Optimal Design and Retrofit of Energy Efficient Buildings, Communities, and Urban Centers

Optimal Design and Retrofit of Energy Efficient Buildings, Communities, and Urban Centers

Moncef Krarti
Professor and Coordinator, Building Systems Program,
University of Colorado, Boulder, CO, USA

ELSEVIER
Butterworth-Heinemann
An imprint of Elsevier

Butterworth-Heinemann is an imprint of Elsevier
The Boulevard, Langford Lane, Kidlington, Oxford OX5 1GB, United Kingdom
50 Hampshire Street, 5th Floor, Cambridge, MA 02139, United States

Library of Congress Cataloging-in-Publication Data
A catalog record for this book is available from the Library of Congress

British Library Cataloguing-in-Publication Data
A catalogue record for this book is available from the British Library

ISBN: 978-0-12-849869-9

For information on all Butterworth-Heinemann publications visit our website at
https://www.elsevier.com/books-and-journals

Publisher: Matthew Deans
Acquisition Editor: Ken McCombs
Editorial Project Manager: Andrae Akeh
Production Project Manager: Anitha Sivaraj
Designer: Christian J. Bilbow

Typeset by Thomson Digital

Front cover credit: Hajer Tnani Krarti

Contents

About the Authors

Moncef Krarti, Professor and Coordinator, Building Systems Program, Civil, Environmental, and Architectural Engineering Department at the University of Colorado, has a vast experience in designing, testing, and assessing innovative energy efficiency and renewable energy technologies applied to buildings. He also directed several projects in designing energy efficient buildings with integrated renewable energy systems. Prof. Krarti has published over 250 technical journals and handbook chapters in various fields related to energy efficiency, distribution generation, and demand side management for the built environment. Moreover, he has published several books on building energy efficient systems. He taught courses related to building energy systems for over 20 years in the US and abroad. As part of his activities as a professor at the University of Colorado, he has been managing the research activities of an energy management center at the University of Colorado with an emphasis of testing and evaluating the performance of mechanical and electrical systems for residential and commercial buildings. He has also helped the development of similar energy efficiency centers in other countries including Brazil, Mexico, and Tunisia. Dr. Krarti has an extensive experience in promoting building energy efficiency technologies and policies overseas, including the development of building energy codes and energy efficiency training programs in several countries, including Tunisia, Sri Lanka, Egypt and collaborative research with over ten countries in Europe, Africa, Asia, and South America.

Moncef Krarti, PhD PE LEED®AP
Professor and Coordinator, Building Systems Program,
University of Colorado, Boulder, CO, USA

Preface

Worldwide, buildings are responsible for over 40% of the total primary energy use and related greenhouse emissions. Through standards and energy efficiency programs, several countries have succeeded in improving the energy performance of new and existing buildings. Designing and retrofitting electrical power systems to be energy efficient have been the primary components of the efforts to reduce energy use consumption by the built environment. Indeed, most energy end-use systems for both residential and commercial buildings such as lighting, air conditioning equipment, and appliances require electrical power to operate. In particular, electricity has to be readily available throughout the building to ensure people can live comfortably and work productively. However, if not designed safely, electrical distribution systems can cause serious injury and even death. Therefore, the main objective when designing and retrofitting electrical distribution systems within buildings is safety for both humans and equipment.

This book outlines the fundamental principles and methods to design safe, flexible, reliable, accessible, and energy efficient electrical power systems for both residential and commercial buildings. In particular, this book presents simplified but effective calculation and analysis methods to design and evaluate safe and energy efficient distribution electrical systems suitable for residential and commercial buildings. These simplified methods are based on well-established engineering principles. In addition, several innovative yet proven energy efficiency technologies and strategies are presented to improve the energy performance of existing electrical systems. The book is designed to be a self-contained textbook aimed at seniors and/or first-year graduate students interested in designing energy efficient distribution electrical systems for buildings. The contents of this book can be covered in a one-semester course for building electrical systems. However, the book can be used as a reference for practitioners and as a text for continuing education short courses. The users of this book are assumed to have a basic understanding of basic electrical circuits including single-phase and three-phase power systems. Basic knowledge of general concepts of engineering economics, and building mechanical systems are also recommended.

The book is organized in 14 self-contained chapters with several worked-out examples, and design case studies. Moreover, several problems are provided at the end of most chapters to serve as review or homework assignment problems for the users of the book. As instructor of a course on building electrical systems at the University of Colorado, I found that the best approach for the students to understand and apply the various design and analysis methods and tools discussed in this book is through group projects. These projects include (1) the design of electrical systems of new residential and commercial buildings and (2) audit and redesign of distribution power systems for existing buildings.

The first chapter provides a basic overview of the basic components of electrical distribution systems specific to both residential and commercial buildings. In addition, the general approach as well as the main objectives of designing building power distribution systems are outlined. In Chapters 2 through 4, a basic review is presented for electric circuits, transformers, and motors. Chapter 5 provides the basic operation of protection devices while Chapter 6 summarizes the design criteria for wiring systems including branch circuits and feeders. Chapters 7 and 8 present the detailed design requirements as well as specific case studies for residential and commercial buildings, respectively. Then, Chapters 9 through 11 outline sequentially the principles of economic analysis, typical energy efficiency measures for electrical systems, and general power quality issues and means to avoid or eliminate them. Chapters 12 and 13 present the components as well typical design procedures for photovoltaic (PV) systems and electrical generation systems. Finally, Chapter 14 introduces optimization-based design methods to integrate renewable electricity generation technologies in designing electrical systems for buildings.

When using this book as a textbook, the instructor should start from Chapter 1 and proceed through Chapter 14 in order. However, some of the chapters can be skipped or covered lightly depending on the time constraints and the background of the students. First, basic design approaches of all components of electrical distribution systems are outlined (Chapter 1). Some of the fundamental principles of electrical power systems are then reviewed (Chapter 2). The characteristics and the basic features of single-phase and three-phase transformers are discussed in details (Chapter 3). Similarly, the operation features and characteristics of single-phase and three-phase electrical motors are summarized (Chapter 4). Then, some of the major components of building electrical distribution systems are presented including protection devices as well as grounding (Chapter 5), and conductors for branch circuits and feeders (Chapter 6). Specific design approaches for specific building types are then outlined with specific applications to well-defined case studies for residential buildings (Chapter 7), and commercial buildings (Chapter 8). To assist in designing energy efficient electrical systems, some economic analysis methods are overviewed (Chapter 9). Then, a wide range of energy efficiency strategies and technologies are discussed to reduce energy use from electrical power systems including transformers, wires, lighting fixtures, motors, and appliances (Chapter 10). Measures and methods to improve the power quality of electricity delivered to building energy systems are also discussed (Chapter 11). Approaches to design systems to generate electricity within buildings are presented including solar PV systems (Chapter 12) and fuel-based generation systems (Chapter 13). Finally, optimization-based analysis approach is introduced and applied to design net zero energy buildings (Chapter 14).

A special effort has been made to use metric (SI) units throughout the book. However, in several instances English (IP) units are also used as they are still the standard set of units used in the United States. A conversion table between the two unit systems (from English to metric and metric to English units) is provided as part of back materials for the book.

I wish to acknowledge the assistance of several people in the conception and the preparation of this book. Special thanks to the input of several of my students at the University of Colorado at Boulder. Finally, I am greatly indebted to my wife Hajer and my three children for their continued patience and support throughout the preparation of this book.

Moncef Krarti

Conversion Factors (Metric to English)

Area	1 m^2	= 1550.0 in^2
		= 10.764 ft^2
Energy	1 J	= 9.4787 × 10^{-4} Btu
	1 GJ	= 0.023884 TOE
	1 GJ	= 9.48043 Therm
	1 GJ	= 0.163456 BOE
	1 XJ	= 0.9478 Quad
	1 kWh	= 3412 Btu
	1 GWh	= 85.9845 TOE
	1 TWh	= 0.0034 Quad
Heat transfer rate	1 W	= 3.4123 Btu/h
Heat flux	1 W/m^2	= 0.3171 $Btu/h{\cdot}ft^2$
Heat generation rate	1 W/m^3	= 0.09665 $Btu/h{\cdot}ft^3$
Heat transfer coefficient	1 $W/m^2{\cdot}K$	= 0.17612 $Btu/h{\cdot}ft^2{\cdot}°F$
Latent heat	1 J/kg	= 4.2995 × 10^{-4} Btu/lb_m
Length	1 m	= 3.2808 ft
	1 km	= 0.62137 mile
Mass	1 kg	= 2.2046 lb_m
	1 ton	= 2204.62 lb_m
Mass density	1 kg/m^3	= 0.062428 lb_m/ft^3
Mass flow rate	1 kg/s	= 7936.6 lb_m/h
Mass transfer coefficient	1 m/s	= 1.1811 × 10^4 ft/h
Pressure	1 Pa = 1 N/m^2	= 0.020886 lb_f/ft^2
		= 1.4504 × 10^{-4} $lb_f/in.^2$
		= 4.015 × 10^{-3} in. water
	1 atm = 100 kPa	= 14.69595 psi = 1 bar
Power	1 kW = 1 kJ/s	= 1.340 HP
		= 3,412 Btu/h
	1 × 10^5 N/m^2	= 1 bar

Refrigeration capacity	1 kJ/h	= 94,782 Btu/h
		= 7.898×10^{-5} ton
	1 kW	= 0.2844 ton
Specific heat	1 J/kg·K	= 2.3886×10^{-4} Btu/lb_m·°F
Temperature	1 K	= (5/9) °R
		= (5/9)(°F + 459.67)
		= °C + 273.15
Temperature difference	1 K	= 1°C
		= (9/5) °R = (9/5) °F
Thermal conductivity	1 W/m·K	= 0.57782 Btu/h·ft·°F
Thermal diffusivity	1 m^2/s	= 38750.0775 ft^2/ h = 645.835 ft^2/min =10.7639 ft^2/s
Thermal resistance	1 C.m^2/W	= 5.7683 °F.ft^2.h/Btu
U-value	1 W/°C.m^2	= 0.17611 Btu/h.°F.ft^2
Volume	1 m^3	= 6.1023×10^4 in^3
		= 35.314 ft^3
		= 264.17 gal
Volume flow rate	1 m^3/s	= 1.2713×10^5 ft^3/h
	1 L/s	= 127.13 ft^3/h
		= 2.119 ft^3/min

Chapter 1

Introduction

Chapter Outline

1.1 GLOBAL TRENDS IN ENERGY CONSUMPTION

In this section, historical energy consumption in the world and select countries as well as regions is first discussed based on reported data for primary energy supply and final energy consumption. Specific trends in current and future energy consumption and energy mix of the building sector are also outlined using available data and analyses. In addition, the economic output per unit energy consumed is evaluated to assess the macroeconomic energy efficiency of select countries and regions in the world.

Optimal Design and Retrofit of Energy Efficient Buildings, Communities, and Urban Centers.
http://dx.doi.org/10.1016/B978-0-12-849869-9.00001-6

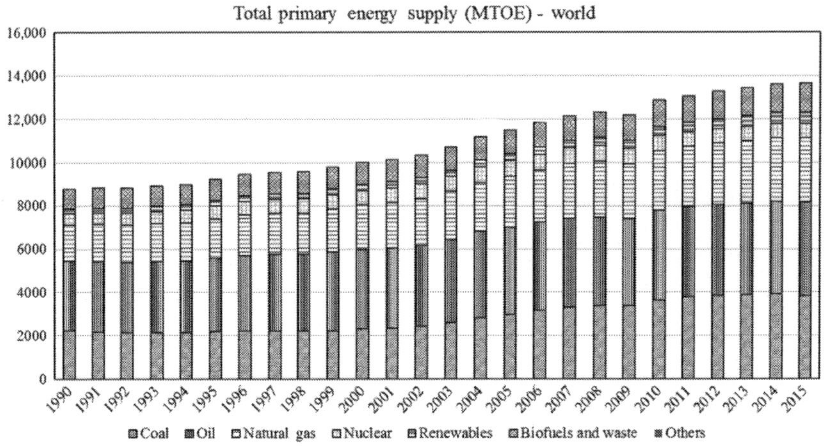

FIGURE 1.1 Annual variation of total primary energy supply (TPES) for the world from 1990 to 2015. *(Source: IEA, 2017).*

1.1.1 Primary Energy Supply

From 1990 to 2015, the total primary energy supply (TPES) in the world increased by 60% while the energy mix remained almost unchanged as noted in Fig. 1.1. Indeed, fossil fuels provided over 80% of the primary energy supply of the world during the last two decades. The specific composition of the world primary energy mix for 2015 is detailed in Fig. 1.2 (IEA, 2017). Specifically, the world relied mainly on oil (32%), coal (28%), and natural gas (22%) to meet its energy needs. Biofuels and waste (10%), nuclear (5%), hydro (2%), and

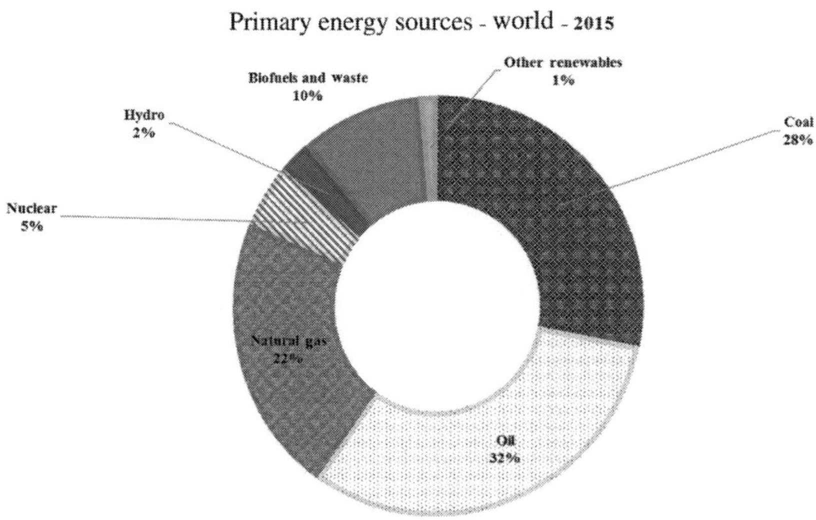

FIGURE 1.2 Resource composition for the world TPES during 2015. *(Source, IEA, 2017).*

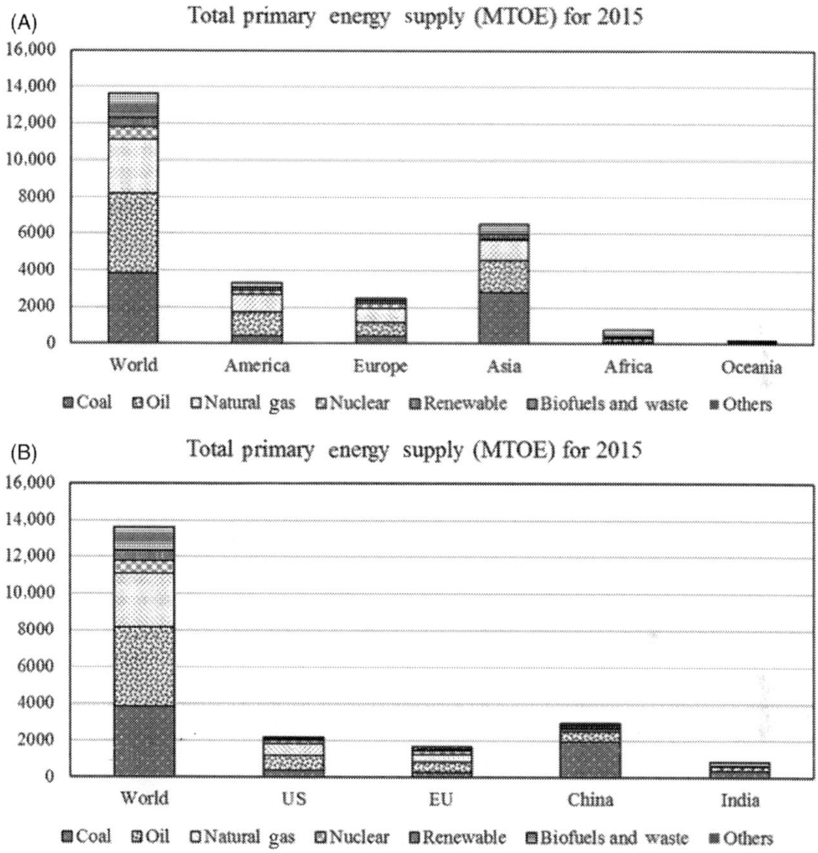

FIGURE 1.3 Primary energy resources for the world compared to those for (A) continents and (B) select economies, 2015. *(Source, IEA, 2017).*

renewable energy resources (1%) were also used to power the world economy during 2015. Fig. 1.3 compares the composition for TPES mix between continents as well as select countries and regions. China and Asia still rely heavily on coal to supply their primary energy needs while most of the other continents and major economies use predominantly oil and natural gas. During 2015, China and the US remained the top energy consumer countries with, respectively, 22% and 16% of the world TPES. Asia, mostly due to China and India, has the highest primary energy consumption among all the continents with over 47% of the world TPES.

As illustrated in Fig. 1.4, the annual world TPES increases at a slightly higher rate than that of the world population growth but significantly lower than the change rate of the world economic output measured by the gross domestic product (GDP). Indeed, the GDP has increased 60% from 1990 to 2015 indicat-

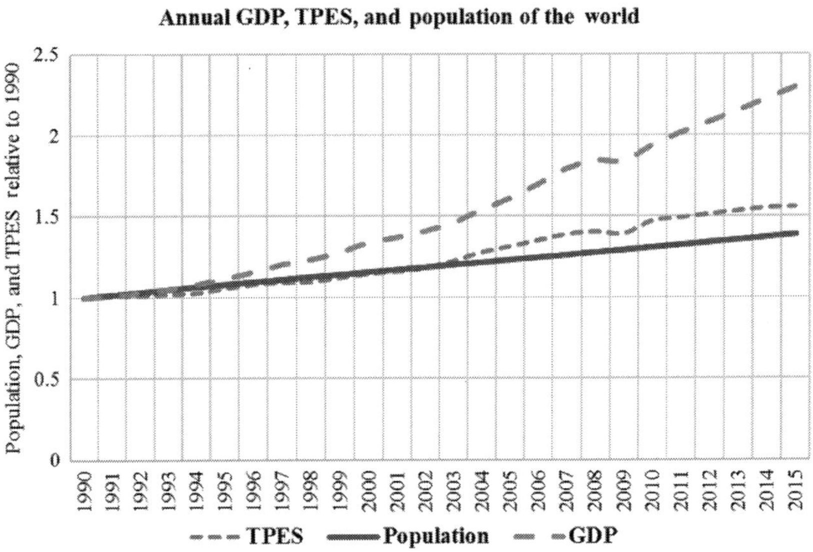

FIGURE 1.4 Annual change of gross domestic product (GDP), TPES, and population of the world indexed to 1990.

ing a decrease in energy intensity and thus an increase in energy efficiency and ultimately an increase in the energy productivity of the overall world economy as discussed in more detail in Section 1.1.4 and Chapter 5.

When normalized by the size of the population, the primary energy supply varies widely between countries as indicated in Fig. 1.5 for the world, US, EU, China, and India from 1990 to 2015 (IEA, 2017). Specifically, the world TPES

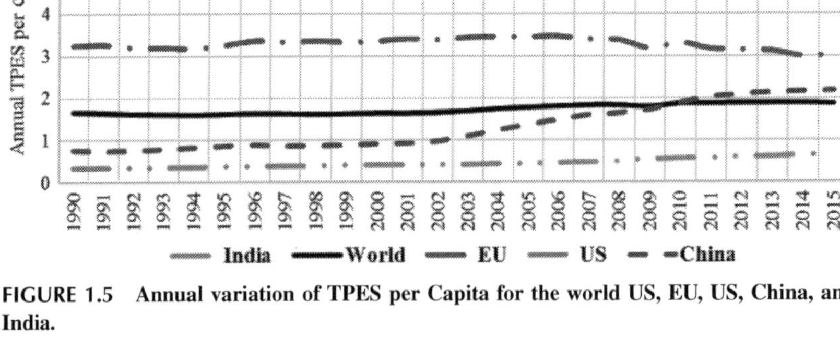

FIGURE 1.5 Annual variation of TPES per Capita for the world US, EU, US, China, and India.

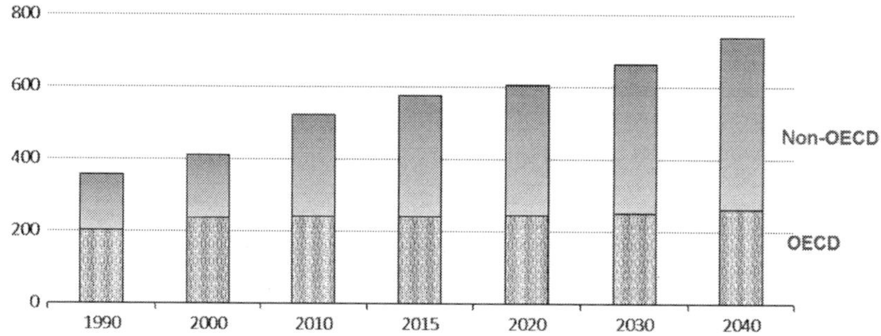

FIGURE 1.6 Predictions of annual TPES in quadrillion of Btu for the world including Organization of Economic Cooperation and Development (OECD) and non-OECD countries. *(Source, EIA, 2017).*

per capita has increased steadily and has reached an average of 2.0 TOE/person in 2015. The average energy use per capita has started lower than that of the world in both China and India but has increased steadily since the 2000s. In fact, TPES per capita in China has passed that of the world average in 2010 and reached 2.2 TOE/person in 2015. The energy use per person in India remains low at 0.75 TOE/person. The US had a significant energy use per person during the entire 1990–2015 period reaching a peak of 8.0 TOE/person in 2000 but declined since to settle at 7.0 TOE/person during 2015. A similar trend is observed for TPES per capita for the European Union (EU) but with a significant decrease since 2008 to be at 3.0 TOE/person for 2015.

Considering the current energy policies of major countries and economies, the US Energy Information Agency (EIA) has projected that future world TPES will continue to grow at a steady pace as illustrated in Fig. 1.6 showing past and future annual TPES for both OECD (i.e., countries part of Organization of Economic Cooperation and Development) and non-OECD countries (EIA, 2017). The EIA projections are based on the reference case with a 3.0% annual economic growth. Specifically, the EIA projects that the world energy consumption will reach 736 quadrillion Btu (i.e., 18,561 MTOE) by 2040, that is a 36% increase relative to the TPES recorded for 2015. As noted in Fig. 1.6, most of the energy consumption increase would be attributed to non-OECD countries including China and India. Using the projection case of high economic growth at 3.3%/year, an additional 40 quadrillion Btu (i.e., 1009 MTOE) is predicted to be consumed worldwide by 2040 relative to the reference case. For the low economic growth at 2.7%/year, the EIA expects that world TPES will be reduced by 29 quadrillion Btu (i.e., 731 MTOE) by 2040 compared to the reference case.

Based on the same EIA predictions as illustrated in Fig. 1.7, fossil fuels will remain the main resources utilized to meet the world's future energy needs. However, the overall share of fossil fuels in the world energy mix is predicted to decline. In particular, while petroleum and other liquids are predicted to remain the largest source of energy, their share in meeting world energy demand is

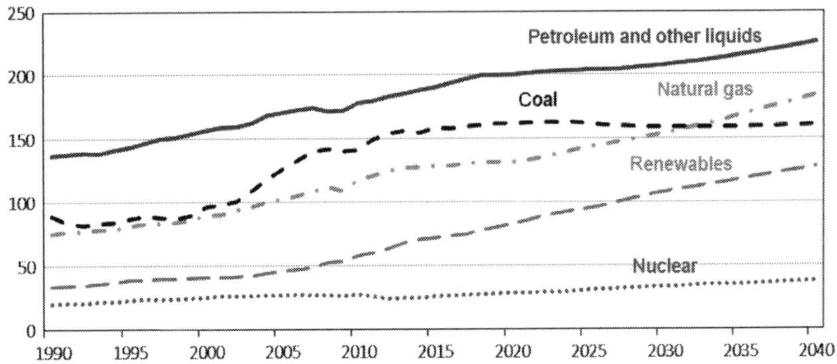

FIGURE 1.7 Predictions of resources of world primary energy consumption expressed in quadrillion of Btu countries. *(Source, EIA, 2017).*

expected to decrease from 33% in 2015 to 31% in 2040. The significant decline in the use of coal is projected to continue while the growth of using natural gas is predicted to increase at 1.4% per year. Renewables are projected to be the world's fastest growing energy resources in the next few decades. In particular, the share of renewables in the world's energy mix is projected to reach 17% by 2040.

1.1.2 Final Energy Consumption

Fig. 1.8 presents the annual variation and resource decomposition of total final energy consumption (TFC) by the world during the 1990–2015 period using reported IEA data (IEA, 2017). As noted for TPES, mostly fossil fuels and primarily oil products remain the main resources utilized to power the various

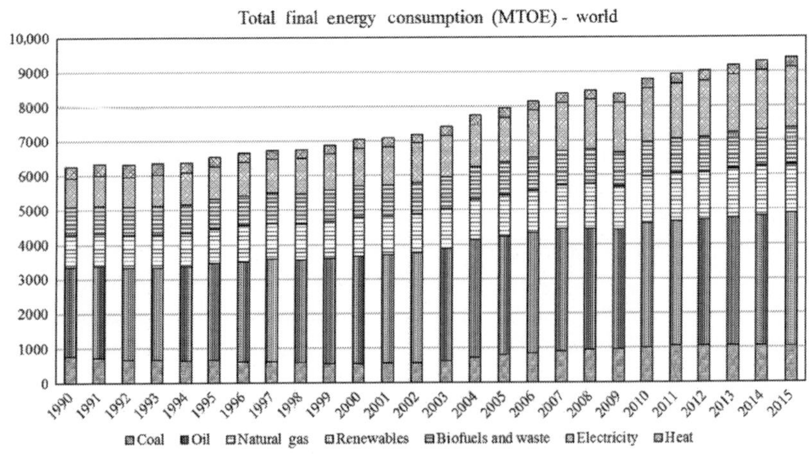

FIGURE 1.8 Annual total final energy consumption (TFC) and energy mix for the world between 1990 and 2015.

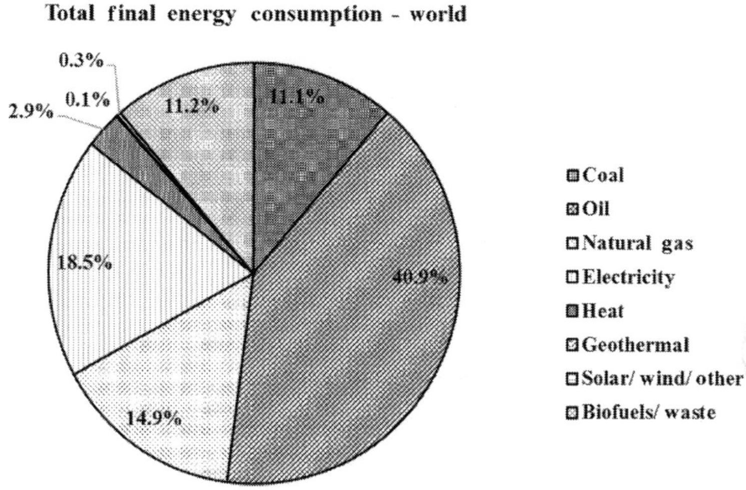

FIGURE 1.9 **Energy mix of TFC for the world during 2015.**

economic sectors of the world. During 2015, as illustrated in Fig. 1.9, the share of oil products was 40.9% of world TFC followed by electricity (18.5%), natural gas (14.9%), biofuels and waste (11.2%), and coal (11.1%). Renewables including geothermal, solar, and wind represented only 0.4% of the TFC in the world during 2015. Moreover, final energy resources are consumed evenly by three sectors: industry, transport, and buildings as illustrated in Fig. 1.10 through the 1990–2015 period. The share in world TFC of the building sector including residential buildings as well as commercial and public services has

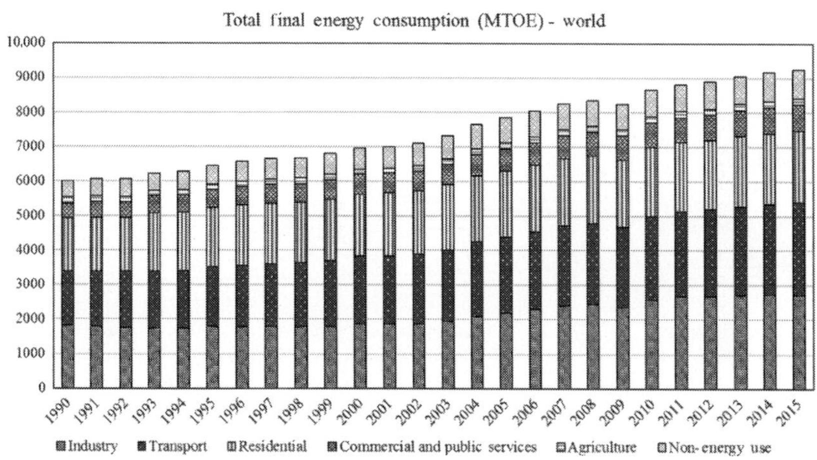

FIGURE 1.10 **Annual world TFC variation by sector during 1990–2015.**

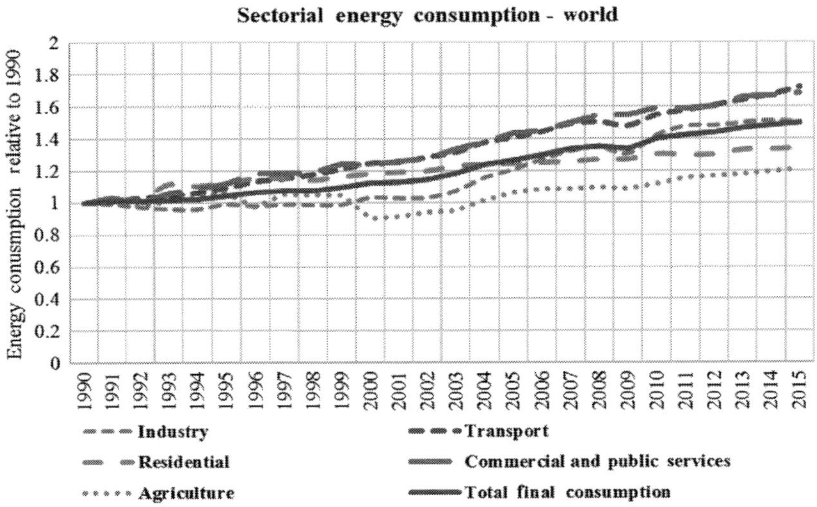

FIGURE 1.11 Annual growth relative to 1990 of world TFC by sector during 1990–2015.

reached 30% during 2015, the highest among all sectors. The growth in the TFC share of the building sector is mostly due to the increase in energy consumption associated with the commercial and public services as indicated in Fig. 1.11. While the overall TFC has increased by 50%, the energy consumed by commercial and service buildings has seen a growth rate of over 70% during the 1990–2015 period. Worldwide, the energy consumed by residential buildings has increased by 34%.

Fig. 1.12 illustrates the annual variation of the TFC as well as the ratio of TFC/TPES reported by IEA for the world during 1990–2015 period (IEA, 2017).

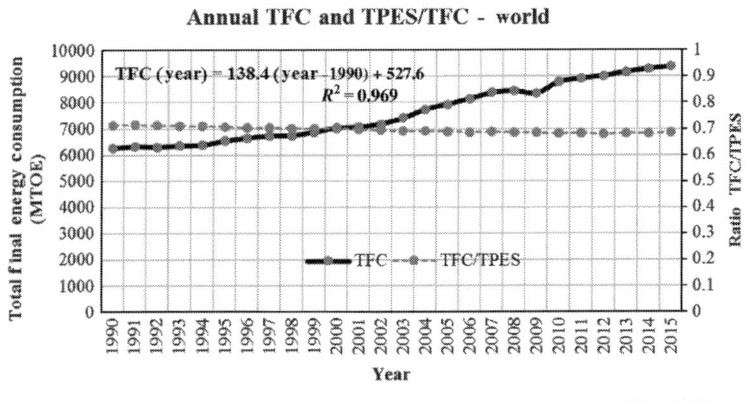

FIGURE 1.12 Annual TFC and ratio of TPES/TFC for the world during 1990–2015.

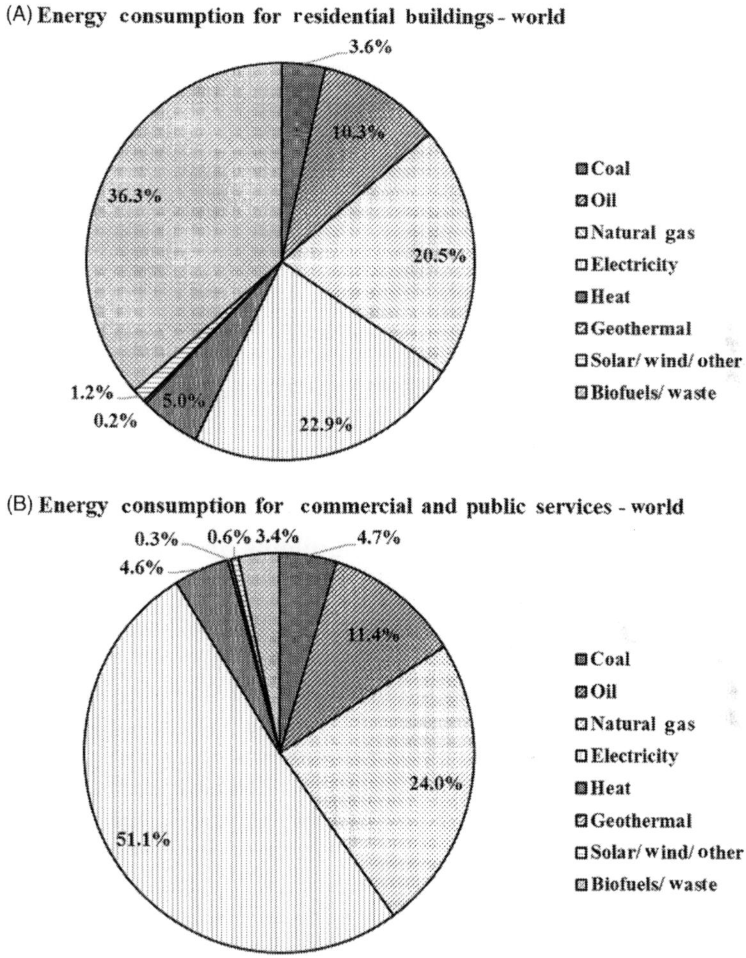

FIGURE 1.13 Share by resource for world's overall energy consumption for (A) residential buildings and (B) commercial and public services during 2015.

As noted in Fig. 1.13, a good linear correlation exists between TFC expressed in MTOE and the year after 1990 provided by the following regression:

$$TFC(year) = 138.4(year - 1990) + 5727.6 \, with \, R^2 = 0.969 \tag{1.1}$$

The ratio of TPES and TFC remains almost constant during the 1990–2015 period slightly decreasing from 0.714 in 1990 to 0.687 in 2015. Example 1.1 illustrates the simple linear regression model of Eq. (1.1) to predict both TFC and TPES for the world in 2040. Using historical data, similar simplified models can be developed to predict future energy consumption for each region and country in the world (Krarti and Dubey, 2017; Krarti et al., 2017).

Example 1.1

Estimate the final energy consumption (TFC) and TPES for the world in 2040 using the simplified model of Eq. (1.1) and assuming that the TPES/TFC ratio is 0.68.

Solution

Using Eq. (1.1), TFC expressed in MTOE for year 2040 can be estimated:

$$TFC(2040) = 138.4(2040 - 1990) + 5727.6 = 12{,}648\,MTOE$$

Thus,

$$TPES(2040) = TFC(2040) \Big/ \left(\frac{TFC}{TPES}\right) = 12{,}648/(0.68) = 18{,}600\,MTOE$$

The model of Eq. (1.1) indicates that TPES is 737 quadrillion Btu, which is close to the EIA predictions based on the reference economic growth case.

1.1.3 Energy Consumption by the Building Sector

The building sector is responsible for the largest share of the world's TFC as indicated in Fig. 1.10. A wide range of resources is utilized to meet the energy demand for residential buildings and the commercial and public services. For the residential buildings, the world utilized during 2015 mostly biofuels and waste (36.3%), followed by electricity (22.9%), and then natural gas (22.5%) as indicated in Fig. 1.13A. For the commercial and public service facilities, the world consumed mostly electricity (51.1%), natural gas (24.0%), and oil (11.4%) as illustrated in Fig. 1.13B. However, the specific energy resources consumed by the building sector vary widely between countries and regions as shown in Fig. 1.14. For instance, biofuels and waste represent the dominant energy resource for residential buildings in India (73%) and to lesser extent in China (28%) but rarely used in the US (4%). On the other hand, natural gas is the energy resource of choice for households in the US (42%) and Europe (36%). For commercial buildings, electricity is the main energy resource for most countries and regions including the US (56%), EU (47%), China (34%), and India (35%).

Moreover, the contribution of the building sector in the national TFC varies significantly between countries and regions. Fig. 1.15 compares the sectorial decomposition of TFC for the world, US, EU, China, and India during 2015. The share of the overall TFC associated to the building sector including residential as well as commercial and public services is 34% for the US, 41% for the EU, 23% for China, and 41% for India during 2015. Within the same region of the world, the share of buildings in the national TFC can vary significantly as noted in Fig. 1.16 for some Middle East and North Africa (MENA) countries. Specifically, Fig. 1.16 shows the annual contribution of the building sector accounting for residential, commercial, and public buildings in national TFC

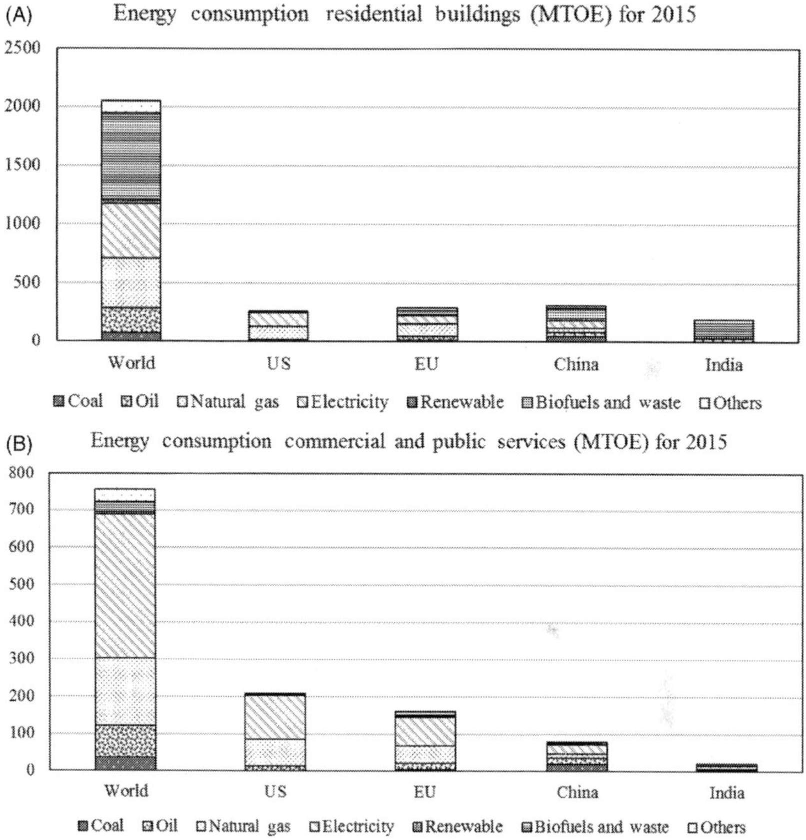

FIGURE 1.14 Resources for energy consumption for world, US, EU, China, and India attributed to (A) residential buildings and (B) commercial and public services during 2015.

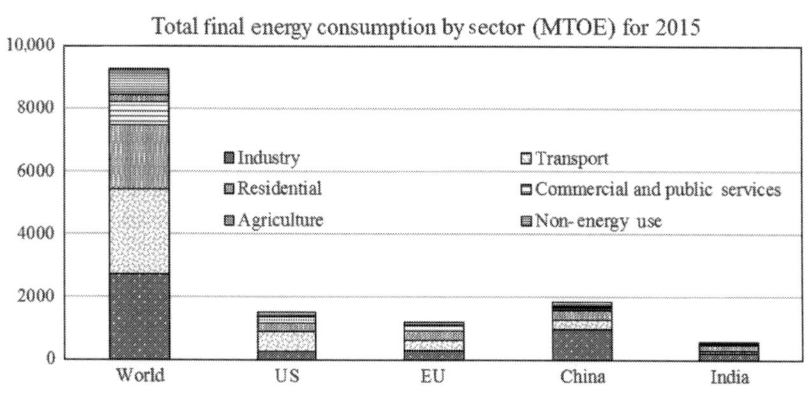

FIGURE 1.15 Sectorial TFC for the world, US, EU, China, and India during 2015.

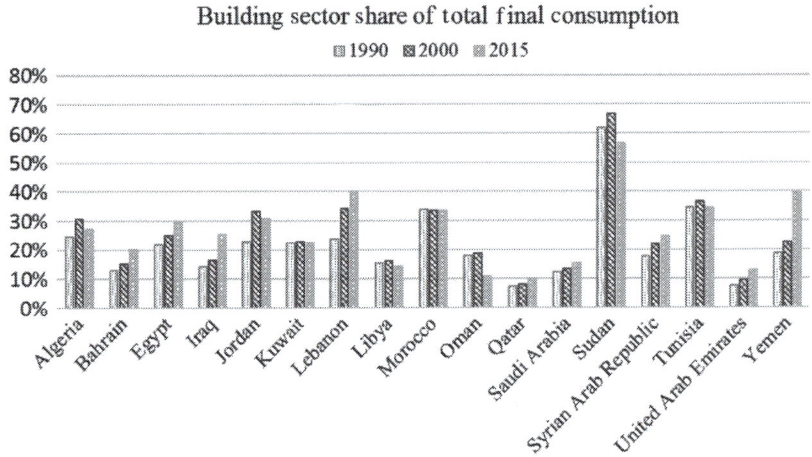

FIGURE 1.16 Share of the building sector in national TFC for select Middle East and North Africa (MENA) countries during 1990, 2000, and 2015.

during the years 1990, 2000, and 2015 for most MENA countries (IEA, 2017). Among the countries listed in Fig. 1.15, the TFC share of the building sector is highest for Sudan (57% in 2015) and is the lowest for Qatar (10% in 2015). For most MENA countries considered in Fig. 1.16, residential buildings consume more energy than commercial/public buildings as clearly indicated by the analysis shown in Fig. 1.17.

EIA projects that the energy consumption attributed to the world building sector, including residential and commercial facilities, will increase by 32%

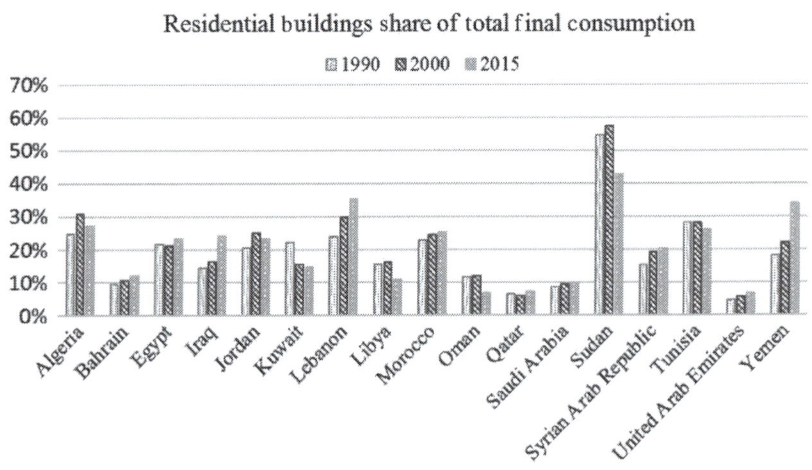

FIGURE 1.17 Share of the residential building sector in national TFC for select MENA countries during 1990, 2000, and 2015.

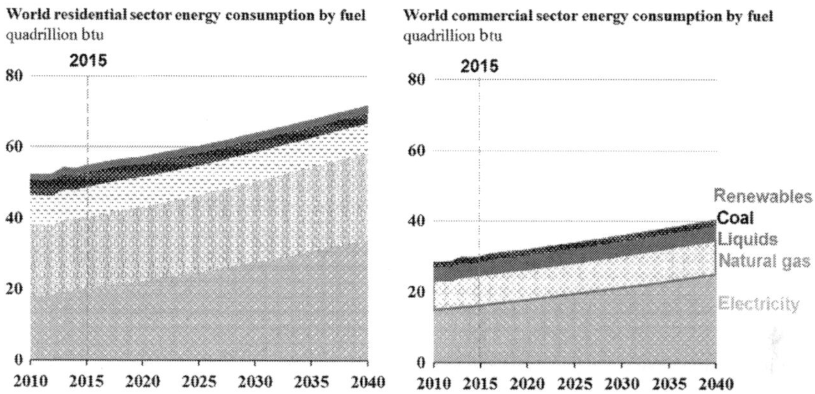

FIGURE 1.18 Projections for energy consumption of the world building sector. *(Source: EIA, 2017).*

between 2015 and 2040 as depicted in Fig. 1.18. It is expected that most of the energy use increase will occur in non-OECD countries especially in Asia including China and India that will experience increasing urbanization rates. Specifically, non-OECD Asia is expected to account for more than 50% of the world increase in the building sector. Between 2015 and 2040, electricity will be the fastest-growing source of energy used in both residential and commercial buildings with an annual growth rate of 2%. As illustrated in Fig. 1.19, the increase in electricity consumption will be driven by improving standards of living in non-OECD countries resulting in higher demand in appliances, lighting, cooling, and commercial services. The increase in energy consumption will offset the impacts of energy efficiency programs applied to buildings, appliances as well as heating and cooling equipment. Moreover, EIA projects that natural

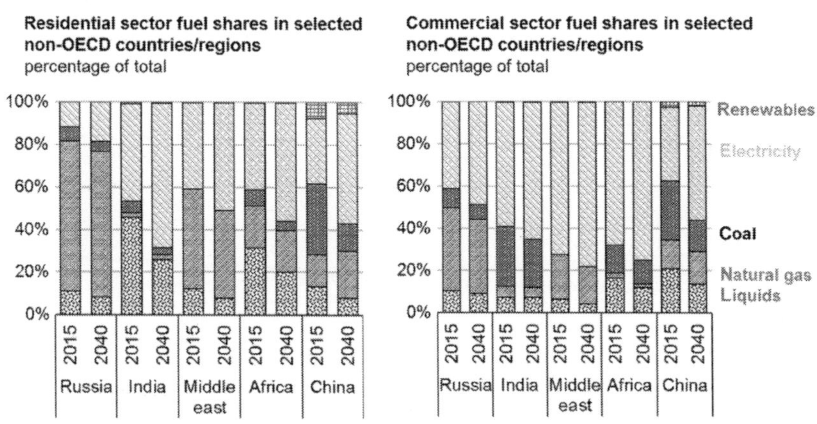

FIGURE 1.19 Distribution of energy resources used in the building sector by non-OECD countries and regions for 2015 and 2040. *(Source: EIA, 2017).*

consumption will increase by 20% during the 2015–2040 period mostly in non-OECD countries. The consumption of coal in the building sector is expected to decline further in the next decades.

1.1.4 Energy Productivity Trends

Defined as the inverse of energy intensity, that is, the economic output per unit of energy consumed, energy productivity has been used as a macroeconomic indicator to assess the balance between economic growth and domestic energy consumption (KAPSARC, 2015). Specifically, energy productivity is often utilized to determine the level of structural diversification between energy-intensive and nonenergy-intensive activities as well as to estimate the overall energy efficiency of an economy. However, as detailed in Chapter 5, energy productivity can be the basis for a framework analysis for energy programs to account for their overall benefits and value added. In particular, the energy productivity-based framework can incorporate the multiple benefits of energy efficiency including economic, environmental, and social. Indeed, an energy productivity framework can include traditional benefits (i.e., reduced energy consumption) as well as nontraditional value added of energy efficiency such as increased asset values, reduced greenhouse gas emissions, and creation of jobs as discussed in more detail in Chapter 5.

Fig. 1.20 presents the annual variation of the GDP per capita in the world, US, EU, China, and India during the 1990–2015 period. It is clear that the economic output per person in the world has steadily increased to reach $14,322/person during 2015. In terms of actual GDP per capita, average value for the world, China, and India have remained significantly less than similar indicators of $34,580/person and $51,592/person reported for the EU and the US,

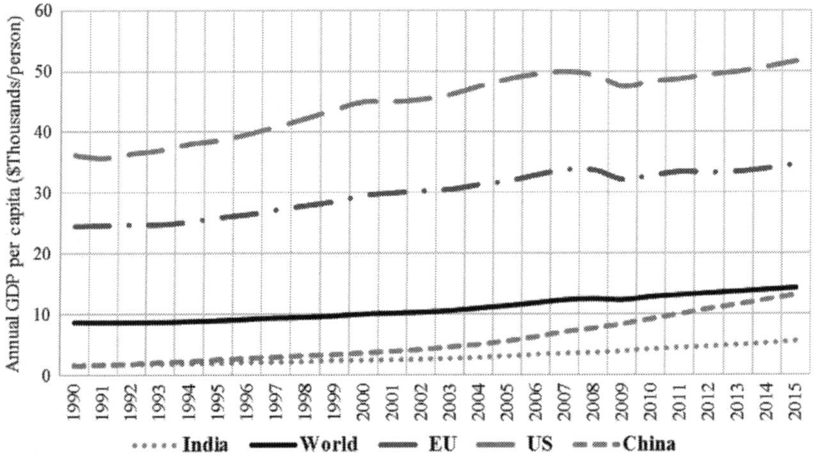

FIGURE 1.20 Annual GDP per capita for the world, US, EU, China, and India during 1990–2015.

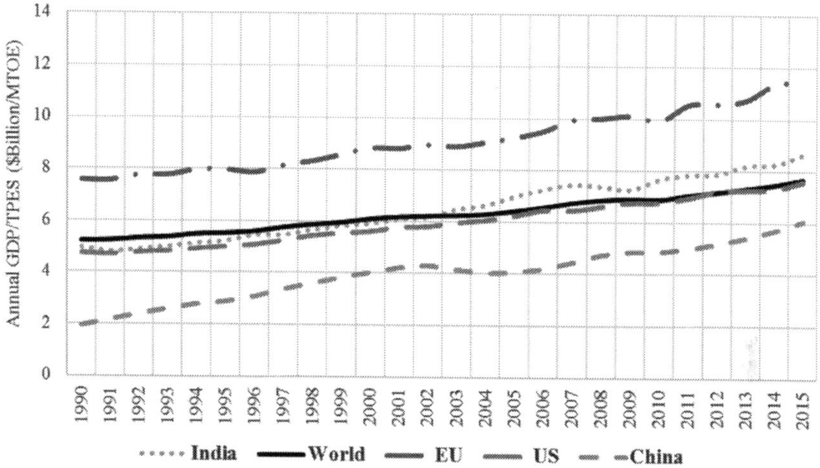

FIGURE 1.21 **Annual energy productivity set as the ratio GDP/TPES, for the world, US, EU, China, and India during 1990–2015.**

respectively. However, the growth rate of the per capita economic output for China, India, and the world has been higher than that for the EU and the US. In particular, the improvement of GDP per capita has increased by 789% for China, 223% for India, and 65% for the world well above the increase reported for the EU (41%) and the US (42%) from 1990 to 2015.

Because of the trends noted for GDP and TPES, the overall energy productivity, defined as the ratio of GDP and TPES, has been increasing in the world as well as the US, EU, China, and India as shown in Fig. 1.21. The energy productivity has steadily increased between 1990 and 2015 in the world as well as the US, EU, China, and India. Specifically, energy productivity for the world has increased by 47.7% from $5.2 billion/MTOE in 1990 to $7.7 billion/MTOE in 2015. During the same period (i.e., from 1990 to 2015), the growth of energy productivity differed between countries and is 51.7% for EU, 60.2% for US, 74.9% for India, and 214.5% for China due to the significant increase in its economic output during the same period.

For several countries, the main driver between the relationship between GDP and energy consumption is the contribution level of the industrial sector to the economy. Indeed, energy-intensive industries are often linked to the early stages of the development of an economy. As the income increases, the economy typically transitions to less intensive industries and sectors such as services. Thus, the more an economy relies on the industrial sector for its development, the more energy it consumes. However, the use of energy-efficient processes and systems can reduce the high-energy dependency to produce wealth. Fig. 1.22 illustrates the annual variation of GDP with total and sectoral final energy consumption for select countries including the US, UK, China, and India over the 1990–2013 period. A strong relationship exists between GDP and TFC

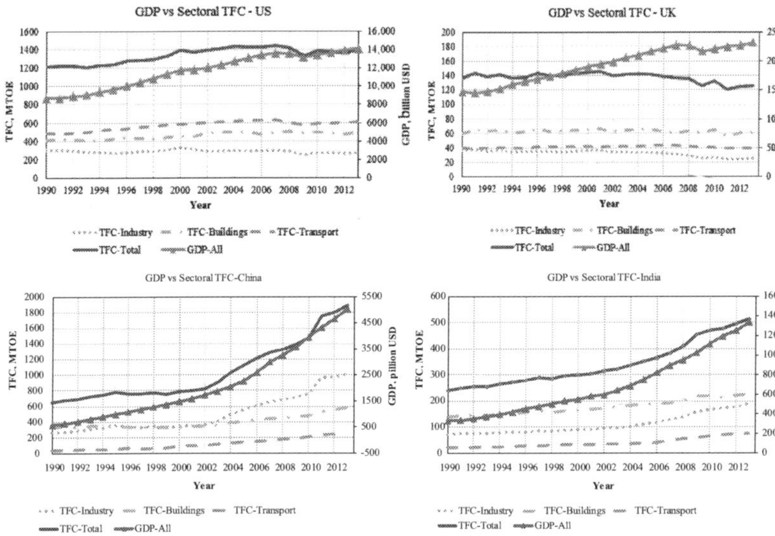

FIGURE 1.22 Annual variation of GDP and energy consumption for industry, transport, and building sectors for US, UK, China, and India.

for China and India, and to a lesser extent for the US. For the UK, however, TFC has decreased since 2000 while GDP continues to increase highlighted by a reduction in energy consumption for all sectors of its economy.

Fig. 1.23 provides the results of a regression analysis between GDP and TFC for the US, UK, China, and India during the period 1990–2013. As expected, good correlations exist between the GDP and TFC for China, India, and the US. In particular, the slope of the linear regression indicates that each unit of energy

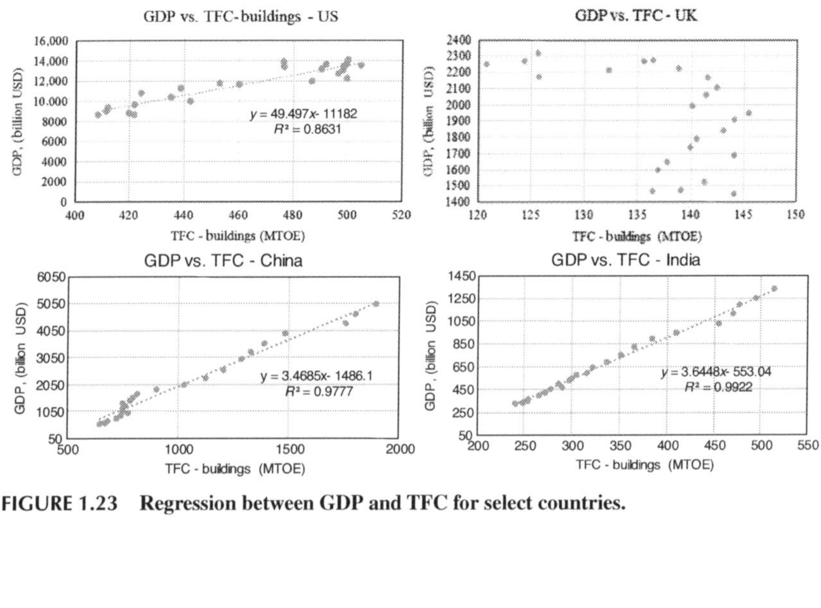

FIGURE 1.23 Regression between GDP and TFC for select countries.

consumed (in MTOE) has increased the GDP by \$49.50 billion for the US, but only by \$3.64 billion for India, and \$3.46 billion for China. The relationship between GDP and TFC is weak for the UK. In fact, the UK follows the expected behavior referred to as the environmental Kuznets curve (EKC) hypothesis.

In econometrics, the EKC hypothesis has been considered by several studies to evaluate the relationship between economic growth and environmental pressure. In particular, Grossman and Krueger (1991) found that there is an inverted U-shaped relationship between income level and environment deterioration. This relation was termed the environmental Kuznets curve or EKC by Panayotou (1993) based on the original work on Kuznets curve indicating an inverted U-shaped correlation between income and inequality. In several studies, the total energy used by a country (expressed typically in terms of energy consumption per capita) is used as an indicator of environmental pressure while GDP (expressed in GDP per capita) is often used to measure the economic growth for the same country (Streimikiene and Kasperowicz, 2016). Some studies considered carbon emissions as an alternative indicator for environmental pressure (Kaika and Zervas, 2013a, 2013b).

According to the EKC hypothesis, the environmental pressure increases with the economic growth until a turning point is reached for a specific income per level. After this turning point, further economic growth leads to improved environmental quality. The results of the EKC analysis are often used to provide policymakers with a better understanding of the link between energy consumption and economic growth and ultimately assist them to formulate sustainable energy policies.

Using the annual energy consumption per capita, $x = E/P$, and the annual economic output expressed GDP per capita, $y = Y/P$, the EKC analysis consist of finding the correlation coefficients generally based on the model defined by Eq. (1.2):

$$e_{i,t} = \alpha_i + \gamma_t + \beta_1 y_{i,t} + \beta_2 y_{i,t}^2 + \beta_3 y_{i,t}^3 + \varepsilon_{i,t} \tag{1.2}$$

where the index i refers to a specific country for which the data and the analysis is carried out, the index t refers to the year when the data are collected, α_i is a constant coefficient specific to the country i, γ_i is a constant coefficient specific to the year t, β_1, β_2, and β_3 are coefficients correlating e to y as noted in Eq. (1.2), and $\varepsilon_{i,t}$ is the error term of the correlation.

Instead of e and y, some studies have used $Ln(e)$ and $Ln(y)$ to estimate the regression coefficients of Eq. (1.2) to avoid zero and negative indicators (Kaika and Zervas, 2013a). Fig. 1.24 illustrates the relation between the e and y as stipulated by the EKC hypothesis and modeled by the correlation of Eq. (1.2). At the macroeconomic scale, the energy consumption increases with the income level as the country is developing from a low-tech farming economy to an energy-intensive industrial economy reaching a turning point. With further development and transition to a less energy-intensive service economy, the energy consumption decreases as the income level continues to increase. For sustainable development, it is desired to increase the economic growth while reducing

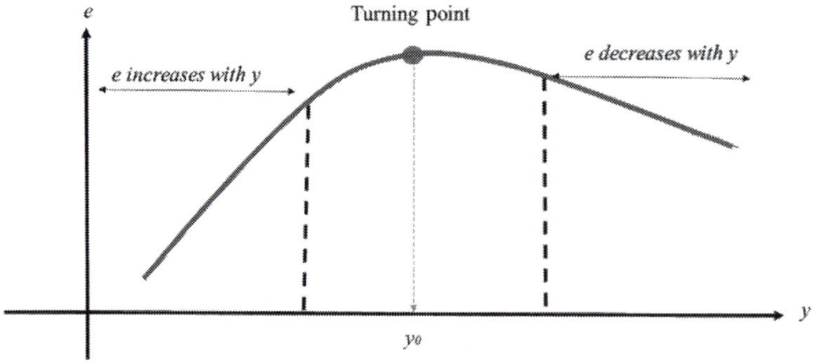

FIGURE 1.24 Typical relationship between energy consumption, *e*, and income level, *y*, following EKC model.

energy consumption as well as carbon emissions, and ultimately improving the environmental quality.

Considering the possible values of the regression coefficients, β_1, β_2, and β_3 in the model of Eq. (1.2), seven scenarios can be obtained but only one of them would satisfy the EKC hypothesis as listed here (Dinda, 2004):

- Scenario 1: if $\beta_1 = \beta_2 = \beta_3 = 0$, there is no relationship between *e* and *y*.
- Scenario 2: if $\beta_1 > 0$ and $\beta_2 = \beta_3 = 0$, an increasing linear relationship exists between *e* and *y*.
- Scenario 3: if $\beta_1 < 0$ and $\beta_2 = \beta_3 = 0$, a decreasing linear relationship exists between *e* and *y*.
- Scenario 4: if $\beta_1 > 0$, $\beta_2 < 0$, and $\beta_3 = 0$, there is an inverted U-shaped relationship that exists between *e* and *y* satisfying the EKC hypothesis as shown in Fig. 1.21. In this case the turning point income level can be estimated by $y_0 = -\beta_1/2\beta_2$ if the polynomial model of Eq. (1.2) is considered and by $y_0 = \exp(-\beta_1/2\beta_2)$ if the logarithmic model is used.
- Scenario 5: if $\beta_1 < 0$, $\beta_2 > 0$, and $\beta_3 = 0$, there is a U-shaped relationship that exists between *e* and *y*.
- Scenario 6: if $\beta_1 > 0$, $\beta_2 < 0$, and $\beta_3 > 0$, there is an N-shaped relationship that exists between *e* and *y*.
- Scenario 7: if $\beta_1 < 0$, $\beta_2 > 0$, and $\beta_3 > 0$, there is an inverted N-shaped relationship that exists between *e* and *y*.

The model of Eq. (1.2) is an empirical model since it requires historical data to determine the correlation coefficients and ultimately the relationship between the energy consumption and the economic output. The model, however, does not allow to identify the specific underlying structural drivers that lead to the relationship. Specifically, the inverse model of Eq. (1.2) cannot be utilized to predict future impacts of the implementation of energy policies in various sectors on the economic output. A deterministic analysis framework using the concept energy productivity is introduced in Chapter 5 to assess the effects of various energy efficiency programs and strategies on both future energy consumption and future economic output at the macroeconomic and microeconomic scales.

1.2 ENERGY EFFICIENCY POLICIES

A study was completed to rate the energy policies of over 111 countries worldwide. The criteria as energy policies and the administrative procedures considered in the analysis are summarized in Table 1.1. The study has found that most countries are making some efforts to improve the accessibility and the sustainability of energy (RISE, 2016). However, several energy policy shortcomings have been identified especially for lesser-developed countries:

- Access to energy and especially electricity in some countries remain low due mostly to the high cost of connecting to the grid. For instance, over 600 million people in the sub-Sahara countries remain without access to electricity. The cost for connecting to the electrical grid in several of these countries can reach \$500 exceeding the worldwide average of \$100. Poor families in some of these African countries live on an income of less than \$2.00 per day.
- Several countries overlook energy efficiency and tend to pursue more aggressively renewable energy. However, implementing energy efficiency

TABLE 1.1 Criteria, Energy Policies, and Administrative Considered in the RISE Study (RISE, 2016)

Criteria	Energy policies and regulations	Administrative procedures
Energy accessibility	• Develop, implement, and monitor energy (electrification) plans • Establish a clear framework for grid electrification, minigrids, and stand-alone systems • Ensure access to affordable energy (electricity)	• Allow grid connection to customers including households • Permit new minigrids
Energy efficiency	• Establish energy efficiency plans • Create entities to overview and regulate energy efficiency plans • Develop and implement mandatory regulations and incentives for energy efficiency (i.e., minimum energy performance standards, building energy codes, energy labeling systems, and carbon pricing and monitoring) • Provide financial mechanisms for energy efficiency	• Establish enforcement mechanisms and procedures
Renewable energy	• Establish legal framework for renewable energy • Provide incentives and regulatory support for renewable energy • Allow grid connection and network access • Provide financial mechanisms for renewable energy	• Permit new renewable energy projects

programs remains globally more cost-effective than investing in renewable energy. The RISE study has noted that only a third of the countries worldwide have made progress in improving the energy efficiency of their economy by labeling energy-efficiency appliances or establishing building energy codes.

- Moreover, several countries are solely relying on grid extension to enhance electricity accessibility to their citizens and are missing the solar revolution. Indeed, development of low-cost PV systems offers today distributed energy resources and decentralized off-grid options that can accelerate the pace of electrification in many remote regions of the world.

1.2.1 Benefits for Improving Energy Efficiency of Buildings

According to the International Energy Agency (IEA, 2013), harnessing the full capacity for energy efficiency in the building sector could boost cumulative global economic output through 2035 by $18 trillion, with related benefits including the following:

- For building owners and occupants: improved durability, reduced maintenance, greater comfort, lower costs, higher property values, increased habitable space, increased productivity, and improved health and safety.
- For governments: improved air quality, reduced societal health costs, improved tax base, increase in GDP, and enhanced energy security. Moreover, energy efficiency can stretch the availability of nonrenewable energy resources, typically the main income source for several countries.
- Utilities: cost and operational benefits associated with decreased customer turnover, improved generation and transmission efficiency, and reduced system capacity constraints.

Moreover, improving energy efficiency of buildings can provide significant job creation potential often higher than that associated with investing in renewable energy. In particular, jobs created by the building energy efficiency sector require broader and often specialized skill sets than those needed for renewable energy. In particular, the job creation potential for improving the energy efficiency of the built environment including retrofitting existing buildings can provide significant social and economic benefits. A study by the Rockefeller Foundation (2012) indicated that retrofitting existing buildings including replacing and upgrading energy consuming systems offer an important investment opportunity with significant economic impacts including high employment potential. Specifically, the study has found that $279 billion can be invested in retrofitting US residential, commercial, and institutional buildings with savings in energy costs of $1 trillion over 10 years (i.e., with an average payback period of less than 4 years) resulting in an annual reduction by 30% in US electricity expenditures. When all the retrofits are undertaken, about 3.3 million jobs can be created throughout the US with the vast majority related to improving

TABLE 1.2 Employment Creation Impacts for Various Energy Sources by $1 Million of Expenditures

Energy source	Direct job creation per $1 million expenditures (number of job-years)	Indirect job creation per $1 million expenditures (number of job-years)	Direct and indirect job creation per $1 million expenditures (number of job-years)
Fossil fuel			
Oil and natural gas	0.8	2.9	3.7
Coal	1.9	3.0	4.9
Energy efficiency			
Building retrofits	7.0	4.9	11.9
Mass transit	11.0	4.9	15.9
Smart grid	4.3	4.6	8.9
Renewables			
Wind	4.6	4.9	9.5
Solar	5.4	4.4	9.8
Biomass	7.4	5.0	12.4

the energy efficiency of residential buildings. A job creation model has been developed by Pollin et al. (2009) based on survey data to estimate the number of direct and indirect full time jobs. Table 1.2 lists the number of job years that can be created from $1 million expenditures in the energy sector for various energy sources using the model of Pollin et al. (2009). For building retrofits, the direct effects consist of jobs created to implement the energy efficiency measures while the indirect effects are associated with the jobs needed to produce and supply energy efficiency equipment and materials. As noted by Pollin et al. (2009), there is an induced job creation associated with the direct and indirect effects, that is, the expansion of employment as the result of spending earnings from people who have acquired direct or indirect jobs. For the energy sector, the level of induced job-year creation is found to be 40% of the level for direct and indirect job-year creation. Thus, the induced job years created from building retrofits are estimated at 4.9 per $1 million of expenditures. Most of the jobs created in building retrofits are in the construction and manufacturing industries with a wide range of pay level and technical specialization including electricians, HVAC (heating, ventilating, and air conditioning) technicians, insulation installers, energy auditors, building inspectors, and construction managers. The total direct and indirect job creation due to $1 million expenditures is 11.9 job-years for the building retrofit while it is only 9.8 job-years for solar systems installation as indicated in Table 1.2.

1.2.2 Barriers for Energy Efficiency

However, securing the expected benefits from energy efficiency can be complex and challenging for several countries. Indeed, the building sector is highly fragmented, conservative, and slow to change. In particular, the building sector faces the following challenges in most countries:

- Multiple groups of stakeholders, with no unifying incentive to prioritize energy efficiency.
- Wide variations in technical education and energy-efficiency knowledge between individuals and groups.
- National and regional differences and inconsistencies in regulatory systems.
- Economic policies for new and retrofit building projects that often impede viable energy efficiency investments.

Several studies and pilot projects have shown that there are four core areas that typically limit the market penetration of energy efficiency:

- Lack of awareness and leadership particularly related to challenges in making the business case foe energy efficiency. Several proven and cost-effective energy efficiency systems are available for buildings. However, most decision-makers are not aware of these systems as well as of recent advances in building energy efficiency technologies.
- Workforce capacity and the need for proper skills and collaboration throughout the value chain to consult, plan, implement, and operate the right solutions. Indeed, several stakeholders are involved in designing, constructing, and operating buildings. It is difficult to achieve high energy efficient buildings without an integrated process between design, construction, and occupancy as well as collaboration effort between various stakeholders.
- Lack of adequate financing models that overcome the split incentives inherent in the building sector and enable value sharing. In particular, initial design decisions of new buildings are typically made by developers and not by owners or occupants who are responsible for paying the energy costs of using the buildings.
- Lack of consistent and long-term policy frameworks (national, subnational, regional, and city), including regulations and incentive schemes. For instance, several countries highly subsidize energy prices and often make investments in energy efficiency that is not cost-effective to building operators and owners. Table 1.3 summarizes the total subsidies for various energy resources reported for the world during 2013–2015. Table 1.4 lists the highest energy subsidy per capita level as well as the subsidization rate reported for countries during 2015.

TABLE 1.3 World Global Subsidies for Energy Resources for 2013–2015 Expressed in Billions of 2015 Real US Dollars (IEA, 2017)

Energy resources	2013	2014	2015
Oil	263,504	242,258	144,161
Electricity	124,578	120,438	100,228
Natural gas	103,020	95,201	76,280
Coal	2016	1622	1181
Total	493,117	459,520	321,850

TABLE 1.4 Top Energy Subsidies Levels by Countries During 2015 (IEA, 2017)

Country	Average subsidization rate (%)	Subsidy per capita ($/person)	Total subsidy as share of GDP (%)
Kuwait	70%	1547.4	5.0%
Saudi Arabia	72%	1542.3	7.4%
Qatar	52%	1508.0	1.8%
Bahrain	52%	1211.9	5.5%
UAE	40%	1082.4	2.9%
Turkmenistan	70%	1021.8	15.4%
Libya	65%	668.0	11.0%
Iran	71%	662.4	13.5%
Venezuela	91%	640.6	8.3%
Trinidad and Tobago	33%	497.4	2.8%
Brunei	30%	422.1	1.5%
Algeria	74%	371.9	8.2%
Argentina	35%	241.1	1.8%
Russia	22%	210.9	2.3%
Uzbekistan	57%	206.8	9.8%
Kazakhstan	25%	180.3	1.8%
Iraq	45%	164.7	3.5%
Egypt	48%	147.0	4.1%
Ecuador	27%	139.0	2.3%
Ukraine	31%	137.7	6.9%

GDP, Gross domestic product.

In the following sections, mechanisms and regulatory strategies to improve energy efficiency of new and existing buildings are described.

1.2.3 Energy Efficiency Programs for New Buildings

The main strategy commonly employed by policy makers to improve the energy efficiency of new buildings is a set of standards and codes. In particular, minimum energy performance standards (MEPS) are often utilized as strategies to ensure only high-efficiency building products and equipment are manufactured, sold, and installed. MEPS can be specific to appliances (i.e., refrigerators, freezers, dishwashers, washing machines, dryers, and ovens), lighting fixtures, and heating and cooling systems. Table 1.5 illustrates an

TABLE 1.5 Labels for Minimum Energy Performance Standards for Refrigerators, Freezers and Air Conditioners, and Washing Machines for KSA (Krarti et al., 2017)

Star rating	Air conditioners (EER = 3.412 COP expressed in Btu/Wh)	Refrigerators/freezers (percent of energy consumption relative to a baseline)	Washing machines (function of energy use per load capacity)
1	<7.5	5%	<2.0
2	7.5–8.5	10%	2.0–2.9
3	8.5–9.0	15%	3.0–3.9
4	9.0–9.5	20%	4.0–4.9
5	9.5–10.0	25%	5.0–5.9
6	10.0–11.5	30%	>6.0
7	11.5–12.4		
7.5	12.4–13.4		
8.0	13.4–14.5		
8.5	14.5–15.6		
9.0	15.6–16.8		
9.5	16.8–18.1		
10	<18.1		

Note: SASO has updated the energy performance standards for washing machines (conforming to star rating 4 and above only), refrigerators (conforming to star rating 1 and above only), and air-conditioners (conforming to star rating 3 and above can only be sold and manufactured).
Source: SASO, 2012. Energy Labelling and Minimum Energy Performance Requirements for Air-Conditioners (Standard No. 3459). Saudi Arabia Standards Organization, Saudi Arabia. SASO, 2013. Energy Labelling Requirements of Household Electrical Clothes Washing Machines (Standard No. 3569). Saudi Arabia Standards Organization, Saudi Arabia. SASO, 2014. Energy Performance Capacity and Labelling of Household Refrigerators, Refrigerator-Freezers, and Freezers (Standard No. 3620). Saudi Arabia Standards Organization, Saudi Arabia.

example of MEPS for appliances set recently in KSA (Krarti et al., 2017). Moreover, the establishment of voluntary or mandatory building energy efficiency codes (BEECs) has been proven to be an effective strategy in several countries to compel building industry stakeholders to design and construct an energy efficient and sustainable built environment. Incentive programs have been considered to encourage building owners and developers to adopt even stricter regulations than applicable MEPS and BEECs to improve the energy performance of new buildings. These incentives can be monetary and nonmonetary as indicated in the following options considered by select countries:

- Full or partial grants (US, Singapore) to cover any additional design costs, or costs beyond meeting BEEC requirements, as well as for demonstration projects involving untested and novel technologies.
- Subsidies for loans or interest rates (Germany, US, Austria, Japan).
- Energy efficiency or green mortgages (US, Mexico).
- Tax benefits (US) using several options such as (1) reductions in import tax duties for energy-efficiency equipment, (2) income tax deductions, and (3) credits for energy efficient appliances and equipment.
- Nonmonetary incentives (US, Korea) through expedited construction permits and/or relaxed zoning restrictions (size, density).
- Awards (US) using Energy Star labels.
- Rating systems LEED (US) and BREAM (UK).

The BEEC establishment process for MEPS and BEECs can be lengthy and involved with development, implementation, enforcement, and revision phases. The following sections outline the type of BEECs and their establishment procedures.

1.2.4 Types of Building Energy Codes

BEECs are also known as energy standards for buildings, thermal building regulations, energy conservation building codes, or simply energy building codes. Energy policy makers and governments utilize BEECs as regulatory mechanisms to reduce energy consumption and environmental impacts of buildings while maintaining acceptable indoor environmental quality. Typically, BEECs consist of mandatory design requirements to improve the energy performance of buildings. Since the 1970s, BEECs have been developed and implemented in several countries. For some countries, it has been shown that BEECs are effective in reducing the energy consumption for buildings. In particular, the implementation of mandatory BEECs has resulted in energy use reduction of households for most EU countries. Indeed, the reduction in energy consumption for residential buildings has ranged from 22% in Germany and the Netherlands to 6% in southern European countries (IEA, 2013).

Two approaches are commonly considered to develop BEECs specific to new buildings:

- *Prescriptive-based approach*: In this case, the BEECs include sets of minimum energy performance requirements for each component of the building: windows, walls, and heating and cooling equipment. Two compliance paths can be considered: (1) each building component has to meet strict minimum energy performance requirements such as minimum thermal performance levels for walls and windows (U-values and SHGC-values), and (2) trade-offs between the energy performance requirements of different components. While the prescriptive-based approaches are simple to implement and enforce, they lack flexibility and do not encourage an integrated design of buildings. Table 1.6 illustrates examples of current BEECs following the prescriptive approach for selected MENA countries (RCREEE, 2015).
- *Performance-based approach*: BEECs based on performance set requirements for the overall building energy consumption. Thus, performance-based BEECs encourage an integrated design approach to account for interactions between building components to optimize the energy performance for the entire building. Several options for the performance-based BEECs can be developed and implemented including the following:
  - Partial performance of subsystems (building envelope using the overall thermal transfer value or OTTV analysis approach)
  - Performance of multiple subsystems (maximum energy use of demand for lighting and air conditioning systems)
  - Total building performance (energy consumption or energy cost) using three options:
    - Fixed budget approach using a fixed energy consumption such as kWh/m^2 (key assumptions are required including operation settings and internal loads).

TABLE 1.6 Prescriptive Requirements for Building Envelope for Selected MENA Countries

Country	Minimal *R*-values Roofs (m$^{2\circ}$C/W)	Minimal *R*-values Walls (m$^{2\circ}$C/W)	Minimal *R*-values Floors (m$^{2\circ}$C/W)
Egypt	1.67	1.0	–
Kingdom of Saudi Arabia	1.75	1.0	–
Lebanon	1.75–2.27	0.48–1.85	0.40–1.82
Tunisia	1.33–1.82	0.83–1.67	–
Jordan	0.51–1.75	0.56	–
Kuwait	2.50	1.75	

MENA, Middle East and North Africa.

TABLE 1.7 Performance Requirements for Office Buildings in Tunisia

Building thermal performance class	*BECTh* value (kWh/m^2 yr)
Class 1	$BECTh \leq 75$
Class 2	$75 < BECTh \leq 85$
Class 3	$85 < BECTh \leq 95$
Class 4	$95 < BECTh \leq 105$
Class 5	$105 < BECTh \leq 125$
Class 6	$125 < BECTh \leq 150$
Class 7	$150 < BECTh \leq 180$
Class 8	$BECTh > 180$

- Custom budget approach using comparative analysis with a reference building (typically, the reference building is the same as the proposed building but complies with a set of prescriptive requirements).
- Points system using a score system for different energy efficiency measures.

Table 1.7 illustrates the performance-based BEEC requirements for office buildings in Tunisia (Ihm and Krarti, 2012). As indicated in Table 1.7, the BEEC requirements are set-based ratings or classes using the annual energy needs allowed for heating and cooling loads (expressed in kWh/m^2 yr). The building has to meet a specific energy rating depending on the climate zone. The energy rating-based BEEC permits an easy revision process of the energy efficiency requirements by simply changing the class levels.

1.2.5 Overview of BEEC Establishment

To be effective in improving the energy efficiency of buildings, BEECs have to go through several phases including development, implementation, enforcement, and revision as summarized in Table 1.8. Indeed, the establishment cycle of BEECs may take over 6 years to be adequately implemented and enforced.

The development process for BEECs varies widely between countries. However, the following steps are typically considered to develop an effective BEEC:

Step 1: Perform a review and an analysis of existing international BEECs.
Step 2: Assemble a working group made up of experts to assist in the BEEC development.
Step 3: Survey representative samples of national building stock through walk-through energy audits and questionnaires.
Step 4: Develop prototypical building energy models as benchmarks for BEEC development.

TABLE 1.8 Typical Phases and Their Duration for Establishing Building Energy Codes

Phases for building energy efficiency code establishment	Year								
	1	2	3	4	5	6	7	8	9
Code development									
Code implementation									
Code enforcement									
Code revision									

Step 5: Collect climatic data, available energy efficiency technologies, and costs of adopting these technologies.
Step 6: Perform technical and economic analyses using state-of-the-art computer simulation tools to assess best energy efficiency measures based on energy savings and cost-effectiveness. The most effective measures are typically included as part of the BEEC requirements.
Step 7: Draft, revise, and finalize BEEC requirements based on public review to include key stakeholders.
Step 8: Adopt BEEC officially as mandatory or voluntary code (selecting if BEEC should be mandatory or voluntary depends often on existing enforcement mechanisms of general building codes including structural, electrical, mechanical, and fire codes).

While designing and developing BEECs, the following features could be considered to maximize their effectiveness:

- *Simplicity and/or flexibility:* BEEC can include both prescriptive and performance approaches (a prescriptive approach is preferred as an initial step for BEEC adoption especially for small buildings).
- *Accuracy:* BEEC has to be technically accurate and consistent (prescriptive requirements should be consistent with performance compliance requirements).
- *Affordability:* BEEC should account for the availability of energy efficiency technologies (the costs of these technologies should be sufficiently low to make them cost-effective).
- *Beneficial impact:* BEEC should provide tangible benefits to owners, occupants, and society (life-cycle cost or LCC analysis should be used as part of the BEEC development).

The implementation approach of a BEEC is often an important step and requires adequate preparation. The following implementation approach is suggested:

- Establish BEEC administration and enforcement structure, typically through a government agency.

- Define a clear BEEC compliance process through easy-to-use forms or computer tools.
- Develop training and building capacity programs for code officials, designers, architects, engineers, contractors, and suppliers.
- Establish outreach programs to the building industry and the public.
- Initiate, complete, and evaluate demonstration programs through incremental financial incentives in the initial phase of BEEC implementation.
- Set firm periods and dates for implementation of the BEEC requirements with clear procedures and mechanisms.
- Identify institutions and mechanisms to evaluate BEEC impact and effectiveness in reducing energy consumption.

The enforcement phase of BEECs is often the most difficult step to complete successfully especially for countries with limited qualified institutions and established regulations and standards related to constructing new buildings. Some of the enforcement options and their associated implementation requirements utilized by some countries are outlined below:

- Government agencies (US): This option requires well-trained inspectors and may result in high operating costs for the government.
- Private agencies (France, Mexico): This option allows a third party, certified by the governmental institutions, to enforce BEEC requirements. A training program for private enforcement agencies needs to be established. The costs incurred by the government to implement this option are generally moderate.
- Self-certification to owners (Germany): This option relies on the builders themselves to ensure compliance with the BEEC requirements typically through "verification of compliance" statements and forms. The costs for implementing this enforcement option are generally low and are mainly associated to the training and certification programs.

1.2.6 Beyond BEEC Requirements

Several countries have set regulations and programs to improve building energy efficiency and design low-energy and sustainable buildings beyond the requirements of BEECs. As will be discussed throughout this book, current technologies allow to design and retrofit cost-effectively buildings to be near zero-energy. The main strategy to push designing and retrofitting high-energy performance buildings beyond BEECs involves rating and labeling systems. Some of the sustainable building rating systems include the following:

- LEED: Leadership in Energy and Environmental Design, is the first rating system for buildings and was developed in 1994 by architects and engineers.
- UK BREEAM: Building Research Establishment's Environmental Assessment Method, is used mostly in the UK and was developed in 1990.

- CASBEE: Comprehensive Assessment System for Building Environmental Efficiency, was developed in Japan in 2001.
- GBTool: Green Building Tool, an international (25 countries) rating system was developed in 1998.
- Green Globes: was developed for Canada using BREEAM and Green Leaf in 2004. It is intended to be an on-line tool.
- Green Mark Rating: Building and Construction Authority (BCA) launched this rating system in 2005 in Singapore for new and existing buildings.

1.2.7 Holistic Design Approach

Fig. 1.25 shows the conventional structure defining the interactions between various members of a design team for a new building. While the architect typically plays the role of a coordinator for the design team, several professionals and engineers have crucial involvement in various phases of the building design and construction. The role and the main duties of each team member as listed in Fig. 1.10 based on the conventional design process are briefly described as follows:

- The architect generally develops the overall aesthetic form and structure of the building including its shell and material requirements while accounting for factors, concepts, and practices that ensure functional, safe, and economical design as per the client specifications. For most current building projects, the architect delivers the design through plans and drawings completed using computer-aided drafting and building information modeling tools. Specifically, the architect coordinates the project and communicates the design specifications to various engineers through computer-aided drafting and/or building information modeling drawings. After the design phase,

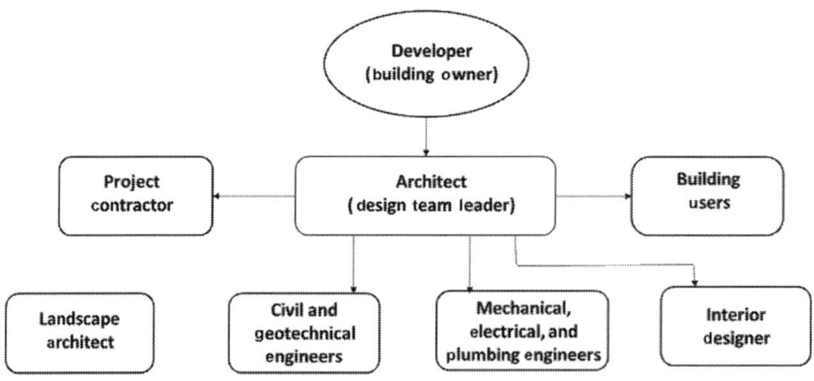

FIGURE 1.25 Interactions between team members of a conventional design process for new buildings.

the architect continues overseeing the project during the construction phase through site visits and some revisions of the design based on any changes to account for the client needs, budget constraints, and other unexpected factors. Through consultations with the architect, engineers provide detailed design specifications for various systems within the building from the conception to the construction phases.

- The mechanical engineer develops design solutions and specifications for energy systems including HVAC, plumbing, and fire protection. Moreover, the mechanical engineer assists the architect in specifying the thermal properties of materials to be used in the building structure as well as any energy systems to improve the energy performance of the building to meet any relevant codes and standards and/or labeling systems.
- The electrical engineer provides design options for power service and distribution as well as lighting, communication, fire detection and alarm, as well as general electrical equipment and associated space requirements. Moreover, the electrical engineer collaborates closely with the architect and the mechanical engineer to meet requirements of any applicable codes and standards and any energy performance targets such as integrating cost-effective energy efficiency technologies including cogeneration and distributed generation systems such as roof-mounted photovoltaic panels.
- The geotechnical or civil engineer evaluates and provides recommendations for the geotechnical specifications of the building site such as ground soil properties and design solutions for earthworks (i.e., grading, drainage, and pavement). In addition, the civil engineer assists in the design of building foundations to ensure that these building elements meet applicable codes and standards. Similarly, the structural engineer assists the architect in the planning, design, and construction of the building structure and its components such as slabs, beams, columns, and foundations. Moreover, the structural engineer specifies the various materials for the building shell following relevant codes and standards.
- The interior designer and the landscape designer work mostly with the client and the architect. The main role of the interior designer is recommending layouts and types of furnishings and decorations within the interior building spaces. The main task of the landscape designer is to propose layouts and types of greenery to be located just outside the building structure. Both designers may also have to communicate their plans and service needs to the mechanical and electrical engineers.

The conventional design process outlined in Fig. 1.25 has several significant shortcomings that often impede designing high performance buildings. In particular, the process limits the involvement of all the stakeholders in the early conception and design phases of the project. For instance, the mechanical engineer is requested to size the mechanical system only after major architectural and structural decisions have been made. Moreover, and due to

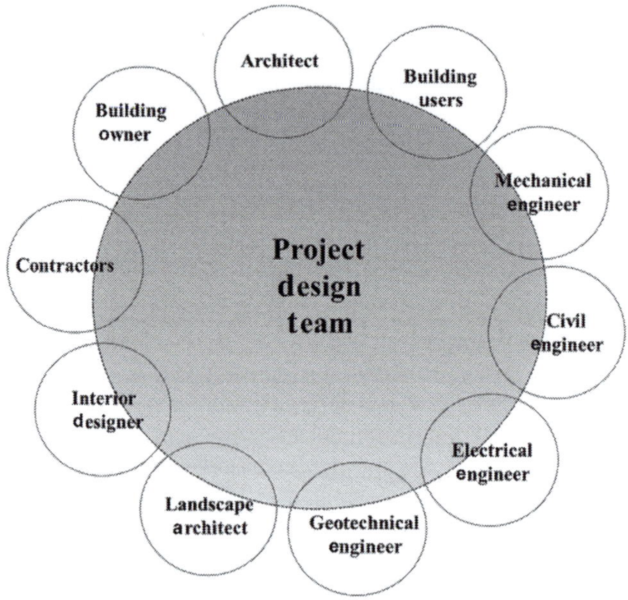

FIGURE 1.26 Interactions between building stakeholders in an integrated design process for new buildings.

limited time and budget given to some systems (i.e., lighting and electrical), innovative and boilerplate design specifications are adopted. Some of these shortcomings can be avoided through an integrated design process as illustrated in Fig. 1.26.

Indeed, the integrated design process allows multiple disciplines and most of the building stakeholders to be engaged and constructively participate in defining and selecting the main specifications of the building. The engagement is facilitated through a series of meetings and discussions. These meetings, also referred to as "charrettes," are especially important in the early design phase to ensure that all relevant building stakeholders including occupants and contractors are able to discuss and share ideas and alternatives. The integrated design process has proven to be effective in achieving higher energy performance and lower costs for sustainable facilities such as LEED-rated or Energy Star-labeled buildings. As will be discussed throughout this book, a wide range of strategies and systems can be utilized to design and retrofit buildings to be more energy efficient. Indeed, several factors affect the overall performance of buildings including mechanical and electrical systems, building occupants, sustainability targets, climate conditions, as well as construction and operation costs. With an integrated design approach, these factors can be considered by the integrated design team members to determine the best synergies and alternatives to achieve the highest levels of energy performance for the new buildings.

1.3 STRATEGIES FOR LARGE SCALE EE PROGRAM IMPLEMENTATION

Three strategies are commonly adopted to develop and implement successfully large-scale energy efficiency programs for existing building stocks:

1. Foundational strategies, also called no-regrets strategies: activities needed to ensure large penetration of the energy efficiency programs.
2. Voluntary strategies: recommended activities to support upgrade projects for all categories of the building stock.
3. Mandatory strategies: needed activities to achieve large-scale upgrade of the existing building stock.

The basic tasks and objectives for each of the listed strategies are described in the following sections.

1.3.1 Foundational Strategies for Large-Scale Energy Efficiency Program

Some of the main phases for foundational strategies to develop large-scale energy efficiency programs for the existing building stock include the following:

- *Phase 1*: Data collection and analysis of energy performance and characteristics of the building stock. In this phase, energy consumption databases for existing buildings should be developed and maintained. The databases can be utilized to establish baselines and benchmarks for the energy performance of existing building stocks. These baselines are then used to develop performance metrics and to document the impacts of any future energy retrofit projects. The primary stakeholders to implement this phase are government agencies, utilities, nonprofit organizations, and building industry professionals. For instance, energy use intensity (EUI) values such as those outlined in Table 1.9 for various building types in France and the US are useful to assess the energy performance of existing buildings within a country or a state (Krarti, 2011).
- *Phase 2*: Support for standards development, compliance, and enforcement. In this phase, programs and standards for improving the energy performance of existing buildings are developed, implemented, and monitored. In particular, administrative structures should be established and the required financial support should be provided for the development and implementation of the programs. Moreover, clear monitoring compliance procedures should be developed to ensure successful enforcement of the implemented energy efficiency standards and programs. Depending on the country, the main tasks for this phase can be completed using government agencies, utilities, building industry professionals, as well as equipment manufacturers and distributors.

TABLE 1.9 Energy Use Intensity (EUI) Values Expressed in kWh/m^2 for Various Building Types in France and the US

Major building activity	France	US
Office	296	300
Education	160	270
Health care	283	610
Lodging	230	325
Food service	416	649
Mercantile and service	264	250
Public assembly	NA	305
Warehouse and storage	NA	147

- *Phase 3*: Education and outreach to building owners and managers. To successfully implement the adopted energy efficiency programs, it is important to educate building owners and managers about the desired goals and the set requirements. In particular, a market research should be carried out to understand customers' attitudes toward energy efficiency specific to existing buildings. Generally, energy-intensive buildings should be first targeted to maximize the benefits of the programs. In addition, a series of informational sessions about the benefits of energy efficiency should be organized to reach building owners, real estate agents, and facility managers. Some of the stakeholders that can assist in the education and outreach efforts include government agencies, utilities, education partners, building industry professionals, and realtor associations.
- *Phase 4*: Training of skilled workforce. This phase is important to ensure the success of any energy efficiency program for existing buildings. Indeed, a skilled workforce is needed to execute any level of building energy retrofits. Due to the wide range of skills required, specific technical training programs have to be developed and delivered possibly through existing educational institutions. These programs should train workforce and contractors in specific energy efficiency technologies suitable for buildings (i.e., high efficacy lighting fixtures, high performance heating and cooling systems, and intelligent controls). These programs should be scheduled regularly to allow practitioners and professionals to update their technical skills. The main stakeholders that can develop and deliver technical and practical training programs include government agencies, utilities, education institutions, and building industry professionals.

Once the foundational strategies for large-scale energy efficiency programs have been developed and tested, two paths for implementation can be considered either voluntary or mandatory. It is recommended to start first with voluntary strategies followed by mandatory strategies.

1.3.2 Voluntary Strategies for Large-Scale Energy Efficiency Program

The development and implementation of voluntary strategies to improve the energy performance of existing buildings include four phases.

- *Phase 1*: Develop a wide range of pathways to achieve energy efficiency. In particular, building owners and managers should have a wide range of options to implement energy efficiency upgrades. For instance, retrofit options could include single energy efficiency measures (i.e., to replace a heating or cooling system), multiple measures (i.e., to upgrade the insulation level in the walls as well as to replace the windows), and deep-retrofits (i.e., to ensure the building meets any current energy-efficiency codes and standards). The contractors need to be informed and educated about available energy efficiency program offerings for the existing building stock as well as about procedures used to document performance metrics of retrofit projects. The main stakeholders for this phase include constructors, building owners, facility managers, realtor associations, and financial institutions.
- *Phase 2*: Develop standardized analysis tools for benchmarking, auditing, and retro-commissioning. These analysis tools are important to facilitate the implementation and monitoring of the energy efficiency programs. Various tools, analysis methods, and protocols are needed depending on the desired objectives including the following:
  - Benchmarking tools to establish baseline metrics for different categories of the existing building stock (offices, hotels, restaurants, retails).
  - Reliable tools and methods for calculating energy use and cost savings for a wide range of building categories and energy efficiency measures.
  - Easy-to-use and cost-effective energy audit protocols and tools.

  The main stakeholders for developing these tools include government agencies, research and educational institutions, building industry professionals, and software developers.
- *Phase 3*: Develop innovative financing mechanisms of energy efficiency upgrades. Building owners and managers often need to have access to financing mechanisms to improve the overall energy performance of their buildings. Several options should be provided to obtain the required capitals through (1) engaging financial institutions to provide loans for energy efficiency projects, (2) encouraging third party private agencies to invest in energy efficiency projects such as energy service companies, and (3) establishing monetary incentives for energy efficiency retrofit projects (such as energy efficiency mortgages and tax reductions). Government agencies,

financial institutions, energy service companies, and nonprofit organizations are typically the main stakeholders for this phase.

1.3.3 Mandatory Strategies for Large-Scale Energy Efficiency Program

To make energy efficiency upgrades for existing buildings mandatory, the following steps can be considered:

- First require minimum energy efficiency levels for public buildings. Governments should provide an example to the private sector by improving the energy efficiency of their buildings.
- Require minimum efficiency levels for rental properties. It is important to ensure that owners provide healthy and energy efficient buildings for renters.
- Require compliance with minimum building energy efficiency for deep retrofits. Typically, deep retrofitted buildings can be required to meet the same energy standards and regulations set for new buildings.

To assist in implementing mandatory programs for existing buildings, several stakeholders can be involved including government agencies, utilities, building industry professionals, property owners, nonprofit organizations, inspectors, building owners, and facility managers.

1.3.4 Approaches for Energy Intensive Buildings

A skilled workforce with a wide range of technical and expertise areas is needed to properly improve the energy performance of existing buildings. Some of the desired expertise areas include the following (Krarti, 2011):

Energy Performance Contracting (EPC): The EPC mechanism has been proven to be effective in financing energy efficiency projects for buildings. Typically, a third party such as energy service companies can evaluate, finance, and implement energy retrofit projects through agreements with building owners or operators. Often these agreements call for proven and sound approaches to measure and verify energy savings incurred from building retrofits. The third party (i.e., energy service companies) would receive payments based on measured and verified energy cost savings achieved by the building retrofit as illustrated in Fig. 1.27.

Demand-Side Management (DSM): This type of management program is typically considered by electric utilities as a mechanism to reduce their customers' peak demands. Moreover, DSM programs are employed by electric utilities to prevent operation or construction of peak generation or spot-market purchases that are often not efficient and are costly. To entice their customers, utilities provide monetary incentives for delivered demand and energy savings.

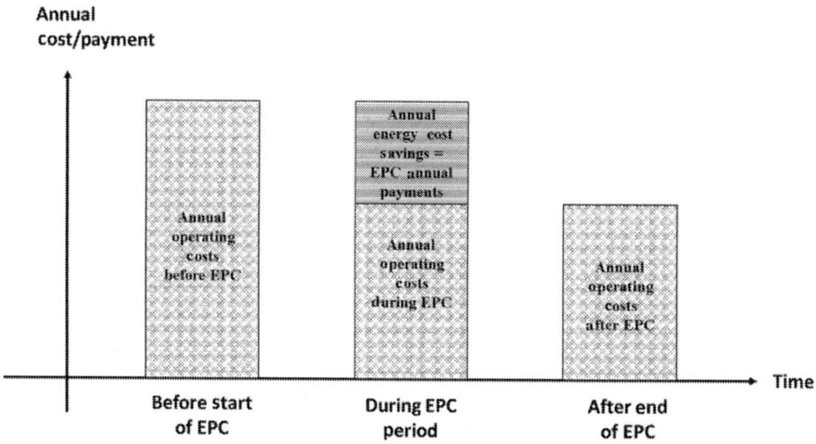

FIGURE 1.27 **Mechanism for energy performance contracting to finance building retrofit projects.**

DSM programs can be targeted for residential and commercial buildings as well as industrial facilities and are typically operated by governmental agencies. There are several DSM programs including the following:

- retrofits (HVAC, lighting, motors, controls)
- new construction (integrated design assistance)
- retro-commissioning (no/low-cost operational)
- load curtailment (on call demand reduction for 3–4 h blocks on peak summer days)
- air-conditioner rebates for efficiency upgrades
- new construction (Energy Star homes)

Commissioning and Retro-Commissioning: For new buildings, commissioning programs can be used to assure that the project concept, design, and installation are appropriate and correct, and that the systems operate properly to the design specifications. For existing buildings, retro-commissioning programs are used to evaluate, troubleshoot, and optimize mechanical systems and their controls to increase system reliability, energy efficiency, occupant comfort, and indoor air quality. Moreover, recommissioning programs can be considered for existing buildings that have already been commissioned or retro-commissioned. In the last decade, continuous commissioning programs have become popular to ensure that there is always an ongoing process to resolve operating problems, minimize comfort complaints, and optimize energy performance for existing buildings. Table 1.10 provides an example to illustrate the difference between commissioning and capital investment projects.

TABLE 1.10 Example to Illustrate the Difference Between Commissioning and Capital Improvement Programs

Option	Capital improvement	Retro-commissioning
Retrofit measure	Replace existing chillers with high-efficiency equipment with rated efficiency from 0.59 kW/ton to 0.43 kW/ton	Identify and implement low-cost improvements for operating the chillers such as fix broken chilled water valves
Energy use savings	160,000 kWh/yr	530,000 kWh/yr
Cost savings	$5700/yr	$19,200/yr
Project cost	$240,000	$3800
Simple payback period	42 years	0.2 year

Measurement and Verification (M&V) Analysis: M&V protocols are required to assess and estimate the energy savings associated to any energy retrofit project especially when energy performance contracts are involved. Several techniques are available to determine the energy savings through the use of inverse or forward models based on short-term or long-term metered data or simple utility data. In particular, the International Performance Measurement and Verification Protocol provides four options for M&V analysis depending on the retrofit project type. In this section, these four options are briefly described including their methods of implementation and common applications (Krarti, 2011):

- Option A: Suitable for energy efficiency measures applied to isolated systems or equipment.
  - Method: Spot measurements, short-term monitoring, simple calculations.
  - Applications: Single energy efficiency measures with known impacts (i.e., lighting fixtures, motors with constant speed).
- Option B: Suitable for single energy efficiency measures.
  - Method: Similar to Option A but only end-use measurements are performed and/or specific values of parameters are identified.
  - Applications: Installations of complex systems (i.e., chillers, boilers, motors with variable speed drives).
- Option C: Suitable for energy retrofit applied to the entire building.
  - Method: Analysis of utility data and/or monitored energy use of entire building.
  - Applications: Acceptable for energy efficiency projects with large expected energy use savings (i.e., weatherization measures including adding thermal insulation to the building envelope). Typical savings of 20% in total building energy consumption.

- Option D: Suitable when detailed simulation and calibrated models are required to estimate energy savings.
  - Method: Modeling of buildings using detailed simulation tools such as eQUEST, EnergyPlus, and others. The energy models have to be calibrated using available metered and/or utility data.
  - Applications: Projects with several energy efficiency measures such as the case for deep retrofits.

Energy Auditing: Energy audit is an important procedure for improving the energy efficiency of existing buildings. Energy auditors require multidisciplinary skills and technical knowledge about all building systems including envelope, lighting, electrical, mechanical, and controls. Moreover, energy auditors should be able to perform a detailed economic analysis to assess the cost-effectiveness of various energy efficiency measures as well as the impact of utility rate selection on annual energy costs. There are four types of energy audits as outlined in the following sections (Krarti, 2011).

Utility Data Analysis: The main purpose of this type of audit is to carefully analyze the operating costs of the facility. Typically, the utility data over several years are evaluated to identify the patterns of energy use, peak demand, weather effects, and the potential for energy savings. In particular, it is important that the energy auditor understands clearly the utility rate structure that are or could be applied to the facility for several reasons including the following:

- Checking the utility charges and ensuring that no mistakes were made in calculating the monthly bills. Indeed, the utility rate structures for commercial and industrial facilities can be quite complex, with ratchet charges and power factor penalties.
- Determining the most dominant charges in the utility bills. For instance, peak demand charges can be a significant portion of the utility bill, especially when ratchet rates are applied. Peak shaving measures can then be recommended to reduce these demand charges.
- Identifying if the facility can benefit from using other utility rate structures or purchase cheaper fuel and reduce its operating costs. This analysis can provide a significant reduction in the utility bills, especially when considering time of use and real-time pricing rate structures.

Moreover, the energy auditor can determine if a facility is prime for energy retrofit projects by analyzing the utility data. Indeed, the energy consumption for buildings can be normalized and compared to known indicators such as EUI commonly used for commercial buildings.

Walk-Through or Level 1 Energy Audit: This audit consists of a short on-site visit to the facility to identify areas where simple and inexpensive actions can provide immediate energy and/or operating cost savings. Some engineers refer to these types of actions as operating and maintenance (O&M) measures. Examples of O&M measures include lowering heating set-point temperatures, replacing broken windows, insulating exposed hot water or steam pipes, and adjusting boiler fuel–air ratio.

Standard or Level 2 Energy Audit: The standard audit provides a comprehensive analysis of the energy systems of the facility. In addition to the activities described for the walk-through audit and the utility cost, the standard energy audit includes the development of a baseline for the energy use of the facility and the evaluation of energy use savings and cost effectiveness of appropriately selected energy efficiency measures. Generally, simplified tools are utilized in the standard energy audit to develop baseline energy models and to predict the energy savings of energy conservation measures. Among these tools are the degree-day methods and linear regression models. In addition, a simple payback period is often utilized as a metric to determine the cost-effectiveness of energy efficiency measures.

Detailed Level 3 Energy Audit: This is the most comprehensive but also time-consuming energy audit type. Specifically, the detailed energy audit includes the use of instruments to measure energy use for the whole building and/or for some energy systems within the building (for instance, by end uses including lighting fixtures, office equipment, appliances, fans, chillers, and boilers). In addition, sophisticated computer simulation programs are typically employed for detailed energy audits to evaluate and recommend energy retrofits for the facility. The computer simulation programs used in the detailed energy audit typically provide the energy use distribution by load type (i.e., energy use for lighting, fans, chillers, boilers, etc.). They are often based on the dynamic thermal performance of the building energy systems and usually require a high level of engineering expertise and training. Techniques available to perform measurements for an energy audit are diverse (Krarti, 2011). However, a rigorous economical evaluation of the energy efficiency measures is generally performed as part of a detailed energy audit. Specifically, the cost-effectiveness of energy retrofits may be determined based on the LCC analysis rather than the simple payback period analysis. As presented in Chapter 5, LCC analysis takes into account a number of economic parameters such interest, inflation, and tax rates.

Table 1.11 summarizes performance data and energy efficiency measures that are generally considered when conducting an energy audit of thermal and electrical systems for existing residential and commercial buildings.

1.4 SUMMARY

This chapter provides an overview of the current and future trends in the energy consumption in the world as well as in various countries and regions. The energy consumption specifically attributed to the building sector is described based on historical data as well as future projections. In addition, the chapter summarizes various energy efficiency strategies and policies suitable for improving the energy performance for new and existing buildings. General technical practice areas required to implement energy efficiency programs are outlined to pinpoint

TABLE 1.11 Performance Data and Energy Efficiency Measures for Existing Residential and Commercial Buildings

Thermal systems	Electric systems
Utility analysis	
Thermal energy use profile (building signature) Thermal energy use per unit area (or per student for schools or per bed for hospitals) Thermal energy use distribution (heating, DHW, process, etc.) Fuel types used Weather effect on thermal energy use Utility rate structure	Electrical energy use profile (building signature) Electrical energy use per unit area (or per student for schools or per bed for hospitals) Electrical energy use distribution (cooling, lighting, equipment, fans, etc.) Weather effect on electrical energy use Utility rate structure (energy charges, demand charges, power factor penalty, etc.)
On-site survey	
Construction materials (thermal resistance type and thickness) HVAC system type DHW system Hot water/steam use for heating Hot water/steam for cooling Hot water/steam for DHW Hot water/steam for specific applications (hospitals, swimming pools, etc.)	HVAC system type Lighting type and density Equipment type and density Energy use for heating Energy use for cooling Energy use for lighting Energy use for equipment Energy use for air handling Energy use for water distribution
Energy use baseline	
Review architectural, mechanical, and control drawings Develop a base-case model (using any baselining method ranging from very simple to more detailed tools) Calibrate the base-case model (using utility data or metered data)	Review architectural, mechanical, electrical, and control drawings Develop a base-case model (using any baselining method ranging from very simple to more detailed tools) Calibrate the base-case model (using utility data or metered data)
Energy conservation measures	
Heat recovery system (heat exchangers) Efficient heating system (boilers) Temperature setback EMCS HVAC system retrofit DHW use reduction Cogeneration	Energy efficient lighting Energy efficient equipment (computers) Energy efficient motors HVAC system retrofit EMCS Temperature setup Energy efficient cooling system (chiller) Peak demand shaving Thermal energy storage system Cogeneration Power factor improvement Reduction of harmonics

DHW, Domestic hot water; *EMCS*, energy management control system; *HVAC*, heating, ventilating, and air conditioning.

to the technical skills required to implement energy efficiency programs for the building sector. Moreover, the concept of energy productivity is introduced using economic output and energy consumption data for the world and select countries.

REFERENCES

Dinda, S., 2004. Environmental Kuznets curve hypothesis: a survey. Ecol. Econ. 49, 431–455.

2017. U.S. Energy Information Administration. International Energy Outlook, Washington, DC, 2016. http://www.eia.gov/ieo (accessed 05.10.17).

Grossman, G., Krueger, A., 1991. Environmental Impacts of a North American Free Trade Agreement. National Bureau of Economic Research, Working Paper WP3914, Cambridge, MA.

IEA, 2013. Capturing the Multiple Benefits of Energy Efficiency, International Energy Agency Report. Accessed August 5, 2017. Available from: http://www.iea.org/publications/freepublications/publication/Captur_the_MultiplBenef_ofEnergyEficiency.pdf.

IEA, 2017. World Energy Balances, Data Services by the International Energy Agency, http://wds.iea.org/ (accessed 07.10.17).

Ihm, P., Krarti, M., 2012. Design optimization of energy efficient residential buildings in Tunisia. Build. Environ. 58 (12), 81–90.

Kaika, D., Zervas, E., 2013a. The environmental Kuznets curve (EKC) theory—Part A: concept, causes and the CO_2 emissions case. Energy Policy 62, 1392–1402.

Kaika, D., Zervas, E., 2013b. The environmental Kuznets curve (EKC) theory. Part B: Critical issues. Energy Policy 62, 1403–1411.

KAPSARC, 2015. Global Shift: the energy productivity transformation. KAPSRAC Energy Workshop Series. The King Abdullah Petroleum Studies and Research Center, Riyadh, http://www.kapsarc.org/.

Krarti, M., 2011. Energy Audit of Building Systems: An Engineering Approach, 2nd edn. CRC Press, Boca Raton, FL.

Krarti, M., Dubey, K., Howarth, N., 2017. Evaluation of building energy efficiency investment options for the Kingdom of Saudi Arabia. Energy 134, 595–610.

Krarti, M., Dubey, K., 2017. Energy productivity evaluation of large scale building energy efficiency programs for Oman. Sustain. Cities Soc. 39, 12–22.

Kuznets, S., 1955. Economic growth and income inequality. Am. Econ. Rev. 45 (1), 1–28.

Panayotou, T., 1993. Empirical Tests and Policy Analysis of Environmental Degradation at Different Stages of Economic Development, Working Paper, Technology and Environment Program. International Labor Office, Geneva.

Pollin, R., Heintz, J., Garrett-Peltier, H., 2009. The Economic Benefits of Investing in Clean Energy. Report from the Political Economy Research Institute (PERI), University of Massachusetts Amherst, Amherst, MA.

RCREEE, 2015. Arab Future Energy Index 1015 – Energy Efficiency. Report by the Regional Center for Renewable Energy and Energy Efficiency (RCREEE), Cairo, Egypt, www.rcreee.org.

RISE, 2016. Regulatory Indicators for Sustainable Energy: A Global Scorecard for Policy Makers. World Bank, Washington, DC.

Rockefeller Foundation, 2012. United States Building Energy Efficiency Retrofits, Market Sizing and Financing Models. Report from Rockefeller Foundation on March 2012.

SASO, 2012. Energy Labelling and Minimum Energy Performance Requirements for Air-Conditioners (Standard No. 3459). Saudi Arabia Standards Organization, Saudi Arabia.

SASO, 2013. Energy Labelling Requirements of Household Electrical Clothes Washing Machines (Standard No. 3569). Saudi Arabia Standards Organization, Saudi Arabia.

SASO, 2014. Energy Performance Capacity and Labelling of Household Refrigerators, Refrigerator-Freezers, and Freezers (Standard No. 3620). Saudi Arabia Standards Organization, Saudi Arabia.

Streimikiene, D., Kasperowicz, R., 2016. Review of economic growth and energy consumption: a panel cointegration analysis for EU countries. Renew. Sustain. Energy Rev. 59, 1545–1549.

Chapter 2

Advanced Building Energy Efficiency Systems

Chapter Outline

2.1 INTRODUCTION

As noted in Chapter 1, buildings represent the highest energy-consuming sector in the global economy using over 50% of the total electricity generated in the world (IEA, 2013). Moreover, it is expected that the energy demand of the building sector will further increase significantly by 2050 with the addition of

Optimal Design and Retrofit of Energy Efficient Buildings, Communities, and Urban Centers.
http://dx.doi.org/10.1016/B978-0-12-849869-9.00002-8

2.5 billion people to the world's population. Specifically, the energy consumption attributed to buildings can rise by 50% by 2050 if no additional policies are taken to improve their energy efficiency performance (IEA, 2013). However, this growth in energy demand can be reduced to just 10% if currently available and proven energy efficiency technologies and strategies are implemented for both new and existing building stocks.

While the amount of energy used by the building sector varies significantly between countries and depend on several factors including climate, population, economic development, and income level, proven energy efficiency technologies and energy policies are globally effective (IEA, 2013). Among the recommended energy efficiency technologies, that are suitable for new and existing buildings, are the following:

- Optimized and advanced envelope systems to reduce undesired heat transmission through windows, walls, roofs, and floors. Indeed, envelope components account for a significant portion of space heating loads: 50% for residential buildings and 60% for commercial buildings (DOE, 2015). Globally, it is estimated that a high-performance building envelope can reduce space heating loads by 20%–30% in cold climates and space cooling loads by 10%–40% in hot climates (IEA, 2013).
- Energy-efficient heating and cooling systems to lower energy requirements to maintain indoor thermal comfort. Heating, ventilating, and air conditioning (HVAC) systems consume 35% of the total energy used by the US building sector (DOE, 2015). The application of low-energy systems such as heat pumps and solar thermal systems is highly recommended to heat and cool buildings. Moreover, the use of distributed generation including renewables and combined heat and power systems can also reduce the end-use consumption for space heating and cooling (IEA, 2013).
- Improved lighting systems and appliances as well as the application of intelligent controls to optimize their operation represent a significant potential to enhance energy efficiency especially for households. Indeed, the use of natural lighting and the adoption of high efficacy lamp technologies can reduce the energy consumption needed to illuminate buildings by 40% by 2050 (IEA, 2013). In the United States, it is estimated that the best currently available technologies can reduce lighting energy use by 60% for residential and commercial buildings compared to the current Energy Star requirements (DOE, 2015).

In this chapter, representatives of the mentioned energy-efficiency technologies are discussed to show their potential in reducing the energy consumption of buildings without affecting occupants' thermal comfort. Additional energy efficiency and renewable energy are presented throughout the book. Some of the technologies can be suitable for individual buildings (Chapter 6), large communities (Chapter 7), and urban centers (Chapter 8).

Typical national energy policies recommended to improve the energy efficiency levels of the building sector consist of the following options:

- Stringent and more rigorous building codes: these codes are developed, implemented, enforced, and continuously updated in several countries. With these codes, supporting infrastructures such as education and training capacities need to be available to ensure that the energy performance of buildings can be continuously improved as more cost-effective technologies are developed and demonstrated. For instance, it has been shown that energy end-use for space heating can be set between 15 and 30 kWh m^{-2} $year^{-1}$ in most cold climates using the currently available HVAC systems (IEA, 2013).
- Stricter minimum performance energy standards: these programs should be set for lighting, appliances, and cooling systems used in buildings. It has been estimated that the application of the best current technologies can reduce the global final energy consumption for lighting, cooking, appliances, and other building equipment by almost 50% by 2050. Moreover, the use of high-performance HVAC systems such as heat pumps should be required to provide space heating and cooling with minimal energy consumption. For instance, moving from electric resistance heaters to heat pumps can increase space heating efficiency by a factor of three (IEA, 2013).
- Establishment of deep renovation programs: these programs can improve significantly the energy performance of existing buildings. Indeed, the global potential of improving the energy efficiency levels of existing building stock is substantial for developed and developing countries. In some countries, the energy consumption of existing buildings can be reduced by 50% using cost-effective measures (Krarti et al., 2017).
- Requirement of zero- or positive-energy buildings: some countries have started to mandate that all new buildings be net-zero energy. The application of holistic integration of energy efficiency with renewable energy technologies is the basis of designing zero-energy buildings. With the lower costs of rooftop photovoltaic (PV) panels, the incremental costs for designing net-zero energy buildings are becoming low and manageable.

In this chapter, the basic principles of energy efficiency for buildings are outlined first. Then, the main indicators and models available to assess thermal comfort within the built environment are presented. Moreover, whole-building analysis methods and tools often used to optimally select suitable energy efficiency measures are described. Finally, three advanced energy efficiency technologies are discussed in detail: (1) dynamic insulation materials (DIMs) as an example of advanced building envelope systems, (2) thermo-active foundations (TAFs) to illustrate a cost-effective application of ground-source heat pumps to heat and cool buildings, and (3) integrated control LED systems to highlight the energy-saving potential of the optimized operation of lighting systems.

2.2 IMPLEMENTATION OF ENERGY EFFICIENCY

2.2.1 Overview of Energy Efficiency Principles

As noted in Fig. 2.1, building loads can be categorized into four: (1) lighting, (2) appliances and equipment, (3) heating and cooling, and (4) domestic hot water. For conventionally designed and operated buildings, energy sources (i.e., electricity, natural gas, or others) are typically delivered by utilities to operate systems that serve building loads. These systems including lighting fixtures, appliances and office equipment, HVAC (heating, ventilating, and air-conditioning) systems, and DHW (domestic hot water) systems, have inherent energy losses due to low efficiencies. Some of the building loads can be handled through passive systems with no energy requirements such as natural ventilation, solar heat gains, and daylighting through apertures.

One of the basic guidelines for building energy efficiency is to reduce building loads while maintaining an acceptable level of indoor thermal comfort. Indeed, indoor thermal comfort should always be maintained and should not be compromised to reduce building loads and ultimately lower energy use and cost. Section 2.3 of this chapter discusses the basic factors that affect indoor thermal comfort as well as some models to estimate metrics for occupant thermal comfort and sensation. Moreover, Chapter 9 outlines the importance of maintaining indoor thermal comfort for work productivity within the built environment. In particular, the impacts of reduced indoor thermal comfort and air quality are quantified in terms of the economic output of the building sector.

To reduce building loads and improve the energy-efficiency levels of new or existing buildings, the following general strategies need to be considered:

- Reduce space heating loads through insulating and improving the air-tightness of the building envelope as well as improving controls.
- Reduce space cooling loads through reducing solar and internal gains as well as the use of the thermal mass of the building.

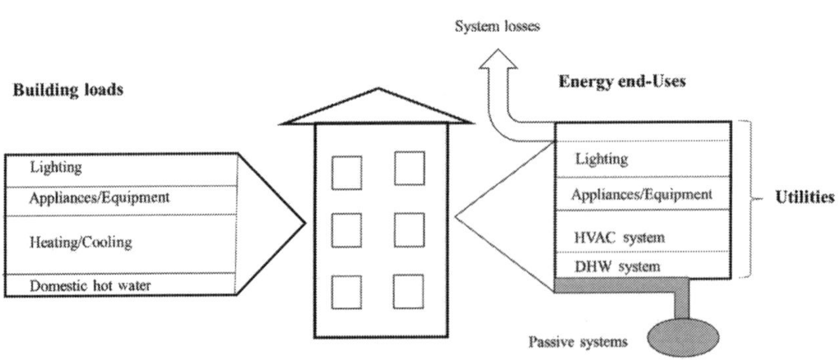

FIGURE 2.1 Building loads and energy consumption end-uses for conventionally designed and operated buildings.

- Decrease energy requirements for ventilation by using natural ventilation or mixed mode ventilation and better controls (i.e., demand ventilation controls).
- Decrease energy use for lighting through the installation of energy-efficient fixtures with high efficacies and integrated controls.
- Reduce energy consumption for heating water by setting the proper temperatures and minimizing distribution losses.
- Decrease energy consumption for office equipment and appliances by selecting energy-efficient systems and setting proper controls for their use.

In addition, systems that serve building loads should be selected to minimize energy consumption. Moreover, energy-efficient systems should be optimally operated to reduce energy losses. Moreover, passive systems should be considered to take advantage of natural energy sources (i.e., solar heat gains, daylighting, and natural ventilation). In summary, an energy-efficient building has low building loads and rely on high-performance systems with reduced energy requirements from the utilities as noted in Fig. 2.2.

To maximize natural energy gains for a building, it is important to consider its surrounding elements and climatic conditions. In hot climates, for instance, solar heat gains should be minimized but natural lighting should be maximized through optimal window sizes and/or improved shading devices. In cold climates, on the other hand, both solar heat gains and natural lighting should be maximized. In most climates, building surrounding elements including, ambient air, ground medium, and sun, can provide heating, cooling, and ventilation using no or little energy consumption through passive systems (natural ventilation, solar direct gains, and sunspaces) and/or low-energy heating and cooling systems (i.e., evaporative cooling systems, earth-tubes, and ground-source heat pumps).

Figs. 2.3 and 2.4 illustrate the concept of using passive systems and low-energy systems for heating and cooling buildings. As noted in Fig. 2.3A, when the outdoor temperatures are below the desired thermal comfort levels during a

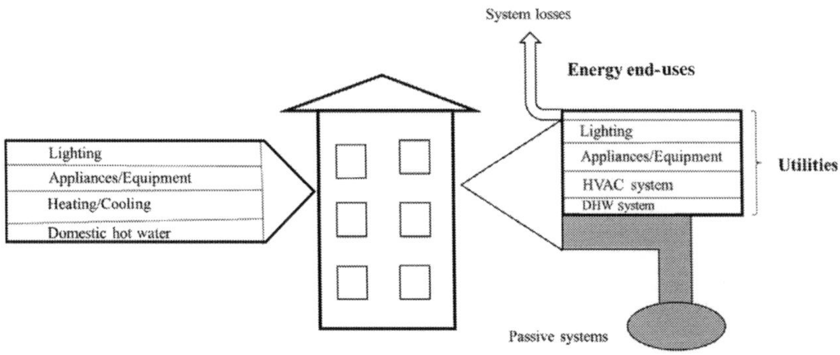

FIGURE 2.2 Building loads and energy consumption end-uses for energy-efficient buildings.

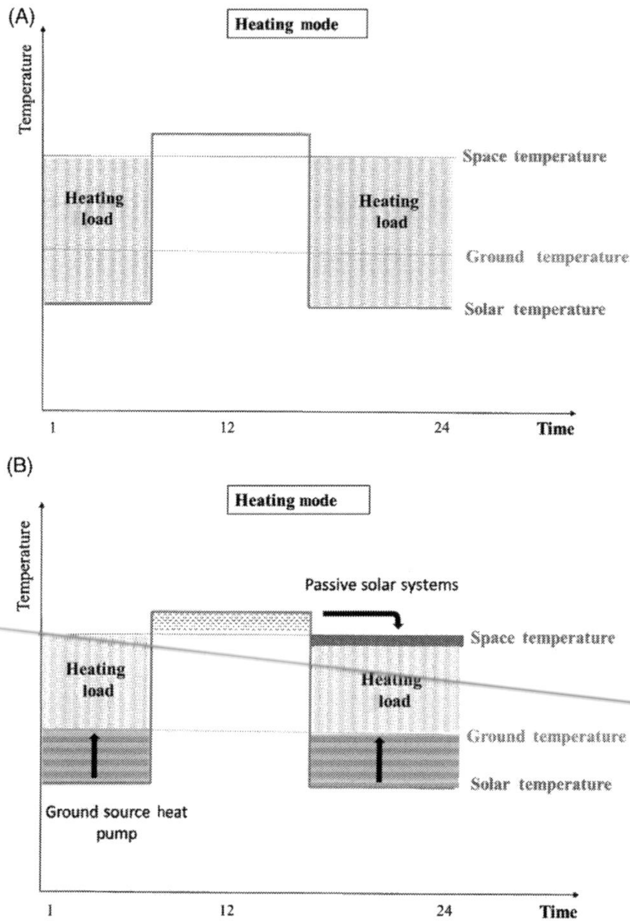

FIGURE 2.3 Building space heating (A) without and (B) with the application of passive- and low-energy systems.

cold day, the building heating load can be significant (shown by the gray area). However, when passive heating systems to store heat gained during the daytime for release during the nighttime as well as low-energy heating systems such as a ground-source heat pump are considered, building heating loads are lowered as noted in Fig. 2.3B. During a warm day, the use of natural ventilation to precool the thermal mass (i.e., floor and walls) as well as a ground-source heat pump can lower the building cooling load as illustrated in Fig. 2.4.

When their loads and energy end-uses are minimized, buildings are considered energy-efficient. Several regulations are typically required for new buildings to ensure that designers meet minimum energy efficiency requirements including insulation and air leakage levels for walls, roofs, floors, and windows as well as efficiency standards for lighting fixtures, appliances, office equipment,

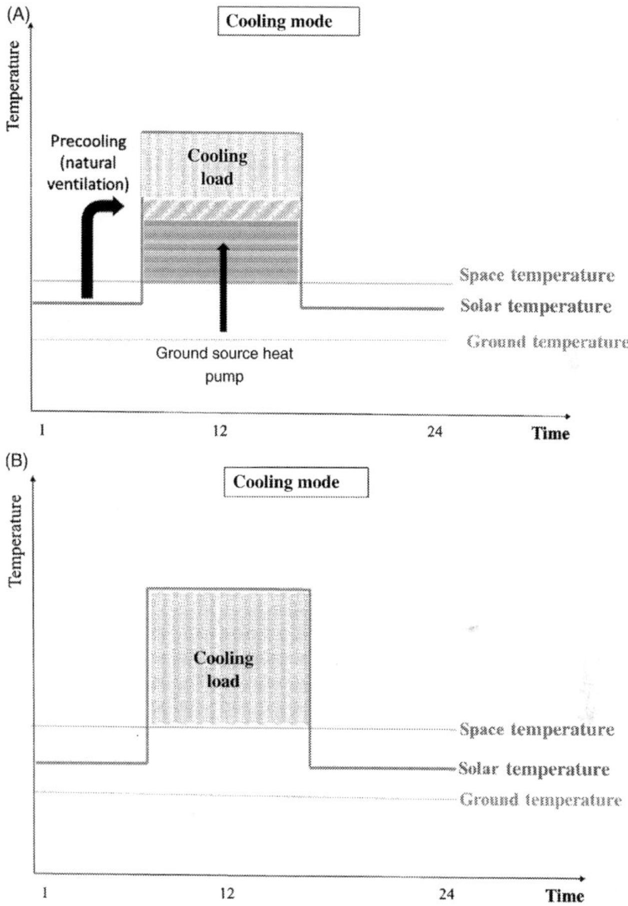

FIGURE 2.4 Building space cooling (A) without and (B) with the application of passive- and low-energy systems.

DHW, and HVAC systems. While not typically mandated, building rating labels and certification programs encourage designers to integrate renewable energy systems to generate electricity and thermal energy needed to meet building loads. In the last decade, the concept of net-zero energy buildings (NZEBs) has become common and mainstream. For a typical NZEB, the energy generated from a rooftop PV panel is either used to meet direct building electricity needs or fed to the utility grid as illustrated in Fig. 2.5. Through net-metering arrangements, buildings can sell as well as buy electricity from utilities, making the grid a large storage system. Options and challenges related to the interconnections of buildings and microgrids with the grid are discussed in more detail in Chapter 4.

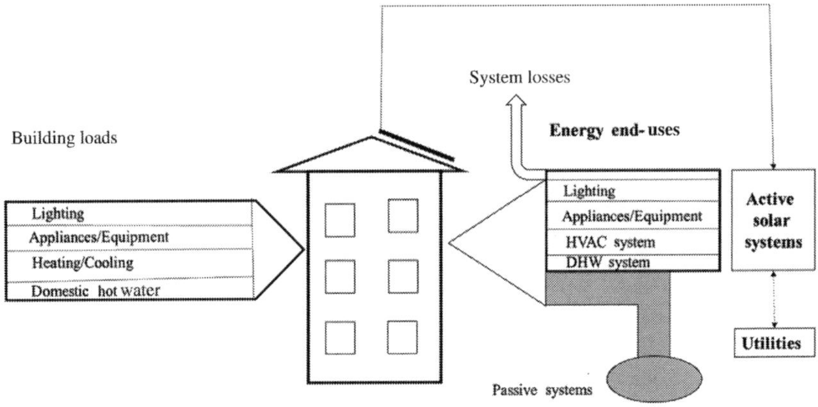

FIGURE 2.5 Energy flow for a net-zero energy building with a rooftop photovoltaic system.

When designing or retrofitting buildings, the basic rule of thumb is to first try to reduce the energy consumption through cost-effective energy efficiency measures. Then the application of renewable energy systems is considered such as rooftop PV panels for achieving NZEBs. Fig. 2.6 compares the energy consumption and generation flows for energy-efficient and NZEBs against a baseline (i.e., conventional design) benchmark.

Different techniques and approaches can be considered in designing and retrofitting buildings to be energy efficient. Some of these approaches are briefly outlined in the following section.

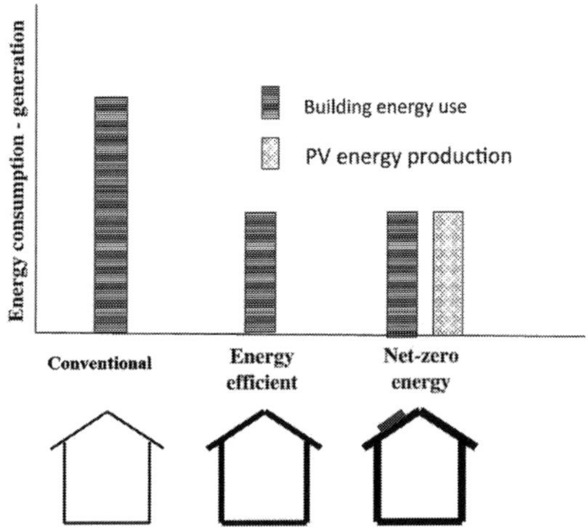

FIGURE 2.6 Energy consumption and generation levels for conventional, energy-efficient, and net-zero energy buildings.

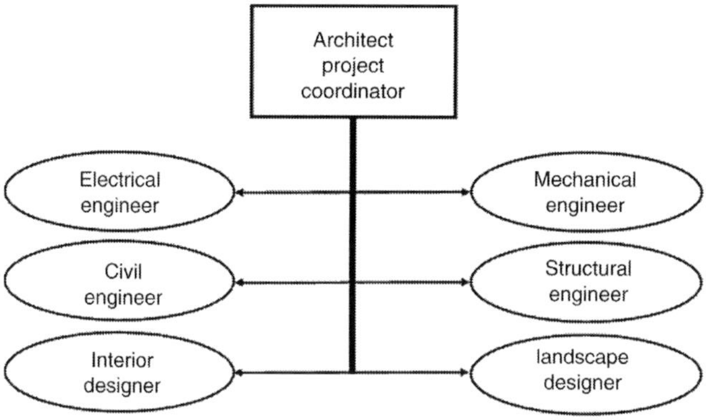

FIGURE 2.7 **Typical design team members for commercial building projects.**

2.2.2 Energy Efficiency Analysis Approaches

One of the main barriers for energy-efficient and sustainable buildings is the fragmentation of the construction industry. Indeed, several stakeholders are involved in financing, designing, constructing, and operating buildings. For instance, Fig. 2.7 illustrates the different stakeholders typically needed to design commercial buildings using the traditional design-bid-build method (Krarti, 2017). In particular, the design team includes an architect acting as the main coordinator of the building project and several engineers responsible for providing specifications of various systems associated with the building. Moreover, several phases are required to complete the design and the construction of a building including schematic design, design development, and construction documentation. To design energy-efficient buildings, significant collaboration between the design team members is required especially in the early design phase. Fig. 2.8 illustrates the impacts on energy efficiency level and associated implementation

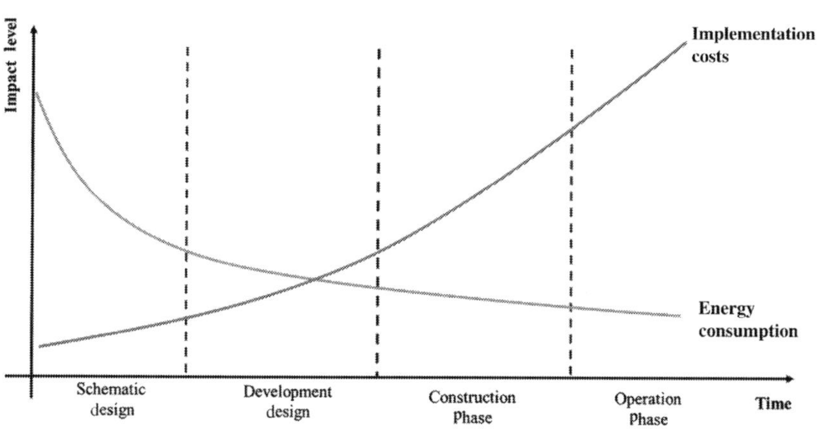

FIGURE 2.8 **Impact on energy efficiency and implementation costs.**

cost during various design phases. As indicated in Fig. 2.8, when the building is constructed and commissioned, it is often too late to improve its energy efficiency unless through basic maintenance and operation measures (Krarti, 2011).

The integrated design approach using the integrated design-build method has been proven to be the most effective delivery process for high-performance buildings (Krarti, 2017; WBCSD, 2008). In the traditional design process, the design team members work in "silo" with minimal cooperation and communication often resulting in suboptimal performance buildings. However, the integrated design approach involves all participants in the early design phase of the building project. In particular, multidisciplinary workshops such as "design charrettes" are arranged to bring together owners, architects, engineers, and other stakeholders (i.e., capital provides, contractors, and potential building users). These stakeholders cooperate across the different specialties to develop the best possible design that meet preset expectations within well-defined constraints. Currently, the integrated design procedure (IDP) is used to design or retrofit buildings that have to meet requirements for building rating labels or certification programs such as LEED or Energy Star. A similar procedure is used when designing or retrofitting buildings that meet high energy-efficiency targets such as those set for zero- or positive-energy buildings. Generally, IDP can achieve improved building performance with lower costs and fewer disruptive changes during the later project stages. Fig. 2.8 clearly shows that the earlier in the process the IDP occurs, the greater the impact on the building energy performance and the lower are the implementation costs (WBCSD, 2008).

To design or retrofit energy efficient buildings, architects and engineers follow several methods and processes. However, two main approaches are typically considered for designing new buildings and retrofitting existing buildings: (1) systems approach and (2) whole-building approach.

2.2.2.1 Systems Approach

The focus of this approach is to consider and evaluate individual or a set of energy efficiency measures. Thus, the impact and the cost-effectiveness of only measures related to specific systems are evaluated such as building envelope (i.e., add wall insulation), lighting systems (i.e., install occupancy sensors), or mechanical systems (i.e., install a heat recovery system). The selection of these measures may be associated with the requirements of building energy-efficiency codes when formulated using prescriptive approaches. In the systems approach, the assessment of the impact of the energy efficiency measures can be carried out using simple analysis techniques including steady-state forward models such as those presented in Section 9.4. Moreover, simplified economic analysis methods are utilized to evaluate the cost-effectiveness and ultimately to recommend energy efficiency measures. A general rule of thumb, energy efficiency measures are considered cost-effective when their simple payback period is less than 5 years. Fig. 2.9A illustrates an example of the energy savings and the cost-effectiveness of select energy efficiency measures considered for a

(A)

Energy conservation measure (ECM)	Recommended	Incremental cost	Electricity cost savings	Simple payback (Yrs)	Rate of return
Additional roof insulation	No	$13,300	$328	40.6	0.00%
HVAC EER upgrade	Yes	$1,387	$1,622	0.9	115.0%
desiccant system	Yes/No	$36,000	$5,422	6.6	14.50%
Central system	No	$70,000	$3,186	22.0	0.00%
Premium motors	Yes	$355	$97	3.7	27%
Lower lighting power density	Yes	$3,302	$1,610	2.1	50.00%
Install PV system+	Yes/No	$254,100	$9,527	9.70	9.20%

(B)

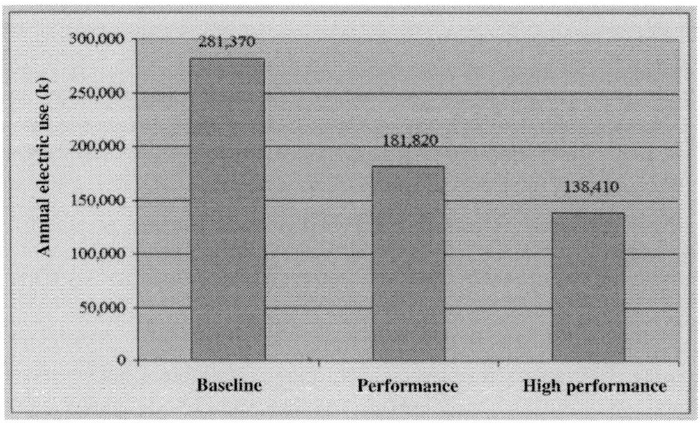

FIGURE 2.9 Example of the application of the systems approach to (A) select cost-effective energy efficiency measures and (B) define high-performance design for a storage facility.

storage facility (Krarti, 2011). Fig. 2.9B compares the energy use costs associated with the baseline building design with two energy-efficient alternatives including a high-performance design option.

2.2.2.2 Whole-Building Approach

This approach is often used for the integrated design process for sustainable buildings. It is the preferred method to achieve maximal levels of energy savings since the approach considers optimized energy efficiency solutions. In most cases, the whole-building approach utilizes detailed computer simulation tools such as those described in Section 9.4. Moreover, the approach requires generally the application of optimization techniques depending on the targets set for the design objectives. Typically, the whole building approach attempts to minimize life-cycle costs of constructing, maintaining, and operating the building to ensure safe, healthy, and comfortable conditions for its occupants. Depending on the cost function, several optimization techniques can be considered as discussed in Chapter 6. Due to thermal interactions between various building

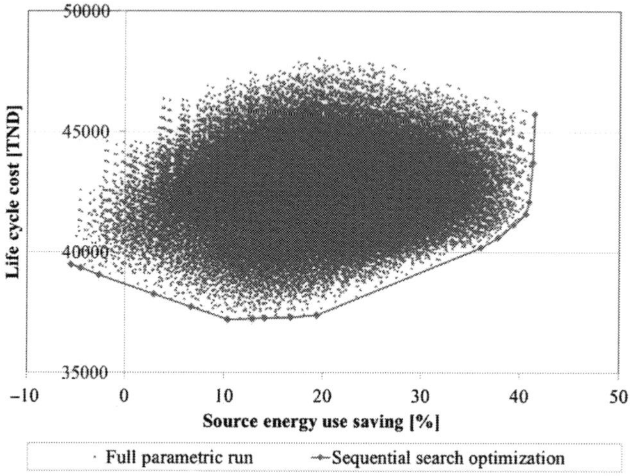

FIGURE 2.10 Comparison of optimal results obtained by the brute-force analysis approach and the sequential search optimization for the 8-EEM package for a villa.

systems, optimal designs or retrofit options may include some energy efficiency measures that when applied individually may not be cost-effective. Thus, a holistic and integrative analysis of several energy efficiency measures is crucial to achieve optimal energy performance for buildings. Fig. 2.10 illustrates the number of combinations formed by eight energy efficiency measures (EEMs) for designing a villa (Ihm and Krarti, 2012). The eight EEMs consist of building orientation, wall insulation, roof insulation, window size, glazing type, lighting level, infiltration rate, and cooling set point. Each EEM has several options (typically 3–7). Each dot in Fig. 2.10 corresponds to one combination of the eight EEMs. The main target of the optimization considered in designing the villa is to minimize life-cycle costs (shown in the *y*-axis) while maximizing the energy use savings (shown in the *x*-axis). In addition to the brute-force technique (i.e., considering all possible combinations of the eight EEMs), Fig. 2.10 also shows the optimal path obtained using the sequential search technique to determine the optimal design solutions for the villa (Ihm and Krarti, 2012). Chapter 6 discusses in more detail the sequential search optimization technique.

2.2.3 Comparative Analysis of Heating and Cooling Systems

In this section, a comparative analysis of several alternatives to heat and cool a residential building is presented to illustrate the impact of HVAC system selection on the energy efficiency level of the building. Both passive and active space heating and cooling options are considered. The analysis is based on a detailed simulation modeling analysis for an apartment building located in Denver, CO (Seo and Krarti, 2010). A 3D rendering of the apartment building is depicted in Fig. 2.11. The passive and active HVAC systems are evaluated to determine

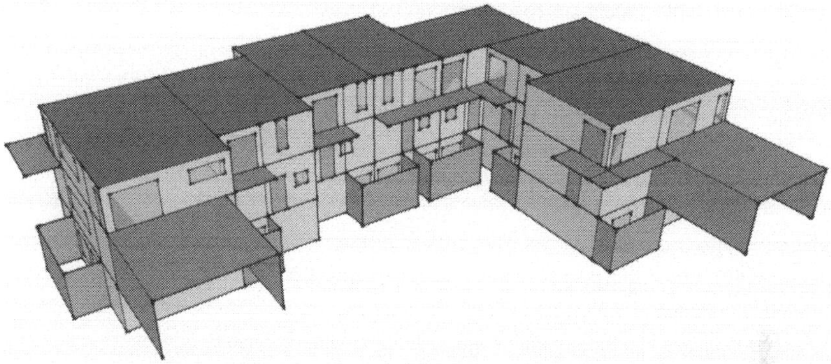

FIGURE 2.11 **Model rendering of apartment building in Denver, CO.**

their performance in terms of energy efficiency and thermal comfort. A brief description of each heating and cooling system is provided in the following sections.

System 1. No HVAC system: The building has no heating or cooling system. Energy is used only for systems such as lighting, appliances, and DHW. This case is considered as a reference case to assess the level of thermal discomfort within the building if no heating and cooling systems are installed.

System 2. Natural ventilation: With this option, the building is cooled using natural ventilation. No mechanical system is used for heating or cooling. For this building, natural ventilation is considered by opening windows only when the outside air temperature is lower than 23°C or when the indoor and outdoor temperature difference is over 1°C.

System 3. Earth tube: This option utilizes heating/cooling energy stored within the ground medium. In this building, an earth tube is used to condition air supplied to each apartment unit in the building. A design flow rate of $0.1\ m^3\ s^{-1}$ (212 cfm) is used for each unit with the following earth tube fan control options:

- Cooled/heated air is provided to each unit using fans.
- No fan operation for the period ranging from October 1 to May 30 due to the very small heat extraction capabilities of the earth tube (the outlet air temperature from the tube is lower than indoor air temperature).

System 4. Radiative walls: Radiative walls are used to heat and cool the apartment unit. Three thermostats per unit are used to avoid significant temperature stratification. It is assumed that the heating season starts October 29 and ends April 17 and that cooling temperature settings are used for the other periods. To operate the system, operative temperature control is used. Simulation analysis to compare various control strategies has indicated that controls based on operative temperature provide more savings in heating/cooling energy use than controls based on mean-air temperature. Chilled water in this system is supplied through a district cooling system. However, the pumping energy and site energy for generating the chilled water are accounted for when evaluating this system.

System 5. DX air conditioner and gas furnace: This HVAC system, most commonly used in the United States to condition residential buildings, consists of a gas-furnace and an electric air conditioner (AC). A COP of 3.0 for the AC and an efficiency of 80% are assumed for this system. This system is considered a benchmark and is used to evaluate the potential energy savings for other HVAC systems.

System 6. Radiative wall with cooling tower: This system is a variation of System 4 with an open type cooling tower (condenser) loop directly connected to the radiative wall system loop to reduce cooling energy consumption.

System 7. Radiative wall with evaporative cooler: This system is also a variation of System 4 with an evaporative cooler used to cool the water supplied to the radiant walls.

System 8. Radiative floors: Instead of radiative walls, radiative floors are utilized to provide heating and cooling throughout each apartment unit of the building. A radiative panel per floor for each unit is considered.

System 9. Heating with radiative floor and cooling with radiative wall: This system uses two separate water loops: (1) a hot water loop to provide heating through the radiative floors as in System 8, and (2) a chilled water loop to provide cooling through radiative walls as in System 4.

In addition to the various alternatives for the HVAC system, other design options are considered including lower DWH usage (low-flow devices), lighting power density (use of LED lamps), and equipment power density (use of high energy efficiency appliances) as well as different floor constructions with higher insulation and mass levels.

Fig. 2.12 summarizes the energy performance of all the heating and cooling systems described earlier and the potential energy savings relative to the conventional air conditioning system (i.e., System 5 with gas furnace and DX cooling). Fig. 2.13 compares the thermal comfort performance of all the systems expressed in terms of the number of hours where the *PMV* (Fanger's predicted mean value) is within −0.5 and 0.5 range for all the apartment units in the building as recommended by ASHRAE Standard 55 for indoor thermal comfort (refer to Section 2.3).

Based on the results provided in Figs. 2.12 and 2.13, the following general observations can be made:

- Without any heating and cooling system, thermal comfort can be achieved 65% of the time. Specifically, the total comfort hours (*PMV* index within −0.5 and 0.5) are 17,000 for System 1 compared to 26,000 achieved by systems with heating and cooling capabilities (including System 5 through

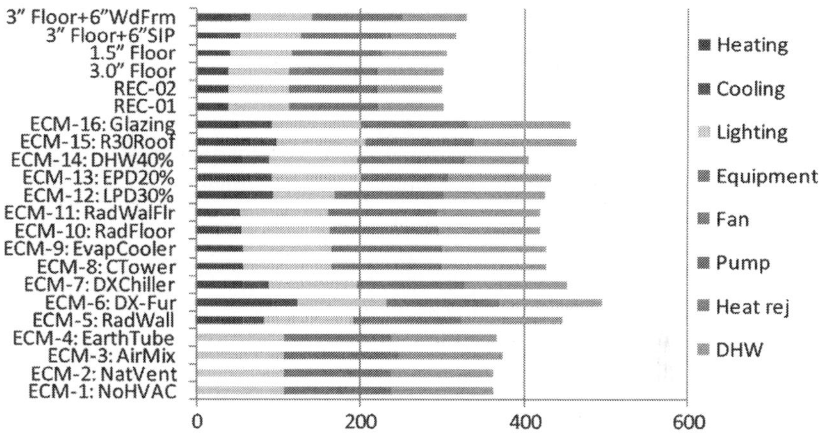

FIGURE 2.12 Summary of annual energy end-use for various heating and cooling systems considered for the apartment building.

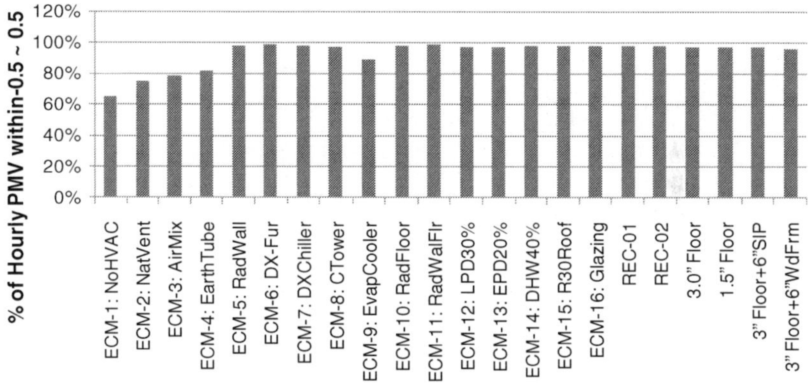

FIGURE 2.13 Summary of annual thermal comfort level for various heating and cooling systems considered for the apartment building.

System 9) out of all 26,280 possible hours (i.e., sum of three floors per unit, i.e., 3*7760 = 26,280).

- System 3 is only effective during the summer since room temperatures in the winter are typically higher than the air temperatures supplied by the earth tube. Detailed estimation for change-over period (from heating to cooling) is needed to better improve the performance of this system.
- The radiative floor system (System 8) saves almost 50% in heating energy use when compared to the radiative wall system (System 4).
- Evaporative cooling with a cooling tower or an evaporative cooler is effective and is highly recommended for cooling for this building located in Denver, CO, characterized by mild and dry climatic conditions during the summer season.

2.3 INDOOR THERMAL COMFORT

2.3.1 Overview of Thermal Comfort

Thermal comfort within the built environment is rather challenging to achieve and maintain since it is affected by several factors. ASHRAE Standard 55, Thermal Environmental Conditions for Human Occupancy, uses the definition that, "Thermal Comfort is that condition of mind that expresses satisfaction with the thermal environment." This definition emphasizes that the judgment of comfort is a cognitive process involving many inputs influenced by physical, physiological, psychological, and other processes. The conscious mind appears to reach conclusions about thermal comfort and discomfort from direct temperature and moisture sensations by the skin, deep body temperatures, and the efforts necessary to regulate body temperatures.

Several thermal comfort indicators do exist including effective temperature, operative temperature, mean radiant temperature, and *PMV*. Most of these indicators are based on heat balance models applied to the human body to indicate that heat generated by metabolic activities ($q_{met,heat}$) must be dissipated by convection (q_{conv}), radiation (q_{rad}), evaporative heat transfer (q_{evap}), and respiration ($q_{resp,sen}$ and $q_{reap,lat}$):

$$q_{met,heat} = q_{conv} + q_{rad} + q_{evap} + q_{resp,sen} + q_{resp,lat} \tag{2.1}$$

A basic overview of the commonly used thermal comfort indicators is provided in the following sections.

Mean radiant temperature: When only convection and radiation heat exchanges are considered, a mean radiant temperature can be defined and estimated using the globe temperature, T_g, air temperature, T_a, and mean air velocity, V, as

$$\begin{aligned} T_{mrt}^4 &= T_g^4 + 0.103 * 10^9 \bar{V}^{1/2} (T_g - T_a) \quad (IP) \\ T_{mrt}^4 &= T_g^4 + 0.247 * 10^9 \bar{V}^{1/2} (T_g - T_a) \quad (SI) \end{aligned} \tag{2.2}$$

Operative temperature: The weighted average of air (T_a) and radiant (T_r) temperatures using convective (h_c) and radiative (h_r) heat transfer coefficients is called the *operative temperature* and is defined as follows:

$$T_{op} = \frac{h_c T_a + h_r T_r}{h_c + h_r} \tag{2.3}$$

Effective temperature: The *effective temperature*, ET^*, is defined as the operative temperature at 50% RH that gives the same total heat transfer as at the actual environmental conditions. The exact relationship to estimate the effective temperature is a function of several variables, including clothing levels, activity levels, and convection and radiation heat transfer coefficients.

Fanger's predicted mean value: The predicted mean vote or *PMV* is the average response of a large number of people. Given the subjective nature of comfort, there is typically a distribution of satisfaction levels among a large group of people. Fanger developed a set of correlations between *PMV* and the body's thermal load, *L*:

$$\begin{aligned} PMV &= 3.155(0.303e^{-0.114M} + 0.028)L \quad US \\ PMV &= (0.303e^{-0.036M} + 0.028)L \quad SI \end{aligned} \tag{2.4}$$

The body's thermal load, *L*, depends on several parameters including clothing level, air velocity, dry-bulb temperature, relative humidity, and mean radiant temperature (Fanger, 1970). Fig. 2.13 shows an empirical relationship between the percentage of people dissatisfied (PPD) with a thermal environment as a function of the *PMV*. As shown in Fig. 2.14, a neutral *PMV* (i.e., *PMV* = 0) indicates that 5% of people are dissatisfied with the indoor thermal comfort.

2.3.2 Pierce Two-Node Model

The Pierce two-node model was developed by Gagge et al. (1970) at John Pierce Foundation Laboratory. The model is a simplified mathematical lumped capacity analysis of the human thermoregulatory system. The model thermally lumps the human body as two isothermal concentric thermal compartments representing the skin and the core. The temperature within a compartment is assumed to be uniform so that the only temperature gradient is between the compartments. This model uses Fanger's model for calculating respiratory heat losses. The core loses energy when the muscles do work. Heat is also lost from the core through respiration. In addition, heat is conducted passively from the core to the skin. Based on the experimental research, the *PMV* modified by effective temperature (PMVET) is proposed similar to the Fanger comfort model (*PMV*) using a nine-point thermal sensation scale (4 to −4). The thermal sensation (TSENS) of sedentary people is represented using an 11-point thermal sensation scale (5 to −5).

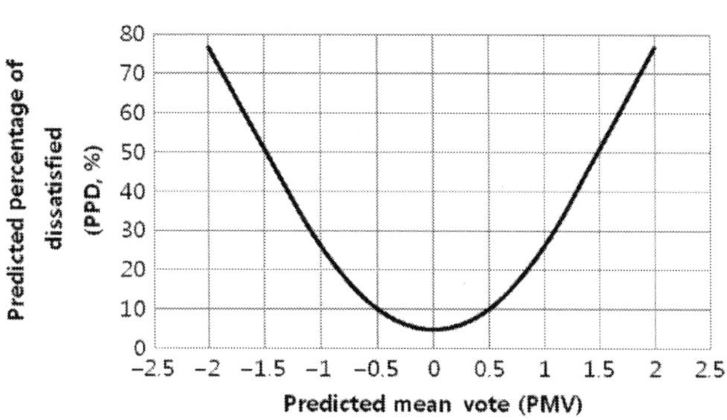

FIGURE 2.14 Correlation between PDD and PMV for the Fanger's model.

2.3.3 KSU Two-Node Model

The KSU two-node model was developed at Kansas State University. It is similar to the Pierce two-node model except that the calculation of thermal sensation (TSV) in the KSU model is different from that used to estimate the TSENS of the Pierce model. The empirical sensation equation was developed from a wide range of experimental conditions ranging from very hot to very cold, activity levels from 1 to 6 *met*, and clothing levels from 0.05 to 0.7 *clo*. Thermal conductance between the core and the skin temperature is applied to calculate thermal sensations in a cold environment. However, the change of skin wittedness is considered in a warm environment. In addition, the thermal sensation vote (*TSV*) has a similar scale as the Fanger comfort model (*PMV*) and Pierce two-node model (TSENS).

2.3.4 ASHRAE Comfort Standard

ASHRAE has developed an industry consensus standard to describe comfort requirements in buildings. The standard is known as ASHRAE Standard 55, Thermal Environmental Conditions for Human Occupancy. The purpose of this standard is to specify the combinations of indoor thermal environmental factors and personal factors that will produce thermal environmental conditions acceptable to a majority of the occupants within the space. Refer to Fig. 2.15 to estimate the acceptable thermal comfort conditions for sedentary activity (let <1.2)

Acceptable operative temperatures depend on the activity of a person (expressed in *met*) and the clothing level (defined by its insulation level or *clo*) as approximated by the following equation:

$$\begin{aligned} T_{op,\text{active}}(°F) &= T_{o,\text{sedentary}} - 5.4(1+clo)(met-1.2) \quad \text{(IP)} \\ T_{op,\text{active}}(°C) &= T_{o,\text{sedentary}} - 3.0(1+clo)(met-1.2) \quad \text{(SI)} \end{aligned} \tag{2.5}$$

2.3.5 Other Thermal Comfort Models

2.3.5.1 Adaptive Predicted Mean Vote

The adaptive predicted mean vote (*aPMV*), sometimes referred to as the corrected *PMV* model, takes into account the behavioral, psychological, and physiological adaptation capabilities of occupants depending on the outdoor conditions. It is often used to reduce the needs for space heating during cold periods and for space cooling during warm periods. An expression for *aPMV* has been established based on the *PMV* value corrected using an adaptive coefficient, β, as follows (Yao et al., 2009):

$$aPMV = (PMV^{-1} + \beta)^{-1} \tag{2.6}$$

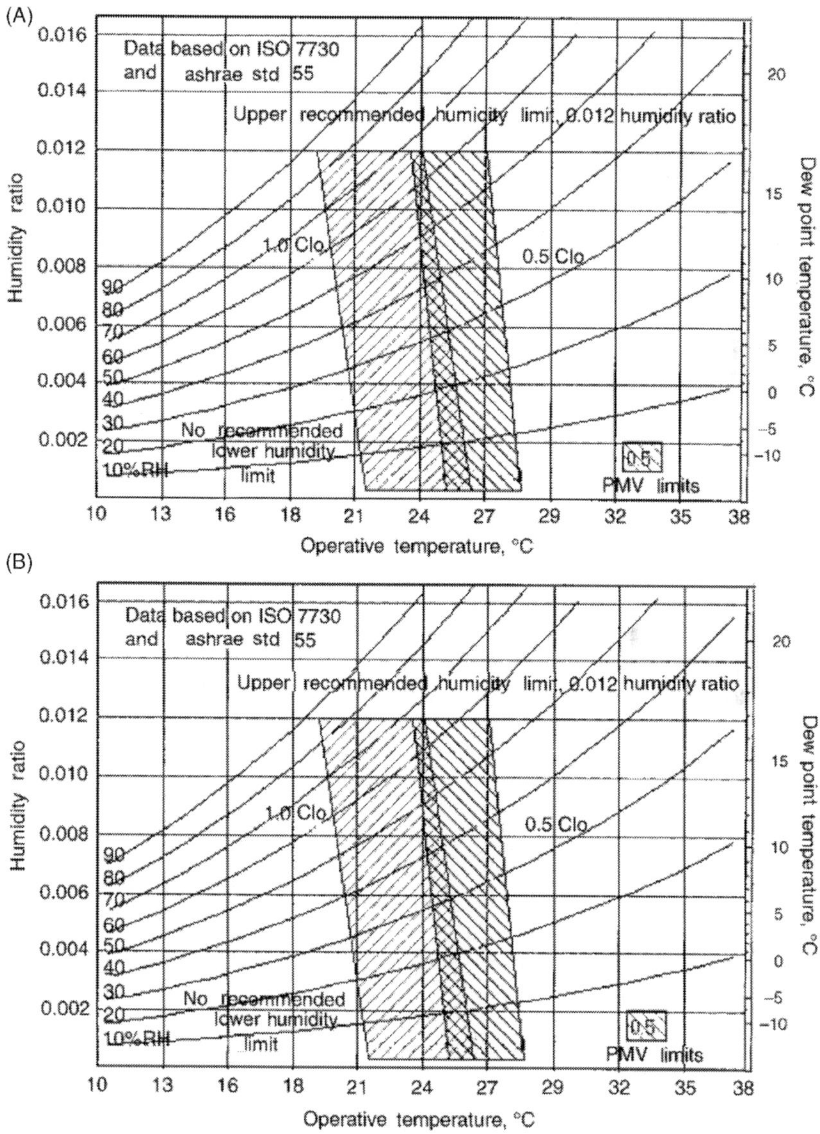

FIGURE 2.15 Psychrometric chart for ASHRAE comfort zones for both heating and cooling seasons.

The adaptive coefficient is typically negative for cold periods and positive for warm periods. Yao et al. (2009) recommended β to be set at -0.125 and 0.293 for, respectively, cold and warm conditions. However, other studies based on field data have found different values for various countries and building types (Enescu, 2017).

2.3.5.2 Extended Predicted Mean Vote

For naturally ventilated buildings using little or no air conditioning and located in hot and humid climates, Fanger and Toftum (2002) have extended the application of the PMV model by adding a correction factor, e_p, as indicated by the following equation:

$$ePMV = e_p \cdot PMV \tag{2.7}$$

The values of the correction factor depend on the climate as well as the number and the expectation level that the surrounding buildings are air-conditioned. Some guidelines for the selection of the e_p value range include the following:

- From 0.5 for locations with few air-conditioned buildings to 0.7 for locations where air-conditioned buildings are common throughout the year with low expectations.
- From 0.7 for locations with few air-conditioned buildings to 0.9 for locations where air-conditioned buildings are common during summer seasons with moderate expectations.
- From 0.9 to 1.0 for locations with most buildings are air-conditioned during warm periods with high expectations.

Several studies have used surveys and measurements to refine or verify the ranges set for the correction factor of the extended PMV model suitable for different regions, period, and countries (Enescu, 2017).

2.3.5.3 New Predicted Mean Vote

The new predicted mean vote (nPMV) model is based on the adaptive thermal comfort approach by reducing the biases between predictions from the PMV model and thermal sensations of occupants found in air-conditioned buildings (Humphreys and Nicol, 2002). The nPMV model can be expressed as follows:

$$nPMV = 0.8(PMV - D_{PMV\text{-ASHRAE}}) \tag{2.8}$$

A regression expression of the function $D_{PMV\text{-ASHRAE}}$ is obtained based on field surveys and measured data as a function of operative temperature (T_{op}), indoor relative humidity (RH_{in}), metabolic rate (met), clothing level (clo), and outdoor mean air temperature (T_{out}) (Humphreys and Nicol, 2002):

$$\begin{aligned} D_{PMV\text{-ASHREA}} &= -4.03 + 0.0949 \cdot T_{in} + 0.00584 \cdot RH_{in} \\ &\quad + 1.201 \cdot met \cdot clo + 0.000838 \cdot T_{out} \end{aligned} \tag{2.9}$$

A similar nPMV model of Eq. (2.9) has been proposed by Kim et al. (2015) using statistical analysis of field measurements.

2.3.5.4 Actual Mean Vote Model

The Actual Mean Vote model is suitable for air-conditioned buildings in tropical climates. This model is defined as the occupants' thermal sensation expressed using the 7-point scale of the ASHRAE Standard 55 model (Hermawan et al., 2015).

2.4 WHOLE-BUILDING ENERGY MODELS

Generally, whole-building energy analysis tools are classified into either forward or inverse methods (Krarti, 2011). In the forward approach as depicted in Fig. 2.16, energy use predictions are based on a physical description of the building systems such as geometry, location, construction details, and HVAC system type and operation. Most of the existing detailed energy simulation tools such as DOE-2, TRNSYS, and EnergyPlus follow the forward modeling approach. In the inverse approach illustrated in Fig. 2.17, the energy analysis model attempts to deduce representative building parameters (such as the building load coefficient (BLC), the building base-load, or the building time constant) using existing energy use, weather, and relevant performance data. In general, the inverse models are less complex to formulate than the forward models. However, the flexibility of inverse models is typically limited by the formulation of the representative building parameters and the accuracy of the building performance data. Most of the existing inverse models rely on regression analysis [such as the variable-base degree-day models (Fels, 1986) or the change-point models (Kissock et al., 1998)] or connectionist approach (Kreider et al., 1997) to identify the building parameters. It should be noted that tools based on forward or inverse approaches are suitable for other applications. Among common applications of whole-building analysis tools are verification of energy savings actually incurred from implementing energy efficiency measures, diagnosis of equipment malfunctions, and efficiency testing of building energy systems (Krarti, 2011).

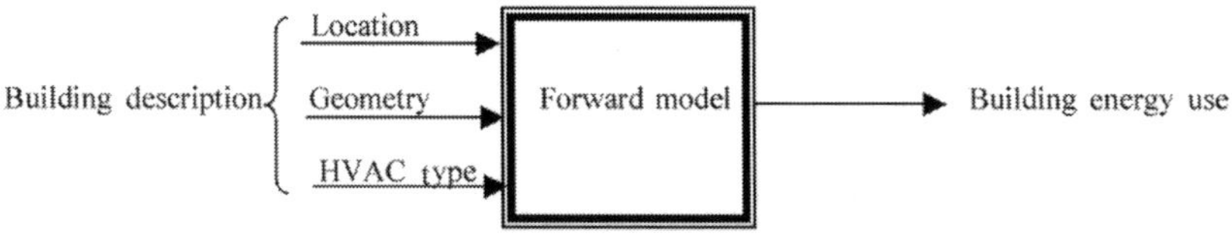

FIGURE 2.16 Basic approach of a typical forward energy analysis model.

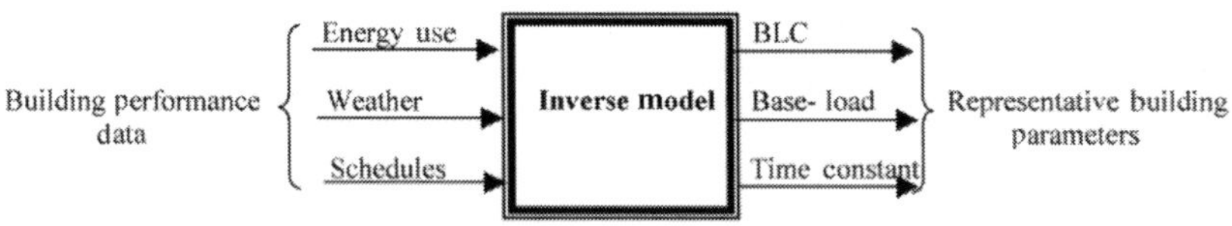

FIGURE 2.17 Basic approach of a typical inverse energy analysis model.

Whole-building energy analysis tools can also be classified based on their ability to capture the dynamic behavior of building energy systems. Thus, energy analysis tools can use either steady-state or dynamic modeling approaches. In general, steady-state models are sufficient to evaluate seasonal or annual building energy performance. However, dynamic models may be required to assess transient and time-dependent effects of building energy systems such as those encountered for thermal energy storage systems and optimal start controls.

In this section, selected energy analysis tools commonly used to predict the energy consumption of buildings are briefly described. As noted earlier, these tools are grouped into two categories:

- Inverse methods: using steady-state and dynamic modeling approaches and include variable-base degree-day methods.
- Forward methods: including either steady-state or dynamic modeling approaches, are often the basis of detailed energy simulation computer programs.

2.4.1 Inverse Modeling Methods

Methods using the inverse modeling approach rely on existing building performance data to identify a set of building parameters. The inverse modeling methods can be valuable tools in identifying measures to improve building energy efficiency. In particular, the inverse models can be used to

- help detect malfunctions by identifying time periods or specific systems with abnormally high energy consumption,
- provide estimates of expected savings from a defined set of energy conservation measures, and
- measure and verify savings achieved by energy retrofits.

Typically, regression analyses are used to estimate representative parameters for the building and/or its systems (such as building load coefficient or heating system efficiency) using measurements or utility data. In general, steady-state inverse models are based on monthly and/or daily data and include one or more independent variables. Dynamic inverse models are usually developed using hourly or subhourly data to capture any significant transient effect such as the case when buildings have high thermal mass levels that can be used to delay or shift cooling or heating loads.

2.4.1.1 Steady-State Inverse Models

These models generally attempt to identify relationships between building energy consumption and selected weather-dependent parameters such as monthly or daily average outdoor temperatures, degree-hours, or degree-days. As mentioned earlier, these relationships can be identified using statistical methods (often using linear regression analysis). The main advantages of steady-state inverse models are the following:

- Simplicity: steady-state inverse models can be developed based on a small data set such as energy data obtained from utility bills.
- Flexibility: steady-state inverse models have a wide range of applications. They are particularly valuable in predicting heating and cooling energy end-uses for residential and small commercial buildings.

However, steady-state inverse models have some limitations since they cannot be used to analyze transient effects such as thermal mass impacts and seasonal changes in the efficiency of the HVAC system. Steady-state inverse models are especially suitable for measurement and verification (M&V) of energy savings accrued from energy retrofits. In this section, only simplified methods based on steady-inverse modeling are briefly presented. These simplified models have been used to determine energy use savings for selected energy efficiency measures and are based on the degree-day method (Krarti, 2011). Using steady-state analysis, building energy use per billing period is correlated to either the average outdoor temperatures or to heating or cooling degree-days (CDDs) (obtained for the billing period). Thus, building energy consumption is estimated for each billing period using the following expressions depending on whether outdoor temperature or degree-day is used.

Eq. (2.10) provides a temperature-base model while Eq. (2.11) outlines a degree-day based model for evaluating building energy use during heating seasons (Krarti, 2011):

$$E_H = 24 \times \frac{BLC}{\eta_H} \times (T_{b,H} - \overline{T}_o) + E_{base,H} \tag{2.10}$$

$$E_H = 24 \times \frac{BLC}{\eta_H} \times DD_H(T_{b,H}) + E_{base,H} \tag{2.11}$$

Eq. (2.12) provides a temperature-base model while Eq. (2.13) outlines a degree-day based model for evaluating building energy use during cooling seasons (Krarti, 2011):

$$E_C = 24 \times \frac{BLC}{COP_C} \times (\overline{T}_o - T_{b,C}) + E_{base,C} \tag{2.12}$$

$$E_C = 24 \times \frac{BLC}{COP_C} \times DD_C(T_{b,C}) + E_{base,C} \tag{2.13}$$

where, $E_{H/C}$ represents the annual building energy use during heating or cooling season; *BLC* is the building loss coefficient; η_H is the average seasonal energy efficiency of the heating system; COP_C is the average seasonal coefficient of performance of the cooling system; $T_{b,H/C}$ is the building balance temperature for heating or cooling energy use; $DD_{H/C}$ is the heating or CDDs (based on the balance temperature); $E_{base,H/C}$ is the base-load for building energy use. It represents the nonheating or noncooling energy use.

Through regression analysis, the balance temperature and the building load coefficient (assuming the heating or cooling system efficiency is known) can be determined. Once building parameters are estimated, the models of Eqs. (2.10)–(2.13) can be used to establish an energy use model for the building and to determine any energy savings attributed to measures that affect one of the three parameters (i.e., balance temperature, building load coefficient, and heating/cooling system efficiency). Fig. 2.18 illustrates the use of the heating model outlined by Eq. (2.11) to correlate the natural gas usage as a function of the heating degree-day (HDD) for a residential building (Kalinic, 2009).

2.4.1.2 Dynamic Models

Steady-state inverse models are only suitable for predicting long-term building energy use (i.e., on monthly or annual basis). Therefore, energy use data are collected for a relatively long time-period (at least one season or one year) to carry out the regression analysis. On the other hand, dynamic inverse models can be used to predict short-term building energy use variations using data collected for a short period of time such as 1 week. Generally, a dynamic inverse model is based on a building thermal model that uses a specific set of parameters. These building model parameters are identified using typically some form of a regression analysis. An example of a dynamic model relating building cooling energy

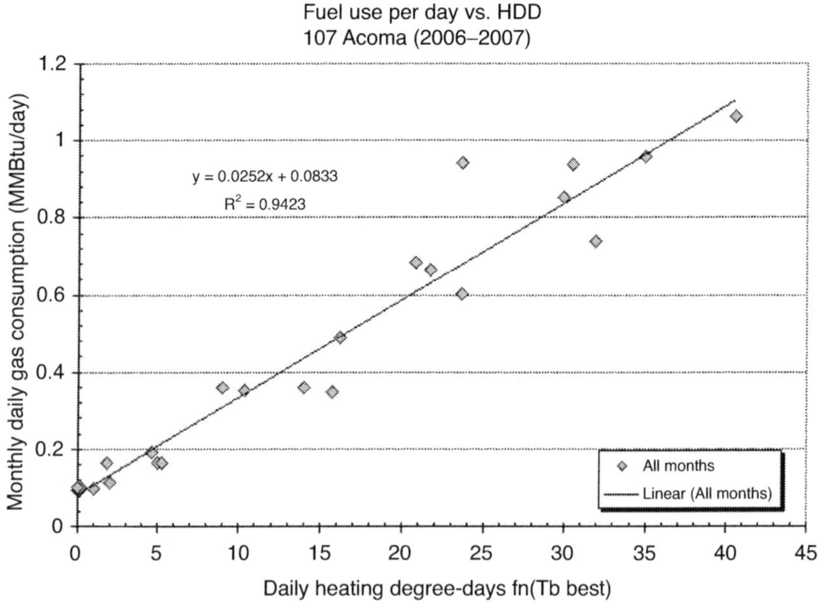

FIGURE 2.18 Analysis of gas consumption as a function of monthly heating degree-days (HDDs).

use to the outdoor air temperatures at various time steps (typically hours) is presented by Eq. (2.14):

$$E_C^n + b_1 E_C^{n-1} + \cdots + b_N E_C^{n-N} = a_0 T_o^n + a_1 T_o^{n-1} + \cdots + a_M T_0^{n-M} \qquad (2.14)$$

Other examples of dynamic inverse models include equivalent thermal network analysis, Fourier series models, and artificial neural networks (Krarti, 2011). These models are capable of capturing dynamic effects such as building thermal mass dynamics. The main advantages of the dynamic inverse models include the ability to model complex systems that depend on several independent parameters. Their disadvantages include their complexity and the need for more detailed measurements to fine-tune the model. Unlike steady-state inverse models, dynamic inverse models usually require a high degree of user interaction and knowledge of the modeled building or system.

2.4.2 Forward Modeling Methods

Forward modeling methods are generally based on the physical description of building energy systems. Typically, forward models can be used to determine energy end-uses as well as predict any energy savings incurred from energy efficiency measures. Selected energy analysis tools that use the forward modeling approach are described in the following sections. For a more detailed discussion, the reader is referred to ASHRAE Handbook of Fundamentals (ASHRAE, 2009).

2.4.2.1 Steady-State Methods

Steady-state energy analysis methods that use the forward modeling approach are generally easy to use since most of the calculations can be performed by hand or using spreadsheet programs. Moreover, most steady-state forward tools are based on one form of the degree-day methods (Krarti, 2011). As described earlier for inverse modeling, the degree-day method uses seasonal degree-day computed at a specific set-point temperature (or balance temperature) to predict energy use required for space heating or cooling. However, degree-days methods are not suitable for predicting space cooling loads especially for massive buildings. In the United States, the traditional degree-day method using a base temperature of 18°C (65°F) has been replaced by the variable-base degree-days method and is applied mostly to residential buildings. In Europe, HDDs using 18°C as the base temperature is often used for residential and commercial buildings.

The variable-base degree-day methods predict seasonal building energy use for space heating as follows (Krarti, 2011):

$$FU = \frac{24 \cdot BLC \cdot f_{DH} \cdot DD_H(T_b)}{\eta_H} \qquad (2.15)$$

where, *FU* represents the fuel use (gas, fuel oil, or electricity depending on the heating system); *BLC* is the building load coefficient including transmission and infiltration losses through the building envelope; f_{DH} is a correction factor to include various effects such night set-back effects, and free heat gains; T_b is the building heating balance temperature; and $DD_H(T_b)$ is the HDDs calculated at the balance temperature T_b.

Variable-base degree-days methods provide generally good predictions of the fuel use for residential buildings dominated by transmission loads. However, they are not recommended for buildings dominated by internal loads and/or with involved HVAC system operation strategies. Example 2.1 illustrates the use of the degree-day method to estimate energy savings from adding roof insulation.

Example 2.1

An apartment building has 1000 m^2 an uninsulated roof with *R*-value of 1.44 W m^{-2}·°C. Determine the payback period of adding insulation (R = 2.0°C·m^2 W^{-1}). The apartment building is electrically heated. The cost of electricity is \$0.10/kWh. The house is located in a location where the HDDs is 2758°C-day (based on 18°C) and is heated 24 h/day, 7 days/week throughout the heating season. Assume that the installed cost of the insulation is \$20/m^2.

Solution

To determine the energy savings due the addition of insulation, we will assume that the annual HDDs before and after the retrofit remained unchanged and are close to 18°C. Using Eq. (2.15) with the retrofitted roof *U*-value to be 0.37 W m^{-2}·°C and heating system efficiency set to be unity (electrical system), the energy savings are calculated to be in kWh/year (the conversion factor of 0.001 is used to convert W into kW):

$$\Delta FU = (24)\cdot(1000)\cdot(1.44-0.37)\cdot(2758)*(0.001) = \frac{72{,}811\,\text{kWh}}{\text{year}}$$

Thus, the simple payback period, *SPP*, for adding insulation on the roof can be estimated to be (refer to Chapter 5)

$$SPP = \frac{(1000\,\text{m}^2)\cdot(\$20/\text{m}^2)}{(72.811\,\text{jWh}/\text{year})\cdot(\$0.10/\text{kWh}^2)} = 2.75\,\text{years}$$

Therefore, the addition of insulation seems to be cost-effective (refer to Chapter 5).

2.4.2.2 Dynamic Methods

Dynamic analytical models use numerical or analytical methods to determine energy transfer between various building systems. These models are based on solutions using resistance-capacitance or RC thermal networks as well as other transient solutions using response factors or heat balance methods to estimate adequately the effects of thermal inertia, due for instance to energy storage in the building envelope and/or its heating system (Krarti, 2011). The main feature of detailed whole-building simulation programs is their capability to account for several parameters that are crucial to accurately predict energy use especially for buildings with significant thermal mass, thermostat setbacks or setups, explicit energy storage, or predictive control strategies. First, a basic RC thermal network model for buildings is presented. Then, some simulation analysis tools commonly used for designing and retrofitting buildings are outlined.

2.4.2.2.1 RC Thermal Network Analysis Technique

An entire building can be represented by a thermal RC network. Using a simple model as shown in Fig. 2.19, two temperature nodes, a single capacitance, and a single resistance can represent building heat transfer (Krarti, 2011). The two temperatures are the indoor and outdoor temperatures; the capacitance represents the entire thermal mass of the building and the resistance represents the heat transfer between the indoor and outdoor temperatures. Additional heat flow paths can be added to account for internal heat gains and heat gain by the HVAC system as noted in Fig. 2.19.

Mathematically, the thermal network shown in Fig. 2.19 can be expressed using the heat balance principle as follows:

$$C\frac{dT_{in}}{dt} = BLC(T_{out} - T_{in}) + Q_{gain} \tag{2.16}$$

where $C = mc_p$ is the capacitance of the building including the mass of all the envelope components (i.e., floor and walls) as well as furniture and interior partitions. Notice that, if the HVAC system is used to maintain the indoor temperature at a constant value, this equation reverts back to a simple and familiar steady-state equation. By rearranging Eq. (2.16), the following expression can be derived:

$$C\frac{dT_b}{dt} = BLC(T_{out} - T_b) \tag{2.17}$$

where T_b is the balance temperature of the building introduced earlier for the variable-base degree-day method as follows:

$$T_b = T_{in} - \frac{Q_{gain}}{BLC} \tag{2.18}$$

The time constant, τ, of the building is then defined as the ratio of the building capacitance, C, over the building load coefficient, BLC:

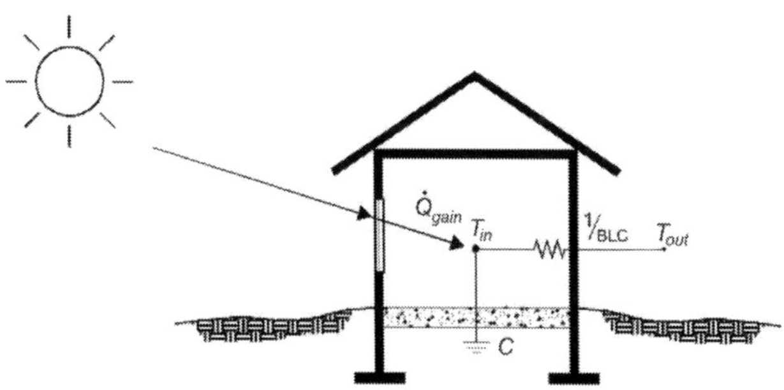

FIGURE 2.19 **Thermal network model for a home.**

$$\tau = \frac{C}{BLC} \tag{2.19}$$

Thus, the differential equation, Eq. (2.17), can be expressed simply as

$$\frac{dT_b}{dt} = -\frac{T_{in} - T_{out}}{\tau} \tag{2.20}$$

If the outdoor air temperature, T_{out}, is constant, the solution to the differential equation is given as noted by Eq. (2.21) with $T_{b,0} = T_b(t = 0)$:

$$\frac{T_b(t) - T_{out}}{T_{b,0} - T_{out}} = e^{-t/\tau} \tag{2.21}$$

Therefore, the indoor air temperature variation with time, $T_{in}(t)$, can be estimated as follows:

$$T_{in}(t) = T_{out} + \frac{Q_{gain}}{BLC} + \left[T_{in}(t=0) - T_{out} - \frac{Q_{gain}}{BLC}\right]e^{-t/\tau} \tag{2.22}$$

In particular, when there is no internal gains (i.e., $Q_{gain} = 0$), the indoor air temperature variation with time, $T_{in}(t)$, can be simply estimated as follows:

$$T_{in}(t) = T_{out} + [T_{in}(t=0) - T_{out}]e^{-t/\tau} \tag{2.23}$$

Example 2.2 illustrates the estimation of the building time constant.

Example 2.2

Consider a home with $BLC = 380$ Btu h^{-1}·°F and a capacitance of $C = 5270$ Btu/°F. During a cold night when the outdoor air temperature is constant at $T_{out} = 0$°F with no internal gains (i.e., $Q_{gain} = 0$), determine the time for the indoor temperature to decrease from 68°F to 50°F.

Solution

First, the time constant of the home is estimated using Eq. (2.19):

$$\tau = \frac{5270}{380} = 13.86\,\text{h}$$

Then, using Eq. (2.23), the time t can be obtained as shown below:

$$t = -\tau * Ln\left[\frac{T_{in}(t) - T_{out}}{T_{in}(t=0) - T_{out}}\right] = 13.86 * Ln\left[\frac{60-0}{68-0}\right] = 1.73\,\text{h}$$

2.4.2.2.2 Detailed Whole-Building Simulation Tools

These tools can predict whole-building energy end-uses and are capable of modeling most, if not all, energy systems present within the building including envelope components (i.e., walls, roofs, floors, doors, and windows) as well as lighting, appliances, and mechanical systems including HVAC equipment. A typical calculation flow chart for a detailed simulation program is presented in Fig. 2.20.

Most detailed computer programs require a high-level of expertise and are generally only suitable for simulating large buildings with complex HVAC systems and involve control strategies that are difficult to model by simplified energy analysis tools. In general, an energy simulation program requires a detailed physical description of the building (including building geometry, envelope construction details, HVAC equipment type and operation, and occupancy and equipment use schedules). Thermal load calculations are based on a wide range of algorithms depending on the complexity and the flexibility of the simulation program. To adequately estimate energy savings from energy efficiency measures for existing buildings, whole-building energy simulation tools have to be calibrated using measured energy data (Krarti, 2011).

While whole-building energy simulation programs are generally capable of modeling most of the building energy systems, they are often not flexible and time-consuming to use, and have inherent limitations. To select the appropriate energy simulation program, it is important that the user be aware of the capabilities of each simulation tool available to him/her. Some of the well-known whole-building energy simulation programs are briefly presented in the following sections:

- DOE-2 (version DOE-2.1) was developed at the Lawrence Berkeley National Laboratory (LBNL) by the US Department of Energy and is widely

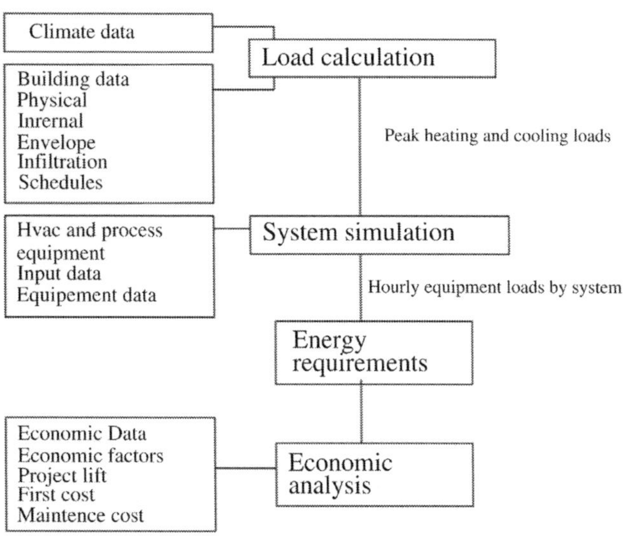

FIGURE 2.20 Flow chart of complete building model.

used because of its comprehensiveness. It can predict hourly, daily, monthly, and/or annual building energy use. DOE-2 is often used to simulate complex buildings. Significant efforts are required to create DOE-2 input files using a programming language called Building Description Language. However, several tools are currently available to facilitate the process of developing DOE-2 input files. Among energy engineers and professionals, DOE-2 has become a standard building energy simulation tool in the United States and several other countries using interfaces such eQUEST.

- EnergyPlus builds on the features and capabilities of DOE-2 and BLAST. It uses integrated solution techniques to correct one of the deficiencies of BLAST and DOE-2—the inaccurate prediction of space and envelope surface temperature variations. Indeed, accurate prediction of space and surface temperatures is crucial to properly analyze energy-efficient systems. For instance, HVAC system performance and occupant comfort are directly affected by space and indoor surface temperature fluctuations. Moreover, EnergyPlus has several features that should aid engineers and architects to evaluate a number of innovative energy efficiency measures that cannot be simulated adequately with either DOE-2 or BLAST. These features include
    - free cooling operation strategies using outdoor air,
    - realistic HVAC systems controls,
    - effects of moisture adsorption in building elements, and
    - indoor air quality with a better modeling of contaminant and air flows within the building.

  Several interfaces for EnergyPlus have been developed over the last few years. A complete list of the interfaces is periodically updated in the EnergyPlus website (EnergyPlus, 2017).
- TRNSYS provides a flexible energy analysis tool to simulate a number of energy systems using user-defined modules. While TRNSYS can be used to model heating and cooling thermal loads for homes, it is widely used for modeling active solar systems including PV and solar hot water heating systems. A good knowledge of computer programming (Fortran) is required to properly use the TRNSYS simulation tool especially to develop models for new systems.

2.5 DYNAMIC INSULATION MATERIALS AND SYSTEMS

DIMs and systems are emerging as a promising new technology to reduce energy consumption required to maintain thermal comfort in buildings (Jelle, 2011). Space heating and cooling of buildings represents a significant portion of the total energy consumption in the United States and other developed countries. In the United States, 48% of total residential energy use and 35% of total commercial energy use are attributed to space heating and cooling (EIA, 2016a).

A traditional building envelope system uses insulation materials with a static *R*-value, which is often selected based on requirements of local building and energy efficiency codes. In contrast, DIM panels are designed so that the rate of heat transfer through a wall can vary over time. The variation in wall thermal properties can be achieved using various approaches including the introduction of inert gases with different thermal conductivity properties (Park et al., 2015) or by mechanically sliding or compressing insulation layers in the wall (Clark et al., 2013). Typically, heat transfer is minimized through the building envelope by selecting walls and roofs with high thermal resistance (i.e., with high *R*-value and low *U*-value). Several advanced materials and systems with low thermal conductivity values have been developed and tested such as vacuum insulation panels, gas-filled panels and aerogels, and reflective insulations (Baetens et al., 2010a, 2010b, 2011; Lee et al., 2016; Schiavoni et al., 2016). However, it is not always beneficial to impose highly insulated building envelope components (Masoso and Grobler, 2008). In fact and in some instances, it is more advantageous to lower the thermal resistance of the building shell. For instance, on a typical summer day, a well-insulated building would effectively limit heat gain due to conduction. However, internal heat gains (due to occupants, equipment, and lighting) and solar heat gain heat (from glazed windows) could remain trapped indoors. To release this unwanted heat, passive strategies, such as stack ventilation or night flushing can be used. Although effective, opening windows is a strategy that is not always available or desirable to utilize for a variety of reasons. In some climates, even during the cooling season, nighttime temperatures may be too low to keep windows open overnight. Hayashi et al. (2014) studied the window-opening behavior of dwellers in nine different detached homes in Japan (both cold and hot-humid climates). Using sensors to monitor and record the opening width and comparing it to the ambient air temperature variation, researchers found that in general, dwellers do not open windows enough to realize the energy savings that is possible with passive cooling (Hayashi et al., 2014). Windows are generally closed at night and when occupants leave during the day. Survey respondents indicated the following reasons for closing windows: outdoor noise and air pollution, crime prevention, and privacy, as well as keeping mosquitoes and other insects out (Hayashi et al., 2014). In a commercial setting, such as an office building, automated strategies have been shown to be significantly more effective than relying on individual controls (Bourgeois et al., 2006; Kapsis et al., 2010).

Using walls and/or roofs with switchable insulation would allow for the building envelope to be well insulated (i.e., having a high *R*-value) when heat transfer between indoors and outdoors is not desirable and to have higher thermal conductance (i.e., having a low *R*-value) when enhanced heat transfer is desirable. Although this technology is still being developed, the idea is not new. In a report published by the Los Alamos National Laboratory in 1982, it was noted that future building efficiency research should focus on control strategies for the transfer and storage of thermal energy (Neeper and McFarland, 1982). In a 2011 review of building insulation materials, Jelle indicates that DIMs show promise as a future technology (Jelle, 2011). In a study on the potential energy savings of DIMs in three different US climates, switchable insulation layers with two different *R*-value settings were considered (Park et al., 2015). In one case, the DIM resistance was altered from a low value of RSI-0.5 to a high value of RSI-2.5 (RSI-0.5/RSI-2.5). In the second case, a low value of RSI-1.0 was combined with a high value of RSI-2.5 (RSI-1.0/RSI-2.5). In all three US climates, it was found that a two-step variable resistance envelope of RSI-0.5/RSI-2.5, was more effective than RSI-1.0/RSI-2.5, and could reduce cooling energy use by an average of 15% and up to 39% in heating-dominated climates, when compared to static insulation layers (Park et al., 2015).

Although very little research has been published on the topic of DIMs, similar technologies exist for other components of the building envelope. For instance, electrochromic widows employ a switching mechanism to vary the transmittance of visible and infrared solar energy (Sbar et al., 2012). In the reported study, using windows that could be switched from 62% visible transmittance (0.47 SHGC) to fully tinted (with less than 2% visible transmittance/0.09 SHGC), potential energy savings were found to be higher than 45% when compared to single pane static glazing and about 20% when compared to ASHRAE 90.1-2007 code-compliant glazing (Sbar et al., 2012). Jonosson and Roos (2010) used computer simulations to evaluate the effectiveness of various control strategies for smart (electrochromic) windows. Although the most energy efficient solution would be to keep windows in their tinted state (low visible transmittance and low SHGC) when there is a need for cooling, and switch to a clear state when there is a need for heating, this solution does not take into account daylight and occupant preference. The same study showed the benefits of adding occupancy sensors so that when people are present, the windows are optimized for daylighting; and when the space is empty, the windows are optimized for energy savings. Other studies have considered coupling ventilation airflow within the building envelope using the concept of breathing walls (Dimoudi et al., 2004; Krarti, 1994; Taylor et al., 1996). A typical breathing wall, referred to sometimes as a dynamic wall, draws a steady stream of filtered air through the wall cavity and into the building to provide clean ventilation and partially conditioned air to the occupants. Several studies indicated that the airflow can effectively vary the thermal resistance of breathing walls depending on the indoor–outdoor temperature difference and the ventilation rate (Qiu and Haghighat, 2007; Solupe and Krarti, 2014). As an extension of breathing walls, recent studies have considered living walls by integrating air and/or water flows as well as phase-change materials (PCMs) within the building envelope (Al-Saadi and Zhai, 2015; El Mankibi et al., 2015; Kosny et al., 2012). Preliminary analysis results indicate that the performance of the living walls is very sensitive to the PCM melting range. With optimal melting temperature close to the heating and cooling set-points, living walls can achieve savings in annual and peak heating and cooling loads of up to 30% (Al-Saadi and Zhai, 2015; El Mankibi et al., 2015). The DIM technology presented in this section does not require any PCM integration or any fluid flow from the outside to the inside of the building.

The aim of this section is to provide an overview of the DIM technology and the potential benefits of its deployment in reducing the heating and cooling requirements for residential and commercial buildings. In Chapter 3, optimized control strategies for DIM are discussed for various climatic conditions. Most of the results outlined in this section are based on published studies.

2.5.1 Description of DIM Technology

Traditional thermal insulation materials, including mineral wool, polyurethane, and expanded and extruded polystyrene (XPS), exhibit low thermal conductivities in the range of 0.02–0.04 W m·K. Recently developed high-performance envelope technologies include vacuum insulation panels (VIPs), gas insulation panels, and nanoinsulation materials. While these state-of-the-art high-performance thermal insulation assemblies have initial conductivities that range between 0.003 and 0.02 W m^{-1}·K, concerns exist regarding the long-term reliability of these insulation material technologies. For example, VIP conductivities are known to rise to 0.008–0.03 W m·K due to water vapor diffusion and other potential mechanisms (e.g., puncturing) (Kalnæs and Jelle, 2014). Similarly, it is well known that high-performance XPS loses its thermal resistance over time due to high vapor diffusivity (Jelle, 2011).

High thermal resistance materials and assemblies are effective in reducing heat transfer through the building envelope when there is a significant indoor–outdoor temperature difference. As noted earlier, building envelopes with high *R*-value can be beneficial for one season of the year but are not necessarily beneficial for the other seasons. For example, in cold climates, high thermal insulation can actually increase building thermal loads—especially during mild periods when it may be beneficial to let heat that is trapped indoors escape outdoors. A number of strategies exist to release stored heat, such as opening windows, night ventilation, or lower conductivity building envelopes.

Some studies have theorized on DIMs and systems in which thermal conductivities of wall assemblies are controllable and tunable, either due to applied electrical voltage to conducting polymers or other electrochromic materials (Jelle et al., 2010; Sadineni et al., 2011). While some prototypes of electrochromic materials are available, these technologies have only been applied to smart windows rather than opaque building envelope systems.

The basic operating principles of some DIM technologies are depicted in Fig. 2.21. In lieu of traditional, static insulation assemblies (e.g., XPS, VIPs, and gas insulation panels), a DIM is a rigid, cellular panel that is placed within the external wall cavities of a building. A possible DIM panel can be an interconnected system of conduits as shown in Fig. 2.21A. Once installed, the bulk thermal diffusivity of the assembly is tailored by the introduction of inert gases with variable conductivities (e.g., carbon dioxide, nitrogen, and helium), thereby affecting the rate of heat transfer through the envelope. As an alternative, a DIM panel can consist of an open cavity with pieces of insulation that can

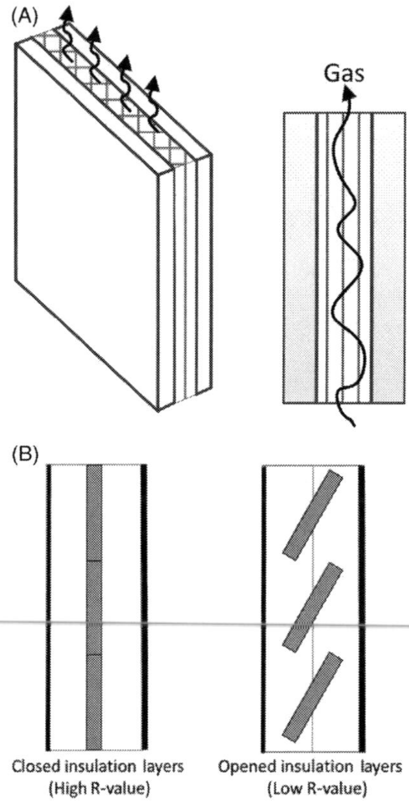

FIGURE 2.21 **Operating principle of dynamic insulation materials (DIMs).**

rotate depending on the desired *R*-value of the panel as outlined in Fig. 2.21B. It is envisioned that the insulation media can be regulated via control strategies based on indoor–outdoor temperature differences and exchanged daily, monthly, or seasonally, depending on the climate and necessary heating and cooling demands.

2.5.2 Model for DIM Systems

One approach to evaluate the energy performance of DIM panels is to use the RC modeling approach. Specifically, a DIM system has been modeled using a 3R2C thermal network model modified to have a virtual node at wall surface exposed to the surroundings (Park et al., 2015). The virtual node has no thermal capacitance so that the surface temperature can react quickly to the surrounding conditions. Fig. 2.22 depicts a thermal network of an exterior wall connecting the indoor space to the outdoor environment. All resistances and capacitances are assumed to be time invariant.

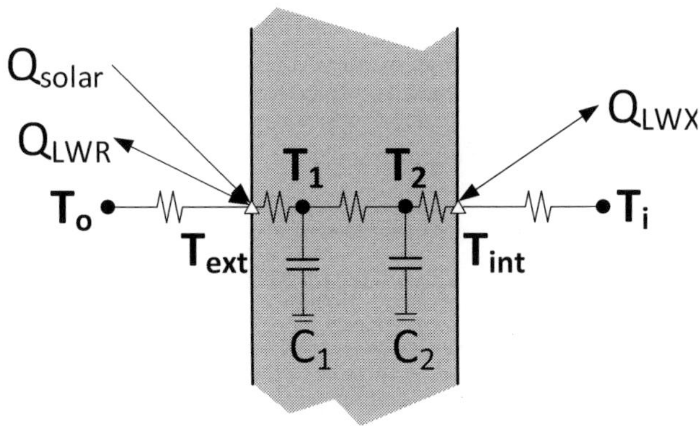

FIGURE 2.22 Description of RC thermal network for a single layer exterior wall.

Convection and long-wave radiation exchanges occur at both sides while short-wave solar radiation is considered only at the outer surface of an exterior wall. The transient heat balance analysis for both nodes yields the following differential equations:

$$\frac{T_{ext}-T_1}{R}A+h_oA(T_o-T_{ext})+Q_{LWR}+Q_{solar}=0 \tag{2.24}$$

$$\frac{T_{ext}-T_1}{R}A+\frac{T_1-T_2}{R}A=C_1\frac{dT_1}{dt} \tag{2.25}$$

$$\frac{T_1-T_2}{R}A+\frac{T_2-T_{\text{int}}}{R}A=C_2\frac{dT_2}{dt} \tag{2.26}$$

$$\frac{T_2-T_{\text{int}}}{R}A+h_iA(T_{\text{int}}-T_i)+Q_{LWX}=0 \tag{2.27}$$

where R is the thermal resistance of materials ($m^2{\cdot}K/W$), C is the thermal capacitance of materials (W/K), A is the area of the surfaces (m^2), T_{out} is outdoor air temperature (°C), T_{zone} is indoor air temperature (°C), T is the surface temperature of the exterior walls (°C), h_o is the convective coefficient at outer surface of the exterior walls ($W\ m^{-2}{\cdot}K$), h_i is the convective coefficient at inner surface of the exterior walls ($W\ m^{-2}{\cdot}K$), Q_{LWR} is net longwave radiation (W), Q_{solar} is exterior solar radiation incident (W), and Q_{LWX} is longwave radiant heat exchange between surfaces (W).

Fig. 2.23 illustrates the RC thermal network used to model a DIM layer in a multilayer exterior wall separating a thermal zone characterized by its indoor temperature, T_i. The material in the middle of the exterior wall (M2) is considered as a DIM. The thermal resistance of M2 is controlled based on the heating or cooling requirement of the thermal zone (Park et al., 2015).

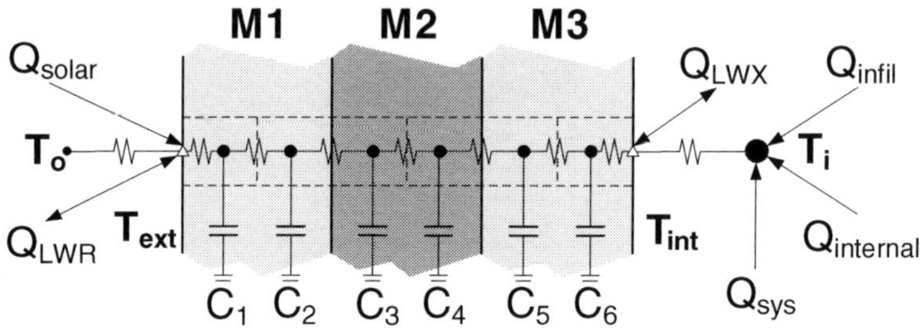

FIGURE 2.23 **Thermal network for a multilayer exterior wall.**

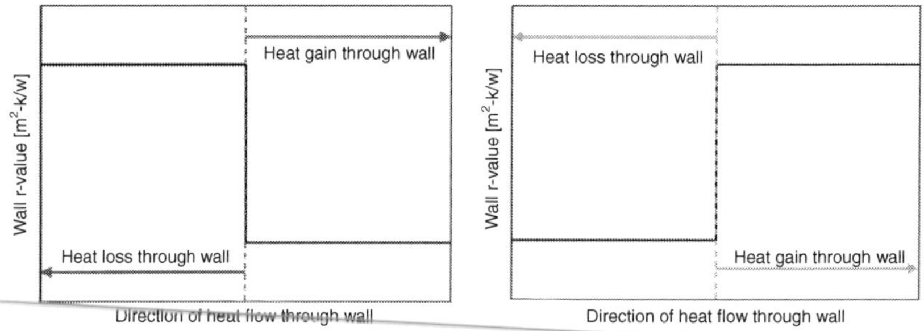

FIGURE 2.24 A two-step *R*-value control strategy for DIM layer during (M2): (A) heating operation and (B) cooling operation.

To control the thermal resistance of M2, the direction of heat flow through the exterior wall has been considered as a control parameter (Park et al., 2015). Then, heat flow direction can be determined by monitoring the inside, middle, and outside surface temperatures of the exterior wall. Fig. 2.24 depicts the control strategy to vary the thermal resistance for the DIM layer and ultimately the exterior wall. First, the controller determines the direction of heat flow through the building envelope. Then, the controller examines temperature distribution inside the exterior wall to determine the appropriate RSI-value (*R*-value in SI unit) of the DIM in the middle layer of the exterior walls. For example, when outside surface temperature is higher than the inside surface temperature in the cooling season, M2 increases its thermal resistance so that it reduces the heat gains through the exterior walls. On the other hand, when outside surface temperature is lower than the inside surface temperature during the cooling season, the thermal resistance of M2 is reduced to release heat from indoors to outdoors through the exterior wall.

Fig. 2.25 summarizes the reported analysis results to assess the potential energy savings associated with deploying of DIMs instead of static insulation layers (Park et al., 2015). The solar heat gains transmitted through windows assist in heating indoor space air, and thus reduce heating thermal loads. However, the addition of heat gains from transmitted solar radiation results in an increase in building cooling thermal loads. The low RSI-value walls are more effective

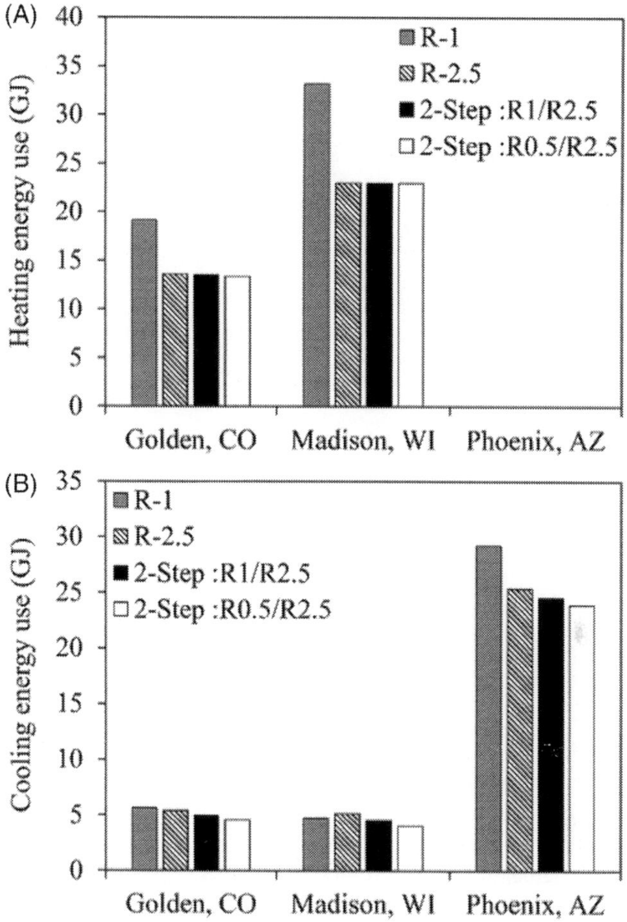

FIGURE 2.25 (A) Annual heating energy use and (B) annual cooling energy use for four DIM control strategies when the building has windows and internal gains in three US locations: Golden, CO; Madison, WI; and Phoenix, AZ.

for releasing heat trapped indoors during cooler nighttime periods than the high RSI-value walls. The two-step control strategies take advantage of the DIM low RSI-value during nighttime and minimize heat gains through the exterior walls during daytime by increasing their thermal resistance. Specifically, it has been found as noted in Table 2.1 that a two-step RSI-1/RSI-2.5 control strategy can reduce space cooling energy use by up to 8% compared to the static RSI-2.5 wall option in Golden, CO, while the two-step RSI-0.5/RSI-2.5 control strategy can reduce space cooling energy use by up to 15% compared to the RSI-2.5 wall option in the same location. Moreover, cooling energy use can be decreased by 11% and 21%, respectively, with two-step RSI-1/RSI-2.5 and two-step RSI-0.5/RSI-2.5 control strategy in Madison, WI. However, the two-step control can

TABLE 2.1 Heating and Cooling Energy Use Savings Compared to Static RSI-1.0 Exterior Walls in Three US Locations: Golden, CO; Madison, WI; and Phoenix, AZ

	Heating Energy Use Savings (GJ)			Savings (%)		
	R-2.5	Two-step RSI-1/ RSI-2.5	Two-step RSI-0.5/ RSI-2.5	*R*-2.5	Two-step RSI-1/ RSI-2.5	Two-step RSI-0.5/ RSI-2.5
Golden, CO	−5.55	−5.65	−5.77	−29.0	−29.5	−30.1
Madison, WI	−10.14	−10.14	−10.15	−30.6	−30.6	−30.6
Phoenix, AZ	n/a	n/a	n/a	n/a	n/a	n/a
	Cooling Energy Use Savings (GJ)			**Savings (%)**		
	R-2.5	Two-step RSI-1/ RSI-2.5	Two-step RSI-0.5/ RSI-2.5	*R*-2.5	Two-step RSI-1/RSI-2.5	Two-step RSI-0.5/ RSI-2.5
Golden, CO	−0.24	−0.67	−1.05	−4.3	−11.8	−18.6
Madison, WI	0.40	−0.18	-0.69	8.5	−3.7	−14.5
Phoenix, AZ	−3.86	−4.70	−5.34	−13.2	−16.0	−18.2

achieve only 6% cooling energy savings and has little impact on space heating energy use in Phoenix, AZ (Park et al., 2015).

The controls for DIMs can be further optimized using a cost function and allowing DIM thermal resistance to vary continuously between the minimal and maximal values (Shekar and Krarti, 2017). Chapter 3 discusses in detail the additional energy savings that DIMs can achieve when operated using optimized controls.

The number of hours when each DIM exterior wall is set at the low thermal resistance over a year has been found to vary depending on the orientation as shown in Fig. 2.26 for three US climates (Park et al., 2015). In particular, the results of Fig. 2.26 indicate that the north wall equipped with a DIM panel is the most actively controlled especially when high internal loads are considered.

2.5.3 Impact of Climate

To determine the potential energy savings that DIM can achieve for various US climates, a broader analysis was completed and reported with 44 different cities representing 15 US climate zones (Menyhart and Krarti, 2017). Table 2.2 summarizes the overall wall *R*-values for both static and variable insulation configurations considered for all 44 cities.

HDDs and CDDs are often used to characterize climate conditions for building energy analysis. Since degree-days vary depending on the weather data used, all

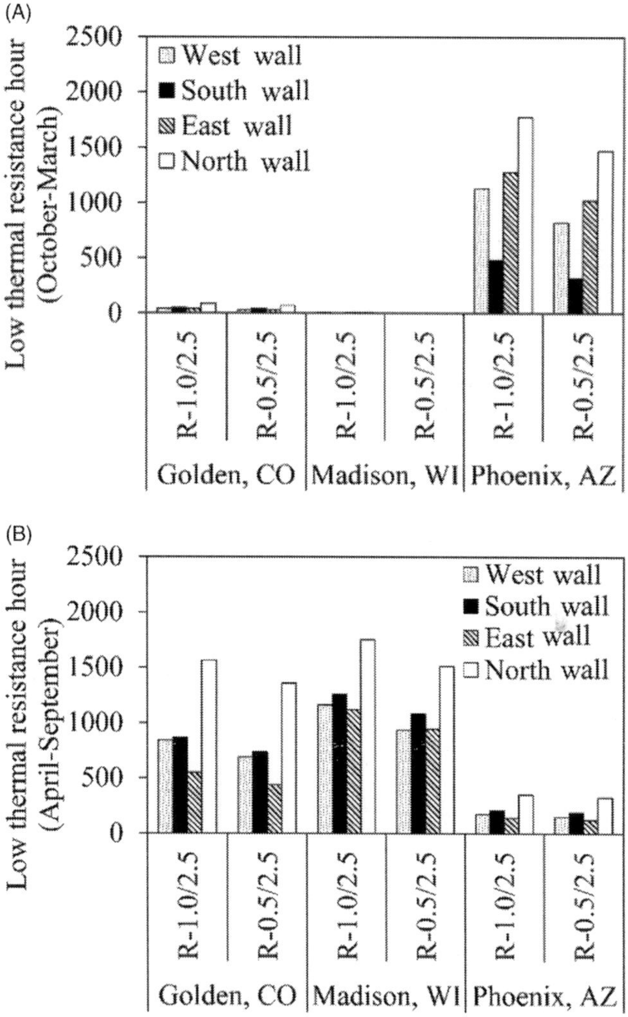

FIGURE 2.26 Number of hours for maintaining low *R*-value for the exterior walls in the building with windows and internal load in three US locations: Golden, CO; Madison, WI; and Phoenix, AZ.

HDDs and CDDs were calculated from TMY3 weather data used in the simulation analysis with a base temperature of 65°F (18°C) (Menyhart and Krarti, 2017).

2.5.3.1 Impact of Heating Degree Days

When comparing the simulated heating savings to HDDs for each location, it is found that there is greater potential for heating energy savings with the deployment of DIMs in colder climates (with high HDD values) as shown in Fig. 2.27 (Menyhart and Krarti, 2017).

TABLE 2.2 Summary of Insulation Levels Tested by Climate Zone

Climate zone	Static insulation			Two-step variable insulation		
	None	Code	High	Med/code	None/code	None/high
1,2	RSI-0.4 (*R*-2.5)	RSI-1.9 (*R*-11)	RSI-3.0 (*R*-17)	RSI-1.4/1.9 (*R*-8.0/11)	RSI-0.4/1.9 (*R*-2.5/11)	RSI-0.4/3.0 (*R*-2.5/17)
3,4,5	RSI-0.4 (*R*-2.5)	RSI-3.0 (*R*-17)	RSI-3.9 (*R*-22)	RSI-1.4/3.0 (*R*-8.0/17)	RSI-0.4/3.0 (*R*-2.5/17)	RSI-0.4/3.9 (*R*-2.5/22)
6,7,8	RSI-0.4 (*R*-2.5)	RSI-3.9 (*R*-22)	RSI-5.3 (*R*-30)	RSI-1.4/3.9 (*R*-8.0/22)	RSI-0.4/3.9 (*R*-2.5/22)	RSI-0.4/5.3 (*R*-2.5/30)

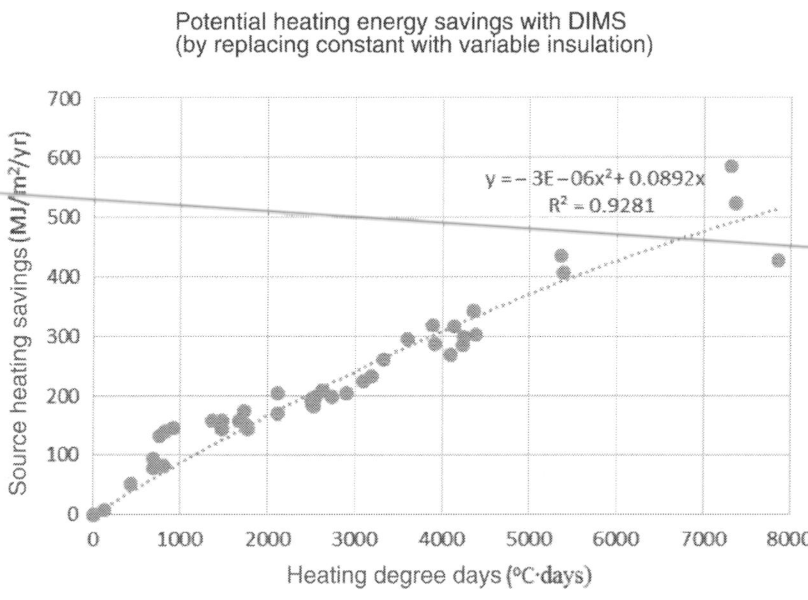

FIGURE 2.27 Net heating energy savings with DIMs by HDD.

Although there is a variation among cities within each climate zone, HDDs is found to be a fairly good indicator of the magnitude of the energy savings that could be realized with the deployment of DIMs for a given location.

2.5.3.2 Impact of Cooling Degree Days

When comparing space cooling energy savings (the difference between two-step: none/high and code-*R*) with CDDs, as shown in Fig. 2.28, the potential for DIMs is found to depend on the baseline used for the comparative analysis to estimate annual energy savings (Menyhart and Krarti, 2017). Specifically,

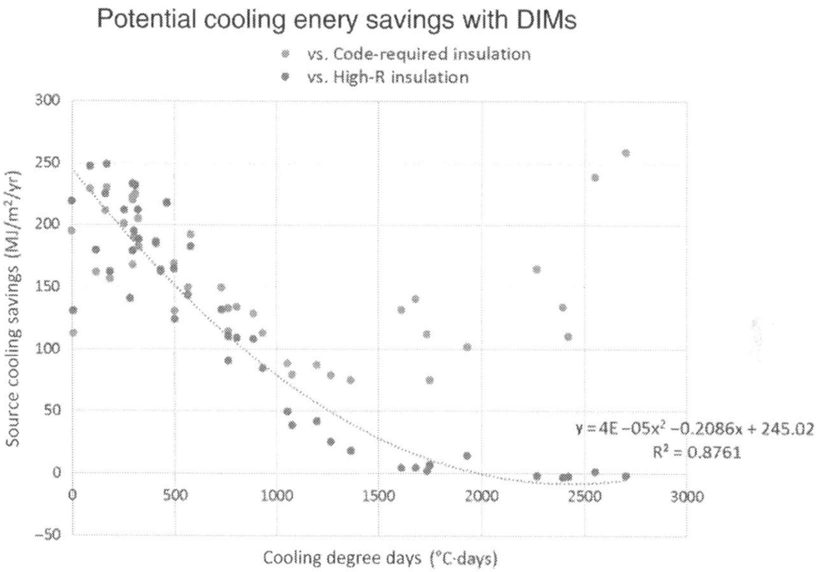

FIGURE 2.28 **Net cooling energy savings as a function of cooling degree-day (vs. high-*R*).**

for warmer climates (high CDD values), two-step controlled DIMs save energy when compared to code-required insulation levels. However, these energy savings could also be achieved by using static insulation with a higher *R*-value. For colder climates, the opposite behavior is found: switching to static insulation with a higher *R*-value does not decrease the cooling load, but a dynamic insulation system would.

2.5.4 Economic Analysis

One important consideration when determining the potential of a new technology is its cost-effectiveness. As part of a cost analysis for the deployment of DIMs, their potential in annual energy cost savings has been estimated (Menyhart and Krarti, 2017). These annual energy cost savings are found to be typically proportional to the reduction in source energy use by a building. Indeed, source energy takes into account the primary fuel used in the production of energy that is delivered by the utility company. Since most cooling equipment in US buildings are served by electricity that is generated by burning fossil fuels, reducing source energy consumption is essential in reversing the effects of climate change. In 2014, it was estimated that 30% of greenhouse gas emissions in the US was attributed to electricity production (EIA, 2016a).

In all US locations considered in the study, it is found that there is potential energy savings with DIMs, when compared with code-required static insulation (Menyhart and Krarti, 2017). However, in several of these locations, similar

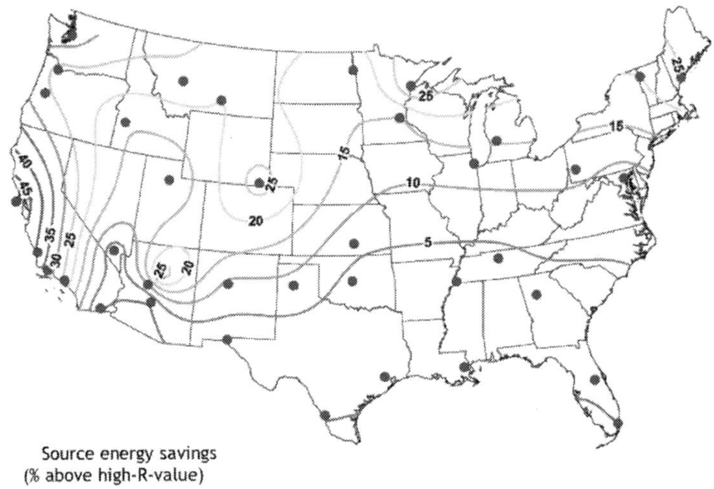

FIGURE 2.29 Source energy savings contour map associated with DIMs applied to US residential buildings.

savings could be realized by investing in a higher level of static insulation, which would be more cost effective than a dynamic insulation system. For DIM technology to be viable in the market place, it would need to outperform static insulation with a higher *R*-value. For this reason, the potential source energy savings were calculated as the difference between using above-code static insulation (i.e., high-*R*) and dynamic two-step insulation (i.e., two-step: none/high-*R*). The US map outlined in Fig. 2.29 shows contours for the percent energy savings associated with the deployment of DIMs on residential buildings by geographical location (Menyhart and Krarti, 2017). The largest potential for total source energy savings (above high-*R*) was 51.2% in San Francisco (57.4% for cooling and 12.9% for heating). For cities in climate zone 1a (Hilo, Honolulu, and Miami), 2a (Houston, New Orleans, and Orlando), and 2b (Imperial, Laredo, and Phonenix) there was only a negligible difference between two-step: none/high and high-*R* (ranging from −0.2% to 1.3% savings). Mostly likely, this is because the diurnal temperature swing during the summer months is not large enough to take advantage of the free cooling that is possible with DIMs.

Another approach to measure the cost effectiveness of deploying DIMs was estimated in terms of site energy cost savings, which directly correlates with a reduction in utility bills for the building owners. Often cost-effectiveness is the primary factor considered by a homeowner when deciding whether or not to invest in energy efficiency measures. It is also used to calculate the return on investment or payback period. The map in Fig. 2.30 shows the potential savings

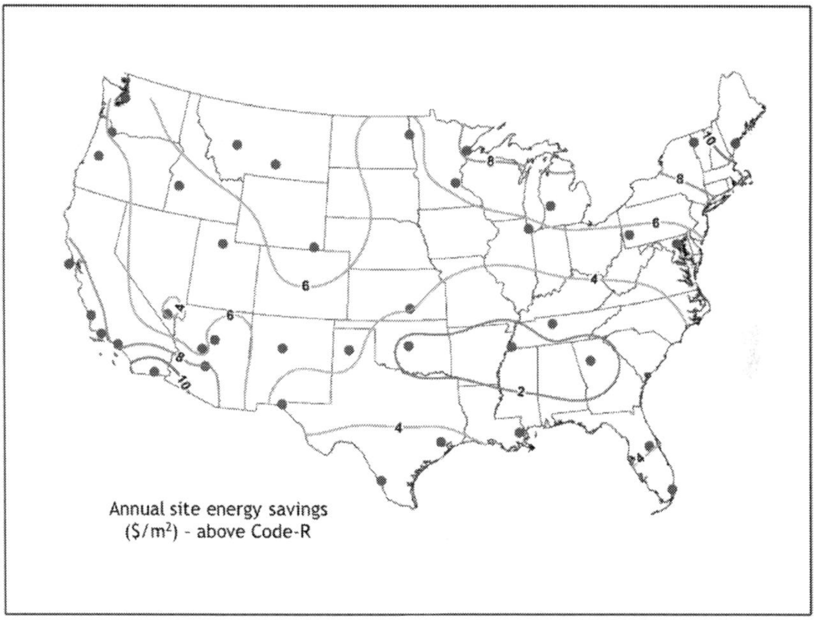

FIGURE 2.30 Energy cost savings contour map associated with DIMs applied to US residential buildings.

per square meter of dynamic insulation systems, in comparison to code-required static insulation, and takes into consideration the current electricity and natural gas rates (EIA, 2016b, 2016c). For electricity, these rates range from a low of \$0.0833/kWh in Oklahoma to a high of \$0.2692/kWh in Hawaii. For natural gas, these rates range from a low of \$0.54/therm in North Dakota to a high of \$3.25/therm in Hawaii.

When considering site energy cost savings, it is found that utility rates have a large impact (Menyhart and Krarti, 2017). In states with high utility rates such as Alaska, California, Maine, and Vermont, DIMs can achieve significant energy cost savings for homes as reflected in the percent cost savings map shown in Fig. 2.30. The largest overall cost savings was \$13.62/m^2 in Fairbanks, AK (\$13.25 from cooling savings and \$0.37 from heating savings). The lowest cost savings tend to be clustered in the Southeast and South Central states—especially Memphis, Atlanta, Oklahoma City, which are all in climate zone 3a. The cost of electricity also tends to be relatively low in these states, with Oklahoma currently being the lowest of the 50 states.

As noted earlier, the best application for DIMs would be in climates where simply adding more insulation is not effective. The additional costs for DIMs would depend primarily on the specific dynamic insulation system adopted and installed. Table 2.3 illustrates the payback period of DIMs applied to residential buildings for various US climates using two scenarios for their

TABLE 2.3 Evaluation for Payback Periods for DIMs Applied to US Homes

Climate zone	City	Site energy cost savings ($/m^2)	Payback period (years)	
			DIMs cost is 150% of static insulation	DIMs cost is 200% of static insulation
3b	El Paso, TX	2.84	3.3	6.6
	Las Vegas, NV	3.60	2.6	5.2
	Los Angeles, CA	7.50	1.3	2.5
3c	San Francisco, CA	8.63	1.1	2.2
	Santa Barbara, CA	8.38	1.1	2.2
	San Luis Obispo, CA	8.96	1.1	2.1
4a	Baltimore, MD	4.97	1.9	3.8
	Nashville, TN	2.77	3.4	6.8
	Wichita, KS	4.21	2.2	4.5
4b	Albuquerque, NM	4.73	2.0	4.0
	Amarillo, TX	4.66	2.0	4.0
	Prescott, AZ	4.55	2.1	4.1
4c	Eugene, OR	6.40	1.5	2.9
	Portland, OR	5.92	1.6	3.2
	Seattle, WA	5.84	1.6	3.2
5a	Chicago, IL	5.33	1.8	3.5
	Grand Rapids, MI	7.31	1.3	2.6
	Pittsburgh, PA	5.65	1.7	3.3
5b	Boise, ID	5.59	1.7	3.4
	Flagstaff, AZ	7.55	1.2	2.5
	Salt Lake City, UT	4.83	2.0	3.9
6a	Burlington, VT	9.40	1.0	2.0
	Minneapolis, MN	6.26	1.5	3.0
	Portland, ME	10.89	0.9	1.7
6b	Billings, MT	6.78	1.4	2.8
	Cheyenne, WY	7.07	1.3	2.7
	Helena, MT	7.02	1.3	2.7
7	Anchorage, AK	11.52	0.8	1.6
	Duluth, MN	8.49	1.1	2.2
	Grand Forks, ND	5.25	1.8	3.6
8	Fairbanks, AK	13.62	0.7	1.4
	Nome, AK	6.84	1.4	2.8

DIMs, Dynamic insulation materials.

additional installation cost relative to the static insulation. For US residential buildings, the average costs of adding wall static insulation is \$1.75 per square foot (\$18.84/m^2). If DIMs could be manufactured for 50% more that static insulation (\$28.26/m^2), the payback period would be less than 3.5 years in all of the cities where it would be recommended (climate zones 3b–8). Even if the cost was double that of static insulation (\$37.68/m^2), the payback period would be less than 5 years in all but three of the cities listed in Table 2.3.

2.5.5 Recommendations

The results of the reported analyses and briefly discussed in this section indicate that although heating energy can be saved through the use of DIMs, especially in warm to mild climates, cooling energy savings are more significant. This is particularly true when accounting for equipment inefficiencies and source energy losses. Moreover, it is found that orientation affects the frequency with which an exterior wall is switched to a lower *R*-value. In colder climates, where DIMs are used to release trapped heat, it is found that the north wall is one that is most frequently switched to a low *R*-value. On the other hand, when the wall *R*-value is lowered to allow heat to enter a building, the south wall is found to be the most active.

When comparing DIMs to code-required static insulation, energy and cost savings can be achieved in buildings located in most US cities. Moreover, the largest energy savings were achieved by using the two-step insulation with the lowest possible low *R*-value and the highest possible high *R*-value. Although seasonal control was only evaluated in select cities, in all cases it was less effective than an hourly controlled DIM system. As will be outlined in Chapter 3, the analysis results indicate that when using dynamic insulation, the greatest energy savings would be achieved by using an automated optimized control system, which responds to actual input from temperature sensors and can take advantage of diurnal temperature swings, rather than relying on seasonal settings and manual control.

In summary, DIMs can be used for enhancing the passive cooling capabilities of buildings especially in mild climates and weather conditions. Moreover, it is expected that DIMs could achieve even more energy savings if intermediate thermal resistance settings are allowed in addition to only high and low *R*-values considered in the analysis reported in this chapter.

HDDs and CDDs were found to be good indicators of potential energy savings. As expected, HDDs correlate well with the potential of heating energy savings associated with DIMs. However, only for the warmest climates, where allowing heat to enter the building during the winter to offset the heating load, DIMs would provide some energy savings relative to adding static insulation. For cities with high CDDs, DIMs can achieve higher cooling savings when compared to code-required insulation levels. For cities with low CDDs, DIMs could provide higher cooling savings when compared to above-code insulation levels.

In the overview outlined in this section, DIMs were only applied for building exterior walls. In climate zones where DIMs have been shown to be effective, future research could be conducted to determine the effect of using variable insulation systems in building roofs. Moreover, DIMs also need to be developed for seamless integration with building envelope systems. To be practical, DIMs must be robust enough to handle the demands of construction sites. To be a viable alternative to static insulation systems, DIM's installation and operation costs must also be comparable with the potential energy cost savings discussed in this section.

2.6 THERMO-ACTIVE FOUNDATIONS

2.6.1 Overview of Thermo-active Foundations

Renewable energy sources can be utilized not only to generate electricity but also to meet heating and cooling needs for buildings. In particular, through the use of heat exchangers embedded in the ground, heat can be extracted from or rejected into the soil medium to maintain thermal comfort within buildings without a significant reliance on any other external energy sources and mechanical systems such as boilers and chillers. Ground source heat pumps or GSHPs are examples of geothermal systems that take advantage of mild and uniform deep ground temperatures to heat and cool residential and commercial buildings. However, high installation costs can make GSHPs using vertical boreholes less cost-effective compared to other conventional air conditioning systems. Therefore, alternative systems and methods have been suggested to reduce the implementation costs of GSHP systems. TAF systems have been shown to represent a viable solution to reduce the excavation costs related to drilling work associated with deep boreholes of vertical GSHPs. TAF systems, also commonly referred to as energy or thermal piles, integrate ground loop heat exchanger pipes within building concrete foundations to partially or fully heat and cool buildings. Thus, TAF systems utilize building foundation elements (i.e., piles) as vertical boreholes to reduce the excavation costs associated with digging work needed for GSHPs. Indeed, a study by Lee et al. (2013) showed that a TAF system integrated with a PHC (prestressed high-strength concrete) pile can reduce the installation cost of the system by over 40%. Therefore, the installation costs associated with TAF systems can be significantly reduced compared to those of GSHPs with vertical boreholes.

The basic operating principles of GSHPs and TAFs are depicted in Fig. 2.31. When compared to GSHPs, TAFs save on the excavation costs of boreholes by integrating heat exchangers within the concrete of the building foundation piles. Table 2.4 compares the basic features of GSHPs and TAFs. However, and unlike the case of GSHPs, no clear design guidelines are available to properly size TAF systems to account for a wide range of parameters such as pile depth, soil type, number of piles, and water table depth.

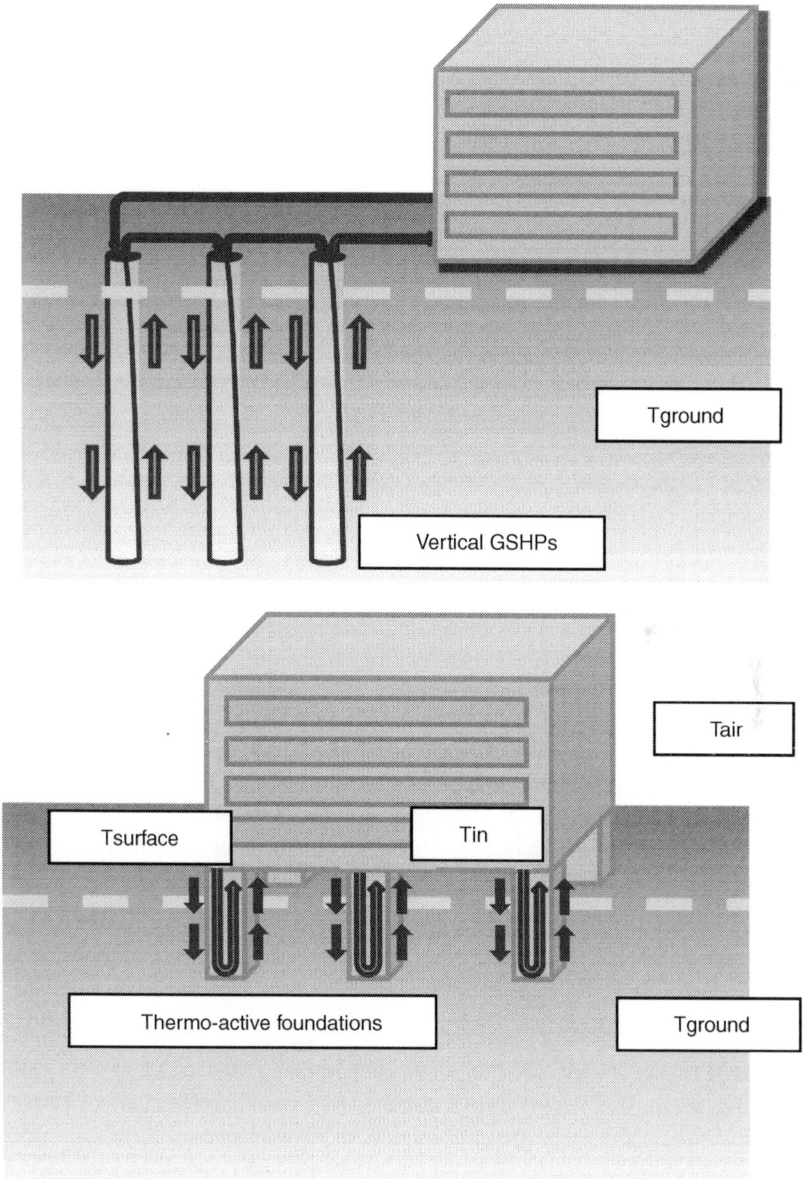

FIGURE 2.31 Schematic comparison of ground source heat pump and thermo-active foundation (TAF) systems.

TABLE 2.4 Comparison of Basic Features Between GSHPs and TAFs

Feature	GSHPs	TAFs
Range of depths	50–600 ft (15–180 m) (ref. ASHRAE std. 32.10)	Various depths depending on building foundations (as shallow as 2–3 m).
Pile configuration (cross-section)	Circular section	Various based on foundation types Circular section Square section Rectangular section
Effective design parameters	Ground temperature Borehole size	Ground temperature Ground surface temperature Foundation size

GSHPs, Ground source heat pumps; *TAFs*, thermo-active foundations.

Limited analyses and research studies have been reported for assessing the thermal performance and best design specifications of TAF systems especially for US climates and building applications (Hamada et al., 2007; Jalaluddin et al., 2011; Laloui et al., 2006; Loveridge and Powrie, 2013; Sekine et al., 2007; Wood et al., 2010). Indeed, the vast majority of existing TAF systems has been installed and evaluated in Europe and Japan. In particular, Brandl (2006) overviewed TAF systems and provided a discussion of the reported thermal and mechanical performance for select case studies in Europe. Brandl noted that concrete, which has good thermal properties, can enhance heat transfer between the ground medium and heat exchanger pipes, and that low-permeability soil and the low hydraulic gradient of groundwater can improve TAF thermal performance. McCartney and Rosenberg (2011), McCartney et al. (2010), and Stewart and McCartney (2013) performed a controlled laboratory experimental analysis using a centrifuge setup to evaluate soil–foundation interactions for a small-scale TAF. It was observed that heat can be transferred effectively to the ground medium through a fluid circulating within heat exchanger pipes embedded in a concrete pile. Kaltreider et al. (2015), Kwag and Krarti (2013), and Rouissi et al. (2011) used the experimental data of Rosenberg (2010), while Kwag and Krarti (2014) utilized testing results obtained by Stewart and McCartney (2013) to validate their numerical thermal models of TAF systems. While Kaltreider et al. (2015) and Rouissi et al. (2012) developed two-dimensional numerical models, Kwag and Krarti (2014) developed a three-dimensional numerical model to assess the thermal performance of TAF systems. The reported analyses using numerical solutions indicated that several physical parameters, such as foundation pile depth, flow velocity, and shank space, can have a significant impact on the ground heat transfer rate between an energy pile and the ground medium. Abdelaziz et al. (2011) also found that the same design parameters affect significantly the thermal performance of TAF systems.

Moreover, Kaltreider et al. (2015) and Rouissi et al. (2012) found that substantial thermal interactions can occur between indoor building environment and TAF system components. Specifically, when compared to the standard foundations (i.e., without embedded heat exchangers), TAF systems can significantly affect ground-coupled building foundation heat transfer and thus can influence heating and cooling thermal loads for air-conditioned buildings. This result indicates that thermal analysis of TAFs differs from that of GSHPs for which no thermal interactions occur between the building and the vertical boreholes.

Currently, several whole-building energy simulation programs employ the thermal response function approach, also called as G-function, to model the thermal performance of GSHPs. The G-function methodology was initially developed by Eskilson (1987) to evaluate the long-term thermal performance of GSHPs. Yavuzturk et al. (1999) later extended the G-function approach to account for the short-time performance of GSHPs. Kwag and Krarti (2014) and Loveridge (2012) applied the G-function methodology to evaluate the thermal performance of TAFs using whole-building simulation tools.

The main objective of this section is to discuss the benefits of using TAFs to heat and cool residential and commercial buildings. The details of the analysis are presented mostly by Kwag (2015) and Kwag and Krarti (2014).

2.6.2 Application of TAFs for Office Buildings

A prototypical floor of an office building with five thermal zones and a total floor area of 463.6 m^2 (4990 ft^2) as shown in Fig. 2.32 was modeled using EnregyPlus to assess the performance of TAF and variable air volume (VAV) systems. The VAV system is coupled with a temperature-based outside air economizer, hot

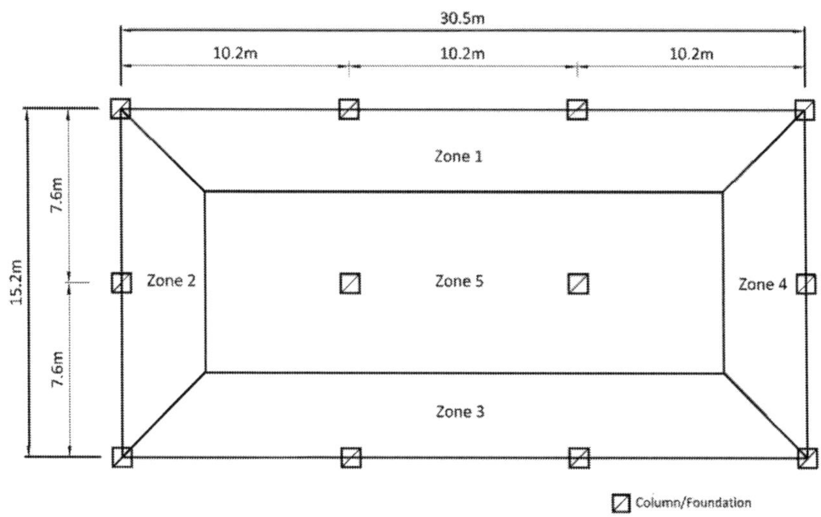

FIGURE 2.32 Location of foundation piles along the office building slab floor.

water reheat coils connected to a hot water boiler, and chilled water cooling coils connected to an electric compression chiller with an air-cooled condenser. The TAF system has a water-to-water heat pump connected to a ground source heat exchanger in addition to a boiler and a chiller as illustrated in Fig. 2.33. Table 2.5 lists the input data to model the VAV and TAF systems.

Kwag and Krarti (2014) evaluated the energy performance of TAF systems for heating and cooling the prototypical office building of Fig. 2.32. Five representative US climate conditions were considered including Chicago, IL; New York, NY; Phoenix, AZ; Denver, CO; and Cheyenne, WY. Table 2.6 outlines the annual average energy consumption to cool and heat the prototypical office building using a standard VAV system considered as the baseline case. The baseline HVAC configuration consists of a VAV system served by a boiler and a chiller as summarized in Table 2.7.

A TAF system is then considered to maintain thermal comfort within the prototypical office building using 12 thermal foundation piles with a depth of

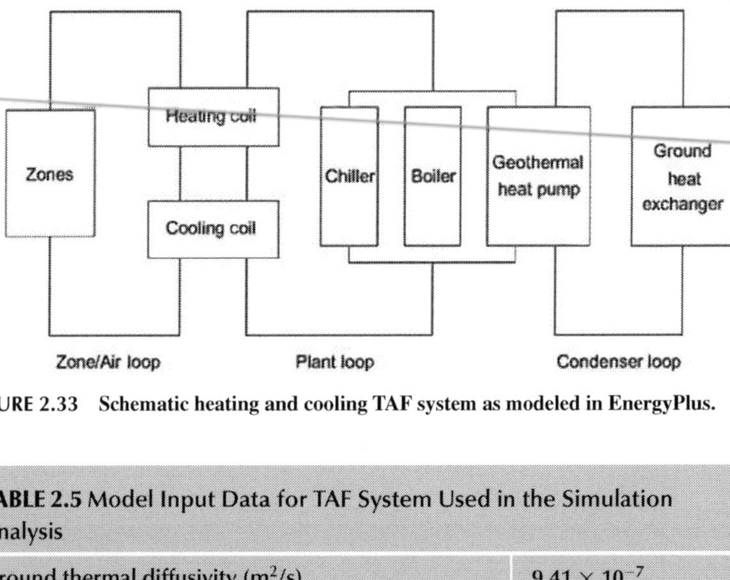

FIGURE 2.33 Schematic heating and cooling TAF system as modeled in EnergyPlus.

TABLE 2.5 Model Input Data for TAF System Used in the Simulation Analysis

Ground thermal diffusivity (m^2/s)	9.41×10^{-7}
Ground thermal conductivity (W/m·K)	1.60
Concrete thermal conductivity (W/m·K)	1.30
Pipe thermal conductivity (W/m·K)	0.36
Foundation depth (m)	20.0
Foundation radius (m)	0.20
Number of thermo-active foundation	12.0

TAF, Thermo-active foundation.

TABLE 2.6 Annual Average Thermal Energy Consumptions of the Baseline Case in Different Climate Conditions

	Chicago, IL	New York, NY	Phoenix, AZ	Denver, CO	Cheyenne, WY
Cooling energy (GJ)	77.483	77.271	83.983	75.510	73.317
Heating energy (GJ)	66.543	49.549	12.044	49.182	64.374

TABLE 2.7 Chiller and Boiler Capacities for a Baseline Case

	Chiller: electric	Boiler: gas
Capacity (kW)	200.0	90.0
Nominal efficiency	COP = 3.2	80%

H = 20 m. The energy consumption for heating and cooling associated with TAF systems is summarized in Fig. 2.34 for all five US climates. As shown in Fig. 2.34, TAF systems reduced the annual heating and cooling energy end-uses of the office building. Because of the relatively small thermal loads of the prototypical small office building, the simulation results show high reduction in cooling energy end-use from 18% (Cheyenne, WY) to 52% (Phoenix, AZ) and heating energy end-use from 40% (Phoenix, AZ) to 52% (Denver, CO). These

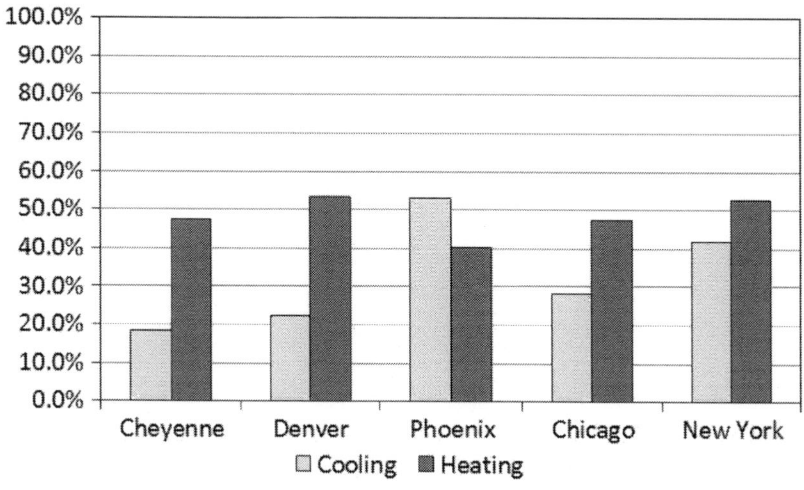

FIGURE 2.34 Percent reduction of cooling and heating energy end-uses associated with a TAF system relative to a VAV system for a prototypical office building in five US climates.

results indicate that TAF systems can significantly reduce cooling and heating energy consumption for office buildings.

2.6.3 Application of TAFs to Residential Buildings

The performance of TAF systems to air condition residential buildings has been also evaluated using an apartment building located in Boulder, CO (Kwag, 2015). The apartment building has three stories with six units per floor. Fig. 2.35 illustrates the prototypical building modeled using EnergyPlus, a whole-building energy simulation program. The net conditioned floor area for the apartment building is 111.5 m^2 (1200 ft^2). The total building floor area including all unconditioned areas is 2899 m^2 (31,210 ft^2). Table 2.8 provides the basic data used to develop the energy model for the residential building.

In the analysis, it was assumed that the baseline heating, air conditioning, and ventilating (HVAC) system includes a central system with an electric chiller

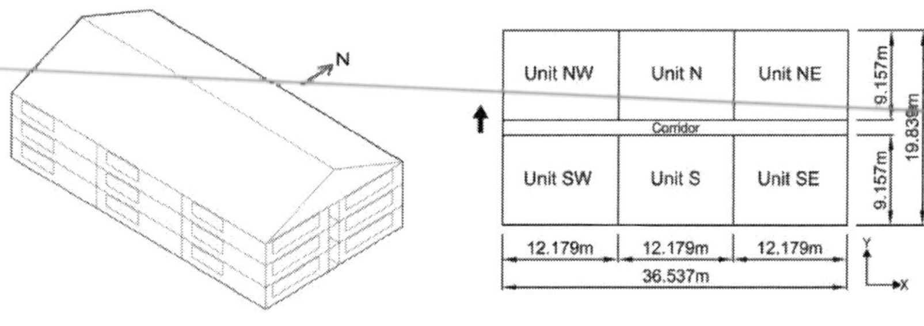

FIGURE 2.35 Apartment building model.

TABLE 2.8 Basic Features of the Prototypical Multifamily Residential Building

Building sizes summary	
Unit size	(x) 12.179 m, (y) 09.157 m, (z) 2.59 m
Corridor size	(x) 36.537 m, (y) 01.525 m, (z) 2.59 m
Building size	(x) 36.537 m, (y) 19.839 m, (z) 7.77 m (11.9 with attic)
Number of floor	Three stories
Number of units per floor	Six units per floor
Building envelope summary	
External roof	0.943 Btu/h·ft^2
External wall	0.065 Btu/h·ft^2
External floor	0.032 Btu/h·ft^2

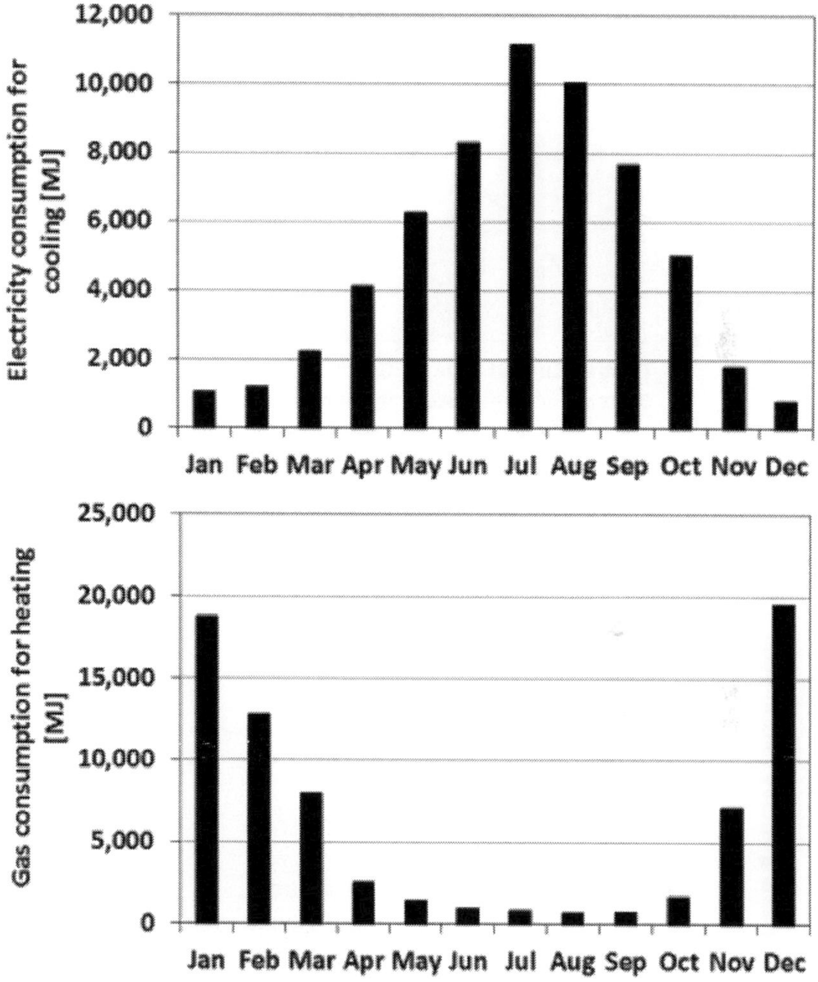

FIGURE 2.36 Monthly energy consumptions for cooling (electricity), and for heating (gas) using the baseline HVAC system.

for cooling and a hot water gas boiler to provide heating as well as domestic hot water for each apartment unit (Kwag, 2015). Fig. 2.36 shows the monthly electricity and gas consumptions of the multifamily residential building served by the baseline HVAC system.

The performance of a TAF system is modeled using EnergyPlus to estimate the energy end-uses to heat and cool the apartment building using the G-function approach. Figs. 2.37 and 2.38 present the cooling and heating end-uses for several heat pump sizes when the TAF system is used to air condition the residential building. The simulation results indicate that TAF systems can significantly reduce the building energy consumption required for heating when

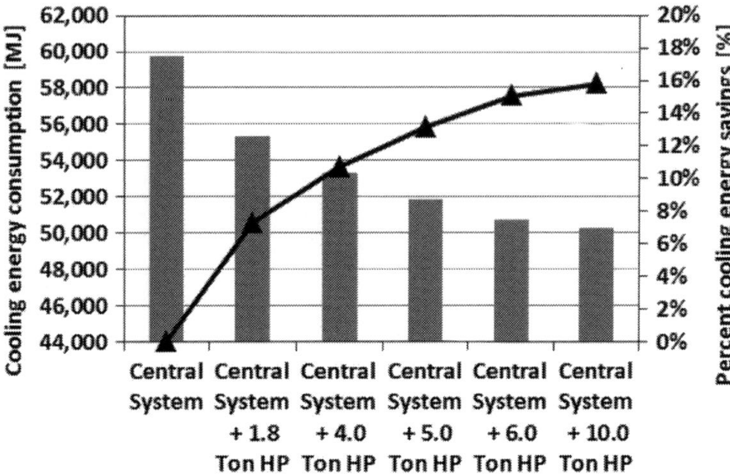

FIGURE 2.37 **Annual cooling energy consumption as a function of heat pump size.**

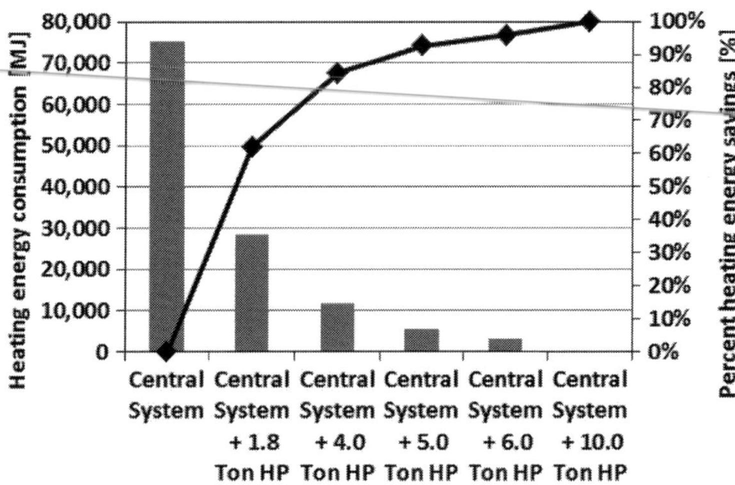

FIGURE 2.38 **Annual heating energy consumption as a function of heat pump size.**

the apartment is located in Boulder, CO, a heating-dominated climate. As the heat pump size is increased, a high reduction in heating and cooling energy end-uses is achieved. Thus, proper design of the heat pump is crucial to optimize the performance of TAF systems.

2.6.4 Recommendations

As alternatives to GSHP systems, TAFs can efficiently heat and cool residential and commercial buildings. The main advantage of TAFs is reduced installation costs compared to GSHP systems with similar benefits. Through the simulation analysis, it was found that TAFs can reduce annual cooling and heating energy use of residential and commercial buildings by 20%–55% for a wide range of US climates. However, and to ensure that TAFs can be widely applied as heating and cooling systems for buildings, design guidelines have to be developed to account for the main design parameters and climatic conditions. Moreover, long-term monitoring of TAF systems in real buildings has to be carried out for a wide range of building types and climates.

2.7 LED INTEGRATED CONTROLS

2.7.1 Background

Light-emitting diodes (LEDs) are commonly used for building lighting applications especially for illuminating interior spaces. LED fixtures are used for new and existing buildings to replace linear fluorescents due to their superior efficacies (lumens/W). The rated lifetimes of LED fixtures are typically at least 50,000 h, significantly higher than the expected life of fluorescent lamps. Other advantages of LEDs include higher controllability (e.g., easier dimming and on/off cycling) and include stringent lighting power density settings as well as control requirements such as automatic shut-off and greater durability.

Moreover, the application of advanced control strategies to operate lighting systems is increasing especially for commercial buildings. Lighting controls range from on/off scheduling based on manual settings, switching on/off based occupancy sensors, and dimming based on daylighting levels. Specifically, integrating LED fixtures with advanced controls capabilities are now becoming more widespread due to their ease of installation and lower costs. In addition, the integrated LED fixture and integrated control system allow for a simple path to building energy code compliance. Indeed, ASHRAE 90.1-2010 and other building codes occupancy sensors in offices, and daylight dimming controls. In particular, the integrated controls also allow for tuning of fixture groups to reduce fixture power from maximum output to medium or low levels, if those settings meet the lighting needs of the indoor space.

In this section, optimal controls for lighting fixtures are first discussed using a laboratory study and various desk location configurations and control strategies. Then, a specific case study for retrofitting lighting systems in two existing office buildings is briefly presented.

2.7.2 Evaluation of Optimal Controls for Lighting Systems

Several studies have shown that up to 60% of electric lighting energy savings can be achieved using properly designed control systems (Seo and Krarti, 2011). However, advanced lighting controls are rarely installed in buildings. For instance, only 10% of US commercial buildings actually utilize daylighting control systems although almost 50% of the buildings have energy-efficient lamps and ballasts (Krarti, 2011). Poor performance of lighting controllers leading to significantly lower energy savings than predicted is one of the common reasons of not installing these control systems. As noted earlier the implementation cost and the ease of installation are also barriers to the widespread use of advanced lighting controls. In this section, the potential energy use savings achieved by an optimal controller is discussed based on laboratory testing analysis and computer simulation results (Khilifi and Krarti, 2006; Seo and Krarti, 2011).

Specifically, the analysis is carried out to evaluate the performance of various optimized daylighting controls for an office room located in Boulder, CO (Khilifi and Krarti, 2006). The room has a floor area of 15.05 m^2 (162 ft^2) with 2.44 m (8 ft) of height. Two windows with tinted glazing were placed in the west façade of the office with a total area of 1.16 m^2 (162 ft^2). The walls as well as the suspended ceiling are white. The floor is covered with a brown carpet. Three electric lighting fixtures in the testing office are installed to provide the desired indoor illuminance level lighting. Each fixture has two separately circuited fluorescent lamps. Continuous indoor measurements were performed over a period of four months. For each day, hourly measurements were monitored from 8:00 am to 6:00 pm throughout the room. All the measurements are performed at desk height (i.e., working plane) of 0.762 m (2.5 ft). During the testing, the door was closed and the electrical lighting was turned off to ensure that space illuminance within the office space is only caused by natural light transmitted from the windows. Fig. 2.39A shows potential positions of working areas while Fig. 2.39B provides the locations of the lighting fixtures existing in the tested office room. Fig. 2.40A–B show the lines for equal illuminance level (at desk

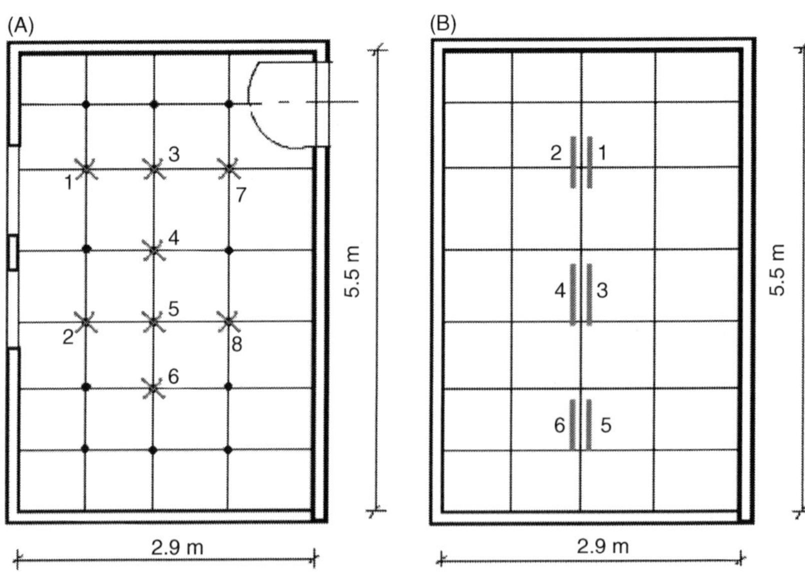

FIGURE 2.39 Floor plan for the tested office room showing (A) selected working zones and (B) locations of lighting fixtures.

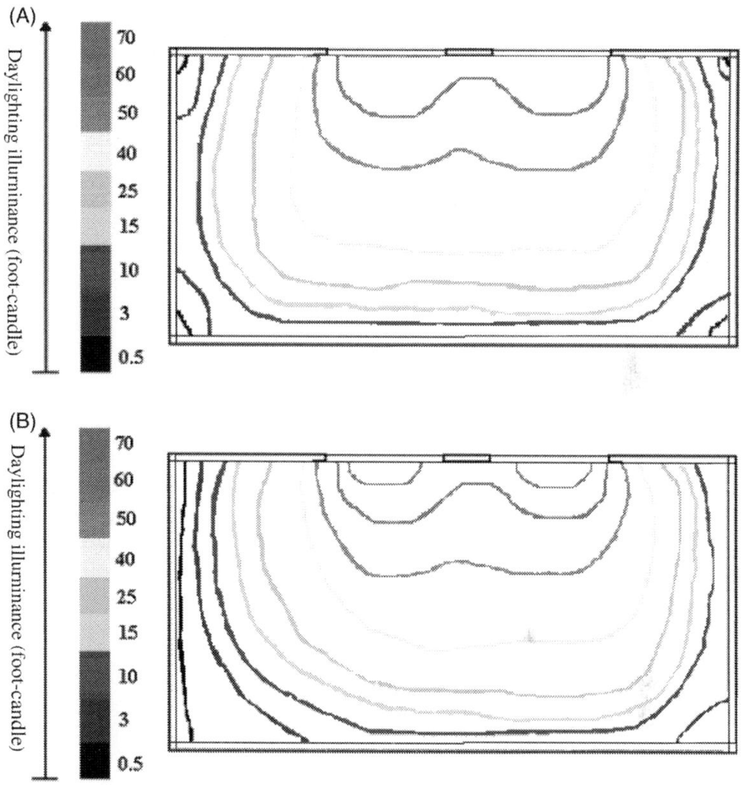

FIGURE 2.40 Iso-illuminance distribution in the tested office room. (A) 10:00 am of March 9 and (B) 4:00 pm of March 9.

height) when the interior shades are fully open, during March 9 (sunny day) at 10:00 am and 4:00 pm, respectively. The 500 lux illuminance level is achieved only near the two windows. Further away from the windows, electrical lighting is required to complement daylighting to accomplish the required 500 lux illuminance level at the work plane.

From the monitoring data, a correlation between outside light and inside light is developed. The relationship is expressed in terms of modified daylighting factors (Khilifi and Krarti, 2006). These factors are defined as ratios of inside horizontal illuminance level measured on a horizontal surface located at the work plane over outside natural illuminance level measured on a horizontal surface (measured by a single sensor located on the roof of the office space). The modified daylighting factors for each potential working zone are listed in Table 2.9. As noted, the demand factors change throughout the day as the position of the sun affects the amount of light reaching each working zone.

The performance of the optimal controller depends on the desk locations and the control types. In the following section, results obtained using optimized controls for various desk location and control combinations are presented.

TABLE 2.9 Modified Daylighting Factors for the Tested Office Room

Time of day	Zone							
	1	2	3	4	5	6	7	8
9–11 am	0.0043	0.0038	0.0032	0.003	0.0027	0.0014	0.0019	0.0017
11 am to 1 pm	0.0061	0.0051	0.0044	0.004	0.0036	0.0018	0.0025	0.002
1–3 pm	0.0075	0.0063	0.0054	0.0053	0.0047	0.0023	0.0034	0.0031
3–5 pm	0.0065	0.0056	0.0047	0.0046	0.0042	0.002	0.0029	0.0027

2.7.2.1 Configuration A: Eight Working Areas

In this configuration, all eight locations (1–8) as outlined in Fig. 2.39A are occupied by desks and are required to have the desired illuminance level of 500 lux. Six separately circuited fluorescent lamps are used to light the office room. The results obtained using a genetic-algorithm (GA) optimization to maximize electrical lighting energy use savings are briefly discussed for various lighting controls (Khilifi and Krarti, 2006).

2.7.2.1.1 ON/OFF Electrical Lighting Control

After the execution of the GA-based optimization, an hourly operation schedule for the electrical lighting system was developed. Table 2.10 gives the operation schedule for all six fluorescent lamps utilizing the ON/OFF lighting controls as well as the electrical power (in Watts) used by the electrical lighting system. The value "1" indicates that the lamp is ON while the value "0" indicates that the lamp should be off.

2.7.2.1.2 Two-Stepped Control

Table 2.11 gives the operation schedule for all six fluorescent lamps utilizing the two-stepped lighting controls as well as the electrical power (in Watts) used by the electrical lighting system.

TABLE 2.10 Operation Schedule for Electrical Lighting Fixtures Using ON/OFF Controls

Time of the day	Lamp number						Energy use (W)
	1	2	3	4	5	6	
9–11 am	1	1	1	0	1	1	352
11 am to 1 pm	1	0	0	0	1	0	140.8
1–3 pm	0	0	0	0	1	0	70.40
3–5 pm	0	0	1	0	0	1	140.8

TABLE 2.11 Operation Schedule for Electrical Lighting Fixtures Using Stepped Controls

Time of the day	Lamp number						Energy use (W)
	1	2	3	4	5	6	
9–11 am	1	0.5	1	0.5	1	0.5	316.80
11 am to 1 pm	0.5	0	0.5	0	0	0.5	105.60
1–3 pm	0	0	0	0	0.5	0	35.200
3–5 pm	0.5	0	0	0.5	0	1	140.80

The value "0.5" means that the lamp should be set at 50% of its full capacity (thus using 50% of electrical input and producing 50% of its rated illuminance output) while "1" means the lamp should be providing 100% of its full illuminance output.

2.7.2.1.3 Dimming Controls

Table 2.12 presents the operation schedule for all six fluorescent lamps utilizing the dimming lighting controls as well as the electrical power (in Watts) used by the electrical lighting system.

The values provided in Table 2.13 correspond to the output and input settings relative to the rated conditions for all electrical lamps.

Energy use savings obtained from GA-based optimization for electrical lighting are determined for the three daylighting controls (i.e., on/off, two-stepped, and dimming) and are summarized in Table 2.13. For each control, energy savings are estimated relative to the case when no daylighting is utilized. As expected, Table 2.13 indicates that the highest energy use savings are achieved by dimming controls during the period between 1 and 3 pm when natural illuminance level is the highest during the day. However, the overall estimated savings in electrical lighting energy use for two-stepped control is 64.59% and is only 4.11% less than the savings of 68.70% achieved by the dimming controls.

TABLE 2.12 Operation Schedule for Electrical Lighting Fixtures Using Dimming Controls

Time of the day	Lamp number						Energy use (W)
	1	2	3	4	5	6	
9–11 am	0.68	0.62	0.79	0.56	0.76	0.92	304.832
11 am to 1 pm	0.34	0.08	0.22	0.26	0.18	0.35	100.672
1–3 pm	0.10	0.04	0.05	0.08	0.17	0.03	33.088
3–5 pm	0.16	0.34	0.15	0.32	0.46	0.67	137.84

TABLE 2.13 Percent Savings in Electrical Lighting Energy Use Based on GA-Based Daylighting Optimization and Three Lighting Controls for Configuration A

Time of the day	Energy savings (%)		
	Electrical lighting control		
	On/off	Two-stepped	Dimming
9–11 am	16.67	25.00	29.90
11 am to 1 pm	66.67	75.00	79.46
1–3 pm	83.33	91.67	93.58
3–5 pm	66.67	66.67	71.84
Entire day: 9 am to 5 pm	58.34	64.59	68.70

2.7.2.2 Configuration B: Working Area Close to Windows

Fig. 2.41 shows another configuration for the working zones (i.e., where desks are located). In this configuration, the desired illuminance level of 500 lux is required to be maintained only in zones 1 through 5. Table 2.14 presents the percent savings in electrical lighting energy use obtained for three control strategies (Khilifi and Krarti, 2006).

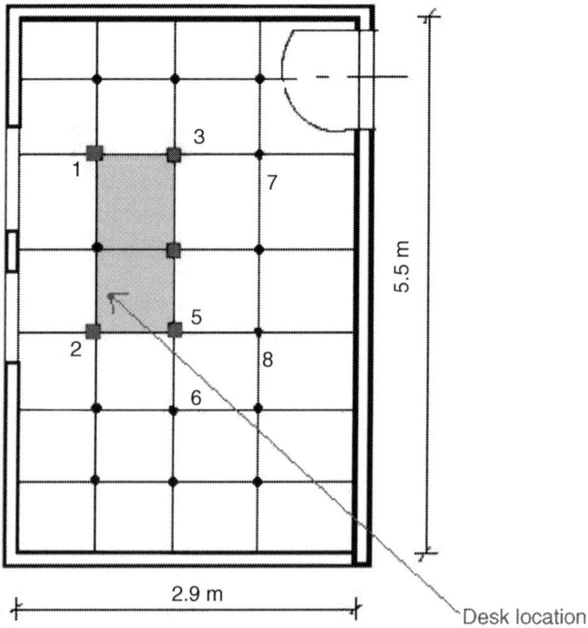

FIGURE 2.41 Working zones for configuration B.

TABLE 2.14 Percent Savings in Electrical Lighting Energy Use Based on GA-Based Daylighting Optimization and Three Lighting Controls for Configuration B

Time of the day	Energy savings (%)		
	Electrical lighting control		
	On/off	Two-stepped	Dimming
9–11 am	33.33	41.67	50.00
11 am to 1 pm	66.67	75.00	80.67
1–3 pm	83.33	91.67	93.50
3–5 pm	83.33	83.33	84.34
Entire day: 9 am to 5 pm	58.34	66.67	72.92

The results of Table 2.14 indicate that the electrical lighting energy use savings obtained for configuration B is higher than that obtained for configuration A for all lighting controls especially during early morning and late afternoon. Indeed, configuration B has fewer locations where the illuminance level need to be maintained at the desired set-point of 500 lux. The overall estimated savings in electrical lighting energy use achieved by the dimming control increased from 68.70% for configuration A to 77.13% for configuration B.

2.7.2.3 Configuration C: Working Area Perpendicular to the Window Wall

Fig. 2.42 shows a configuration for the working zones (i.e., where a desk placed). In this configuration, the desired illuminance level of 500 lux is required in only zones 2, 5, 6, and 8. Table 2.15 presents the percent savings in electrical lighting energy use obtained from GA-based daylighting optimization for the three daylighting controls (i.e., on/off, two-stepped, and dimming) specific to the working zones of configuration C.

Due to the location of the desk and thus the working zones, the electrical lighting energy use savings obtained for configuration C are lower than those obtained for configuration B for all lighting controls especially during early morning and late afternoon. For instance, the overall estimated savings in electrical lighting energy use achieved by the dimming control is reduced from 77.13% for configuration B to 72.09% for configuration C.

2.7.3 Recommendations

A GA optimization approach is shown to be effective in controlling the operation of lighting fixtures taking advantage of incoming daylight. Specifically, the

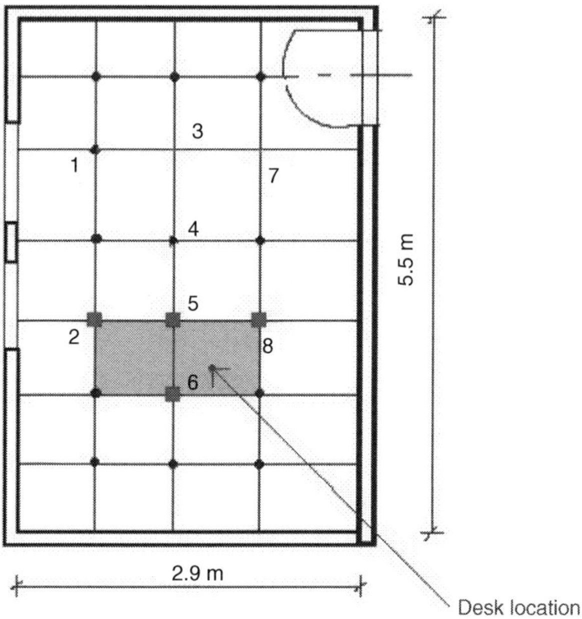

FIGURE 2.42 Working zones for configuration C.

TABLE 2.15 Percent Savings in Electrical Lighting Energy Use Based on GA-Based Daylighting Optimization and Three Lighting Controls for Configuration C

Time of the day	Energy savings (%)		
	Electrical lighting control		
	On/off	Two-stepped	Dimming
9–11 am	16.67	25.00	32.83
11 am to 1 pm	66.67	75.00	78.17
1–3 pm	83.33	91.67	93.00
3–5 pm	66.67	66.67	84.34
Entire day: 9 am to 5 pm	58.34	58.34	64.59

GA-based optimal controller is combined with on–off, stepped, and dimming controls to reduce energy use associated with electrical lighting while maintaining desired illuminance levels for specific working zones. The performance of the GA-based daylighting controls is tested for various working zone configurations in an office space located in Boulder, CO. In particular, the potential savings in electrical lighting energy use are estimated or all lighting controls and can range from 50% to 70%.

The potential energy savings are found to depend on the lighting control and on the position of the working zones relative to the windows. As expected, the highest energy savings are obtained for working zones located closer to the windows. Moreover, the results indicate that dimming controls achieve the highest energy savings especially during early morning and late afternoon hours. The GA optimization approach can be utilized to not only maintain illuminance level in specific working zones but to ensure desired uniformity in spatial indoor illuminance distribution throughout the office space. Moreover, the GA-based optimization can be used to operate the shading devices for the windows to reduce glare from daylighting.

2.7.4 Field Study Analysis

In this section, the results of a reported field study are summarized to demonstrate the performance of integrated control LED fixtures (Shackelford et al., 2015). Specifically, the field study includes two sites where the existing lighting systems have been retrofitted using integrated control LED units. Each integrated control LED unit includes the LED lamp, driver, daylighting controls, and occupancy sensors. The units are designed to replace T-8 or T-12 fluorescent fixtures using the existing wiring system. The calibration of the units is performed automatically to facilitate their installation and commissioning. The two sites of the study are briefly described:

Site-1: The first site is part of a 30-story building located in downtown Atlanta, GA. Specifically, the lighting system for a 1198 m^2 (12,900 ft^2) space has been upgraded from two-lamp fixtures with an average area density of 8.7 m^2 (94 ft^2) per fixture to integrated control LED units. The space consists of large open office areas with two private offices, two conference rooms, and one breakroom.

Site-2: In the second site, integrated LED units have replaced the existing lighting system for one floor of a 28-story building, located in downtown Chicago, IL. The retrofitted floor consists of a large open office space with some private offices and conference rooms with a total area of 1835 m^2 (19,750 ft^2). The existing lighting system is made up of mostly three-lamp fixture parabolic troffers with an average area density of 7 m^2 (76 ft^2) per fixture.

Measurements of energy use and photometric characteristics of the lighting system before and after the retrofit have been carried out for several months in both sites averaging 6 weeks for the preretrofit period and 5 months for the postretrofit period. The results of the testing and monitoring are summarized in Table 2.16 for both sites (Shackelford et al., 2015). Based on photometric measurements at the study locations, average illuminance and efficacy at the work

TABLE 2.16 Summary of the Field Test Results for Two GSA Sites

Indicators	Site-1	Site-2
Mean illuminance		
Preretrofit	432 lux (40.1 fc)	341 lux (31.7 fc)
Postretrofit	470 lux (43.7 fc)	429 lux (39.9 fc)
Percent improvement	9.0%	25.9%
Work plane efficacy		
Preretrofit	60.8 Lm W^{-1}	44.7 Lm W^{-1}
Postretrofit	99.3 Lm W^{-1}	78.8 Lm W^{-1}
Percent improvement	63.3%	78.5%
Lighting power density		
Preretrofit	7.1 W m^{-2} (0.66 W ft^{-2})	7.4 W m^{-2} (0.69 W ft^{-2})
Postretrofit	4.7 W m^{-2} (0.44 W ft^{-2})	5.4 W m^{-2} (0.50 W ft^{-2})
Percent reduction	33.3%	27.5%
Energy use intensity		
Preretrofit	19.2 kWh m^{-2} y^{-1} (1.78 kWh ft^{-2} y^{-1})	27.6 kWh m^{-2} y^{-1} (2.56 kWh ft^{-2} y^{-1})
Postretrofit	11.4 kWh m^{-2} y^{-1} (1.06 kWh ft^{-2} y^{-1})	10.5 kWh m^{-2} y^{-1} (0.98 kWh ft^{-2} y^{-1})
Percent savings	40.5%	61.7%

plane have improved in both sites as summarized in Table 2.16. In particular, the work plane efficacy results for both sites have shown improvements of 63.3% in site 1 and 78.5% in site 2.

The annual energy use of the lighting system before and after the retrofit has been calculated based on field measurements. Significant energy use savings have been estimated for both sites due to the retrofit of the lighting systems using the integrated control LED fixtures. Specifically, the annual energy consumption associated with lighting has been reduced by 40.5% in site 1 and 61.7% in site 2. Part of these savings are attributed to the improvement in the fixture efficacy since the lighting power density in both sites have been reduced by 33.3% in site 1 and 27.5% in site 2. However, most of the savings are due to improved controls especially in site 2 where on–off settings from the occupancy sensors and dimming reduction due to daylighting controls have reduced lighting energy consumption by 41% (Shackelford et al., 2015).

It should be noted that the study has indicated that the average cost of the lighting retrofit in both sites is estimated to be about \$35.42/m^2 (\$3.29/ft^2). This cost results in a simple payback period for the lighting retrofit that varies between 10 and 15 years depending on the electricity prices. The cost of the integrated control LED fixtures is estimated to be only \$8.83/m^2 (\$0.82/ft^2) for new constructions and therefore the payback is reduced to be in the range of 2.5–4 years.

2.8 SUMMARY

In this chapter, the basic energy efficiency principles for designing and retrofitting buildings are presented. These principles are applicable to any building type and any climatic conditions without affecting indoor thermal comfort. The reduction of building thermal loads, the use of energy efficiency equipment, and reliance on renewable energy systems are the basis of improving the energy performance of buildings. In addition to an overview of models and indicators of indoor thermal comfort, the chapter outlines suitable thermal analysis methods and techniques to assess and improve the energy performance of buildings. In particular, whole-building simulation tools are often used to develop integrated and optimized energy efficiency solutions for designing and retrofitting buildings. Finally, three representative energy efficiency technologies are described to highlight some of the advanced and innovative strategies that should be considered to efficiently heat and cool buildings.

PROBLEMS

2.1 Determine the heating base-load, the balance temperature, and the BLC of a house located in Boulder, CO using an inverse modeling technique based on the monthly utility bills provided in the table. For the analysis assume that the house is heated 24 h/day and that the seasonal efficiency of the gas-furnace is about 78%. It should be noted that the house is generally unoccupied during the period of May 25 to July 24 (for 2014 and 2013).

Date of bill	Billed days	Therms used	Outdoor temp. (°F)
Dec. 20, 2014	34	230	33
Nov. 21, 2014	30	71	51
Oct. 17, 2014	29	50	63
Sep. 19, 2014	31	21	72
Aug. 20, 2014	28	18	77
Jul. 24, 2014	35	6	79
Jun. 20, 2014	30	10	65
May 22, 2004	30	61	56
Apr. 18, 2004	30	118	45
Mar. 22, 2014	30	163	38

Date of bill	Billed days	Therms used	Outdoor temp. (°F)
Feb. 16, 2014	27	186	26
Jan. 23, 2014	35	230	32
Jan. 02, 2014	30	199	32
Nov. 17, 2013	29	132	41
Sep. 19, 2013	29	16	73
Sep. 5, 2013	31	4	78
Jun. 20, 2013	30	9	67
May 24, 2013	31	53	56
Apr. 25, 2013	30	87	49
Mar. 22, 2013	27	107	41
Feb. 24, 2013	32	159	36

Provide any pertinent comments on the thermal performance of this house.

2.2 Consider the following characteristics for a warehouse:

Roof area:	10,000 ft^2
Existing roof *R*-value:	5
Degree-days (winter):	6000
Degrees-hours (summer):	17,000
Fuel cost:	\$8/MMbtu
Boiler efficiency:	0.70
Electric rate	\$0.09/kWh
Air-condition requirement:	0.7 kW/ton

It is proposed to add *R*-19 insulation in the roof.

(1) Comment on the potential energy savings of adding insulation.
(2) If the insulation costs \$0.35/ft^2, determine the payback of adding insulation.

2.3 Determine the annual energy savings due to replacing single-pane window (*R*-1) by a double-pane window (*R*-2). The total area of the glazing is 200 ft^2. Assume the following:

Heating degree-days:	6000
Cooling degrees-days:	500
Fuel cost:	\$8/MMBtu
Boiler efficiency:	0.80
Electric rate:	\$0.12/kWh
Refrigeration requirement:	0.70 kW/ton
Insulation cost:	\$0.225/ft^2

2.4 Do you recommend to add *R*-20 insulation to a 25,000 ft^2 uninsulated roof (*R*-5) when you consider a 25-year life cycle with a rate of return of 8%? Assume the following:

Heating degree-days:	5000
Cooling degrees-hours:	10,000
Fuel cost:	$8/MMBtu
Boiler efficiency:	0.70
Electric rate:	$0.10/kWh
Refrigeration requirement:	0.75 kW/ton
Insulation cost:	$0.225 ft^2

2.4 Model a one-floor 42-ft by 24-ft ranch house using any detailed simulation tool such eQUEST, EnergyPlus, TRNSYS, and ESP-r. Use the following assumptions to model the house:

- The whole house is one thermal zone with a floor-to-ceiling height of 9 ft.
- Attic is unheated and is modeled as a separate zone with a 3-ft height.
- The roof is flat.
- The temperature is kept at 70°F (heating) and 75°F (cooling) throughout the year.
- Assume an infiltration rate of 0.75 ACH.
- Assume that the DWH usage is 0.2 gallon/person.
- The wall has *R*-11 fiberglass insulation and the ceiling has an insulation of *R*-19. Use material library to select the proper construction materials.
- The house is heated and cooled through a central residential system with a gas furnace efficiency of 0.80.
- The house is located in Denver, CO.
  1) Determine the design heating and cooling loads for the house. Select the size of the required gas furnace (Btu/h) and the air conditioning system (in tons).
  2) Determine the total annual gas and electricity usage for the house. Compare the gas usage for heating (only) predicted by simulation to that calculated from the DD method assuming $T_b = 65°F$. What is should be T_b for the house so the DD method matches the simulation prediction for the annual heating gas usage.
  3) Determine the energy savings (using the simulation tool) and the cost-effectiveness (use both simple payback period and LCC analyses) of each of the following energy conservation measures (assume the house located in Denver, CO):

     (i) The infiltration rate is cut to 0.25 ACH with a total cost of $200.
     (ii) The insulation is increased to *R*-19 the walls with a cost of 0.25/ft^2.
     (iii) The insulation is increased to *R*-30 in the ceiling with a cost of $0.35/ft^2.

(iv) Install high efficiency furnace with a seasonal efficiency of 0.90 with an incremental cost of $750.

For the economic analysis, assume 25 years and 5% discount rate. Assume the electricity cost is $0.08/kWh and $0.95/therm.

Briefly discuss if the results make sense to you (explain why).

REFERENCES

Abdelaziz, S.L., Olgun, C.G., Martin, II, J.R., 2011. Design and Operation Considerations of Geothermal Energy Foundations, GeoFrontiers 2011. ASCE, Dallas, TX, p. 450–459.

Al-Saadi, S.N., Zhai, Z., 2015. Systematic evaluation of mathematical methods and numerical schemes for modeling PCM-enhanced building enclosure. Energy Build. 92, 374–388.

ASHRAE, 2009. Handbook of Fundamentals. American Society of Heating, Refrigerating and Air-Conditioning Engineers, Inc., Atlanta, GA.

Bourgeois, D., Reinhart, C., MacDonald, I., 2006. Adding advanced behavioural models in whole building energy simulation: a study on the total energy impact of manual and automated lighting control. Energy Build. 38 (7), 814–823.

Baetens, R., Jelle, B.P., Thue, J.V., Tenpierik, M.J., Grynning, S., Uvsløkk, S., 2010a. Vacuum insulation panels for building applications: a review and beyond. Energy Build. 42, 147–172.

Baetens, R., Jelle, B.P., Gustavsend, A., Grynninga, S., 2010b. Gas-filled panels for building applications: a state-of-the-art review. Energy Build. 42, 1969–1975.

Baetens, R., Jelle, B.P., Gustavsen, A., 2011. Aerogel insulation for building applications: a state-of-the-art review. Energy Build. 43, 761–769.

Brandl, H., 2006. Energy foundations and other thermo-active ground structures. Geotechnique 56 (2), 81–122.

Clark, W.W., Schaefer, L.A., Knotts, W.A., Mo, C., Kimber, M., 2013. Patent US20130081786 – variable thermal insulation. Google Books [Online]. Available: https://www.google.com/patents/US20130081786.

Dimoudi, A., Androutsopoloulus, A., Lykoulis, S., 2004. Experimental work on a linked, dynamic and ventilated wall component. Energy Build. 36 (5), 443–453.

DOE, 2015. Quadrennial Technology Review: An Assessment of Energy Technologies and Research Opportunities. Department of Energy Report, Washington, DC.

Eskilson, P., 1987. Thermal Analysis of Heat Extraction Boreholes. University of Lund, Department of Mathematical Physics, Lund, Sweden, (Doctoral Thesis).

EIA, 2016a. U.S. Energy Information Administration – EIA – Independent Statistics and Analysis, 2016. http://www.eia.gov.

EIA, 2016b. U.S. Price of Natural Gas Sold to Commercial Consumers (Dollars per Thousand Cubic Feet). Available from: http://www.eia.gov.

EIA, 2016c. Average Retail Price of Electricity. Available from: http://www.eia.gov/.

El Mankibi, M., Zhai, Z., Al-Saadi, S.N., Zoubir, A., 2015. Numerical modeling of thermal behaviors of active multi-layer living wall. Energy Build. 106, 96–110.

EnergyPlus, 2017. EnergyPlus Documentation. US Department of Energy, Energy Efficiency for Buildings, Building Technologies Program, Washington, DC, Available from: https://energyplus.net/documentation.

Enescu, D., 2017. A review of thermal comfort models and indicators for indoor environments. Renew. Sustain. Energy Rev. 79, 1353–1379.

Fanger, P.O., 1970. Thermal Comfort. Danish Technical Press, Copenhagen.

Fanger, P.O., Toftum, J., 2002. Extension of the PMV model to non-air-conditioned buildings in warm climates. Energy Build. 34 (6), 533–536.

Fels, M., 1986. Special issue devoted to measuring energy savings: the scorekeeping approach. Energy Build 9 (12), 127–136.

Gagge, A.P., Stolwijk, J.J., Nishi, Y., 1970. An effective temperature scale based on a simple model of human physiological regulatory response. ASHRAE Trans 70 (1), 247–262.

Hamada, Y., Nakamura, M., Kubota, H., Ochifuji, K., 2007. Field performance of an energy pile system for space heating. Energy Build. 39 (5), 517–524.

Hayashi, M., Yoshinori, H., Sugawara, M., 2014. Dwellers' habit of opening windows in detached houses in cold and hot-humid climate of Japan. J. Environ. Protect. 5, 663–671.

Hermawan, A., Prianto, E., Setyowati, E., 2015. Thermal comfort of wood–wall house in coastal and mountainous region in tropical area. Proc. Eng. 125, 725–731.

Humphreys, M.A., Nicol, J.F., 2002. The validity of ISO-PMV for predicting comfort votes in every-day thermal environments. Energy Build. 34, 667–684.

IEA, 2013. Transition to Sustainable Buildings: Strategies and Opportunities to 2050. International Energy Agency Report, https://www.iea.org/publications/freepublications/publication/Building2013_free.pdf.

Ihm, P., Krarti, M., 2012. Design optimization of energy efficient residential buildings in Tunisia. Build. Environ 58 (12), 81–90.

Jalaluddin, Miyara, A., Tsubaki, K., Inoue, S., Yoshida, K., 2011. Experimental study of several types of ground heat exchanger using a steel pile Foundation. Renew. Energy 36, 764–771.

Jelle, B.P., Gustavsen, A., Baetens, R., 2010. The path to the high performance thermal building insulation materials and solutions of tomorrow. J. Build. Phys. 34 (2), 99–123.

Jelle, B.P., 2011. Traditional, state-of-the-art and future thermal building insulation materials and solutions – properties, requirements and possibilities. Energy Build. 43, 2549–2563.

Jonosson, A., Roos, A., 2010. Evaluation of control strategies for different smart window combinations using computer simulations. Solar Energy 84 (1), 1–9.

Kalinic, N., 2009. Measurement and Verification of Savings from Implemented Energy Conservation Measures (MS Report). University of Colorado, Boulder, CO.

Kalnæs, S.E., Jelle, B.P., 2014. Vacuum insulation panel products: a state-of-the-art review and future research pathways. Appl. Energy 116, 355–375.

Kaltreider, C., Krarti, M., McCartney, J., 2015. Heat transfer analysis of thermo-active foundations. Energy Build. 86 (1), 492–501.

Kapsis, K., Tzempelikos, A., Athienitis, A.K., Zmeureanu, R.G., 2010. Daylight performance evaluation of a bottom-up motorized roller shade. Solar Energy 84 (12), 2120–2131.

Khilifi, A., Krarti, M., 2006. Genetic algorithm based controls for daylighting. In: Proceedings of ASME Solar Energy Conference. July 13–18, Denver, CO.

Kim, J.T., Lim, J.H., Cho, S.H., Yun, G.Y., 2015. Development of the adaptive PMV model for improving prediction performances. Energy Build. 98, 100–105.

Kissock, K., Reddy, T.A., Claridge, D.E., 1998. Ambient-temperature regression analysis for estimating retrofit savings in commercial buildings. ASME J. Sol. Energy Eng 120 (3), 168–176.

Kosny, J., Kossecka, E., Brzezinski, A., Tleoubaev, A., Yarbrough, D., 2012. Dynamic thermal performance analysis of fiber insulations containing bio-based phase change materials (PCMs). Energy Build. 52, 122–131.

Krarti, M., 2017. Design of Energy Efficient Electrical Systems for Buildings. CRC Press/Taylor & Francis, New York, NY, p. 494.

Krarti, M., Dubey, K., Howarth, N., 2017. Evaluation of building energy efficiency investment options for the Kingdom of Saudi Arabia. Energy 134, 595–610.

Krarti, M., 1994. Effect of airflow on heat transfer in walls. J. Solar Energy Eng. 116 (1), 35–42.

Krarti, M., 2011. Energy Audit of Building Systems: An Engineering Approach, second ed. CRC Press, Boca Raton, FL.

Kreider, J.F., Blanc, S.L., Kammerud, R.C., Curtiss, P.S., 1997. Operational data as the basis for neural network prediction of hourly electrical demand. ASHRAE Trans 103 (2), 926–934.

Kwag, B.C., Krarti, M., 2013. Performance of thermoactive foundations for commercial buildings. ASME Solar Energy Eng. J. 135 (4), 040907-040907-10.

Kwag, B.C., Krarti, M., 2014. Development of a response factor model for thermo-active building foundation. In: Proceedings for ASME 2014 International Mechanical Engineering Congress and Exposition. November 10–14, Montreal, Canada.

Kwag, B.C., 2015. Thermal Performance and Design Guidelines of Thermo-active Foundations. University of Colorado, Boulder, CO, (PhD Dissertation).

Laloui, L., Nuth, M., Vulliet, L., 2006. Experimental, numerical investigations of the behaviour of a heat exchanger pile. Int. J. Numer. Anal. Methods Geomech. 30 (8), 763–781.

Lee, J.U., Kim, T., Leigh, S.B., 2013. Thermal performance analysis of a ground-coupled heat pump integrated with building foundation in summer. Energy Build. 59, 37–43.

Lee, S.W., Lim, C.H., Bin Salleh, E., 2016. Reflective thermal insulation systems in building: a review on radiant barrier and reflective insulation. Renew. Sustain. Energy Rev. 65, 643–661.

Loveridge, F., 2012. The Thermal Performances of Foundation Piles Used as Heat Exchangers in Ground Energy Systems. University of Southampton, Faculty of Engineering and the Environment, (Doctoral thesis).

Loveridge, F., Powrie, W., 2013. Temperature response functions (G functions) for single pile heat exchangers. Energy 57, 554–564.

McCartney, J.S., Rosenberg, J.E., Sultanova, A., 2010. Engineering performance of thermo-active foundation systems. In: Goss, C.M., Kerrigan, J.B., Malamo, J., McCarron, M.O., Wiltshire, R.L. (Eds.), GeoTrends: The Progress of Geological and Geotechnical Engineering in Colorado at the Cusp of a New Decade (GPP 6). ASCE, pp. 27–42.

McCartney, J.S., Rosenberg, J.E., 2011. Impact of heat exchange on the axial capacity of thermo-active foundations. In: Han, J., Alzamora, D.E. (Eds.), In: Proceedings of Geo-Frontiers 2011 (GSP 211). ASCE, Reston, VA, pp. 488–498.

Menyhart, K., Krarti, M., 2017. Potential energy savings from deployment of Dynamic Insulation Materials for US residential buildings. Build. Environ. doi: 10.1016/j.buildenv.2016.12.009, 2016.

Masoso, O.T., Grobler, L.J., 2008. A new and innovative look at anti-insulation behaviour in building energy consumption. Energy Build. 40 (10), 1889–1894.

Neeper, D.A., McFarland, R.D., 1982. Some Potential Benefits of Fundamental Research for the Passive Solar Heating and Cooling of Buildings. Los Alamos National Laboratory, Los Alamos, NM.

Park, B., Srubar, W.V., Krarti, M., 2015. Energy performance analysis of variable thermal resistance envelopes in residential buildings. Energy Build. 103, 317–325.

Qiu, K., Haghighat, F., 2007. Modeling the combined conduction-air infiltration through diffusive building envelope. Energy Build. 39, 1140–1150.

Rouissi, K., Krarti, M., McCartney, J.S., 2011. Analysis of thermo-active foundation using U-tube heat exchangers. ASME Solar Energy Eng. J. 134 (2), 154–161.

Rosenberg, J., 2010. Centrifuge Modeling of Soil Structure Interaction in Thermo-Active Foundations, MS Thesis report. University of Colorado, Boulder, CO.

Sbar, N.L., Podbelski, L., Yang, H.M., Pease, B., 2012. Electrochromic dynamic windows for office buildings. Int. J. Sustain. Built Environ. 1 (1), 123–139.

Sadineni, S.B., Madala, S., Boehm, R.F., 2011. Passive building energy savings: a review of building envelope components. Renew. Sustain. Energy Rev. 15 (8), 3617–3631.

Schiavoni, S., D'Alessandro, F., Bianchi, F., Asdrubali, F., 2016. Insulation materials for the building sector: a review and comparative analysis. Renew. Sustain. Energy Rev. 62, 988–1011.

Sekine, K., Ooka, R., Yokoi, M., Shiba, Y., Hwang, S., 2007. Development of a Ground-Source Heat Pump System With Ground Heat Exchanger Utilizing the Cast-In-Place Concrete Pile Foundations of Buildings. , ASHRAE Trans DA-07-061, p. 558–566.

Seo, D., Krarti, M., 2010. Energy Analysis of Cecil Residential Complex (Final Report). Building Systems Program, University of Colorado, Boulder, CO.

Seo, D., Krarti, M., 2011. Development of an optimal daylighting controller. Build. Environ. 46 (5), 1011–1019.

Shackelford, J., Robinson, A., Rubinstein, F., 2015. Retrofit Demonstration of LED Fixtures with Integrated Sensors and Controls. Lawrence Berkeley National Laboratory Report prepared for US General Services Administration, Berkeley, CA.

Shekar, V., Krarti, M., 2017. Control strategies for dynamic insulation materials applied to commercial buildings. Energy Build. 154. doi: 10.1016/j.enbuild.2017.08.084.

Solupe, S., Krarti, M., 2014. Assessment of air infiltration heat recovery and its impact on energy consumption for residential buildings. Energy Convers. Manage. 78, 316–323.

Stewart, M.A., McCartney, J.S., 2013. Centrifuge modeling of energy foundations under cyclic heating and cooling. ASCE J. Geotech. Geoenviron. Eng. 1–11. doi: 10.1061/(ASCE)GT.1943-5606.0001061.

Taylor, B.J., Cawthorne, D.A., Imbabi, M.S., 1996. Analytical investigation of the steady-state behavior of dynamic and diffusive building envelopes. Build. Environ. 31 (6), 519–525.

WBCSD, 2008. Energy Efficiency in Buildings: Business Realities and Opportunities. Summary Report by World Business Council for Sustainable Development, http://wbcsdpublications.org, (accessed 25.08.17).

Wood, J.C., Liu, H., Riffat, S.B., 2010. An investigation of the heat pump performance and ground temperature of a piled foundation heat exchanger system for a residential building. Energy 25, 4932–4940.

Yavuzturk, C., Spitler, J.D., Rees, S.J., 1999. A transient two-dimensional finite volume model for the simulation of vertical U-tube ground heat exchangers. ASHRAE Trans. 105 (2), 475–485.

Yao, R., Li, B., Liu, J., 2009. A theoretical adaptive model of thermal comfort – adaptive Predicted Mean Vote (aPMV). Build. Environ. 44, 2089–2096.

Chapter 3

Control Strategies for Building Energy Systems

Chapter outline

3.1 INTRODUCTION

Currently, almost all new buildings have some control systems to manage the operation of various building equipment including HVAC systems. More elaborate control systems can simultaneously operate several mechanical and electrical equipment dispersed throughout the facility. In particular, these energy management control systems (EMCS) can be used to reduce and limit the energy demand of the entire facility. In the last decade, most of the advances in the HVAC equipment are due to the modern electronic controls, which are now cheap, flexible, and reliable.

The development of EMCS is mostly attributed to the introduction of computerized building automation systems. In fact, energy management represents one of several tasks performed by an integrated building automation system (IBAS). Among other tasks of the IBAS include fire safety, vertical

Optimal Design and Retrofit of Energy Efficient Buildings, Communities, and Urban Centers.
http://dx.doi.org/10.1016/B978-0-12-849869-9.00003-X

transportation control, and security regulation. Advanced IBAS include logic for interaction between lighting, HVAC, and security systems. For instance, if an automated occupancy sensor detect presence of people in specific spaces during late hours (during night or weekends), the information can be used to adjust indoor temperature (for comfort) and to reinstate elevator service (to ensure that people can leave the building). Moreover, EMCS can provide facility operators with recommendations on maintenance needs (such as lighting fixtures replacement) and alarms for equipment failures (such as motors when they burn out).

The use of energy-efficient equipment does not always guarantee energy savings. Indeed, a good management of the operation of this equipment is a significant factor to reduce whole building energy use. Generally, building energy loads are continuously changing with time due to fluctuations in weather and changes in equipment use and occupancy. Thus, effective energy management requires the knowledge of the facility loads. Two approaches are typically applied:

- *Load Tracking*: The operation of equipment is modulated to respond to the actual needs in the facility. As an example, the compressor in a centrifugal chiller may change speed to match the cooling demand. The actual needs of a facility can be determined by continuous monitoring. As an example, the load on the chiller can be estimated if the chiller water flow and chilled water supply and return are monitored.
- *Load Anticipation*: In some applications, the needs of a facility have to be predicted to be able to modulate adequately the operation of equipment. For instance, in cooling plants with a thermal energy storage system, it is beneficial to anticipate the future cooling loads to be able to decide when and how much to charge and discharge the storage tank. Load prediction can be achieved by analyzing the historical pattern variations of the loads.

Using monitored data and other parameters characterizing the building, energy control systems enable operators and managers to operate efficiently HVAC and lighting systems to maintain comfort levels. In the following sections, building energy control systems and some of their applications are presented and discussed.

3.2 BASIC CONTROL PRINCIPLES

3.2.1 Overview of Control System Components

Control systems are used to match equipment operation to load requirements by changing system variables. A typical control system includes four elements:

1. Controlled variable is the characteristic of the system to be controlled (for instance, the indoor temperature is often the controlled variable in HVAC systems).

TABLE 3.1 Types of Sensors Used for HVAC Systems Controls

Variable to sense	Types of sensors
Temperature	• Bimetallic strip • Sealed bellows • Bulb and capillary sensors
Pressure	• Variable resistance • Capacitance
Liquid flow	• Orifice • Venturi tube • Flow nozzles • Vortex shedding sensors • Positive displacement flow sensors • Turbine based flow sensors • Magnetic flow sensors • Ultrasonic flow sensors
Air flow	• Hot wire anemometers • Pitot static tube
Relative humidity	• Resistance relative humidity sensor • Capacitance relative humidity sensor • Temperature condensation • Condensation and wetting • Quartz crystal relative humidity sensor
	• Hydrostatic sensors • Ultrasonic sensors • Capacitance sensors

2. Sensors that measure the controlled variable (for instance thermocouple can be used to measure indoor temperatures). Table 3.1 lists some of the sensors commonly used to control HVAC systems.
3. Controllers that determine the needed actions to achieve the proper setting for the controlled variable (for instance, the damper position of the variable air volume (VAV) box terminal can be modulated to increase the air supply to increase the indoor temperature of the zone if it falls below a set point).
4. Actuators are the controlled devices that need to be activated to complete the actions set by the controllers (to vary the air supplied by a VAV box, the position of the damper is changed by an actuator through direct linkages to the damper blades).

Generally, two categories of control systems can be distinguished including closed loop and open loop systems. In a closed loop system (also known as a feedback control system), the sensors are directly affected by (and thus sense) the actions of the actuators. A typical control of a heating coil is an example of a closed loop system. However, in an open loop system (also called a feedforward

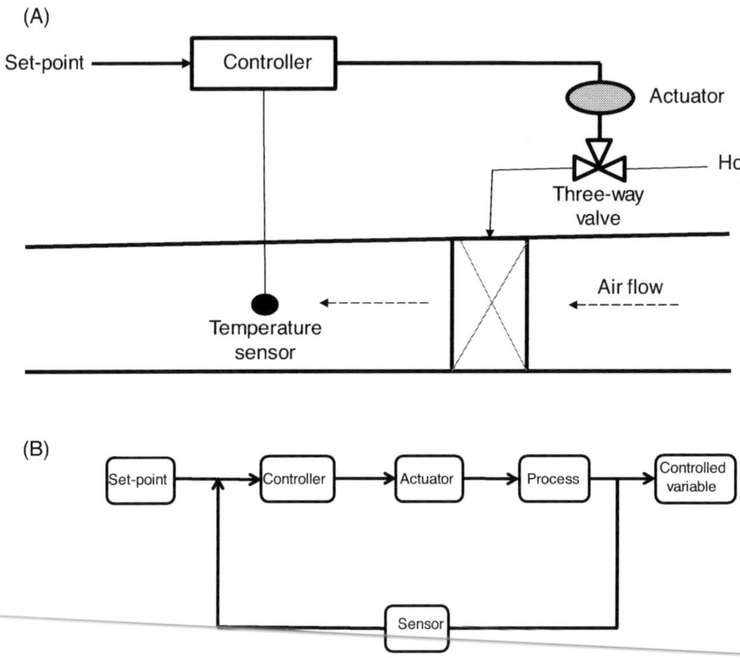

FIGURE 3.1 Typical representations for a heating coil control system. (A) Basic closed-loop control for a heating coil. (B) Equivalent control diagram for the heating coil.

control system), the sensors do not directly sense the actions of the controllers. The use of a timer to set the temperature of the heating coils would be an example of an open loop system since the time may not have a direct connection with the thermal load on the heating coils.

Fig. 3.1 shows the various components and terms discussed earlier as well as an equivalent control diagram for a closed-loop control system for a heating coil.

3.2.2 Transfer Function Analysis

In HVAC controls, the analysis of control systems can be carried out using transfer functions connecting input to output signals. For instance, Fig. 3.2 shows the relationship between the input function $u(t)$ and the output $y(t)$ after a process function, $g(t)$, typically referred to as a response function.

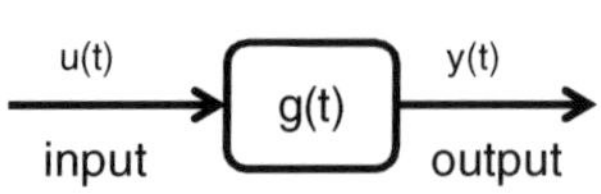

FIGURE 3.2 Time domain input–output relation through a response function *g*.

For linear processes, the output function, $y(t)$ can be estimated from the input function, $u(t)$, using the following expression:

$$y(t)=\int_{-\infty}^{\infty}u(t-\tau)g(\tau)d\tau \tag{3.1}$$

As a notation, the relation of Eq. (3.1) can be written simply as

$$y(t)=u(t)*g(t) \tag{3.2}$$

For instance, in case the process can be represented by a constant response function, that is, $g(t) = K$. Then, from Eq. (3.1), the output function is proportional to the input signal as shown by Eq. (3.3):

$$y(t)=K\cdot u(t) \tag{3.3}$$

In general, the relationship between the input and output involve linear differential functions such as the relationship provided by Eq. (3.4):

$$a\frac{d^2y}{dt^2}+b\frac{dy}{dt}+cy=\alpha\frac{du}{dt}+\beta u \tag{3.4}$$

To find the relationship between the input and output functions, and thus define the function $g(t)$, the Laplace transform analysis is used. The Laplace transform function for any function $f(t)$ is defined as $F(s)$ as follows:

$$F(s)=\int_{0}^{\infty}f(t)e^{-st}\,dt \tag{3.5}$$

Through the Laplace transform, any linear differential equation can be simplified. For instance, Eq. (3.4) is expressed in the Laplace domain as follows:

$$as^2Y(s)+bs\,Y(s)+cY(s)=\alpha sU(s)+\beta U(s) \tag{3.6}$$

Thus,

$$Y(s)=\frac{\alpha s+\beta}{as^2+bs+c}U(s) \tag{3.7}$$

Therefore, the response function $G(s)$ in the Laplace transform can be written as

$$G(s)=\frac{Y(s)}{U(s)}=\frac{\alpha s+\beta}{as^2+bs+c} \tag{3.8}$$

Fig. 3.3 illustrates the input–output relationship through the function G in the Laplace transform domain. The function G is called a transfer function to find the output $Y(s)$ from the input $U(s)$ by simple multiplication operator in the Laplace transform domain:

$$Y(s)=G(s)\,U(s) \tag{3.9}$$

FIGURE 3.3 **Transform domain input–output relation through a response function *s*.**

The output function $Y(s)$ can then be converted into the time domain, $y(t)$, using the inverse of the Laplace transform as shown by Eq. (3.10):

$$f(t) = \int_0^{\infty} G(s) e^{st}\, ds \tag{3.10}$$

Table 3.2 summarizes for typical response functions their associated Laplace transforms as well as transfer functions.

When two signals are used as input variables, u_1 and u_2, first the sum of the output signals, X, from two transfer functions, G_1 and G_2, are added before the resultant is applied to a third transfer function, G_3, to obtain the output signal, y, as shown in Fig. 3.4 using the Laplace transform analysis.

Using the transfer functions G_1, G_2, and G_3, the output $y(s)$ can be determined from the $U_1(s)$ and $U_2(s)$ as indicated by Eqs. (3.11) and (3.12):

$$Y(s) = G_3(s)\ X(s) \tag{3.11}$$

and

$$X(s) = G_1(s)\ U_1(s) + G_2(s) U_2(s) \tag{3.12}$$

Thus, the output can be directly obtained from the input signals U_1 and U_2:

$$Y(s) = G_1(s)\ G_3(s)\ U_1(s) + G_2(s)\ G_3(s)\ U_2(s) \tag{3.13}$$

The output from the two input variables is simply the sum of the output signals from each input variable obtained by the transfer functions $G_1\ G_3$ and $G_2\ G_3$.

TABLE 3.2 Laplace Transforms for Selected Response Functions

Time domain response function	Laplace transform function, $G(s)$	Input–output relationship
$\delta(t)$ (impulse function at $t = 0$)	1	$y(t) = u(t)$
e^{-at}	$\frac{1}{s+a}$	$\frac{dy}{dt} + a y = u$
$\sin(\omega t)$	$\frac{\omega}{\omega^2 + s^2}$	$\frac{d^2y}{dt^2} + \omega^2 y = \omega u$
$\delta(t - T)$ (impulse function at $t = T$)	e^{sT}	$y(t) = u(t-T)$
$H(t)$ (Integrator)	$\frac{1}{s}$	$y(t) = \int_0^t u(\tau) d\tau$

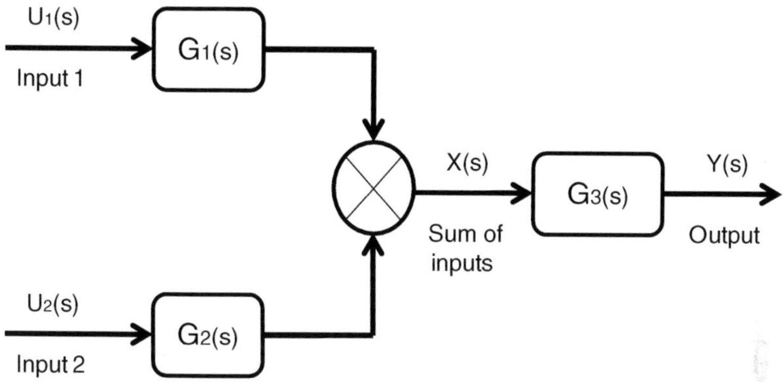

FIGURE 3.4 Interconnection between two input signals and the output functions.

Another common interconnection is when there is a feedback loop from the output to the input to estimate the error function as shown in Fig. 3.5 and expressed in terms of transfer functions

$$E(s) = U(s) - G_2(s)\,Y(s) \tag{3.14}$$

and

$$Y(s) = G_1(s)\,E(s) \tag{3.15}$$

Thus,

$$Y(s) = G_1(s)U(s) - G_1(s)G_2(s)Y(s) \tag{3.16}$$

Therefore,

$$Y(s) = \frac{G_1(s)}{1 + G_1(s)G_2(s)}U(s) \tag{3.17}$$

Then, the error transfer function can be estimated by substituting Eq. (3.17) into Eq. (3.14):

$$E(s) = \frac{1}{1 + G_1(s)G_2(s)}U(s) \tag{3.18}$$

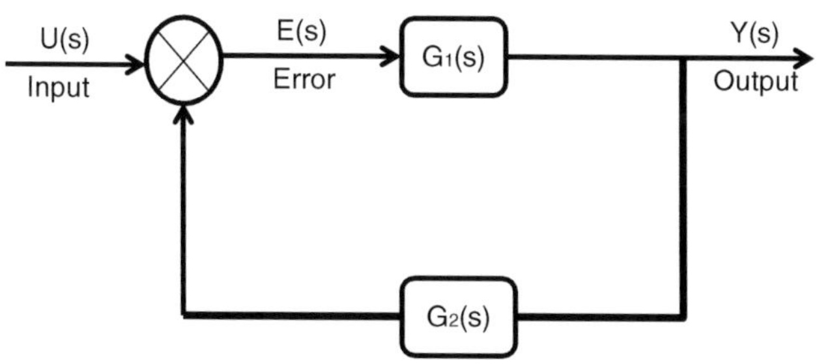

FIGURE 3.5 Error signal obtained input and the output signals.

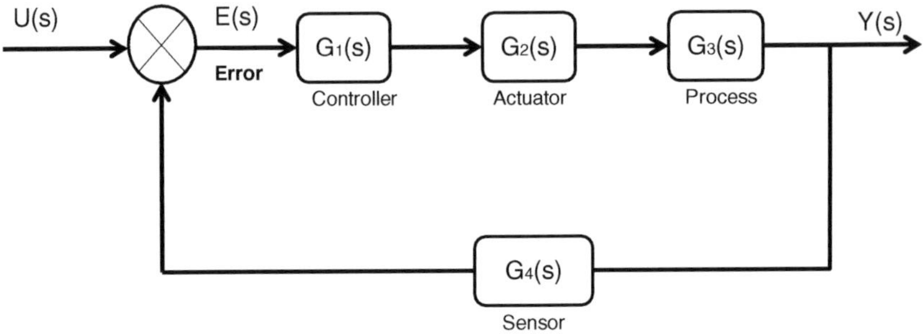

FIGURE 3.6 Transfer function diagram for input and the output signals for the control process of Fig. 3.1.

As an application of the two interconnected transfer functions shown by Figs. 3.4 and 3.5, the transfer function for any linear control scheme can be developed. For instance, the transfer function diagram for the control process of Fig. 3.1 is shown in Fig. 3.6.

The input–output transfer function for the control process can be derived to be

$$Y(s) = \frac{G_1(s)G_2(s)G_3(s)}{1 + G_1(s)G_2(s)G_3(s)G_4(s)} U(s) \tag{3.19}$$

Moreover, the error transfer function can be shown to have the following expression:

$$E(s) = \frac{1}{1 + G_1(s)G_2(s)G_3(s)G_4(s)} U(s) \tag{3.20}$$

In the following section, common control modes for HVAC systems are described in detail.

3.2.3 Control Modes

Each control system can use different control modes to achieve the required objectives of the control actions. Four control modes are commonly used in operating HVAC systems.

3.2.3.1 Two-Position

This control mode allows only two values (on–off or open–closed) for the controlled variable and is best suited for slow-reacting systems. Fig. 3.7A shows the effect of two-position control on the time variation of the controlled variable (such as the air temperature due to on–off valve position in a heating coil). To avoid rapid cycling, a control differential can be used. Due to the

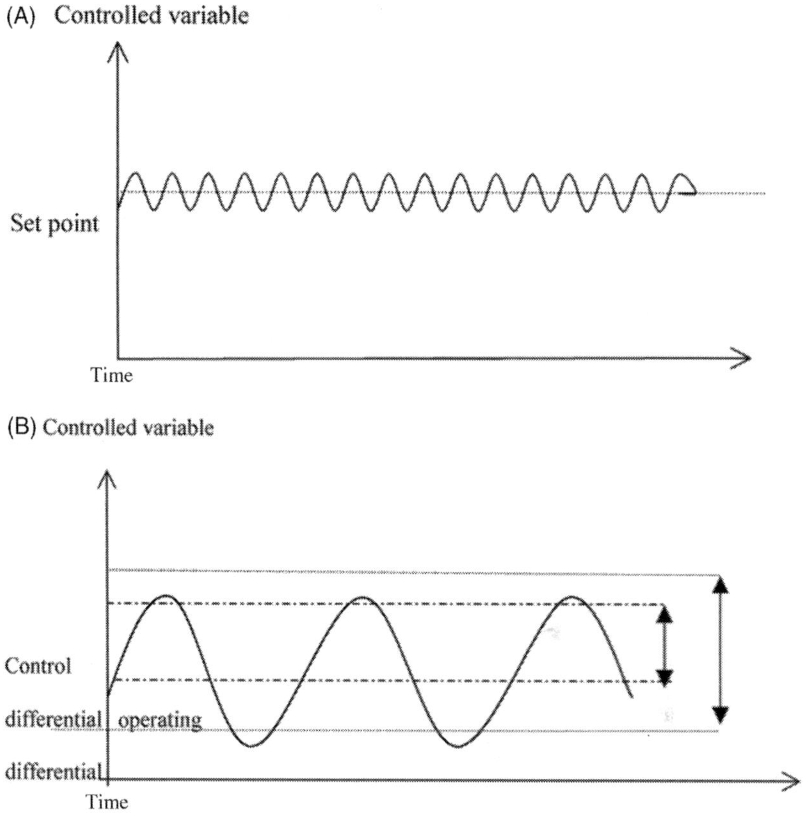

FIGURE 3.7 Effect of two-position control on the time variation of a controlled variable. (A) Two-position action when no control differential is used (rapid cycles). (B) Two-position action with a control differential.

inherent time lag in the sensor response and to the thermal mass of the HVAC system, the controlled variable fluctuates with an operating range (called operating differential) with higher amplitude than the control differential. Thus, the operating differential is always higher than the control differential as illustrated in Fig. 3.7B.

Examples of two-position controls are domestic hot-water heating, residential space-temperature controls, and HVAC system electric preheat elements.

3.2.3.2 Proportional

This mode has a linear relationship between the incoming sensor signal and the controller's output. The relationship is established within an operating range for the sensor signal. The set point of a proportional controller is the sensor input, which results in the controller output to be at the midpoint of its range.

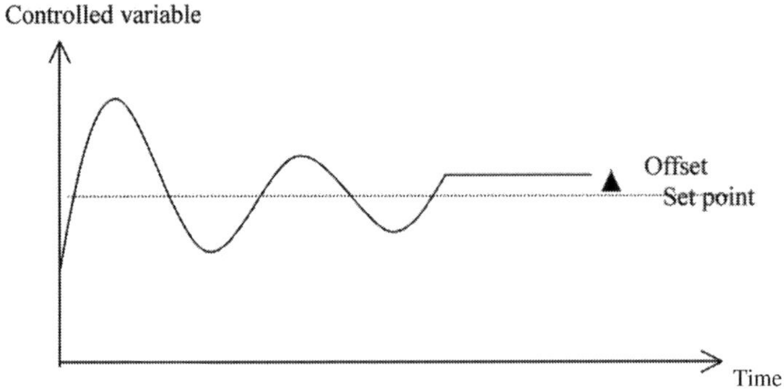

FIGURE 3.8 **Proportional control effect on the time variation of a controlled variable.**

Mathematically, the controller output, u, is given by the following equation for a proportional control:

$$u = K_p e + u_0 \tag{3.21}$$

The offset or error, e, is the difference between the set point and the value of the controlled variable. The proportionality constant, K_p, is called the proportional gain constant. The controller bias, u_0, is the value of the controller output when no error exists. As depicted by Eq. (3.21), the proportional control is not capable of reducing the error since an error is required to produce any controller action. Therefore, the controlled variable fluctuates within a throttling range as depicted in Fig. 3.8.

It should be noted that when the gain constant is very large, an unstable system can be obtained. Generally, proportional controllers are used with slow stable systems that have small offset. In Example 3.1, the proportional gain constant for a controller is estimated.

Example 3.1

A hot water heating coil has a set-point of 35°C with a throttling range of 10°C. The heat output of the coil varies from 0 to 50 kW. Assuming that a proportional controller is used to maintain the air temperature set-point, determine the proportional gain for the controller and the relationship between the output air temperature and the heat rate provided by the coil. Assume steady-state operation.

Solution

Using Eq. (3.21), the relationship between the heat rate, Q, and the error in the air temperature at the coil outlet can be put in the following form:

$$Q = K_p(T_{setpoint} - T_{air}) + Q_o$$

Since

(i) When the heat rate $Q = Q_{min} = 0$ kW, the coil outlet air temperature is $T_{air} = T_{min} = 35°C - 5°C = 30°C$, and

(ii) When the heat rate $Q = Q_{max} = 50$ kW, the coil outlet air temperature is $T_{air} = T_{max} = 35°C + 5°C = 40°C$

The proportional gain K_p can be determined as follows:

$$Q_{max} - Q_{min} = K_p(T_{min} - T_{max})$$

or

$$K_p = [Q_{max} - Q_{min}] / [T_{min} - T_{max}] = -50\,kW / 10°C = -5\,kW / °C$$

Similarly, the constant Q_o can be determined from

$$Q_{min} = K_p(T_{setpoint} - T_{min}) + Q_o$$

or

$$Q_o = Q_{min} - K_p(T_{setpoint} - T_{min}) = 0 + 5\,kW / °C * (35°C - 30°C) = +25\,kW$$

Therefore, the relationship between the heat rate output and the air temperature for the heating coil is

$$Q = -5\,(T_{setpoint} - T_{air}) + 25$$

Thus, as long as the heat rate is different from Q_o = 25 kW, the quantity ($T_{setpoint} - T_{air}$), which is the error in the proportional control equation, cannot be equal to zero.

Generally, proportional controllers are used with slow stable systems that have small offset.

3.2.3.3 *Integral*

This control mode is typically incorporated with a proportional control mode to provide an automatic means to reset the set point to eliminate the offset. The combination of the proportional and integral actions is called "proportional-plus-integral" or simply PI control. Mathematically, the PI control can be expressed as follows:

$$u = K_i \int e \cdot dt + K_p e + u_0 \qquad (3.22)$$

where K_i is the integral gain constant (also known as the reset rate) and has the effect of adding a correction to the controller output whenever an error exists. For HVAC systems, the typical K_p/K_i ratio is less than 60 min.

The PI control can be applied to fast-acting systems that require large proportional bands for stability. Typical applications include mixed-air controls, heating or cooling coil controls, and chiller-discharge controls.

3.2.3.4 *Derivative*

This control action is used to speed up the response of the system in case of sudden changes. The derivative control mode is included in a combination of proportional-plus-integral-plus derivative (PID) control modes for fast-acting systems that tend to be unstable such as duct static-pressure controls. The mathematical model for the PID control is given by Eq. (3.23):

$$u = K_d \frac{de}{dt} + K_i \int e \cdot dt + K_p e + u_0 \tag{3.23}$$

where K_d is the derivative gain constant. The derivative term generates a corrective action proportional to the time rate of change of the error. The ratio K_d/K_p is typically less than 15 min for most HVAC applications. If the system has a uniform offset, the derivative term has little effect. The use of PID controls is typically less common than the PI controls for HVAC systems since no rapid control responses are needed.

To illustrate the action of P, PI, and PID control modes, Fig. 3.9 compares the response of the system to an input step change. As expected the proportional control results in an offset and the controlled variable does not reach the set point. The correction term due to the PI control slowly forces the controlled variable to reach the set point value. Finally, the derivative term of the PID control provides a faster action to allow the controlled variable to attain the set point.

In addition to these conventional control modes, other intelligent controllers have been investigated in various engineering fields including HVAC equipment controls as will be discussed in the following section.

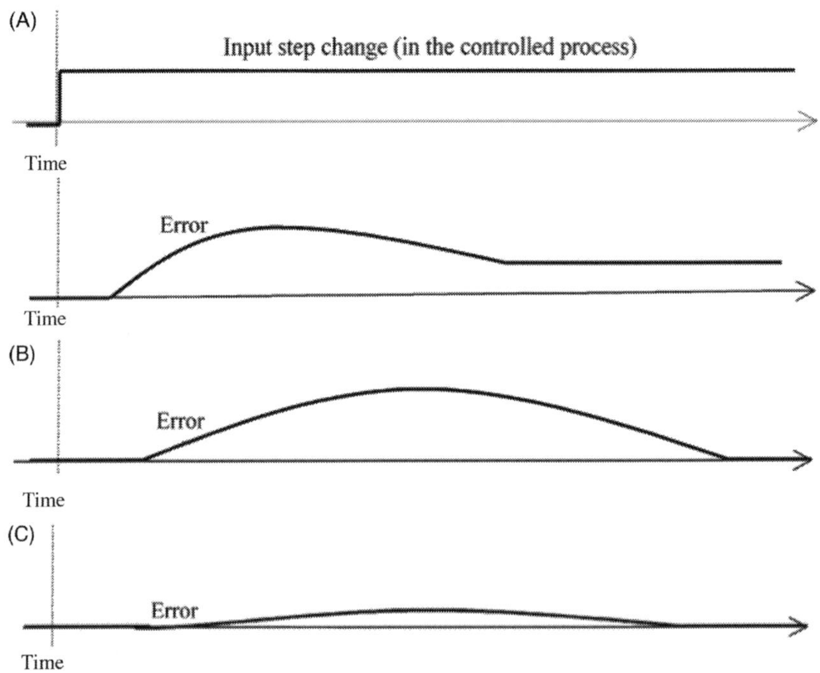

FIGURE 3.9 Comparison of the reaction of three control modes to an input step change. (A) Proportional (P) control action. (B) Proportional-plus-integral (PI) control action time. (C) Proportional-plus-integral-plus-derivative (PID) control action time.

3.2.4 Intelligent Control Systems

In the future, it is expected that intelligent control systems will be commonly available to operate HVAC systems. A number of research studies are applying intelligent system methodologies so that control systems have human-like capabilities such as pattern recognition, adaptation, learning, reasoning, and associative memory to operate complex systems that conventional control techniques cannot handle. Various intelligent control environments have been considered for HVAC systems including the following:

- *Expert Controls*: Rule-based expert systems are based on human-like reasoning and can be powerful for solving practical engineering problems. Indeed, expert control systems have been applied to self-tuning, structured knowledge–based adaptive control, fault diagnosis, and scheduling and planning. However, rule-based expert systems are not yet suitable to process numerical knowledge and thus to provide precise solutions. The development of an expert system environment that can deal with qualitative and quantitative knowledge, and to automatically explore knowledge, is still in the realm of future expert control research.
- *Fuzzy Controls*: Since the development of fuzzy logic in the 1960s, fuzzy control has been one of the most attractive strategies in controlling complex systems with imprecise and/or uncertain knowledge of system information and behavior. Fuzzy logic can be applied to system modeling, estimation, optimal control, and adaptive control with the requirements of only system fuzzy knowledge and input/output data. Applications of fuzzy logic in expert systems are becoming attractive to establish intelligent expert control systems with fuzzy knowledge representation and fuzzy reasoning.
- *Artificial Neural Networks*: Similar to biological neural networks, artificial neural networks consist of a large number of simple nonlinear processing elements, typically called nodes or neurons, which are interconnected with adjustable weights. Well-trained neural network models can provide qualitative and quantitative knowledge and has powerful functions in learning and self-organization. These features make neural networks more suitable in dealing with numerical data than expert systems.

3.2.5 Types of Control Systems

To achieve the actions of the control systems just discussed, several types of energy sources are used. In particular, the following types of control systems are used in HVAC applications:

- Pneumatic devices are used with low-pressure compressed air at 0–20 psig. Pneumatic systems are common in older installations.
- Electric devices using 24–120 V or even higher voltage sources.
- Electronic devices with low direct current voltages varying from 0 to 10 V. These devices are being installed in new commercial buildings especially with direct digital control (DDC) systems.

- Hydraulic systems when large forces are required with pressure larger than 100 psi.
- Self-generated energy derived from the change of state of the controlled variable or from the energy available in the process plant.

For HVAC retrofit applications, it is recommended that DDC systems be considered. Indeed, currently developed DDC systems use the latest digital technology including features such as intelligent controllers, high-speed communication networks, and sophisticated control algorithms. All these features allow more energy-efficient control strategies to be implemented. Moreover, digital devices present additional advantages compared to pneumatic or electric devices as outlined next:

- Little or no maintenance is required for digital devices.
- Calibration of digital devices can be performed through remote instructions issued over a network. Some digital devices have the advantage to be continuously self-calibrating.
- Better accuracy is obtained from digital controls compared to pneumatic or electric devices.

While DDC devices have been used since the 1980s, it is only recently that manufacturers have developed systems that house interposing devices (such as relays, transducers, and hard-wired logic) in the same package with the electronic devices. These new DDC systems are suitable for retrofitting applications since it is now economical to convert existing pneumatic and electric analog controls to DCC systems. The best candidates for DDC retrofitting are air handling units, heat exchangers, distribution pumps, and cooling towers. In general, the larger the equipment size, the faster the payback period. Replacing the controls for small HVAC equipment such as package unitary systems (including unit ventilators, heat pumps, and fan coils) may not be cost-effective.

Retrofit of pneumatic controllers can be made using electronic-to-pneumatic (E/P) transducers to convert signals so electronic and pneumatic control components can be combined in the same control loop. For instance, the controller and the sensor in an existing pneumatic HVAC control system can be converted into electronic devices while the actuator can remain pneumatic. The pressure output of the E/P transducer should match the electric signal. The use of E/P transducers allows retrofit of control systems with minor interruption in the operation of the controlled system.

3.3 SUPERVISORY CONTROLLERS

Supervisory controllers can interact with all the subsystems inside a building to effectively meet building needs with appropriate operator input. Typically, supervisory controllers are called building automation system, building management systems, energy management system, or more commonly energy monitoring and control systems (EMCS).

3.3.1 Basic Components of an EMCS

To control and operate equipment for heating, ventilating, and air conditioning, or for lighting and process equipment, an energy monitoring (or management) and control system (EMCS) can be used. A typical EMCS is configured into a network that includes sensors and actuators at the bottom level, microprocessor controllers in the middle, and a computer at the top with a modem to allow remote monitoring and control of the building energy systems. For a typical commercial building, an EMCS system can be cost-effective in reducing the energy use for HVAC and lighting systems.

Energy management is often just one element of an IBAS, which regulates security, fire safety, lighting, HVAC systems, and elevators. Advanced IBAS systems include logic for interaction between various systems such as HVAC, lighting, and security systems. Indeed, the automated occupancy count information obtained for different spaces in a facility can be used to adjust indoor temperature settings, reduce or turn-off lights, and ensure proper elevator operation.

The size of an EMCS system is typically classified based on the total number of points connected to a system. Five size categories are generally considered for EMCS systems:

1. Large EMCS systems with more than 2000 points
2. Medium EMCS systems with 500–2000 connected points
3. Small EMCS systems with 200–1000 points
4. Small centralized EMCS systems with 50–500 points
5. Micro EMCS systems with less than 100 points

A typical EMCS system is depicted in Fig. 3.10 and includes a central control unit (CCU), a processing memory, storage devices, input–output devices, a central communications controller (CCC), data transmission medium, a field interface devices (FIDs), multiplexers, instruments, and controls. A brief description of the major components of an EMCS system is presented in the following sections.

- The CCU consists generally of a computer with memory for the operating system software, command software, and implementation of application algorithms. In particular, computations and logical decision functions for central supervisory control and monitoring are performed by the CCU. However, data and programs are stored in and retrieved from the memory or storage devices such as magnetic tape and disk systems. Typically, the CCU has input/output ports for specific equipment such as printers and terminals.
- The CCC is typically a computer with enough memory to execute specific programs required to reformat, transfer, and perform error checks on data coming from the CCU or the FIDs. The CCC may have back-up capabilities in case of CCU failure.
- FIDs consist generally of computer devices with memory, input/output, communications, and power supply. The FIDs record the monitored and controlled

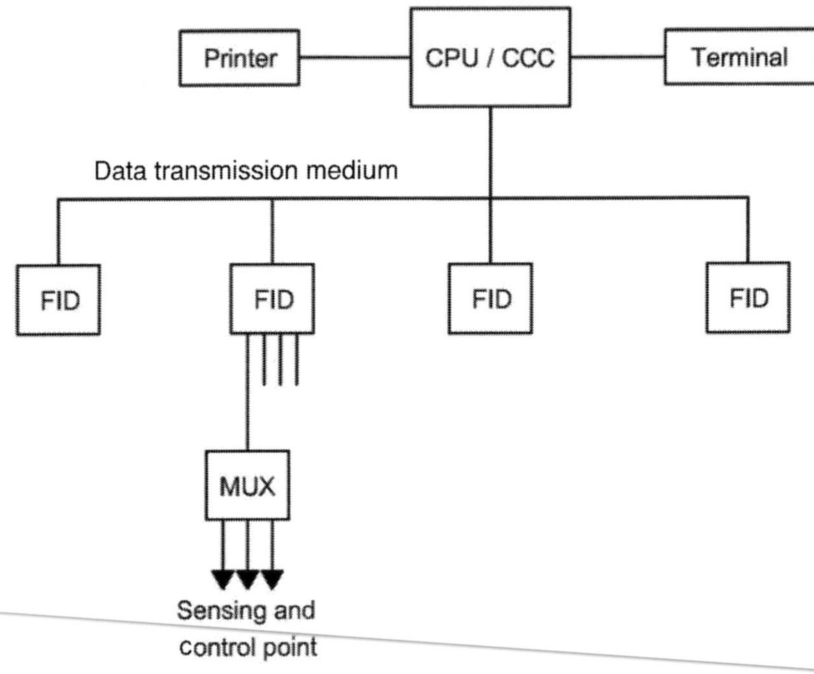

FIGURE 3.10 Typical EMCS components and configuration.

data, perform calculations and logical operations, accept and process system commands. They should be capable of operating in case of CCU failure.

- A multiplexer is a device that communicates between the data environment and its associated FID. The multiplexer is functionally part of the FID and can thus be in the same enclosure.
- The data transmission medium is the communications link that allows transferring data between the CCU and its associated FIDs such as telephone lines, optical fibers, or coaxial cables.

3.3.2 Typical Functions of ECMS

Several control operations and functions can be performed by an EMCS including the following:

1. Equipment operation such start–stop, on–off, and open–close controls.
2. Alarm functions such as abnormal equipment status, high or low parameter values (temperature, pressure, refrigerant level, etc).
3. Computer programming and table look-up functions for energy management and equipment use optimization including enthalpy economizer controls, chiller plant optimization, load shedding based on demand monitoring, lighting control by zones.

4. Monitoring of operation conditions such as temperatures, pressures, energy end-uses.

Other functions and applications of EMCS are being developed and implemented in some facilities. Some of these applications are discussed in the following sections:

EMCS can be used with expert systems to make intelligent operating decisions based on stored and accumulated knowledge. For instance, integrated EMCS/expert systems can be used to diagnose faults and inefficiencies in HVAC and lighting systems. Typically, an EMCS/expert system would be programmed with basic information about the building equipment and their design operation characteristics. Over time, actual operating data would be recorded and used to train the expert system and thus improve its diagnostic accuracy. If a fault were detected, the EMCS/expert system would notify the building operator of the probable problems and suggest remedial actions or simply initiate automatically countermeasures in case of serious problems. As an example, if the indoor temperature for a specific space is noted to be rising significantly beyond the throttling range, the EMCS/expert system would examine all the relevant sensor readings (such as airflow rates, fan energy use, and cooling water flow and temperature). Then, the EMCS/expert system would indicate using its knowledge base that the probable cause for the problem is faulty fan operation or leakage in the chilled water pipe (70% faulty fan and 30% chilled water pipe leakage).

Another function of EMCS that has been implemented in several buildings especially in the last decade is the maintenance of acceptable indoor air quality (IAQ) levels. Indeed, the cost of sensors for monitoring air pollution compounds such carbon monoxide (CO), carbon dioxide (CO_2), and volatile organic compounds have decreased sufficiently enough to be incorporated in commercial and institutional buildings. For instance, CO_2 demand controlled ventilation has been implemented in several spaces including classrooms, conference rooms, theaters, and auditoriums.

3.3.3 Design Considerations of an EMCS

When an EMCS is recommended for a facility, it is important to consider some practical issues to ensure a successful design and operation. In particular, the EMCS is recommended to do the following:

- Allow redundancy into the control systems in case of system failure. In particular, it is important that the local control devices are able to manage and operate the system for a reasonable period of time when the CCU malfunctions or needs servicing.
- Provide clear information about the system operation. For instance, color graphic displays are preferred to numerical data for the use of facility operators and managers.

- Perform a thorough commissioning of all the components and functions of the EMCS under various operating conditions (peak cooling and peak heating modes as well as part-load conditions).
- Train the operating staff to use the EMCS to their best advantage. In particular, the operators should understand the EMCS capabilities and benefits.

The selection of an EMCS for a given facility depends typically on the required functions and on economic considerations. The desired functions and controls from an EMCS are based on several aspects including the building type, HVAC system zoning, occupancy profiles, accuracy requirements, and the objectives of the owners. For instance, owner-developers may be more interested in having an EMCS with the lowest first cost rather than a system that provides a high-quality environment, which may be one of the main objectives of owner-occupants. Moreover, high-rise buildings tend to have centralized systems with built-up fans and associated controls. On the other hand, low-rise buildings typically use package rooftop fan systems with local controls.

The cost benefits of the EMCS can be evaluated after identification of all the desired controls. Some of the important factors that affect the cost of an EMCS include

- system size and number of control points,
- degree of automation in the control functions, and
- accuracy requirements for sensors and controls.

To ensure that the control systems conserve energy and save operating cost, it is important to follow basic principles in their design including the following:

1. The energy-consuming equipment should be operated only when needed. For instance, heating temperature set-points should be set back during unoccupied periods. The heating equipment should be operated only to maintain these set-back temperatures (typically between 50°C and 55°C to prevent freezing damages for various components of the HVAC system).
2. The simultaneous heating and cooling should be avoided. Proper zoning and HVAC system selection can minimize (if not eliminate) the need for providing heating and cooling at the same time.
3. The outdoor air intake should be controlled. In the United States, only minimum requirement ASHRAE Standard 62-1999 for ventilation need to be supplied to the building when no economizer cycles are used.
4. The heating and cooling should be provided efficiently. In particular, only actual heating and cooling load requirements should be met. In addition, free cooling/heating or low-cost energy sources should be considered first to maintain comfort within the building.

Finally, it should be noted that energy control systems should be designed to be simple and easy to operate and maintain.

3.3.4 Communication Protocols

A communication protocol consists of a set of rules that have to be applied to exchange data between two parts of an EMCS. Most manufacturers of building automation systems have their proprietary communication protocols. Therefore, building owners are forced to purchase only equipment from the original manufacturer if he or she wants to expand the existing automation system. Due to the lack of interoperability between communication protocols from various manufacturers, it is almost impossible to take advantage of a facility-wide approach to control optimization and energy savings. A single manufacturer does not provide all the best possible control strategies for electrical demand limiting, heating and cooling optimization, and similar energy savings options. Moreover, several control subsystems in a building including HVAC controls, lighting/daylighting interface, fire alarm and life safety, security, and communication systems are generally manufactured by different companies. Integrating all these subsystems is a difficult task without a common communication protocol.

As a solution to the limitations and the difficulties inherent to proprietary protocols, a number of open communication protocols have been developed in recent years. These open protocols have some interoperability capabilities and thus allow building owners and managers to keep the door open to competition on any future expansion projects. One of the most widespread and widely accepted open protocols is the BACnet, a data communication protocol for Building Automation and Control networks (Bushby and Newman, 1991). BACnet is a nonproprietary open protocol standard that supports various communication networks ranging from high-speed Ethernet local area networks to low cost networks. Since ASHRAE developed BACnet, no one specific company or consortium has an advantage or an influence on the future development of the standard. Any changes for BACnet are published for public review and comment after discussions on an ASHRAE open committee that includes representatives from industry, academia, and government.

To design a BACnet device, a manufacturer needs to identify the BACnet objects and services required to achieve the intended functionality for the device. A BACnet object is a standard data structure defined with a set of properties and data types. In its current version, the BACnet standard defines 20 objects such as loops, tables, schedules, commands, and programs. BACnet services are the programmed actions that use the data objects to achieve the function of the device. Services, defined by the current version of the BACnet standard, include alarm and event services (to notify of any alarm and event), file access services (to read and write files), object access services (to read or write the properties of objects), remote device management services (to troubleshoot and maintain devices), and virtual terminal services (to allow interaction between a terminal and the device). Moreover, the BACnet device has to conform to a set of specifications using a series of conformance classifications. Each conformance classification adds functional services to the device. Thus, each BACnet device design would have a protocol implementation conformance statement prepared by the manufacturer to identify the BACnet options available in the device.

Over the last two decades, several manufacturers have already developed BACnet control devices. The integration and the application of these devices in real installations have been thoroughly reported and documented.

3.4 MODEL-BASED CONTROLS

Model-based controls or MPCs are flexible control techniques since they can handle time-varying external as well as internal constraints and disturbances. Indeed, a typical MPC uses a model for the system to be controlled to predict certain future states and develop control strategies to minimize a cost function over the prediction horizon in the presence of any disturbances and constraints. Only the first set of the computed control strategies is applied. Then, the entire process is repeated as the prediction horizon is moved.

The optimization cost function can take the form of tracking error, control effort, energy cost, demand cost, power consumption, or a combination of these objectives (Afram and Janabi-Sharifi, 2014). Constraints can be placed on the rate and range limits of the actuators and the manipulated and controlled variables (e.g., upper and lower limits of indoor temperatures, supply airflow rate limits, and damper position limits). External and internal disturbances acting on the system are considered in the search of optimal control strategies such as those associated with weather conditions and occupant activities.

MPCs are used in supervisory and local controls for various HVAC systems. Some of the applications of MPCs include

- optimize energy and cost savings,
- increase peak load shifting capability, and
- improve IAQ and thermal comfort.

Fig. 3.11 presents the flowchart of a model for a building used to control passive and active thermal energy storage (TES) systems (Hajiah and Krarti, 2012). In particular, the model includes four main modules: (1) input data, (2) building models, (3) control strategies, and (4) controlled building. Through the input data module, the user has to provide the required data (building features such construction details, internal gains, utility rates, and operating schedules as well as recorded weather data), to first estimate thermal loads and energy end uses using the building models module and then to determine the optimal control strategies using the control strategies module. The actual implementation of the control strategies and associated performance is typically evaluated and measured through the controlled building module. The simulation environment shown in Fig. 3.11 is model-based and can be applied to a wide range of building types and operating conditions.

Fig. 3.12 illustrates a predictive MPC implementation when predictors of weather and building loads are considered (Krarti and Henze, 2006).

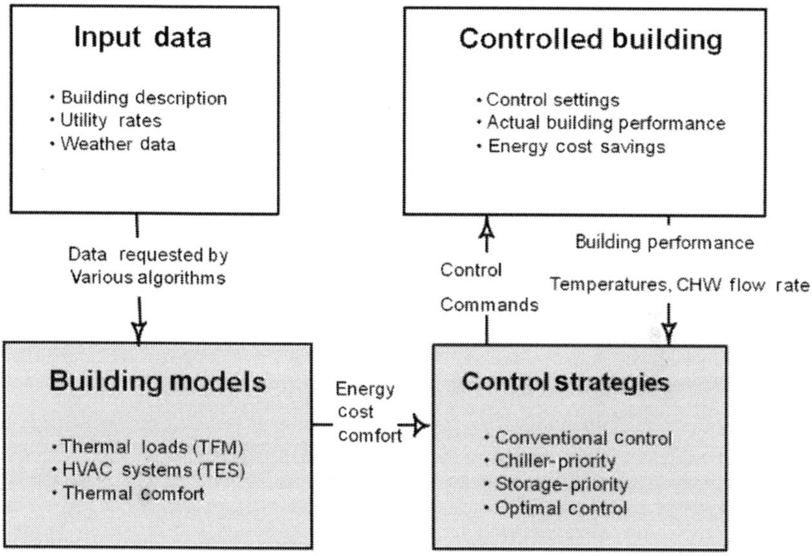

FIGURE 3.11 **Flowchart of a MPC simulation environment for MPC.**

Specifically, a sequential approach to model-based control is shown in Fig. 3.12 following modules: short-term forecasting, optimization, and postprocessing and control command implementation. Generally, a weather station provides the current weather data to the short-term weather predictor. This predictor provides an improved forecast for the next planning horizon to the optimal controller, which adjusts the control variables in the model according to Fig. 3.13 until convergence is reached. The optimal solution is passed to a postprocessor

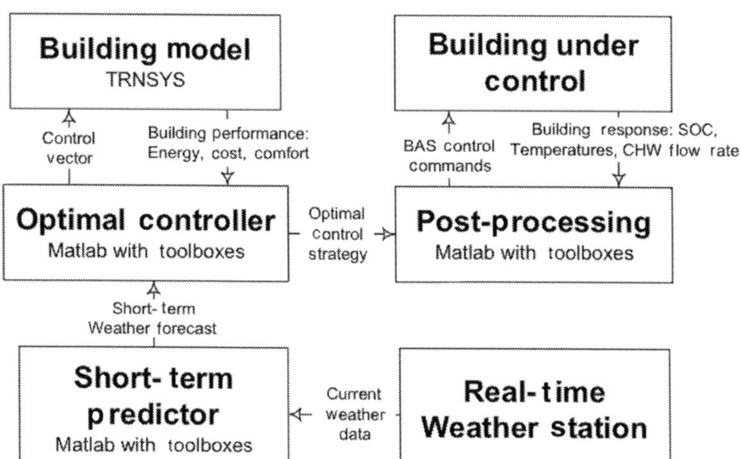

FIGURE 3.12 **A schematic for a real-time predictive optimal model-based control.**

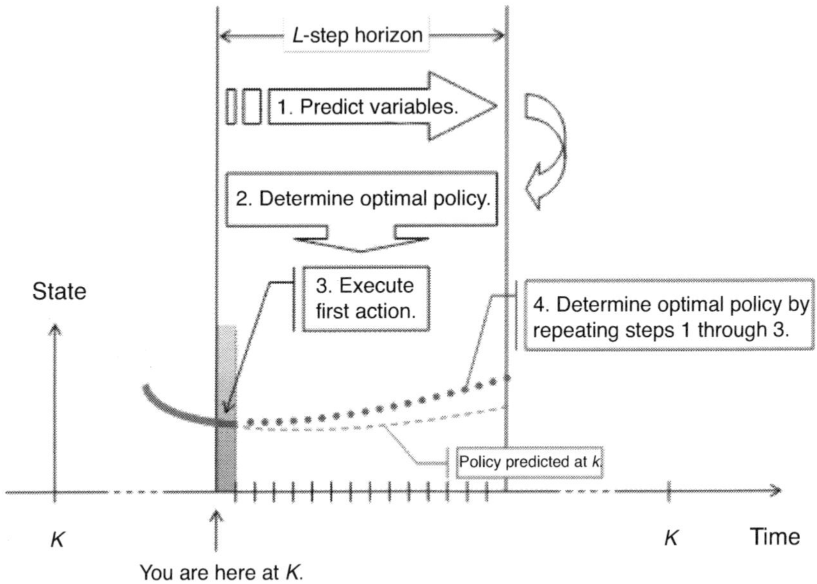

FIGURE 3.13 A basic implementation procedure for a closed-loop predictive MPC.

that interprets the optimal results and turns them into commands understood by the building automation system of the facility under control.

Several predictor models have been tested to provide the most accurate weather forecasts (Henze et al., 2004). Some of the short-term weather forecast models include bin predictor, random walk, and harmonic predictor; linear parametric models such as the auto-regression integrated moving average; and nonlinear models such as artificial neural networks. Typical weather predicted variables include ambient air dry-bulb temperature, relative humidity, global solar radiation, and direct normal solar radiation.

Different building energy systems models have been considered with varying complexity ranging from data-driven, also known as black box or inverse models, to physics-based models, also referred to as forward models (Afram and Janabi-Sharifi, 2014). Forward models are based on fundamental engineering principles and known parameters of building energy systems such as building envelope components as well as lighting, electrical, and mechanical equipment. Forward models range from RC networks that utilize lumped thermal capacitances and resistances to detailed simulation analysis tools such as EnergyPlus and TRNSYS (Krarti and Henze, 2006). On the other hand, inverse models include artificial neural networks, support vector machine algorithms, and statistical models (such as autoregressive moving average as well as Box and Jenkins or BJ models). Often the prediction data-driven models are verified against those from forward models under limited ranges of applications.

A wide range of optimization techniques can be used for MPC of building energy systems depending on the cost function and the constraints (Afram and Janabi-Sharifi, 2014; Henze et al., 1997). The optimization techniques include linear programming or Simplex method, quadratic programming, dynamic programming, mixed integer programming, evolutionary algorithm, particle swarm optimization, and the genetic algorithm (GA). Other meta-heuristic optimization techniques have been also considered for specific MPC applications such as simulated annealing, differential evolution, ant colony optimization, bee algorithm, the Tabu search, the Harmony search, firefly algorithm, cuckoo search, artificial immune systems, the memetic algorithm, the cross-entropy method, and the bacterial foraging method.

For a predictive MPC application as shown in Fig. 3.13, the optimal controller carries out an optimization over a predefined planning horizon L (generally $L = 24$ h) and of the generated optimal strategy only the first action is executed. At the next time step the process is repeated. The final control strategy of this near-optimal controller over a total horizon of K steps is thus composed of K initial control actions of K optimal strategies of horizon L, where $L < K$. Fig. 3.13 illustrates the four-step procedure involved in determining the predictive optimal control strategy. By moving the time window of L time steps forward and updating the control strategy after each time step, a new forecast is introduced at each time step and yields a policy (*thick dotted line*), which is different from the policy found without taking new forecasts into account (*thin dashed line*). Since we optimize over a planning horizon of L hours, we can only minimize an approximate cost function, which allows for the determination of a near-optimal strategy, whose cumulative utility cost approaches the desired target at the end of the billing period.

3.5 CONTROL APPLICATIONS

Energy management and control systems (EMCS) can be used to perform several functions and tasks. Early building automation systems were limited to simple functions such as simple on/off programming including duty cycling and load shedding. Currently, more complex functions and controls can be achieved by EMCS. Some of these controls are now available in standard program packages such as the following:

- Duty cycling for motor loads to provide sequential shut-down for short periods for equipment such as supply fans of small air handling units. However, for large motors, frequent duty cycling is not generally recommended due to adverse effects on belts, bearings, and motor drives.
- Demand shedding to limit electrical loads. However, it is generally difficult to identify loads that can be shed without affecting the building performance especially for HVAC systems that are not generally needed when electrical demand is high. Therefore, equipment not related to HVAC systems such plug

loads or lighting fixtures are typically considered for demand shedding. In particular, programmable lighting controls can be combined with other energy-saving lighting measures including dimming and occupancy sensors.
- Partial space conditioning to allow systems to cool only a small portion of the building by controlling supply-air dampers serving various zones. With the use of variable frequency drives, it is now possible to adjust fan speed to match small loads for most central fan systems.

For buildings, supervisory controllers have evolved to perform several functions including the following:

- Maintain acceptable indoor quality environment in occupied spaces of buildings including thermal comfort, air quality, and visual comfort.
- Reduce energy consumption and thus energy and operating costs for buildings.
- Add some flexibility by allowing the use of some equipment while others are turned off either for regular maintenance or for repair.
- Assist in diagnostic analysis of faults specific to energy systems hardware through a series of adjustments in control strategies.

The applications for supervisory controllers to improve the energy efficiency of building energy systems vary significantly and can range from single energy system to the entire building's set of equipment and subsystems. The following sections present some of these applications spanning a wide range of building energy systems.

3.5.1 Building Envelope Systems

As noted in Chapter 2, innovative and dynamic materials with variable properties are currently being considered for building envelopes. Among these materials are the dynamic insulation materials or DIMs that can vary their thermal properties based on external signals. In this section, different control options for DIMs are presented to determine the best schedule for DIM settings associated with building walls (Park et al., 2015). Specifically, the following objective function and constraints for optimization analysis has been proposed (Shekar and Krarti, 2017):

$$f = \min\left(\sum_{0}^{24\,\mathrm{h}} E_{\mathrm{cooling}} * \mathrm{rate}_{\mathrm{cooling}} + E_{\mathrm{heating}} * \mathrm{rate}_{\mathrm{heating}}\right) \tag{3.24}$$

$$\text{for}: kl \le k \le ku$$

where k is thermal conductivity values to be set for the DIM for all the building walls, kl and ku are lower bound and upper bounds of thermal conductivity values, $\mathrm{rate}_{\mathrm{cooling,heating}}$ are utility rates for electricity and natural gas, respectively, E_{cooling} is energy use associated with the cooling system, and E_{heating} is energy use associated with the heating system.

The optimization objective function of Eq. (3.24) is set to minimize the total building heating and cooling energy cost during one day by varying the thermal conductivity value (and thus *R*-value) settings of the DIM for each building wall separately. The thermal conductivity values of each wall have an upper and lower bounds, which have been set to be 2.0 W/m·K and 0.032 W/m·K, respectively (Shekar and Krarti, 2017). For a fixed thickness of 0.075 m (3 in) insulation, the upper and lower bounds of thermal conductivity correspond to the DIM layer having a low resistance of 0.04 m^2 K/W to a high resistance of 2.37 m^2 K/W. The GA-based optimizer is considered to evaluate all potential DIM layer *R*-values between 0.04 m^2 K/W and 2.37 m^2 K/W to identify the best hourly DIM settings for a 24-h period for all the building walls (Shekar and Krarti, 2017). The optimal selection would result in the least combined heating and cooling energy costs as noted by Eq. (3.24). For simplicity, energy and energy demand penalties have been excluded in the reported study from the objective function (Shekar and Krarti, 2017). However, the GA-optimization framework can handle any energy and demand charges and additional constraints (such as desired thermal comfort levels).

Other important parameters required for the GA-based optimization implementation include the number of variables to optimize (n_{vars}) and the population size. In particular, the number of variables can affect significantly the search effort and thus the computational time needed to identify the optimal DIM settings. It should be noted that the parameter n_{vars} defines the number of daily periods during which the DIM settings are to be switched. For instance, specifying n_{vars} as 6 implies that the DIM settings are optimized for every 4-h period during one day (24 h/6 = 4 h). The population size specifies the number of options the GA-based optimizer utilizes during each iteration to reach the optimum solution. A population size of 50 implies that the GA analysis involves 50 sets of DIM settings during each iteration. Based on a sensitivity analysis of the GA optimization performance, it was found that a population size of 200 is sufficient for determining the optimal DIM settings (Shekar and Krarti, 2017). However, the analysis has indicated that the optimization is very sensitive to change in the parameter n_{vars}.

Since GA-based optimization over an entire year is computationally time-consuming, the analysis considers representative days for each season to evaluate the impact of optimal DIM control for all the building walls. Representative days are selected from the application of the two-step control strategy. In particular, three representative days have been selected for Golden, Colorado include one for the swing season (one day in May), one for the heating season (one day in December), and one for cooling season (one day in July). To assess the benefits of GA-based optimal controls, the total heating and cooling energy use and cost savings using DIM walls are estimated relative to a statically but highly insulated office building. For the cost analysis, average US energy prices of \$0.10/kWh electricity and \$8.9/1000 ft^3 natural gas are considered based on

Energy Information Administration (EIA) average energy costs for commercial buildings (EIA, 2016a,b).

In the following sections, the optimal controls obtained for DIM walls for an office building are described for representative days (Shekar and Krarti, 2017).

3.5.1.1 Swing Season Analysis

Fig. 3.14 shows optimized *R*-value settings obtained for the DIM layers applied to the south, west, north, and east walls for three days: May 6th, 16th, and 24th (Shekar and Krarti, 2017). The optimization results show that DIM *R*-value settings change significantly for each wall orientation during May 16th and 26th but only slightly during May 6th. Specifically, DIMs for all walls are

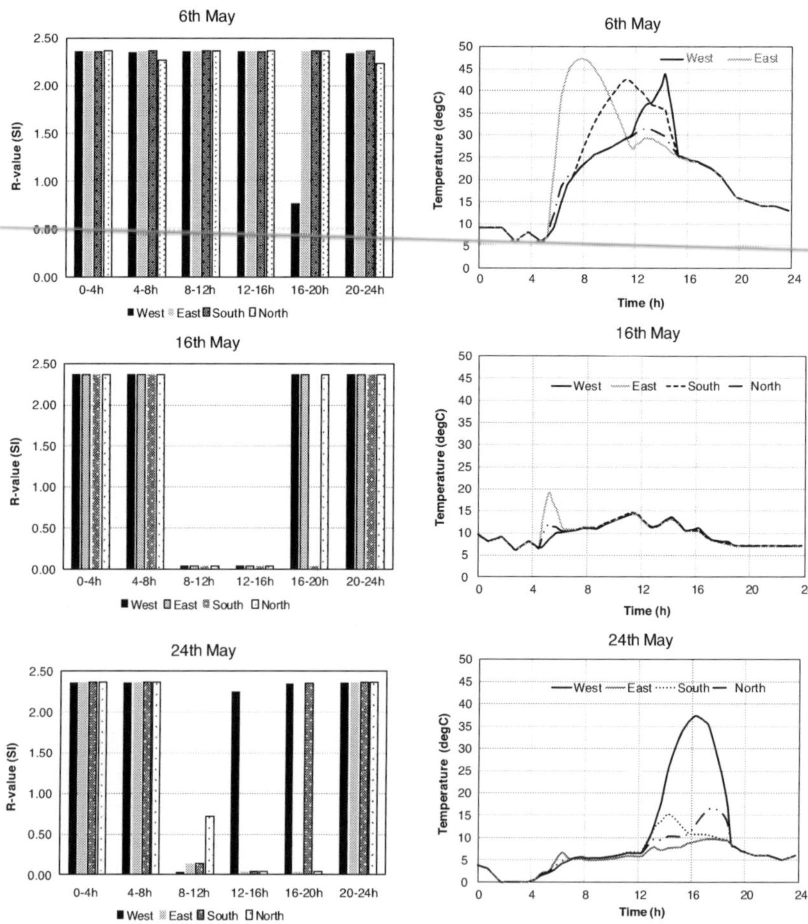

FIGURE 3.14 GA-based optimized DIM insulation *R*-value settings obtained using 4-h periods and sol-air temperatures during May 6th, 16th, and 24th for an office building in Golden, CO.

set to switch to a low R-value for the entire daytime period between 8:00 and 16:00 on May 16th. Similarly, the low R-value settings are selected for all the walls during the 8:00–12:00 period on May 24th but only for the east and north walls during the 4-h periods 12:00–16:00 and 16:00–20:00. It should be noted that the period of 8:00–16:00 coincides with the time of the day when the building is occupied and has high internal gains. Indeed, heat gains associated with equipment, lighting, and people are at their maximum during this period. Due to these high internal heat gains, GA-based optimizer opts to set the DIM layers at their lowest thermal resistance value to allow the excess heat gains trapped indoors to be released outdoors through the building envelope. For May 6th, unlike the case during May 16th and 24th, the DIM layers for all the walls are set at the high thermal resistance R-value during all the periods. To better understand this selection by the GA-based optimization, sol-air temperature hourly variations for the three days in May are also included in Fig. 3.14. It should be noted that sol-air temperature combines the effects of the ambient temperature and the incident solar radiation on the walls. Since during May 16th and 24th, the sol-air temperature remains low during the occupied hours (8:00–16:00), the GA-based optimizer decides to set the DIM layers at their lowest R-value for the outdoors to act as a sink for indoor heat gains. When the sol-air temperature is high, the DIM layers are set at their high R-value to prevent additional heat gains from outdoors as is the case for May 6th during the occupied period. Therefore, and even though the internal heat gains in the building are high, the walls with DIM layers are not able to release this excess heat as the sol-air temperature during the same time is also high. During these periods, the GA-based optimization determines that the best option is to have a highly insulated building envelope to ensure that the cooling thermal load is not increased further by the high sol-air temperatures. This same behavior can also be noted for the west wall on May 24th. Indeed, the increase in the sol-air temperature on the west-facing wall after 12:00 prevents the optimizer to set the DIM layer for that wall to a low R-value while the DIM layers in the other walls have been switched to lower R-value settings.

As noted in the results shown in Fig. 3.14, DIMs have the ability to vary the thermal resistance of the walls in response to mainly building internal loads and external ambient conditions. This feature can allow DIMs to be utilized to lower the building indoor temperatures during unoccupied periods and thus freely precool the building thermal mass. The advantage of this free precooling is to maintain lower cooling thermal loads during the daytime and hence lower energy use and cost. To investigate this additional benefit of DIMs, the office building has been passively precooled by allowing its indoor temperature to float during unoccupied periods. The impact of this precooling strategy associated with the flexible settings of DIMs is illustrated in Fig. 3.14. In particular, the time variation of total building thermal cooling load and the indoor mean air temperature (MAT) are shown during May 16th. It is evident that the MAT variation obtained by the GA-based optimization during unoccupied periods is

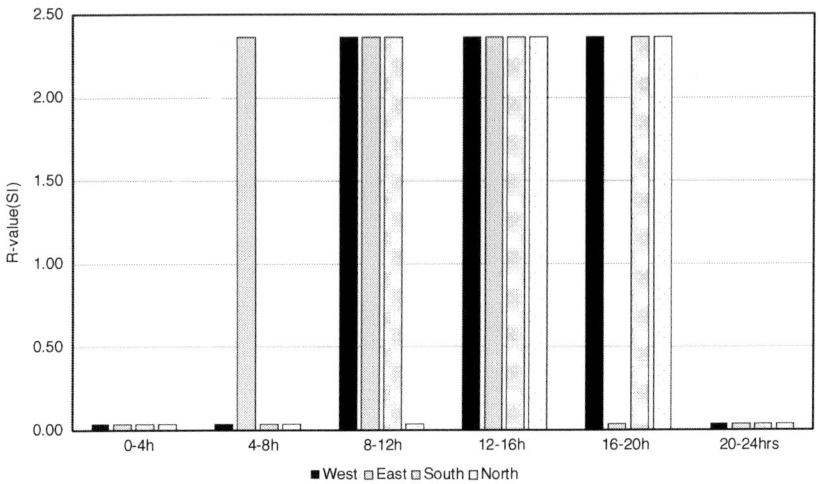

FIGURE 3.15 ***R*-value settings obtained by GA-based optimization for May 6th when the office building is located in Golden, CO.**

lower when compared to that obtained when the building walls are statically insulated with a high *R*-value. Moreover, the DIM-based precooling strategy lowers the total building thermal cooling load during the occupied daytime period.

The same precooling strategy analysis has been carried out during May 6th, which is characterized by high sol-air temperature during the daytime (Shekar and Krarti, 2017). The optimal *R*-value settings for the DIM layers for May 6th are summarized in Fig. 3.15 for all four wall orientations. Based on the GA-based optimization results, the DIM layers have to be set at the low *R*-value during the unoccupied periods but need to have the high *R*-value during the daytime period since the sol-air temperatures for all walls are high as depicted in Fig. 3.16. As a result, the building is precooled during the unoccupied period lowering its daytime thermal cooling load.

3.5.1.2 Heating Season Analysis

The impact of optimal settings for DIM layers has been evaluated during the heating season for an office building located in Golden, Colorado (Shekar and Krarti, 2017). Specifically, two days from December have been selected to assess the performance of optimally controlled DIM layers applied to all the exterior walls of the small office building. In particular, December 3rd and 16th were used for the GA-based optimization analysis with one mild day (December 3rd) and one extremely cold day (December 16th). The use of highly insulated walls is typically considered the most effective approach during the entire heating season of Golden, Colorado to reduce heat loss from residential and commercial buildings. However, the results of the GA-based optimization indicate that when the external conditions (i.e., sol-air temperatures) are mild as

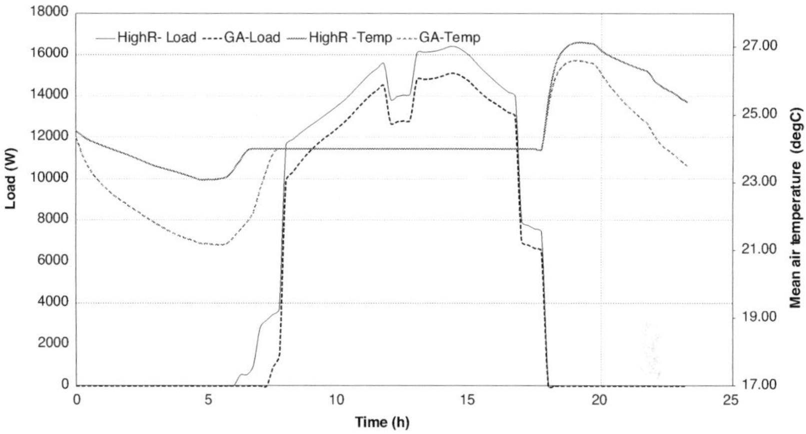

FIGURE 3.16 Thermal loads and temperature settings obtained by GA-based optimization during May 6th when the office building is located in Golden, CO.

is the case on December 3rd, DIM layers with low *R*-value settings can be utilized to release internal heat gains from the office building during the occupied period. Passive precooling ability has not been considered in the optimization analysis as it would not be practical to use this feature in a heating month. The thermal resistance *R*-value settings for DIMs obtained for December 3rd by the GA-based optimization are shown in Fig. 3.17. It can be seen that the *R*-value is lowered for the DIM layers of the west and north walls during the occupied period. For December 3rd, the GA-based optimization has saved as much as 16% of total office building heating and cooling energy costs.

On the other hand, and for a heating-dominated climate such as that of Golden, Colorado, mild weather conditions are seldom experienced during the winter season. A more representative day for Golden, Colorado winter weather conditions is December 16th, which has extreme cold temperatures. Fig. 3.18 represents the time-variation of the sol-air temperature experienced by each building wall during December 16th in Golden, Colorado. It can be seen that all the walls are subject to very low sol-air temperatures ranging from 5°C to −15°C throughout the day. For these extreme outdoor conditions, the GA-based optimization maintains high *R*-values for all the walls with DIM layers during the entire day of December 16th.

3.5.1.3 Cooling Season Analysis

Moreover, the effectiveness of DIMs has been assessed during the cooling season for an office building located in Golden, Colorado (Shekar and Krarti, 2017). First, the analysis is carried out for the entire month of July using the two-step simplified control strategy. It was found that the potential for energy use cost savings due to the application of DIMs is rather limited to very few days in July. The DIM energy saving potential is further evaluated using the GA-based

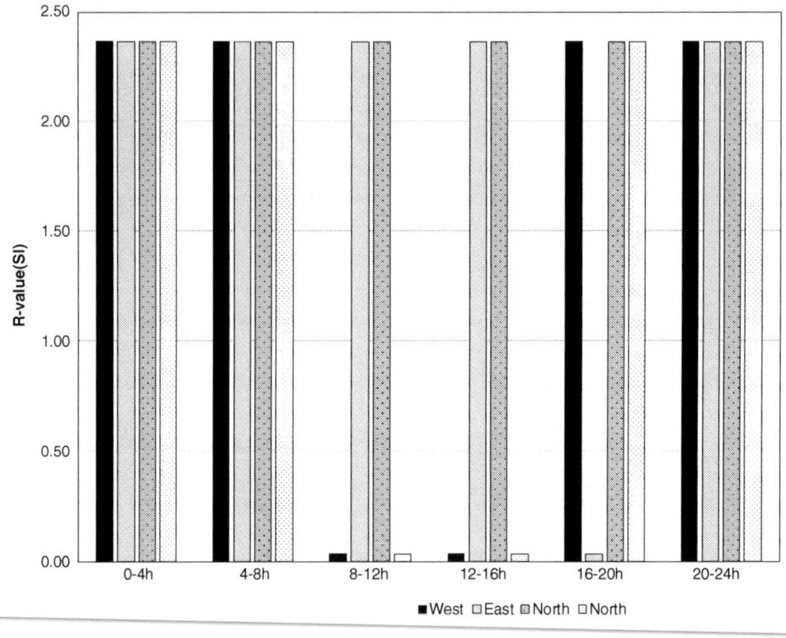

FIGURE 3.17 ***R*-value settings obtained by GA-based optimization for December 3rd when the office building is located in Golden, CO.**

optimization for two days in July: a hot day represented by July 8th and a mild day represented by July 24th. The percent energy cost savings for both days are estimated to be 5% for July 8th and 18% for July 24th. It should be noted that the passive precooling ability of the DIM layers has been considered for the optimization analysis performed for the two days in July. While July 8th represents a typical hot day in July, the 5% energy cost saving potential associated with

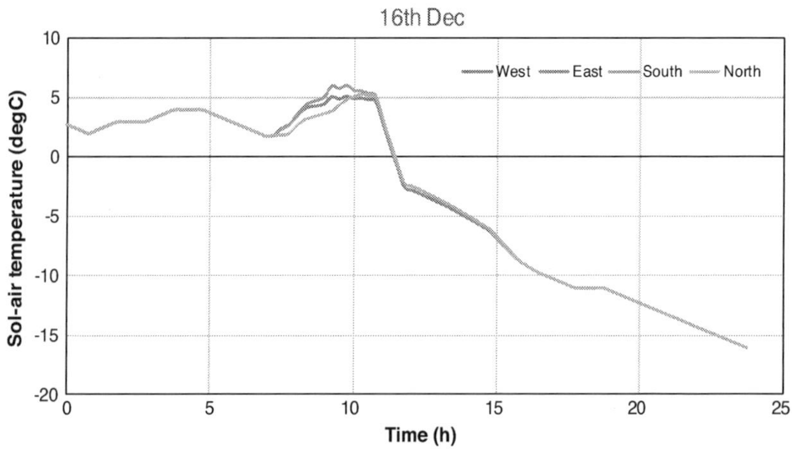

FIGURE 3.18 Hourly variation of sol-air temperatures (December 16th Golden, CO).

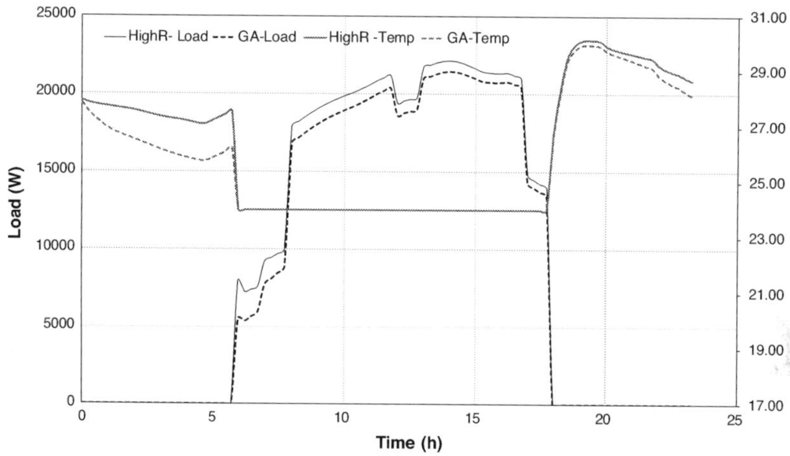

FIGURE 3.19 Thermal loads and temperature settings obtained by GA-based optimization during July 8th when the office building is located in Golden, CO.

the use of DIMs is actually due to the ability of the building to passively and freely precool during the unoccupied periods. The DIM operation settings and their impact during July 8th are better illustrated by the time-variation of thermal cooling load and MAT as well as optimal *R*-value settings as shown in Figs. 3.19 and 3.20, respectively.

3.5.1.4 Comparative Analysis of Control Strategies

While the opportunity of energy use and cost savings associated with the application of DIMs is evident from the results obtained using GA-based optimization, it is important to compare these results to those obtained from the simplified two-step control (Menyhart and Krarti, 2016; Park et al., 2015).

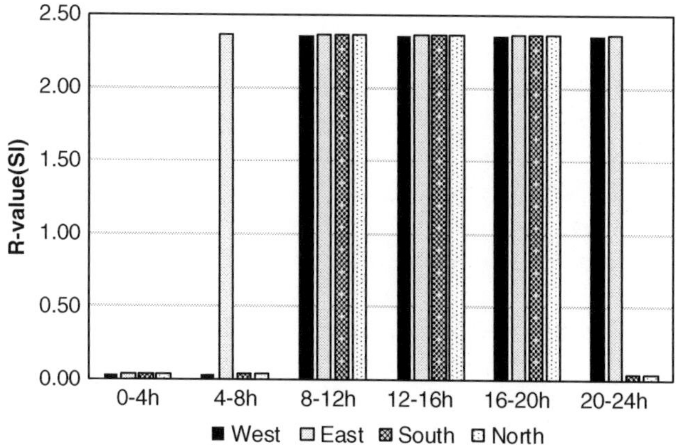

FIGURE 3.20 *R*-value settings obtained by GA-based optimization for July 8th when the office building is located in Golden, CO.

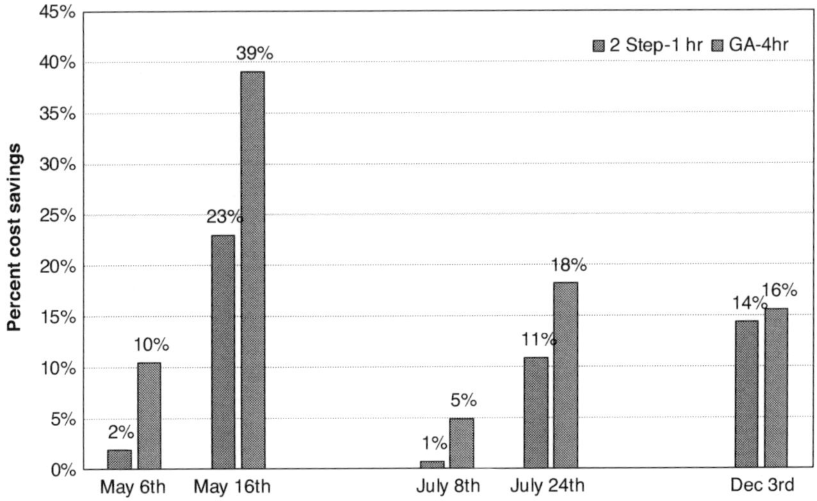

FIGURE 3.21 Percent energy cost savings from DIMs obtained using GA-based optimization and those achieved by two-step controls for five representative days when the office building is located in Golden, CO.

Fig. 3.21 provides a summary of the comparative analysis results for five representative days (i.e., two days for the swing season, one day for the heating season, and two days for the cooling season) (Shekar and Krarti, 2017). The comparative analysis is carried out using 4-h periods for the GA-based optimization with the ability to passively precool the building and 1-h periods for the two-step control. It should be noted that the GA-based optimization results using 1-h and 4-h periods were found to provide the same energy use and cost savings associated with the application of DIM layers for all the exterior building walls.

It is clear from the results outlined in Fig. 3.21 that the GA-based optimization outperforms the two-step controls for all the five days considered in the analysis. This finding is expected since the GA-based optimization algorithm actually searches for the best DIM settings to reduce the total building energy use costs as defined by Eq. (3.24). The significant differences in the potential energy cost savings clearly indicate that the two-step approach, while simple to implement, is not the most effective control technique to operate the DIM layers. It should be also noted, however, that during the heating season, both techniques provide similar performance and lead to almost identical energy cost savings. Further analysis shows that the main difference in the performance between the two control options during both the swing and cooling seasons is the ability of the GA-based optimization to take advantage of cool outdoor temperatures to adjust the DIM settings to freely precool the building thermal mass during unoccupied hours.

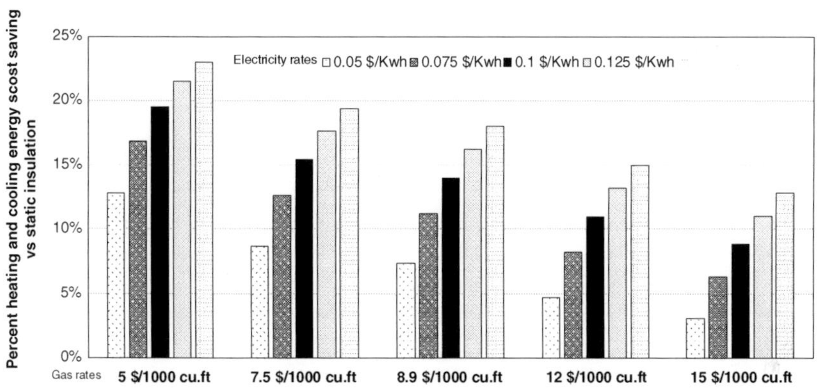

FIGURE 3.22 Impact of electricity and natural gas prices on the energy cost savings that DIMs can achieve for an office building located in Golden, CO.

3.5.1.5 Sensitivity Analysis

It has been shown that the performance of DIMs applied to the exterior building walls can be affected by several design and operating parameters including utility rates (i.e., energy and demand charges), floor aspect ratios, climates, and wall constructions (Park et al., 2015). A series of sensitivity analyses is carried out using the small office building to assess the performance of DIMs under various design and operating conditions as briefly described in the following sections.

3.5.1.5.1 Impact of the Utility Rates

The energy prices used for the reported analyses are based on EIA average rates obtained for the US commercial buildings (EIA, 2016a,b). To evaluate the impact of the energy prices on the behavior of the DIMs, both the electricity rates and natural gas rates are varied from the considered baseline values. The analysis to assess the impact of energy prices is performed for December 3rd when the office building located in Golden, Colorado, experiences heating and cooling thermal loads. The results of the sensitivity analysis are summarized in Fig. 3.22.

Since electricity is mostly consumed during the daytime by the cooling system, the DIM settings are adjusted to release the internal heat gains to the outdoors to lower the required cooling energy. Hence, as the electricity price increases, the controller increasingly activates the DIMs regardless of the natural gas prices as can be noted from the results shown in Fig. 3.22. Similarly, the heating requirements for the office building occur mainly during the unoccupied period on December 3rd, the DIM layers are generally set at high R-value. Thus, while any increase in gas prices increases the heating energy costs, it forces the controller to maintain high R-value for the walls for longer periods and thus reduce the overall energy cost saving potential for the DIMs for any given electricity price as is apparent in Fig. 3.22.

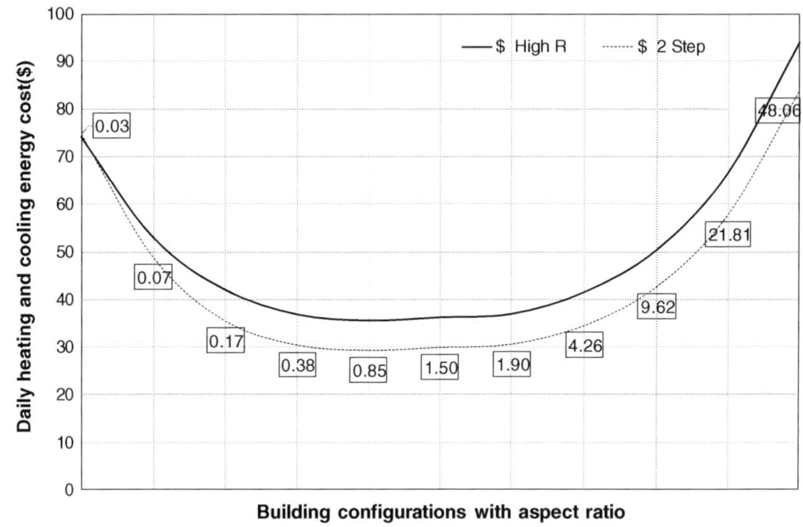

FIGURE 3.23 Variation of heating and cooling energy costs as a function of the office building floor aspect ratio (indicated in the *boxes*) during December 3rd in Golden, CO.

3.5.1.5.2 Impact of the Building Shape

The effect on DIM performance of changing the building shape has also been evaluated for an office building. In particular, the shape is varied by changing the building floor aspect ratio defined as the ratio of the length to the width of the building footprint. By keeping the building orientation and floor area fixed, the floor aspect ratio is varied from a wide floor to a long floor. The sensitivity results obtained for December 3rd are illustrated in Figs. 3.23 and 3.24 when the office building is located in Golden, Colorado. In particular, Fig. 3.23 shows the variation of the combined heating and cooling energy costs as a function of the floor aspect ratio. Fig. 3.24 represents the percent

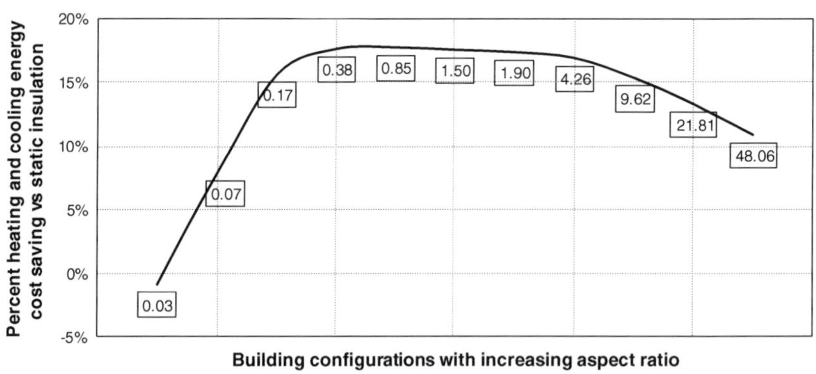

FIGURE 3.24 Variation of percent energy cost saving as a function of the office building aspect ratio (indicated in the *boxes*) during December 3rd in Golden, CO.

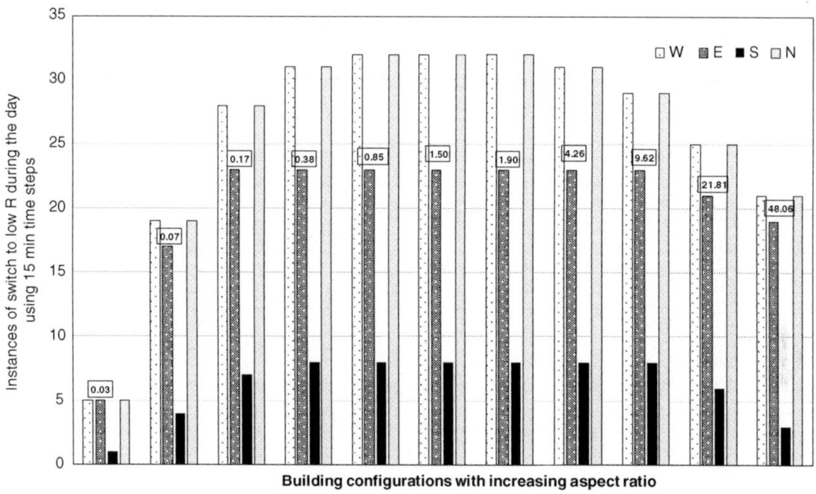

FIGURE 3.25 Frequency of DIM settings to low *R*-value as a function of the office building aspect ratio.

energy cost savings due to DIMs as a function of the building floor aspect ratio. Both figures clearly indicate the percent energy cost saving that can be achieved by DIMs compared to statically insulated walls is the highest when the building floor aspect ratio is close to 1, which corresponds to the most compact building form (i.e., a square).

Fig. 3.25 shows the frequency (in terms of the number of hours) when a low *R*-value is set for the DIM layers as a function of the building floor aspect ratio for all wall orientations. Fig. 3.25 clearly indicates that DIMs are most active for a compact building with a floor aspect ratio close to unity for the four wall orientations. It should be noted that the DIM layers are set differently depending on the wall orientation DIMs facing north and west are the most active. This high activity for both walls is associated to their low sol-air temperature experienced during the day.

3.5.2 Electrical and Lighting Systems

3.5.2.1 Motor Duty Cycling Controls

Frequent turning on and off HVAC systems (in particular fan motors) may be actually detrimental and may not be cost-effective over the life cycle of the equipment due to added maintenance and repair costs. However, on/off cycling of motor loads can be performed safely without long-term damages when minimum on and off times are respected. The National Electrical Manufacturers Association (NEMA) provides a set of recommendations for minimum on/off times of duty cycling of motor loads [2]. Some of these recommendations are summarized in Table 3.3. It is highly recommended, however, that the motor manufacturers be directly consulted to determine their suggested minimum off-time and allowable number of starts per hour.

TABLE 3.3 Allowable Number of Start Per Hour and Minimum Off-Time for Motor Loads

Motor size HP (kW)	Two-pole motors		Four-pole motors		Six-pole motors	
	Maximum starts/h	Minimum off-time (seconds)	Maximum starts/h	Minimum off-time (seconds)	Maximum starts/h	Minimum off-time (seconds)
2.0 (1.5)	11.5	77	23.0	39	26.1	35
5.0 (3.75)	8.1	83	16.3	42	18.4	37
7.5	7.0	88	13.9	44	15.8	39
15.0	5.4	100	10.7	50	12.1	44
20.0 (15.0)	4.8	110	9.6	55	10.9	48
25.0 (18.75)	4.4	115	8.8	58	10.0	51
30.0 (22.5)	4.1	120	8.2	60	9.3	53
40.0 (30.0)	3.7	130	7.4	65	8.4	57
50.0	3.4	145	6.8	72	7.7	64

Source: NEMA, 1994. Energy Management Guide for Selection and Use of Polyphase Motors, Standard MG-10-1994, National Electrical Manufacturers Association, Rosslyn, VA.

To fully benefit from duty cycling energy cost savings (due mostly to reduction in demand charges), two methods can be used. In the first method known as parallel duty cycling, all the motors are cycled on and off at the same time. This method can provide energy cost savings when the duty period is less than the demand period (typically 15 min in most utility rates). However, when the duty period exceeds the demand period, there is no reduction in demand changes if all the motors are cycled on and off at the same time. In this case, it is recommended to use the second method of duty cycling called staggered duty cycling, which alternates the on and off times of the motors.

Most mechanical equipment manufacturers recommend extended duty cycling periods (higher than the typical demand period of 15 min). Therefore, staggered duty cycling approach should be considered in most applications to ensure the safety of HVAC equipment while reducing operating costs.

For example, a 33% reduction in the annual energy use and costs due to staggered duty cycling of three identical two-pole fan motors, each rated at 30 kW (40 hp) if the motor has a cycle of 20 min on and 10 min off as the minimum duty cycle. Indeed, at any given time, only two out of the three motors are operating. Example 3.2 illustrates the calculation procedure for energy cost savings incurred from staggered duty cycling measure.

Example 3.2

Determine the reduction in the annual energy costs due to staggered duty cycling of three identical fan motors (each rated at 30 kW [40 hp]). The motor manufacturer specifies that a cycle of 20 min on and 10 min off as the minimum duty cycle. The utility monthly demand charge is $10/kW. First, determine the recommended NEMA duty period assuming two-pole motors.

Solution

(a) Using Table 3.3, the allowable number of starts per hour for 40 hp, two-pole motors is 3.7 starts/h. Thus, each start should last:

Start period = (60 min/h)/(3.7 starts/h) = 14.4 min/start

The duty period is thus about 15 min. Since the duty period is the sum of off-time and on-time, and since the minimum off-time is about 2 min (130 s based on Table 10.2), the maximum on-time allowable by NEMA standard is 13 mins. Thus, the manufacturer on-time is longer than that recommended by NEMA.

(b) The reduction in the electrical demand peak due to staggered duty cycling approach is one-third of the total demand of all three motors (or 1/3 * 90 kW = 30 kW). Indeed, at any given time, only two out of the three motors are operating. Thus, the annual savings in electrical demands charges is given as follows:

$$\Delta \text{kW} = 12 * 30\,\text{kW} = 360\,\text{kW/year}$$

Therefore, the staggered duty cycling control provides an annual energy cost savings of $3600.

3.5.2.2 *Daylighting*

The use of daylighting controls to reduce the reliance on electrical lighting systems to maintain visual comfort within indoor spaces is one effective approach to significantly reduce the energy consumption in commercial buildings. A typical daylighting control system is illustrated in Fig. 3.26, including photo-sensors to measure the illuminance level at various locations within the daylight space, controllers to determine the required luminance output for the electrical lighting fixtures, and a set of control strategies. Two common conventional control strategies are used for daylighting systems including stepped and dimming controls as outlined in Fig. 3.27.

3.5.2.2.1 Energy Savings from Daylighting Controls

Several studies indicated that daylighting can offer a cost-effective alternative to electrical lighting for commercial and institutional buildings. Through sensors and controllers, daylighting can reduce and even eliminate the use of electrical lighting required to provide sufficient illuminance levels inside office spaces. A simplified calculation method has been developed by Krarti et al. (2005)

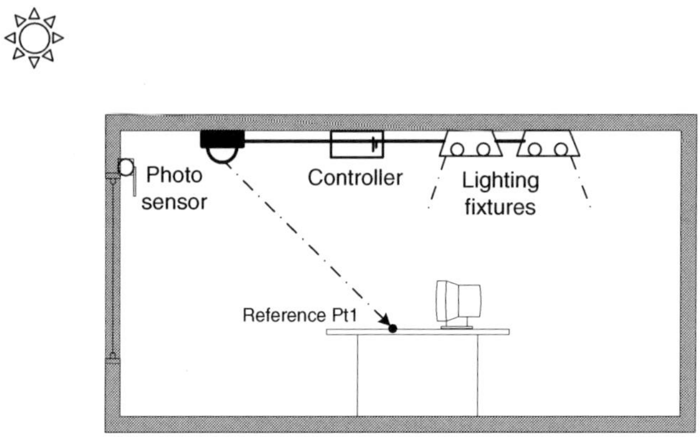

FIGURE 3.26 Typical daylighting control system for an indoor space.

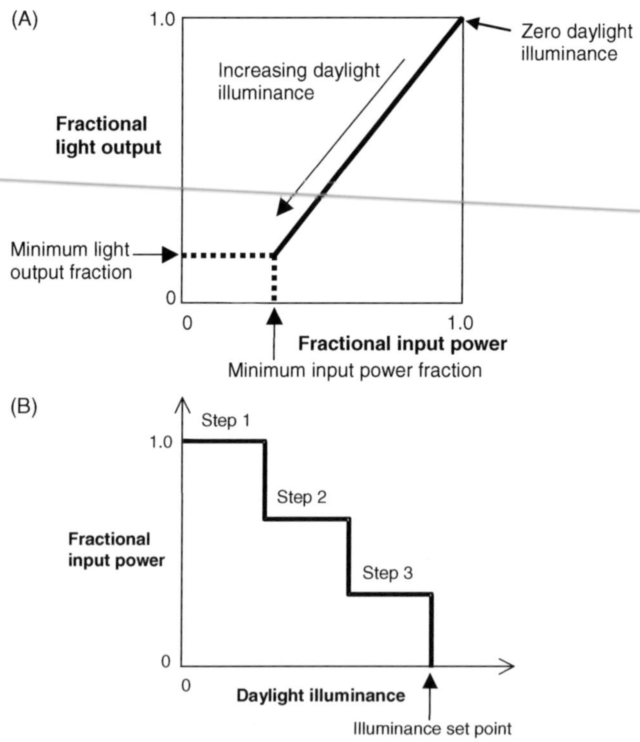

FIGURE 3.27 Conventional (A) dimming and (B) stepped control strategies for daylighting systems.

to estimate the reduction in the total lighting energy use due to daylighting with dimming controls for office buildings. The method has been shown to apply for office buildings in the United States as well as in Egypt (Al Mohimen et al., 2005). The simplified calculation method is easy to use and can be used as a predesign tool to assess the potential of daylighting in saving electricity use associated with artificial lighting for office buildings.

To determine the percent savings, f_d, in annual use of artificial lighting due to implementing daylighting using daylighting controls in office buildings, Krarti et al. (2005) found that the following equation can be used:

$$f_d = b\left[1 - \exp\left(-a\tau_w A_w / A_p\right)\right]\frac{A_p}{A_f} \qquad (3.25)$$

A_w/A_p is window to perimeter floor area. This parameter provides a good indicator of the window size relative to the daylit floor area. A_p/A_f is perimeter to total floor area. This parameter indicates the extent of the daylit area relative to the total building floor area. Thus, when $A_p/A_f = 1$, the whole building can benefit from daylighting. a and b are coefficients that depends only on the building location and is given by Table 3.4 for various sites throughout the world. τ_w is the visible transmittance of the glazing.

TABLE 3.4 Coefficients *a* and *b* of Eq. (3.25) for Various Locations Throughout the World

Location	*a*	*b*	Location	*a*	*b*
Atlanta	19.63	74.34	Casper	19.24	72.66
Chicago	18.39	71.66	Portland	17.79	70.93
Denver	19.36	72.86	Montreal	18.79	69.83
Phoenix	22.31	74.75	Quebec	19.07	70.61
New York City	18.73	66.96	Vancouver	16.93	68.69
Washington D.C.	18.69	70.75	Regina	20.00	70.54
Boston	18.69	67.14	Toronto	19.30	70.48
Miami	25.13	74.82	Winnipeg	19.56	70.85
San Francisco	20.58	73.95	Shanghai	19.40	67.29
Seattle	16.60	69.23	K-Lumpur	20.15	72.37
Los Angels	21.96	74.15	Singapore	23.27	73.68
Madison	18.79	70.03	Cairo	26.98	74.23
Houston	21.64	74.68	Alexandria	36.88	74.74
Fort Worth	19.70	72.91	Tunis	25.17	74.08
Bangor	17.86	70.73	Sao Paulo	29.36	71.19
Dodge City	18.77	72.62	Mexico91	28.62	73.63
Nashville	20.02	70.35	Melbourne	19.96	67.72
Oklahoma City	20.20	74.43	Roma	16.03	72.44
Columbus	18.60	72.28	Frankfurt	15.22	69.69
Bismarck	17.91	71.50	Kuwait	21.98	65.31
Minneapolis	18.16	71.98	Riyadh	21.17	72.69
Omaha	18.94	72.30			

Example 3.3 outlines the calculation procedures to estimate the annual energy savings associated with installing daylighting dimming controls for an office building.

Example 3.3

Problem

Consider a three-story office building 50-m by 30-m located in Denver, Colorado with a window-to-wall ratio of 0.25 with low-E glazing, which has a visible transmittance of 0.41. The building has a lighting density of 10 W/m^2. Determine the energy and cost savings implementing a dimming daylighting control system through the perimeter zones (5-m wide) of the building. Assume that without daylighting controls, the lighting system is operated 8 h/day, 5 days/week, 50 weeks/year. The cost of electricity is 0.10/kWh. Assume the typical wall height is 3 m per floor.

Solution

First, the various areas used in Eq. (3.25) are computed on a per floor basis:

$$A_f = 50\ \text{m} \times 30\ \text{m} = 1500\ \text{m}^2;$$

$$A_p = 2 \times (50 + 30 - 10) \times 5\ \text{m}^2 = 700\ \text{m}^2;$$

$$A_w = 0.25 \times 2 \times (50 + 30) \times 3\ \text{m}^2 = 120\ \text{m}^2;$$

Based on Eq. (3.25) using the coefficients $a = 19.36$ and $b = 72.8f$, the percent reduction in electrical lighting associated with dimming daylighting controls installed in all perimeter zones for one floor is

$$f_d = 72.86 \times [1 - \exp(-19.36 \times (0.41)120 / 700)]\frac{700}{1500} = 25.3\%$$

Therefore, the savings in annual electrical energy for the entire building associated with lighting system are computed as follows:

$$\Delta \text{kWh} = 25.3\% * 3 * \left(1500 \times 10 \times \cdot 8.5.50 \cdot \frac{1}{1000}\right) = 22{,}753\ \text{kWh/year}$$

Thus, the annual energy cost saving is \$2275/year.

As an alternative to conventional controls, Seo et al. (2011) have developed an optimal daylighting controller. The optimal controller relies on an outdoor sensor to regulate the operation of the indoor lighting fixtures as outlined in Fig. 3.28. Moreover, the optimal controller considers the actual lamp performance data to estimate the spatial illuminance distribution due to the electrical lighting system within an indoor space. Moreover, the optimal controller can

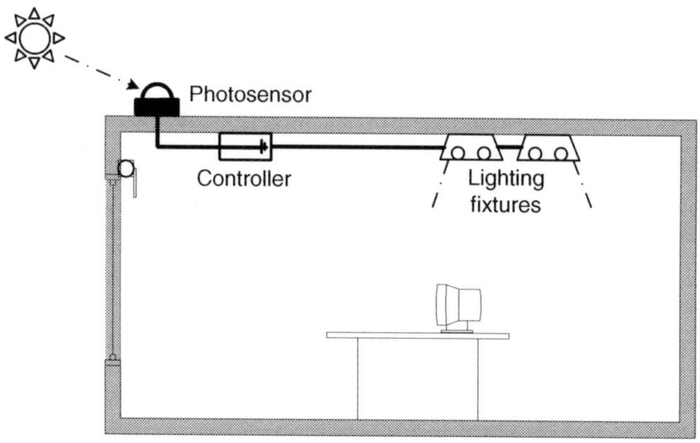

FIGURE 3.28 Daylighting control system for an indoor space using an outdoor photo-sensor.

determine the optimal settings for each fixture or lighting circuit within the space to maintain the desired indoor illuminance levels at specific reference points such as workplaces. In addition, the proposed daylighting controller can utilize only one outdoor photo-sensor or can forgo totally any sensing system to maintain the desired indoor illuminance levels. Indeed, once the lighting system for an indoor space is designed, the performance of the lighting system can be predicted using detailed lighting simulation tools.

To illustrate the application of the lighting and daylighting simulation tool presented by Seo et al. (2011) to optimize the design of lighting circuiting layouts in buildings, a space with five possible lighting circuit layout options are considered as illustrated in Fig. 3.29. The lighting circuiting options in Fig. 3.29 indicate how the lamps are physically wired. Each lighting circuit has its own controller. For instance, the lamps in Option 1 are all wired to the same circuit and are connected to one controller. Thus, in Option 1, the entire space is one lighting zone controlled by one controller/photo-sensor system. In Option 4, there are three lighting circuits associated to three lighting zones. Each lighting zone is controlled by one controller. It should be noted that each lamp illuminance output affects lighting level in all three zones. For the analysis carried out in this section, it is assumed that the photo-sensor (i.e., the reference point) in each lighting zone is located in the middle of the zone (see Fig. 3.29).

Table 3.5 compares lighting energy use savings obtained using optimized daylighting controls with various window orientations and circuiting layouts for a building space located in Golden, Colorado. As indicated in Table 3.5, the lighting energy savings due to daylighting controls are the highest (67.2%) in Layout 5 when the windows are facing south. When the windows are facing north, lighting energy savings ranging from 0.2% to 40% are obtained depending on the circuiting layouts. For east- and west-orientated windows, lighting

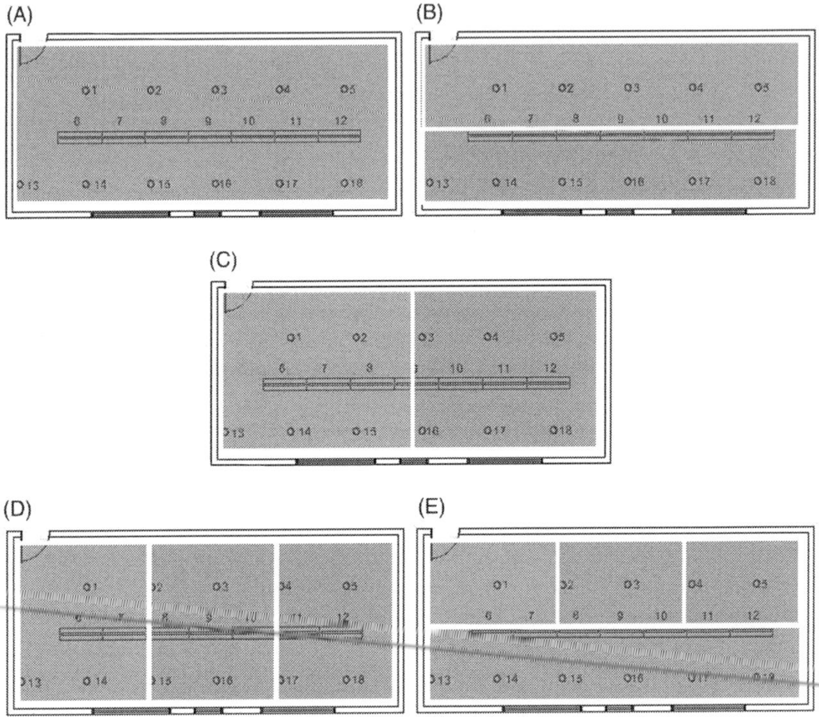

FIGURE 3.29 Lighting circuiting layout options for optimal daylighting controller. (A) Layout Option 1. (B) Layout Option 2. (C) Layout Option 3. (D) Layout Option 4.

energy savings of 18% to 50% can be achieved. The results outlined in Table 3.5 indicate that Layout 5 provides the most energy savings up to 40% relative to Layout 1. For Layout 2, the lighting energy savings relative to Layout 1 range from 1.4% to 6.6% depending on the orientation of the windows. It should be noted that daylighting control performance for Layouts 4 and 5 are similar for east and west orientations. However, Layout 5 outperforms Layout 4 for south and north orientations (Fig. 3.30).

3.5.3 Optimal Start of HVAC Systems

After the energy crisis of the 1970s, engineers found that building utility bills can be reduced by 12%–34% by merely implementing an occupied thermostat setback (Bloomfield and Fisk, 1977). During the cooling season, the unoccupied zone temperature set-point is raised while during the heating season, the set-point is lowered. In some mild climates, the indoor temperatures are allowed to float during night periods rather than defined by a night set-point. However, the winter night setback is typically set between 13°C (55°F) and 15.5°C

TABLE 3.5 Summary of Parametric Simulation Results for Four Window Orientations and Five Circuiting Layouts (Golden, CO)

	Layout1	Layout2	Layout3	Layout4	Layout5
(a) Lighting energy saving relative to no daylight control					
West	28.5%	33.2%	38.1%	49.4%	50.4%
North	0.2%	2.5%	8.4%	29.7%	40.2%
East	18.6%	19.8%	29.7%	43.1%	45.2%
South	46.5%	49.9%	57.5%	61.6%	67.2%

	Layout2	Layout3	Layout4	Layout5
(b) Lighting energy saving relative to Layout 1				
West	6.6%	13.4%	29.2%	30.7%
North	2.3%	8.2%	29.6%	40.0%
East	1.4%	13.6%	30.1%	32.6%
South	6.3%	20.5%	28.2%	38.7%

(60°F). Thus, building indoor temperature is colder than its occupied set-point in the early mornings. Due to the building thermal mass, the heating system has to be turned on earlier than the scheduled occupied time to achieve thermal comfort when people first enter the building.

The amount of time a building takes to recover from its night setback to its occupied set-point is usually referred to as the building recovery time. The length of the recovery period depends on several factors including outdoor ambient temperatures, indoor temperatures, and building thermal characteristics. Therefore, the recovery period can vary daily throughout the heating season especially in climates with sudden changes in the outdoor temperatures. However, the recovery time is typically set to be the same throughout the entire heating season by building operators to simplify the start controls of the heating system. This recovery time is defined as the earliest time the heating system needs to be started for the coldest day of the year. While, this approach may achieve thermal comfort at the start of the occupancy periods throughout the heating season, it does not ensure optimal start times for the heating system especially during mild winter mornings.

With an ECMS, algorithms can be developed to determine the optimum start times and thus the best recovery periods. Several algorithms have been suggested in the literature. Typically, the recovery times are adjusted daily based on outdoor ambient temperatures and initial building zone temperatures (which may not be necessarily close to the set-point temperatures during unoccupied periods). In the following sections, some of the simplified algorithms for estimating building recovery times are presented.

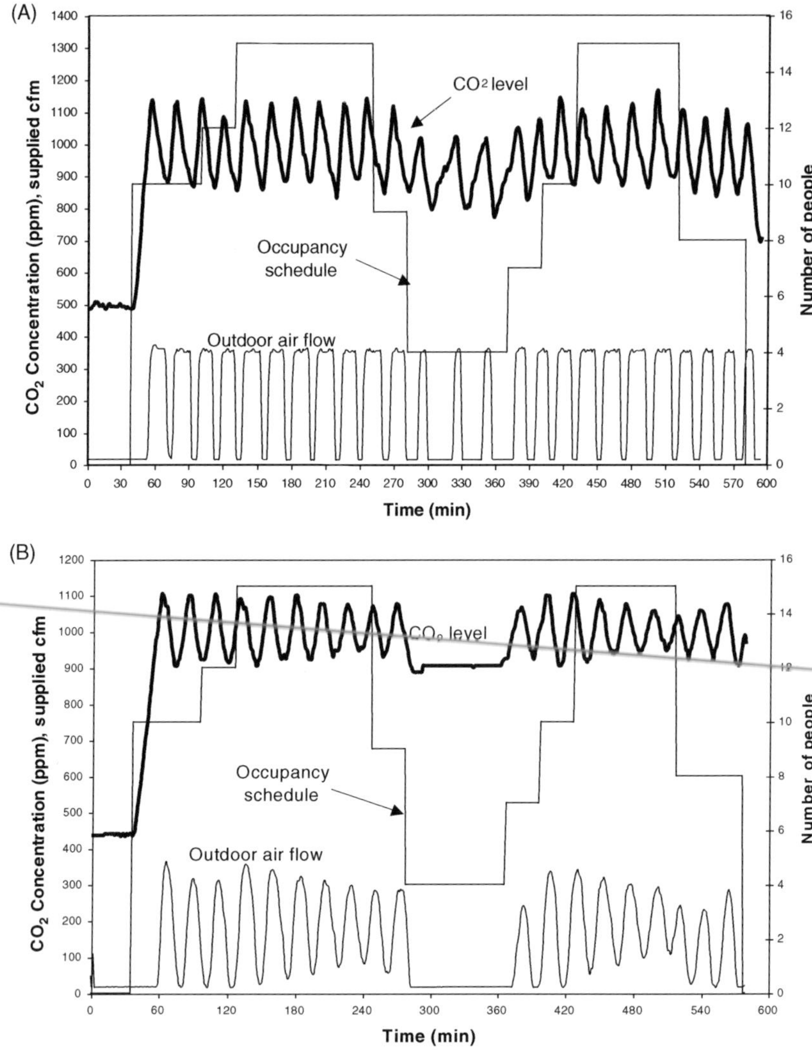

FIGURE 3.30 Outdoor air flow and CO2 variations using DCV controls. (A) ON/OFF DCV control. (B) PID DCV control.

Method 1: A linear relationship between recovery times, τ, and outdoor ambient temperatures, T_{amb}:

$$\tau = a_0 + a_1 T_{amb} \tag{3.26}$$

where

$$a_0 = \tau_{max} + \frac{\tau_{max} T_{amb,max}}{T_{amb,zero} - T_{amb,max}}$$

and

$$a_1 = -\frac{\tau_{max}}{T_{amb,zero} - T_{amb,max}}$$

τ_{max} is the maximum recovery period, $T_{amb,max}$ is the outdoor ambient temperature during the time when the maximum recovery period is obtained, $T_{amb,zero}$ is the outdoor ambient temperature during the time when the recovery period is zero.

It should be noted that the approach presented by Eq. (3.26) is relatively easy to implement since it does not require any regression analysis to determine the relation coefficients a_0 and a_1.

Method 2: A linear relationship between the recovery time, τ, and both outdoor ambient temperature, T_{amb}, and initial zone temperature, $T_{zone,initial}$

$$\tau = a_0 + a_1 T_{amb} + a_2 T_{zone,initial} \tag{3.27}$$

where the coefficients a_0, a_1, and a_2 are determined based on a regression analysis. This approach is first developed and implemented by Jobe and Krarti (1997). To determine the regression coefficients, it is recommended that data for at least five days be used.

Method 3: A quadratic relationship between the recovery time, τ, and both outdoor ambient temperature, T_{amb}, and initial zone temperature, $T_{zone,initial}$, using a weighting function in the form of

$$\tau = a_0 + w a_1 T_{amb} + (1 - w) \cdot (a_2 T_{zone,initial} + a_3 T^2_{zone,initial}) \tag{3.28}$$

where the weighting parameter w is defined as follows:

$$w = 1000^{-[(T_{zone,initial} - T_{setnight})/(T_{zone,final} - T_{setnight})]}$$

with $T_{setnight}$ the night (or unoccupied) setback temperature and $T_{zone,final}$ is the occupied set-point temperature.

The approach presented by Eq. (3.28) is proposed by Seem et al. (1989) based on results from computer simulations.

A comparative analysis performed by Jobe and Krarti (1997) between the three approaches indicated that all three methods can reduce the start-up time for the heating system and thus save energy compared to the common approach that relies on setting the maximum recovery time throughout the entire heating season. Table 3.6 summarizes the daily average reduction time in recovery period for two educational buildings located in Colorado using the three approaches discussed earlier.

It is clear from the results presented in Table 3.6 that the approach described as Method 2 provides the highest recovery time reduction while providing adequate thermal comfort within the two buildings. For the two buildings used in the comparative analysis, the recovery time was found to vary rather linearly with outdoor ambient temperatures. Therefore, and since Method 3 places more emphasis on indoor air temperatures (refer to Eq. (3.28)), it may not be adequate for the considered buildings.

TABLE 3.6 Daily Average Reduction in Recovery Periods for Two Buildings Located in Colorado

Building Method #	Maximum recovery period (min)	Predicted recovery period (min)	Reduced start-up time (min)	Percent reduction from maximum recovery period (%)
Building 1				
Method 1	90	64	26	29
Method 2	90	53	37	41
Method 3	90	77	13	14
Building 2				
Method 1	75	19	56	74
Method 2	75	19	56	74
Method 3	75	28	47	62

3.5.4 HVAC Equipment Operation

3.5.4.1 *Outdoor Air Intake Controls*

Due to the increase in the number of occupant complaints regarding poor IAQ and the increase in buildings diagnosed with sick building syndrome, the control and measurement of outside air intake rates has come to the forefront of attention of many HVAC engineers and designers. The majority of HVAC system designers today rely on the ASHRAE "Ventilation Rate Procedure" described in ASHRAE Standard 62 (2004), *Ventilation for Acceptable Indoor Air Quality*. ASHRAE Standard 62 specifies minimum ventilation rates as a function of building use and occupancy to provide adequate IAQ for conditioned spaces.

Unfortunately, the necessary monitoring equipment and control logic to maintain minimum outdoor intake rates are often nonexistent or are used improperly if they are installed. Consequently, several commercial buildings, and in particular those with VAV systems, have been found to have inadequate ventilation (Sterling et al., 1992). The use of appropriate airflow measurement or VAV control techniques is critical to maintain minimum outside air intake rates. Specifically, theoretical and experimental analyses have been performed to determine the accuracy of various techniques for outside airflow measurement and control applicable to VAV systems (Krarti et al., 1999).

In economizer systems, the size of the outside air duct must be large enough to safely provide 100% of the design flow. This large size, however, results in very low airflow velocities during minimum outside air intake rate mode, which can make measurement difficult with pressure-based airflow measurement devices.

Table 3.7 summarizes the results of an experimental comparative analysis performed by Krarti et al. (2000) to evaluate some of the control techniques for

TABLE 3.7 Summary of Comparative Results for the Control and Measurement Techniques Tested by Krarti, M., Schroeder, C., Jeanette, E., Brandemuehl, M., 2000. Experimental analysis of measurement and control techniques of outside air intake rates in VAV systems. ASHRAE Trans. 106(2), 234–255.

					Outside air intake rate measurements					
					Averaging Pitot-tube array			Electronic thermal anemometry		
System description	Measurement control[a]	Case	Set-point (cfm)	Validity (%)	Mean (cfm)	Stdev (cfm)	RMS (cfm)	Mean (cfm)	Stdev (cfm)	RMS (cfm)
Fixed damper position	-NA-	1-A	1600	14	656	658	1,150	682	564	1048
Fixed damper position	-NA-	1-B	2400	23	1410	678	1,199	1407	680	1199
Fixed damper position	NA	1-C	3200	26	2178	819	1,309	2124	870	1513
Plenum pressure control	NA	2-A	1600	100	1630	69	75	1544	70	89
Plenum pressure control	NA	2-C	3200	100	3288	94	129	3279	88	118
Direct control with economizer duct	P	3-A	1600	100	1635	38	52	1546.05	46	71
Direct control with economizer duct	P	3-C	3200	100	3192	50	51	3225.24	47	53
Direct control with economizer duct	E	4-A	1600	94	1695	55	110	1637.00	55	67
Direct control with economizer duct	E	4-C	3200	100	3228	49	57	3263.10	50	80
Volume tracking	E	5-A	1600	0	2427	439	936	2436.06	458	953

(Continued)

TABLE 3.7 Summary of Comparative Results for the Control and Measurement Techniques Tested by Krarti, M., Schroeder, C., Jeanette, E., Brandemuehl, M., 2000. Experimental analysis of measurement and control techniques of outside air intake rates in VAV systems. ASHRAE Trans. 106(2), 234–255. (*cont.*)

					Outside air intake rate measurements					
					Averaging Pitot-tube array			Electronic thermal anemometry		
System description	Measurement control[a]	Case	Set-point (cfm)	Validity (%)	Mean (cfm)	Stdev (cfm)	RMS (cfm)	Mean (cfm)	Stdev (cfm)	RMS (cfm)
Direct control with dedicated duct	P	6-A	1600	100	1639	39	55	1634.15	47	58
Direct control with dedicated duct	P	6-B	2400	100	2430	41	51	2457.15	51	77
Direct control with dedicated duct	E	7-A	1600	100	1643	36	56	1640.01	43	58
Direct control with dedicated duct	E	7-B	2400	100	2404	39	40	2428.30	42	50
Injection fan	P	8-A	1600	100	1621	31	37	1621.83	36	42
Injection fan	P	8-B	2400	100	2429	28	40	2440.38	36	54
Injection fan	E	9-A	1600	100	1622	38	44	1617.09	36	40
Injection fan	E	9-B	2400	100	2418	33	37	2426.78	32	42
Direct control	C	NA[b]	1600	75	1632	137	141	1605[c]	119[c]	119[c]

[a] *P, averaging pitot-tube array; E, electronic thermal anemometer; C, CO_2 concentration balance.*
[b] *A different system setup was used for testing the concentration balance measurement technique.*
[c] *Value is for CO_2 concentration balance measurement technique, not electronic thermal anemometry.*

outdoor air intake under repeatable laboratory conditions. Specifically, Table 3.7 provides the average value, the standard deviation, the root mean square of the outdoor air intake flow rate, and the validity of each measurement and control method tested in a laboratory set-up. In particular, three measurement techniques are used to determine the air flow rates: (1) averaging pitot-tube array station (P), (2) electronic thermal anemometer (E), and (3) CO_2 concentration balance technique (C). For more details on these measurement techniques, the testing set-up, and the experimental results, the reader is referred to Krarti et al. (2000).

The percentages listed in Table 3.6 in the column labeled "validity" were calculated from Eq. (3.29):

$$\text{validity} = \frac{n_v}{n} \tag{3.29}$$

where n_v is the number of valid data points.

Each test presented in Table 3.7 is subject to errors from the airflow measurement and the control technique used. Each 10 second data point, x_i, recorded during testing was considered valid if it met the following two conditions:

1) $|x_i - \text{set point}| \le (\text{set point} \cdot 10\%)$ and

2) $\frac{e_i}{x_i} < 15\%$ where e_i is the predicted error for the airflow measurement in the laboratory.

The first condition attempts to account for the accuracy of the control technique by requiring the data point to be within 10% of the set point. The second condition attempts to account for the accuracy of the airflow measurement technique by requiring the predicted error of the data point to be less than 15%.

In summary, accurate measurement and control of outside air intake rates in VAV systems is possible when careful attention is paid to proper installation and operation of system equipment. In systems where uniform airflow profiles exist, the use of an averaging Pitot-tube array or an electronic thermal anemometry, depending upon the expected velocities, for the direct measurement of outside airflow rates allows for direct control of minimum outside air intake rates. When these conditions are not met, the installation of a separate, dedicated minimum outside air duct, or the use of the concentration balance airflow measurement technique provide adequate alternatives. However, calculating the outside airflow rate using a temperature balance will not provide accurate results for all building operating conditions. Plenum pressure control in systems where measurement of the outside airflow rate is not possible should provide adequate control of minimum outside air intake rates. The traditional constant-air-volume (CAV) control strategy of a fixed minimum outside air damper position, and the more robust volumetric fan tracking technique are not capable of accurately controlling outside airflow rates in VAV systems.

3.5.4.2 Demand Ventilation Controls

To ensure that ventilation is provided only to present occupants, demand controlled ventilation or DCV controls have been developed. Strategies for the use of CO_2-based DCV controls have been evaluated by several studies. Most of these studies (Emmerich et al., 1994; Knoespel et al., 1991; Kusuda, 1976; Ogasawara et al., 1979; Sorensen, 1996; Vaculik and Plett, 1993) have focused on the potential energy savings of the CO_2-based demand controlled ventilation systems, with CO_2 concentration reported as a measure of the IAQ performance. Strategies with CAV and VAV HVAC systems have been developed. In computer simulation modeling, researchers compared the performance of several ventilation strategies, including constant outdoor airflow at the ASHRAE Standard 62-1989, minimum outdoor airflow at the typical rate with temperature-based economizer, DCV with a step-flow control algorithm, DCV with a step-flow control and a temperature-based economizer, and DCV with on–off control. The on–off control employed an algorithm in which outdoor airflow is set at 100% if the high CO_2 set-point is exceeded and at 0% if the CO_2 concentration drops below the low set-point.

Other researchers used DCV control strategies and/or variations of these strategies in real buildings. In summary, control strategies applied in fields so far include the following:

1. The ASHRAE Standard 62 (20 cfm/person in office buildings).
2. The on/off control strategy (Davidge, 1991; Woods et al., 1982).
3. The proportional control strategy (Donnini et al., 1991; Fleury, 1992; Haghighat and Donnini, 1992).
4. The step-flow control strategy (Fehlmann et al., 1993; Zamboni et al., 1991).

Alalawi and Krarti (2002) have evaluated the effect of various CO_2-based demand controlled ventilation strategies on HVAC system energy consumption under controlled and repeatable test conditions. Three control strategies were evaluated: (1) the base case control strategy that controls the outside/recirculated damper to provide a fixed outside air fraction equivalent to 20 cfm/person, (2) the ON/OFF control strategy that opens and closes the damper as a function of the CO_2 level deviation from the set-point, and (3) a PID controller that controls the damper position according to the error involved in the control process. Another control strategy was also evaluated with the damper controlled proportionally to the zone CO_2 concentration level.

The main findings of the Alawi and Krarti experimental analysis are as follows (Alalawi and Krarti, 2002):

- ASHRAE minimum outdoor air ventilation rate (20 cfm/person) is needed to maintain an acceptable IAQ of less than 1000 ppm of CO_2 concentration within an office space. The experiments performed at lower ventilation rates showed that these rates do not guarantee that the CO_2 concentration level is kept below the limit (1000 ppm).

TABLE 3.8 Comparative Analysis of Various DCV Control Strategies

One-zone controls of CO_2 and temperature set-points			
Parameter	**Base case**	**ON/OFF**	**PID**
Total Volume of air (ft^3)	108101	55794	65248
Percent reduction in volume of air	–	48.4	39.6
Maximum CO_2 level (ppm)	860	956	983
Duration CO_2 exceeds limit (min)	0	2.1	35.1
Average outside temperature (F)	80.9	81.9	84.2
Two-zone controls of CO_2 and temperature set-points			
Parameter	Base case	ON/OFF	PID
Total volume of air (ft^3)	245974	299602	190858
Percent reduction in volume of air[a]	–	−21.8	22.4
Maximum CO_2 level (ppm)	855	1007	965
Duration CO_2 exceeds limit (min)	0	72.3	13.8
Average outside temperature (F)	80.5	68.2	80.3

[a]A (−) sign indicates an increase in the volume or energy consumption.

- A PID controller can maintain acceptable indoor CO_2 level with the minimum outdoor air requirements, followed by the proportional control and lastly the ON/OFF control. A reduction of about 33% can be obtained when applying the PID controller instead of the ON/OFF control strategy. Fig. 3.14 compares the performance of both ON/OFF and PID DCV controls to maintain acceptable IAQ within an office space.
- The one-zone controls to maintain indoor CO_2 concentration and temperature settings indicate that the ON/OFF control strategy provides the minimum outdoor air requirements followed by the PID controller (as designed and implemented in their study) with a percent reduction of about 48.4% and 39.6%, respectively. The two-zone experiments, however, showed that the PID provided the lowest requirement of outside air by about 22.4%. Table 3.8 summarizes the experimental results to assess the performance of various DCV control strategies (Alalawi and Krarti, 2002).

3.5.4.3 Radiant Slab Heating Systems

Radiant floor panel heating systems are widely used in several European and Asian countries. They consist of embedded hot water coils in floor slabs of residential and commercial buildings to provide space heating as illustrated in Fig. 3.31 (ASHRAE, 2012). Control strategies of radiant floor heating systems on slab-on-grade foundation are more challenging than those utilized to operate conventional

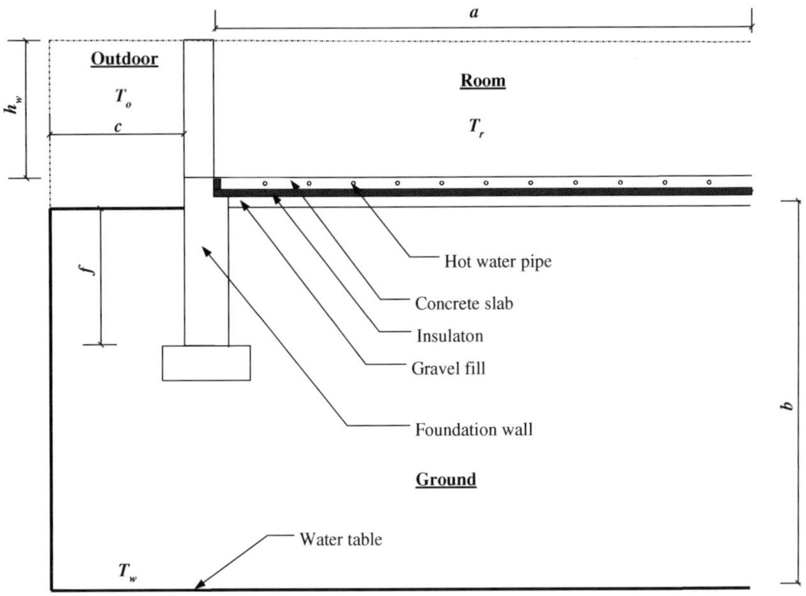

FIGURE 3.31 Ground-coupled heat transfer for a typical hot water radiant floor heating.

hot air heating systems. Due to the inherent time lag of ground-coupled heat transfer, the control of a radiant floor heating system can be difficult for maintaining indoor space temperature and thermal comfort within an acceptable range.

Conventional control strategies for radiant floor heating panels typically maintain space temperature using mostly temperature or heat flux modulation techniques. Temperature modulation control sets the supply water temperature to be proportional to outdoor temperature or to the difference between a desired set-point and room air temperatures. While flux-modulation control attempts to ensure that a slab delivers heat to the space at a rate per unit area proportional to the difference between the room air temperature and the slab surface temperature. The heat flux provided to the slab is proportional to the difference between the supply and return water temperatures. Typically, three temperatures are considered for the radiant slab systems to maintain indoor thermal comfort: MAT, mean radiant temperature (MRT), and operative temperature (OT). Based on a detailed analysis of Ihm and Krarti (2005), Fig. 3.32 presents the total heating energy use and the number of hours when thermal comfort levels exceed the acceptable range (i.e., $-0.5 \leq \text{PMV} \leq +0.5$) for the three conventional controls during a cold day in Denver, Colorado. As shown in Fig. 3.32, total heating energy use is generally proportional to the increasing set-point temperature for all three control strategies. Depending on the control strategy, lower set-point temperature can save some heating energy use at the expense of a reduction in indoor thermal comfort. To further evaluate the performance of control strategies, the set-point temperatures are selected based on both achieving low energy use

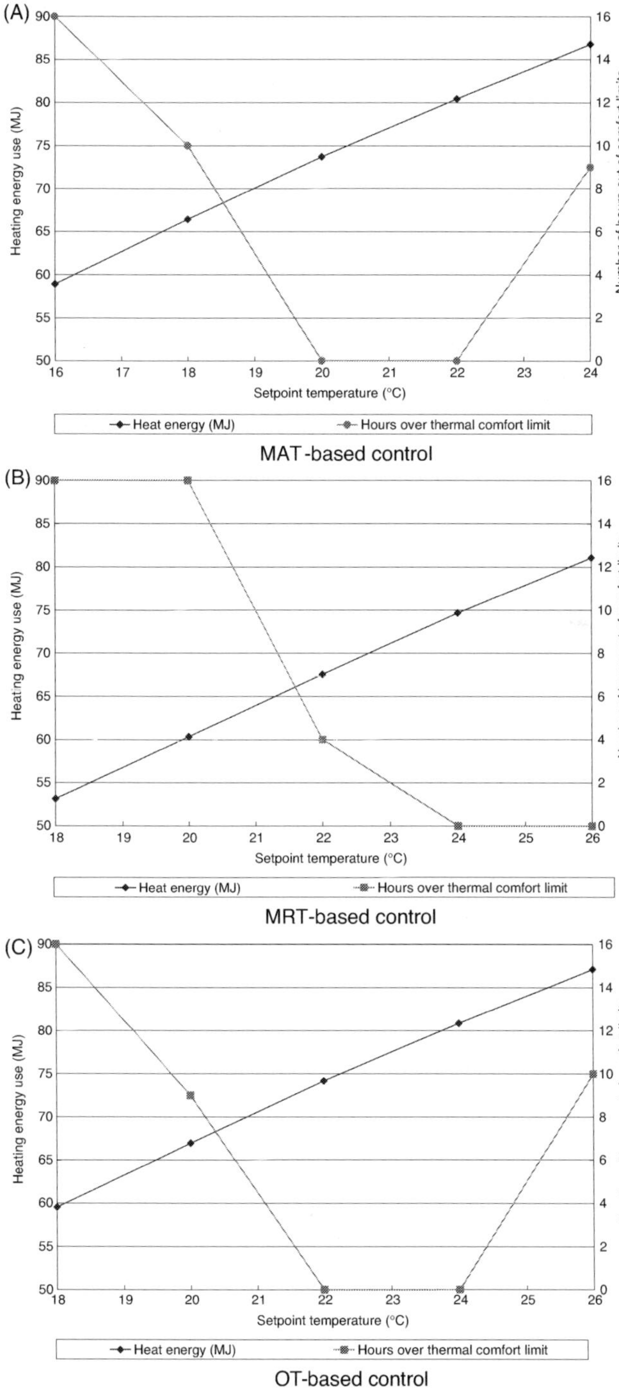

FIGURE 3.32 Comparative of the three conventional controls for radiant slab heating system. (A) MAT-based control. (B) MRT-based control. (C) OT-based control.

and satisfying building thermal comfort. Therefore, the set-point temperatures are chosen as 20°C for MAT-based control, 24°C for MRT-based control, and 22°C for OT-based control for further analysis.

Due to the dynamic characteristics of the radiant floor systems, optimal controls could be more appropriate than conventional controls to reduce energy consumption while maintaining building thermal comfort. The controlled variable to optimize building energy use is selected as time-varying indoor set-point temperature during one day to minimize the total daily operating cost taking into account the constraint of maintaining indoor space temperatures within a range defined by upper and lower temperature limits.

For the optimal control strategies, the objective function is constrained by the condition that thermal comfort has to be maintained during the occupied hours within acceptable levels. That is, the space indoor temperature should be kept within predefined indoor set-point temperature limits. Simmonds (1993) used optimal controls to demonstrate that additional energy savings could be achieved by HVAC systems if the control is based on maintaining the Fanger's predicted mean vote (*PMV*) index rather than the dry bulb temperature within a thermal comfort range. In general, a *PMV* value within ±0.5 is acceptable for thermal comfort according to ASHRAE (2014). Thus, desired *PMV* values can be used as constraints of the objective function to keep proper indoor thermal comfort conditions during the occupied period using the following condition:

$$-0.5 \leq PMV \leq +0.5 \tag{3.30}$$

Thus, the optimal control seeks to minimize the total energy use of radiant floor heating systems for any 24-h period as a function of the building thermal characteristics, HVAC system, and outdoor environment with Eq. (3.30) as a constraint. In the analysis discussed by Ihm and Krarti (2005), the objective function to be minimized is defined as the total energy cost of heating the building constrained by thermal comfort (*PMV*) as expressed by the following equation:

$$C = \sum_{k=1}^{24} \left[E_k + P(PMV_k) \right] \tag{3.31}$$

where C is total energy cost during a day combined penalty, k is hour, E_k is total energy use in a building at hour k, P is penalty when its *PMV* violates the thermal comfort range, and PMV_k is thermal comfort index (*PMV*) at hour k.

Fig. 3.33 shows the set-point temperature and MAT variation for optimal and MAT-based controls when the slab thickness is 5 cm. The predicted set-point temperatures by optimal control vary at each hour while those of conventional MAT-based control are set to be 20°C during the occupied hours of the day (Ihm and Krarti, 2005). The total heating energy use associated with in optimal control (46.9 MJ) is less than that required by the MAT-based control (70.0 MJ). It should be noted that the maximum heating energy for the optimal control is higher than that for the MAT control. However, the optimal control uses more effectively the

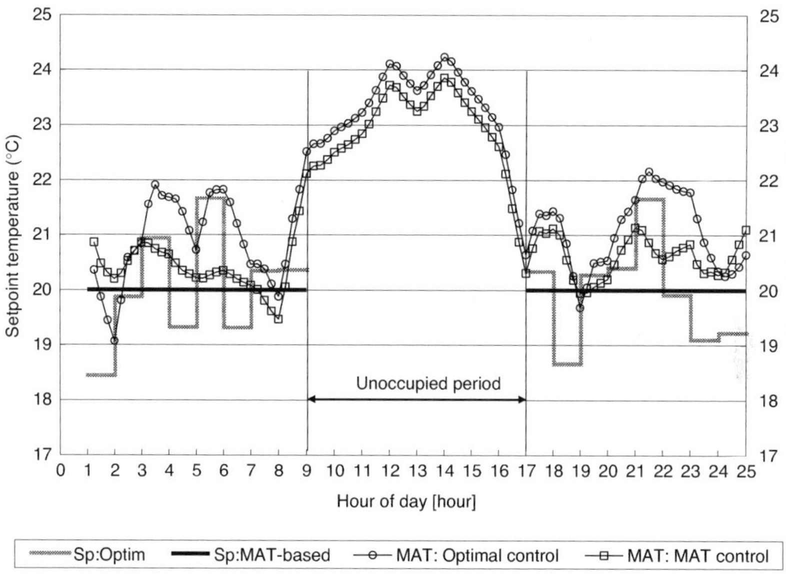

FIGURE 3.33 Comparison of setpoint temperature and MAT variation for both optimal and MAT-based controls for a 5-cm-thick slab on January 21 in Denver, CO.

high thermal mass of slab-on-grade floor and the ground to reduce total heating energy while maintaining acceptable thermal comfort levels for the occupants.

Several thermal comfort models have been proposed to predict the thermal sensation of people using personal, environmental, and physiological variables that influence thermal comfort. EnergyPlus incorporates three thermal comfort models: (1) Fanger comfort model using the *PMV* developed by Fanger (1970), (2) Pierce two-node model using the *PMV* modified by effective temperature (PMVET) developed by J. B. Pierce Foundation (Berglund, 1978), and (3) the KSU two-node model using the thermal sensation vote developed by Kansas State University (Berglund, 1978). The performance of the optimal controls for radiant slab systems depend on the thermal comfort model considered in the optimization. Table 3.9 summarizes the total building heating energy savings attributed to optimal controls relative to the MAT-based control. It indicates

TABLE 3.9 Comparison of the Performance for MAT-Based and Optimal Controls for Three Thermal Comfort Models

Comfort model	Optimal control total heating energy (MJ)	Savings (%)
Fanger	54.7	26.9
KSU	52.2	29.5
Pierce	57.7	7.7

TABLE 3.10 Comparison of the Performance for MAT-Based and Optimal Controls for Various Values of Slab Thickness

Slab thickness (cm)	MAT control total heating energy (MJ)	Optimal control total heating energy (MJ)	Percent saving in total building energy (%)
5	70.0	46.9	30.1
10	74.0	54.7	26.9
15	74.5	59.8	22.7

that optimal controls can reduce total building energy by 26.9% based on the Fanger model, by 29.5% based on KSU two-node model, and by 7.7% based on the Pierce two-node model. The lower savings obtained for the Pierce model is expected since it has a 9-point thermal sensation scale (4 to −4) rather than the 7-point thermal scale (3 to −3) of Fanger and KSU models.

The potential savings associated with optimal controls depend on the thermal mass of the slab floor. Table 3.10 summarizes the total building heating energy savings attributed to optimal controls relative to MAT-based controls for various slab thickness. The results indicate that optimal controls can reduce total building energy by 30.1% for a 5-cm slab, by 26.9% for a 10-cm slab, and by 22.7% for a 15-cm slab.

3.5.4.4 *Cooling/Heating Central Plant Optimization*

Cooling and heating central plants offer several opportunities to reduce energy operating costs through optimal or near-optimal controls for individual equipment (local optimization) and for the entire HVAC system (global optimization). While optimal controls have been developed and implemented for various components of cooling and heating central plants, global optimization remains considerably a complex endeavor and only few strategies have been suggested and tested.

In this section, some of the local optimal control strategies are discussed. Moreover, operating strategies for entire cooling/heating plants are briefly discussed.

3.5.4.4.1 Single Chiller Control Improvement

Before replacing an existing chiller, it may be more cost-effective to consider other cooling alternatives or simple operating strategies to improve cooling plant energy performance. In particular, a significant improvement in the overall efficiency of a chiller can be obtained through the use of automatic controls for the following:

- Supply chilled water at the highest temperature that meets the cooling load.
- Decrease the condenser water supply temperature (for water-cooled condensers) when the outside air wet bulb temperature is reduced.

Example 3.4 illustrates typical energy cost savings due to improved controls for a single chiller cooling plant.

Example 3.4

A centrifugal chiller (having a capacity of 500 kW and an average seasonal COP of 4.0) operates with a leaving chilled water temperature of 4.5°C. Determine the cost savings incurred by installing an automatic controller that allows the leaving chilled water temperature to be set on average 2.5°C higher. Assume that the COP of the chiller is increased by 8% and that the number of equivalent full-load hours for the chiller is 1500/year and that the electricity cost is $0.07/kWh.

Solution

The energy use savings can be calculated using Eq. (9.12) with $SEER_e = 4.0$; $SEER_r = 4.0*1.08 = 4.32$; $N_{h,C} = 1500$; and $Q_C = 500$ kW; $LF_C = 1.0$ (assume that the chiller is sized correctly):

$$\Delta E_C = 500\,\text{kW} * 1500\,\text{h / year} * 1.0 * \left(\frac{1}{4.0} - \frac{1}{4.32}\right) = 13{,}890\,\text{kWh / year}$$

Therefore, the annual energy cost savings are $970/year.

3.5.4.4.2 Controls for Multiple Chillers

When a central cooling plant consists of several chillers, a number of control alternatives exist to meet a building cooling load. Effective controls would select the best alternative for operating and sequencing the chillers to minimize the cooling plant operating costs.

Simple guidelines can be followed to operate multiple chillers at near-optimal performance. Typically, chiller operating variables such as chilled water temperature and condenser water flow rate are adjusted to ensure optimal controls. Some of the near-optimal control guidelines to operate electrically driven central chilled water systems are summarized (ASHRAE, 2011):

- Multiple chillers should be controlled to supply identical chilled water temperatures.
- For identical chillers, the condenser water flow rates should be controlled to provide identical leaving condenser water temperatures.
- For chillers with different capacities but similar part-load performance, each chiller should be loaded at the same load fraction. The load fraction for a given chiller can be set as the ratio of its capacity to the sum total capacity of all operating chillers.

To determine the optimal chiller sequencing, a detailed analysis is generally needed to account for several factors including the capacity and the part-load performance of each chiller and the energy use associated with all power-consuming devices such as distribution pumps.

3.5.4.4.3 Controls for Multiple Boilers

The use of an array of small modular boilers provides a more energy efficient heating system than a single large boiler especially under part-load operation conditions. Indeed, each of the modular boilers can be operated close to its peak capacity and thus its highest energy efficiency. To optimally operate multiple boilers, it is important to know when to change the number of boilers on-line and/or off-line. The mere addition of a second boiler on-line when one boiler cannot handle the load may not provide the minimum operating cost. Indeed, the increase of firing rate (due to additional heating load) on any given boiler can cause a decrease in thermal efficiency due to higher flue-gas temperatures and thus higher thermal losses. However, the addition of a second boiler on-line increases the standing losses due to auxiliaries and the thermal losses through the added casing and piping of the second boiler. Therefore, a detailed analysis is needed to determine the changeover points for the multiple boilers. These changeover points depend on the characteristics of each boiler (ASHRAE, 2011).

3.5.5 HVAC Systems With Thermal Storage

3.5.5.1 Precooling of Building Thermal Mass

The storage capabilities of a building structure can be used to shift a portion of on-peak cooling loads to off-peak periods and thus reduce electrical demand and energy charges. This measure can be achieved by precooling the building thermal mass. Reported studies have shown that when an effective control strategy is used, up to 35% in energy cost savings can be achieved when an effective control strategy is used to determine when and how much to precool the building (Braun, 1992; Morris et al., 1992). Some additional savings in operating costs can be achieved if free cooling is used for precooling when cool outdoor air temperature is introduced to the building during the night using air handling fans. In some cases, the cost of operating air handling fans may be less than the reduction in operating costs for mechanical cooling during occupied periods.

Precooling building thermal mass is an example of the application of the EMCS to reduce operating costs. Precooling of the building thermal mass can be effective at lowering building operating costs. This strategy can have a large impact when chillers have high loads during periods of high occupancy and high outdoor temperatures (which typically coincide with on-peak periods in rate structures). By reducing the on-peak cooling load it is possible to reduce chiller energy use during these critical periods, thereby reducing energy costs.

The precooling control strategy is illustrated in Fig. 3.34 and compared with conventional night setup control. Precooling starts with a constant temperature set point t_{pre}, typically during the night. The warm-up period is then used to reset the zone air temperature set point so that the cooling system turns off without calling for heating. During this time, the zone air is warmed by lighting and equipment loads. The occupied set point, t_{occ}, is set at the low end of the

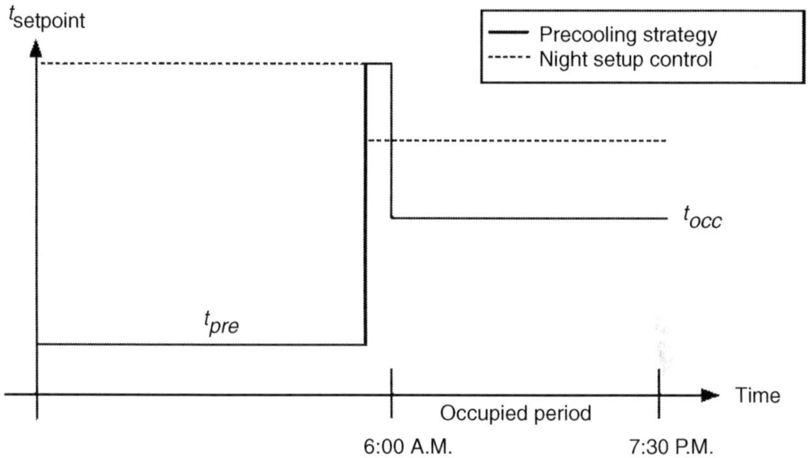

FIGURE 3.34 Typical indoor air temperature set points for precooling controls.

comfort region so that the building mass charge is held as long as cooling capacity is available. This set point is maintained until the limit on cooling capacity is reached. After this point, temperatures in the zones float up and the building thermal mass provides additional cooling.

Based on long-term simulation analysis, the annual energy cost savings associated with precooling has been estimated for various time-of-use utility rates (Morgan and Krarti, 2006). For time-of-use rates, on-peak to off-peak ratio for energy and demand charges is defined as follows:

R_e: ratio of on-peak to off-peak energy charges:

$$R_e = \frac{\text{Peak Energy Rate(\$ / kWh)}}{\text{Off} - \text{Peak Energy Rate(\$ / kWH)}} \tag{3.32}$$

R_d: ratio of on-peak to off-peak demand charges:

$$R_d = \frac{\text{Peak Demand Rate(\$ / kW)}}{\text{Off} - \text{Peak Demand Rate(\$ / kW)}} \tag{3.33}$$

Figs. 3.35 and 3.36 show the variation of the annual energy cost savings for a typical office building located in four US locations due to a 4-h precooling period as a function of R_d and R_e, respectively. The office building has a heavy thermal mass of 105 lbm/ft^2 (513.7 kg/m^2) and the time-of-use rate has an 8-h on-peak period (Morgan and Krarti, 2006).

Using detailed simulation analysis for 16 US climates, Katipamula et al. (2010) estimated the potential energy savings associated with precooling of commercial buildings. The results of the simulation analysis are summarized in Table 3.11 for most of the building types in all 16 climates. The potential savings using precooling strategies can be substantial for most building types and

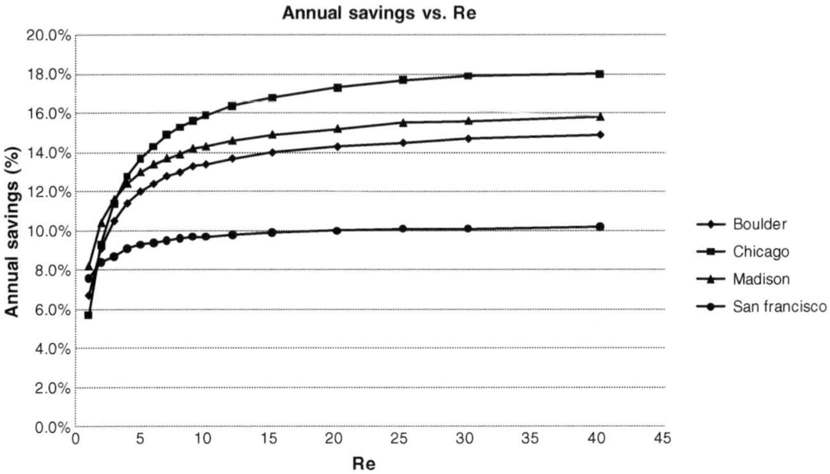

FIGURE 3.35 **Annual energy cost savings due to precooling relative to conventional controls as a function of R_e.**

climates where economizer cooling potential is modest and the annual sensible load is dominant.

3.5.5.2 Passive and Active Thermal Energy Storage Systems

There are two common approaches to store cooling thermal energy in buildings: active and passive systems. The active systems consist of ice or chilled water storage tanks, commonly known as TES systems, which are charged at night and discharged during the day. The passive systems utilize the thermal mass of

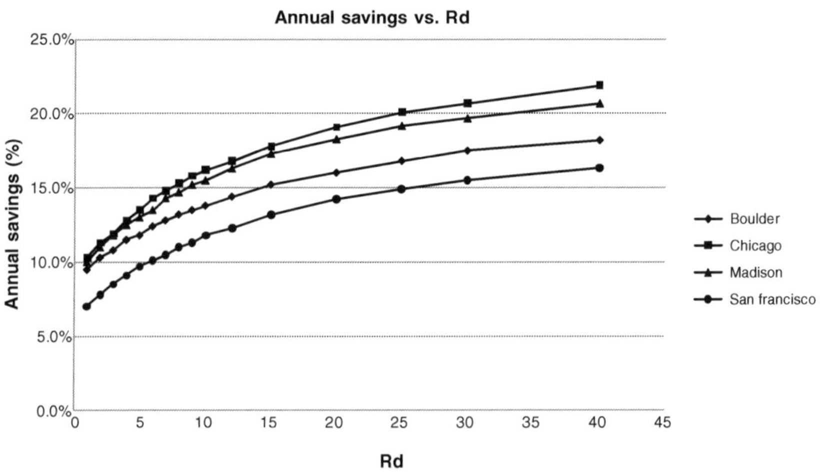

FIGURE 3.36 **Annual energy cost savings due to precooling relative to conventional controls as a function of R_d.**

TABLE 3.11 Precooling Energy Savings Potential for 16 US Climates

Building type	Minimum (%)	Maximum (%)	Average (%)
Office			
small	59	77	70
medium	13	52	37
large	21	61	40
Retail			
standalone	56	73	66
strip mall	45	63	56
Primary school	46	55	51
Secondary school	32	49	43
Hotel, large	16	57	44
Supermarket	59	78	68
Warehouse	50	81	69
Outpatient	65	83	78
Hospital	48	76	64

the building materials to precool the building at night when the electrical rates are low. Both active and passive systems have been used to shift some of the cooling loads from on-peak to off-peak utility rate periods. Several benefits can be incurred from reducing on-peak electrical demands. From a building owner's point of view, the prime motivation for load shifting is to avoid high energy rates and to reduce the overall demand charges. From the power utility's perspective, however, the benefit of active or passive storage systems is their effectiveness in reducing peak electricity demand (Kintner-Meyer and Emery, 1995).

Some studies have considered the use of passive and active TES systems to reduce or even eliminating on-peak cooling demands and thus reduce demand charges associated with operating the cooling system (Henze and Krarti, 1999; Liu and Henze, 2006). Optimal control takes advantage of the building thermal capacitance to minimize the operating cost over a specified period of time such as a day while meeting some required constraints. Building loads are shifted to off-peak hours by proper adjustment of space temperature set points (thermostat settings) throughout the day.

The control strategies for passive and active TES systems can include the following:

- Conventional controls without load shifting (i.e., no precooling of building thermal mass and no utilization of ice storage system).
- Precooling of building thermal mass (typically unoccupied periods) to reduce peak cooling loads (Braun, 1990).

- Chiller-priority control strategy through the operation during on-peak periods of first the chiller (with a capacity smaller than the peak cooling load) and then the ice storage tank to meet any thermal load above the chiller capacity (Henze and Krarti, 1999).
- Storage-priority control strategy through the operation during on-peak periods of first the ice storage and then the chiller to meet cooling thermal loads. The storage-priority control is set so the ice tank is not discharged prematurely (Braun, 1992; Henze and Krarti, 1999, Morgan and Krarti, 2006).
- Optimal control is that control trajectory that minimizes the combined energy and demand charges over the simulation period. The objective function to be minimized is the total cost (i.e., including energy and demand) of the cooling system. The total cost of the cooling system is represented mathematically as follows (Henze and Krarti, 1996, 1999; Hajiah and Krarti, 2012):

$$C = r_{d,0}P_{\max,0} + r_{d,1}P_{\max,1} + \sum_{k=0}^{K} r_{e,\gamma(k)}P(k)\Delta t \tag{3.34}$$

where C is the combined cost including both energy and demand charges, $r_{d,0}$ is off-peak demand charge \$/kW, $r_{d,1}$ is on-peak demand charge \$/kW, $P_{\max,v}$ is peak demand incurred in rate period v ($v = 0$ for off-peak, and $v = 1$ for on-peak), r_e is energy charge \$/kWh, K is total number of hours in the simulation period, $P(k)$ is total cooling and noncooling plant power consumption at k, $\gamma(k)$ is 0 if k is during off-peak period, and 1 if k falls during on-peak period, and Δt is time interval of 1 h.

For the case where only energy charges are minimized, the cost function is reduced to

$$C_{\text{energy}} = \sum_{k=0}^{K} r_{e,\gamma(k)}P(k)\Delta t \tag{3.35}$$

On the other hand, if only the demand charges are minimized, the cost function becomes

$$C_{\text{demand}} = r_{d,0}P_{\max,0} + r_{d,1}P_{\max,1} \tag{3.36}$$

Thus, the optimal control that seeks to minimize the combined energy and demand charges has the following cost function:

$$C = C_{\text{energy}} + C_{\text{demand}} \tag{3.37}$$

The electrical energy cost represents the total cooling plant power consumption multiplied by the energy charge \$/kWh. The electric demand cost on the other hand is generally based on the customer's maximum electrical power over 1 month. The maximum electrical power $P_{\max}$ is defined as the maximum value of electrical power consumed in the building during the on-peak period of the day. Because the simulation period of this research study is assumed to be a day

TABLE 3.12 Cost Savings Obtained for Optimal TES Control, Optimized Precooling, and Optimal Control Using Building Thermal Mass and Ice STORAGE system

Control type	Energy cost ($)	Demand cost ($)	Total cost ($)	Savings in total cost (%)
Conventional (no TES system)	106.1	34.4	140.4	–
Optimized precooling	91.3	28.2	119.5	14.9
Optimized ice storage system	96.0	29.3	125.3	10.8
Optimized precooling and ice storage system	81.0	20.0	101.0	28.1

(i.e., 24-h period), the electrical demand cost (i.e., demand charge multiplied by P_{max}) is discounted by a factor of 30 to reflect the charges associated to one day in a typical month (i.e., average demand daily cost).

Under controlled laboratory testing conditions, Hajiah and Krarti (2012) have performed a series of experimental analysis to compare the performance optimal controls for passive and active TES systems against that of conventional controls. The results of the experimental analysis for the three control strategies are summarized in Table 3.12. The results indicate significant cost savings can be achieved using the optimal controls compared to the conventional control strategy especially for the demand cost optimal control. Indeed, when charges are optimized, 13.2% savings in demand charges are obtained with 9.9% savings in total charges. When total charges are minimized, more than 3.8% cost savings can be achieved.

The ice storage system optimal controller achieved total cost savings of 10.8% relative to the conventional control (system operating during occupancy hours only with no precooling). However, higher savings of 28.1% in the total daily operating cost are obtained with the use of the combined optimal controller (i.e., use of precooling and ice storage) relative to the conventional control. This result illustrates clearly the advantage of using both passive thermal energy storage system (i.e., building thermal mass) and an active thermal energy storage system (i.e., ice storage tank) to minimize total building energy costs.

Using an HVAC laboratory with two full-zones, three control strategies have been implemented and tested. The experimental results have been compared to the simulation predictions as shown in Fig. 3.37 (Hajiah and Krarti, 2012). In general, fair agreement is obtained between the simulation predictions and experimental results for the three control strategies. For all the three tested control strategies, the measured savings are within 10% of the predicted cost savings. The measured energy, demand, and total costs for control strategies are generally slightly higher

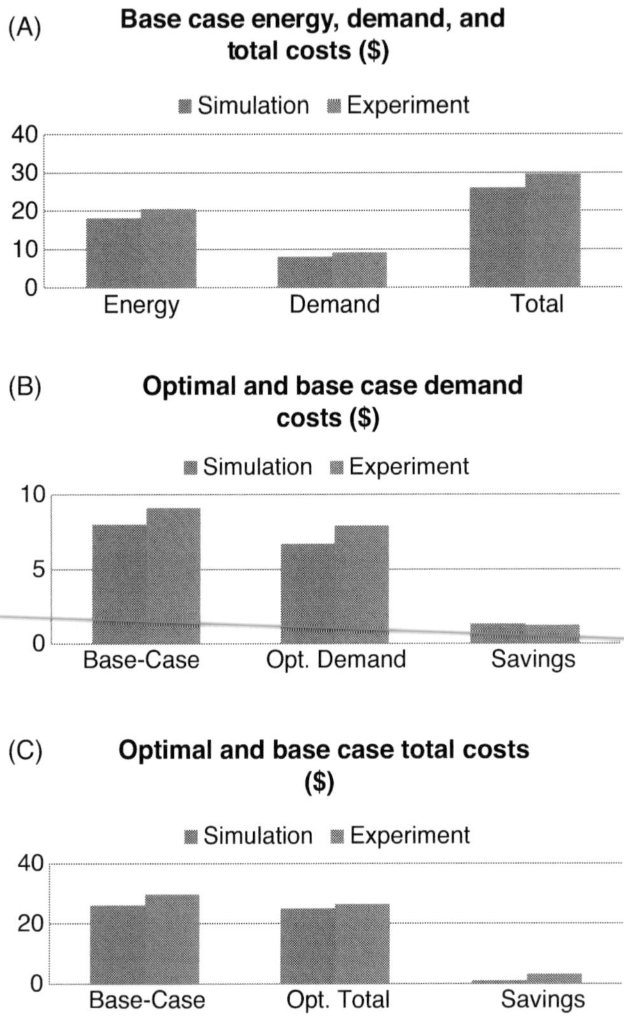

FIGURE 3.37 Comparison between experimental results and simulation predictions for three control strategies. (A) Conventional control. (B) Demand cost optimal control. (C) Total cost optimal control.

than the results obtained from the simulation analysis. The lower predicted values for energy and demand costs are most likely due to the fact that the actual energy efficiencies of various energy-consuming devices such as fans, pumps, and chiller may be lower than the assumed values in the simulation models.

Model predictive controls applied to commercial buildings require short-term weather forecasts to optimally adjust set points in a supervisory control environment. Predictive control strategies for passive and discrete thermal

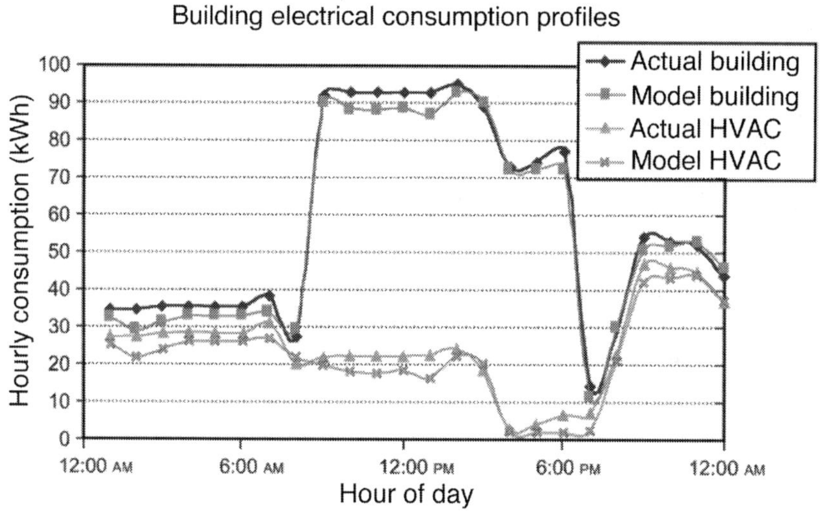

FIGURE 3.38 Building electricity use profiles for 6 h predictive optimal control.

storage systems have been field-tested. Specifically, Morgan and Krarti (2010) investigated the performance of various control strategies for combined passive and discrete TES systems in a Colorado elementary school equipped with an ice storage system. The predictive optimal control strategies were developed using an EnergyPlus-based simulation environment (Zhou et al., 2005). The simulation environment was found to be effective in defining and implementing predictive optimal controls for passive and discrete TES systems for the buildings. Figs. 3.38 and 3.39 show examples of the field testing results for the building energy use performed by Morgan and Krarti (2010).

3.5.5.3 Neural Network Controls of Thermal Energy Storage Systems

Several model-based optimal TES control strategies have been proposed in the literature (Braun, 1992; Drees and Braun, 1994; Henze and Krarti, 1999). These optimal control strategies are generally obtained through optimization techniques and computer simulations and are difficult to implement in the field. Recently, an optimal TES controller using self-learning neural networks has been proposed and tested in full-scale laboratory conditions (Massie, 1998).

The NN-based optimal controller is developed to minimize the operating cost of the cooling plant within a 24-h period. In addition to NN-based equipment models for the chiller and the TES system, the NN-based optimal controller utilizes a training network and a predictor network. The equipment models are based on feed-forward networks while the training and the predictor networks use recurrent networks that compute output vectors sequentially in time. The training network learns the relationship between the controlled

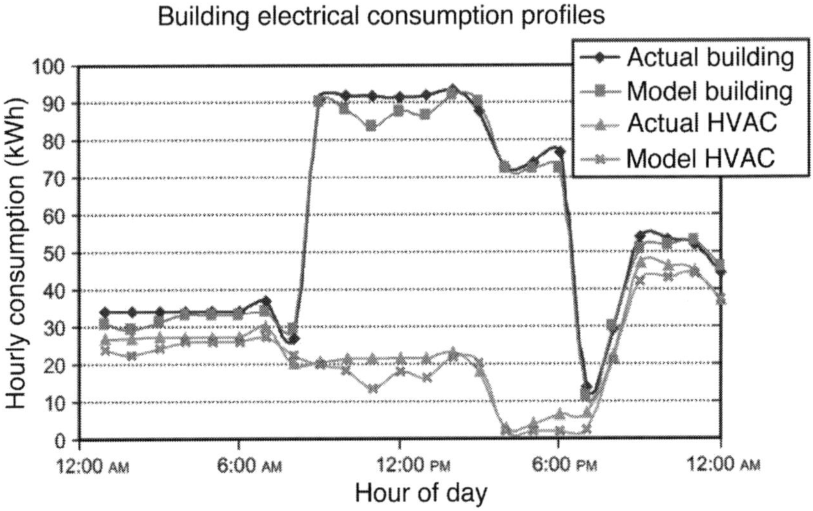

FIGURE 3.39 Building electricity use profiles for 24 h predictive optimal control.

and uncontrolled variables and the cooling plant operating conditions (such as electrical power consumption for the chiller and charging/discharging rates for the TES system). The training network weights are then passed to the predictor network to act as the activation functions. The main goal of the predictor network is to find acceptable set-points for the control variables that minimize operating cost. Massie (1998) used two control variables to control the TES system: (1) chiller set-point temperature and (2) the primary loop three-way valve position. Then, charging/discharging rates for the TES system can be determined using values for the chiller set-point temperatures and the three-way valve positions

Using laboratory conditions, the NN-based optimal controller has been implemented and found to be capable of operating an ice-storage system for least cost under various utility rate structures (Massie, 1998). Fig. 3.40 illustrates the performance of an NN-based optimal controller for a typical office building under a strong price incentive. Specifically, the utility rate is characterized by demand and energy charge ratios of 5 (i.e., ratios of on-peak to off-peak charges). The on-peak period starts at 8:00 and ends at 19:00. As depicted in Fig. 3.40, the noncooling electrical load between 7:00 and 22:00 is dominant compared to the cooling electrical load associated with the cooling plant operation. Due to the strong incentive for load shifting, the NN-based optimal controller begins charging the ice storage system at the start of the off-peak period (i.e., 19:00.) by lowering the chiller set-point temperature to its lowest value (26°F [−3.3°C]). The charging process continues until the end of the off-peak period (i.e., 8:00). Between 8:00 and 12:00, the NN-controller opts to use the chiller to fully meet the building cooling load instead of

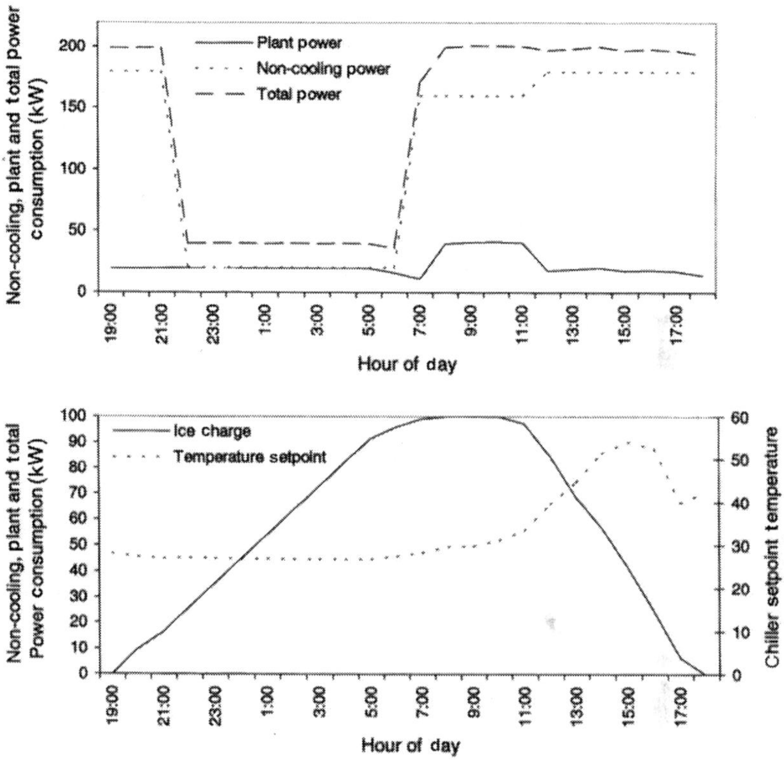

FIGURE 3.40 NN-based optimal control strategy for a typical office under strong price incentive (Massie, 1998).

discharging the ice storage system. This action is justified by the fact that the noncooling load profile allows for additional power consumption (from the chiller) during the period between 8:00 and 12:00 without any increase in the on-peak demand charges. Starting from 12:00 when the noncooling load returns to its maximum value, the NN-controller utilizes the ice storage system as well as the chiller to meet the building cooling load.

3.6 SUMMARY

In this chapter, an overview of the basic components and applications of controls suitable for building energy systems is presented. In particular, the energy cost savings incurred by various functions of ECMS have been illustrated through selected examples and applications. In addition to being knowledgeable of the currently available control systems and applications, building energy managers should be aware of the developments in the intelligent control systems especially those applicable to building energy systems.

PROBLEMS

3.1 A chiller water heating coil has a set-point of 12°C with a throttling range of 2°C. The cooling capacity of the coil varies from 0 to 10 kW. Assuming that a proportional controller is used to maintain the air temperature set-point, determine the proportional gain for the controller and the relationship between the output air temperature and the cooling rate provided by the coil. Assume steady-state operation.

3.2 Determine the reduction in the annual energy costs due to staggered duty cycling of four pump motors (each rated at 15 kW [20 hp]). The motor manufacturer specifies that a cycle of 15 min on and 10 min off as the minimum duty cycle. The utility monthly demand charge is $15/kW. First, determine the recommended NEMA duty period assuming four-pole motors.

3.3 Consider a ranch house with 100 m^2 located in Phoenix, Arizona with a total window area of 15 m^2 with double-pane glazing, which has a visible transmittance of 0.35. The home has 20 incandescent lamps each rated at 60 W.

1. Estimate the energy use and cost savings when the incandescent lamps are replaced by 13-W CFLs
2. Determine the energy use and cost savings due to daylighting harvesting strategies for both options with the existing incandescent lamps and with CFLs.

Assume that without daylighting harvesting, each lamp is operated 4 h/day when daylighting can be harvested, 6 days/week, 50 weeks/year. The cost of electricity is 0.12/kWh.

3.4 Determine the energy and cost savings of daylighting harvesting to a thin two-story building (each floor is 10 m by 20 m) with a window-to-floor area ratio of 30%. The building has clear windows (i.e., visible transmittance = 0.78) and is located in Denver, Colorado. The lighting system has a density of 0.1 W/m^2 and operates 3000 h/year. Assume that the cost of electricity is $0.10/kwh.

REFERENCES

Afram, A., Janabi-Sharifi, F., 2014. Theory and applications of HVAC control systems: a review of model predictive control (MPC). Build Environ. 72, 343–355.

Alalawi, M., Krarti, M., 2002. Experimental analysis of demand ventilation controls. ASHRAE Trans. 108 (2), 105–135.

Al Mohimen, M., Hanna, G., Krarti, M., 2005. Analysis of daylighting benefits for office buildings in Egypt. ASME Solar Energy Eng. J. 127, 366–370.

ASHRAE, 2004. Ventilation for Acceptable Indoor Air Quality, Standard 62-2004. American Society of Heating, Refrigerating and Air-Conditioning Engineers, Inc, Atlanta, GA.

ASHRAE, 2011. Handbook of HVAC Applications, American Society of Heating. Refrigerating and Air-Conditioning Engineers, Inc, Atlanta, GA.

ASHRAE, 2012. Handbook of Systems and Equipment, American Society of Heating. Refrigerating and Air-Conditioning Engineers, Inc, Atlanta, GA.

ASHRAE, 2014. Handbook of Fundamentals, American Society of Heating. Refrigerating and Air-Conditioning Engineers, Inc, Atlanta, GA.

Berglund, L., 1978. Mathematical models for predicting the thermal comfort response of building occupants. ASHRAE Trans. 84 (1).

Bloomfield, D.P., Fisk, D.J., 1977. The optimization of intermittent heating. Build Environ. 12, 43.

Braun, J.E., 1990. Reducing energy costs and peak electrical demand through optimal control of building thermal storage. ASHRAE Trans. 96 (2).

Braun, J.E., 1992. A comparison of chiller-priority, storage-priority, and optimal control of an ice system. ASHRAE Trans. 98 (1), 893–902.

Bushby, S.T., Newman, H.M., 1991. The BACnet communication protocol for building automation systems. ASHRAE J. 33 (4), 14–18.

Davidge, B., 1991. Demand controlled ventilation systems in office buildings. In: Proceedings of the 12th AIVC Conference Air Movement and Ventilation Control within Buildingspp. 157–171.

Donnini, G., Haghighat, F., Hguyen, V., 1991. Ventilation control of indoor air quality, thermal comfort, and energy conservation by CO_2 measurement. In: Proceedings of the 12th AIVC Conference Air Movement and Ventilation Control within Buildingspp. 311–331.

Drees, K.H., Braun, J.E., 1994. Modeling and Control of Area-Constrained Ice Storage Systems, Report No. 1796-2 HL 94-21. Johnson Controls, Inc.

EIA, 2016. U.S. Price of Natural Gas Sold to Commercial Consumers (Dollars per Thousand Cubic Feet). Available from: http://www.eia.gov.

EIA, 2016b. Average Retail Price of Electricity 2013. Available: http://www.eia.gov/.

Emmerich, S., Mitchell, J., Beckman, W., 1994. Demand-controlled ventilation in a multi-zone office building. Indoor Environ. 3, 331–340.

Fanger, P.O., 1970. Thermal Comfort. Danish Technical Press, Copenhagen.

Fehlmann, J., Wanner, H., Zamboni, M., 1993. Indoor air quality and energy consumption with demand controlled ventilation in an auditorium. Proceedings of the 6th International Conference on Indoor Air Quality and Climate 5, 45–50.

Fleury, B., 1992. Demand controlled ventilation: a case study. Proceedings of the 13th AIVC Conference Ventilation for Energy Efficiency and Optimum Indoor Air Quality, 343–346.

Haghighat, F., Donnini, G., 1992. IAQ and energy-management by demand controlled ventilation. Environ. Technol. 13, 351–359.

Hajiah, A., Krarti, M., 2012. Optimal controls of building storage systems using both ice storage and thermal mass–Part I: simulation environment. Energy Convers. Manage. 64, 499–508.

Henze, G.P., Krarti, M., 1996. Ice storage system controls for the reduction of operating cost and energy use. ASME Solar Eng. J. 120 (4), 275–281.

Henze, G.P., Dodier, R.H., Krarti, M.M., 1997. Development of a predictive optimal controller for thermal energy storage systems. Int. J. HVAC&R Res. 3 (3), 233–264.

Henze, G.P., Krarti, M., 1999. The impact of forecasting uncertainty on the performance of a predictive optimal controller for thermal energy storage systems. ASHRAE Trans. 105 (1).

Henze, G.P., Kalz, D., Felsmann, C., Knabe, G., 2004. Impact of forecasting accuracy on predictive optimal control of active and passive building thermal storage inventory. Int. J. HVAC&R Res. 10 (2), 153–178, American Society of Heating, Refrigerating, and Air-Conditioning Engineers, Atlanta, Georgia.

Ihm, P., Krarti, M., 2005. Optimal control strategies for heated radiant floor systems. ASHRAE Trans. 111 (1).

Jobe, T., Krarti, M., 1997. Field Implementation of optimum start heating controls. In: Proceedings for ASME Solar Eng 305.

Katipamula, S., Armstrong, P.R., Wang, W., Fernandez, N., Cho, H., Goetzler, W., Burgos, J., Radhakrishnan, R., Ahlfeldt, C., 2010. Cost-Effective Integration of Efficient Low-Lift Baseload Cooling Equipment: FY08 Final Report. PNNL-19114.

Kintner-Meyer, M., Emery, A.F., 1995. Optimal control of an HVAC system using cold storage and building thermal capacitance. Energy Build. 23 (1), 19–31.

Knoespel, P., Mitchell, J., Beckman, W., 1991. Macroscopic model of indoor air quality and automatic control of ventilation system. ASHRAE Trans. 97, 1020–1030.

Krarti, M., Brandemuehl, M., Schroeder, C., 1999. Techniques for measuring and controlling outside air intake rates in variable air volume systems. Final Report for ASHRAE RP-980, American Society of Heating. Refrigerating, and Air Conditioning Engineering, Atlanta, GA.

Krarti, M., Schroeder, C., Jeanette, E., Brandemuehl, M., 2000. Experimental analysis of measurement and control techniques of outside air intake rates in VAV systems. ASHRAE Trans. 106 (2), 234–255.

Krarti, M., Erickson, P., Hillman, T., 2005. A simplified method to estimate energy savings of artificial lighting use from daylighting. Build. Environ. 40, 747–754.

Krarti, M., Henze, G.P., 2006. Real-time predictive optimal control of active and passive building thermal storage inventory. Final Report, DOE Award DE-FC-36-03GO13026. Washington, DC.

Kusuda, T., 1976. Control of ventilation to conserve energy while maintaining acceptable indoor air quality. ASHRAE Trans. 82, 1169–1181.

Liu, S., Henze, G.P., 2006. Experimental analysis of simulated reinforcement learning control for active and passive building thermal storage inventory–Part 1: theoretical foundation. Energy Build. 38 (2), 142–147.

Massie, D., 1998. Optimal Neural Network-Based Controller for Ice Storage Systems, (Ph.D. dissertation). University of Colorado.

Menyhart, K., Krarti, M., 2016. Potential energy savings from deployment of dynamic insulation materials for US residential buildings. Build. Environ.doi: 10.1016/j.buildenv.2016.12.009.

Morgan, S., Krarti, M., 2006. Impact of electricity rate structures on energy cost savings of pre-cooling controls for office buildings. Energy Build., 38.

Morgan, S., Krarti, M., 2010. Field testing for optimal controls of TES systems. ASHRAE Trans. 116 (1).

Morris, F.B., Braun, J.E., Treado, S.J., 1992. Experimental and simulated performance of optimal control of building thermal storage. ASHRAE Trans. 100 (1), 402.

NEMA, 1994. Energy Management Guide for Selection and Use of Polyphase Motors, Standard MG-10-1994. National Electrical Manufacturers Association, Rosslyn, VA.

Ogasawara, S., Taniguchi, H., Sukehira, C., 1979. Effect of energy conservation by controlled ventilation: case study in a department store. Energy Build. 2, 3–8.

Park, B., Srubar, W.V., Krarti, M., 2015. Energy performance analysis of variable thermal resistance envelopes in residential buildings. Energy Build. 103, 317–325.

Seem, J.E., Armstrong, P.R., Hancock, C.E., 1989. Comparison of seven methods for forecasting the time to return from night setback. ASHRAE Trans. 95 (2), 439.

Seo, D., Ihm, P., Krarti, M., 2011. Development of an optimal daylighting controller. Build. Environ. 46 (5), p.1011.

Shekar, V., Krarti, M., 2017. Control strategies for dynamic insulation materials applied to commercial buildings. Energy Build 154, 305–320.

Simmonds, P., 1993. Thermal comfort and optimal energy use. ASHRAE Trans. 99 (1), 1037–1048.

Sorensen, B., 1996. Simulation of a small VAV plant. Proc. Indoor Air 2, 199–204.

Sterling, E.M., Collet, C.W., Turner, S., 1992. Commissioning to avoid indoor air quality problems. ASHRAE J. 34 (10), 28.

Vaculik, F., Plett, E., 1993. Carbon dioxide concentration-based ventilation control. ASHRAE Trans. 99, 1536–1547.

Woods, J., Winakor, G., Maldonado, E., Kipp, S., 1982. Subjective and Objective Evaluation of a CO_2-controlled variable ventilation system. ASHRAE Trans. 88, 1385–1393.

Zamboni, M., Berchtold, O., Filleux, C., Fehlmann, J., Drangsholt, F., 1991. Demand controlled ventilation–an application to Auditoria. Proceedings of the 12th AIVC Conference Air Movement and Ventilation Control within Buildings, 143–145.

Zhou, G., Ihm, P., Krarti, M., Liu, S., Henze, G.P., 2005. Integration of an internal optimization module within EnergyPlus. IBPSA Proc.

Chapter 4

Utility Rate Structures and Grid Integration

Chapter Ouline

4.1 INTRODUCTION

Energy cost is an important part of the economic viability of several energy conservation measures. Therefore, it is crucial that an energy auditor and/or building manager understands how energy costs are determined. Generally, a considerable number of utility rate structures do exist within the same geographical location. Each utility rate structure may include several clauses and charges that make following the energy billing procedure sometimes a complicated task. The complexity of utility rate structures is becoming even more acute with the deregulation of the electric industry. However, with new electric utility rate structures [such as real-time-pricing (RTP) rates], there can be more opportunities to reduce energy costs in buildings.

At the beginning of this chapter, the primary energy sources consumed in the United States are described. The presentation emphasizes the energy use

Optimal Design and Retrofit of Energy Efficient Buildings, Communities, and Urban Centers
http://dx.doi.org/10.1016/B978-0-12-849869-9.00004-1

and price by end-use sectors including residential, commercial, and industrial applications. In the United States, buildings and industrial facilities are responsible for, respectively, 36% and 38% of the total energy consumption. The transportation sector, which accounts for the remaining 26% of the total US energy consumption, uses mostly fuel products. However, buildings and industries consume predominantly electricity and natural gas. Coal is primarily used as an energy source for electricity generation due its low price.

Throughout the chapter, various features of utility rate structures are outlined. More emphasis is given to the electrical rate structures since a significant part of total energy cost in a typical facility is attributed to electricity. However, the price rate structures of other energy sources are discussed. The information provided in this chapter is based on recent surveys of existing utility rate structures. However, the auditor should be aware that most utilities revise their rates on a regular basis. If detailed information is required on the rates available from a specific utility, the auditor should contact the utility directly.

At the end of the chapter, an overview of microgrids and the integration of renewable energy generation is discussed. In particular, types of microgrids are outlined with their operation modes. Some of the challenges and procedures are presented when the interconnection of microgrids with the utility grid is considered.

4.2 ENERGY RESOURCES

The sources of energy used in the US include coal, natural gas, petroleum products, and electricity. Electricity can be generated from either power plants fueled from primary energy sources (i.e., coal, natural gas, or fuel oil) or from nuclear power plants or renewable energy sources (such as hydroelectric, geothermal, biomass, wind, photovoltaic, and solar thermal sources).

In the US, the energy consumption has fluctuated in response to significant changes in oil prices, economic growth rates, and environmental concerns especially since the oil crisis of the early 1970s. Specifically, the US energy consumption increased from 73 quadrillion British thermal units (Btu) in 1982–97 quadrillion Btu in 2016 but has stabilized since early 2000 (EIA, 2017). Table 4.1 summarizes the changes in the US energy consumption by source from 1982 to 2016.

It is clear from the data summarized in Table 4.1 that the total consumption of fossil fuels has not increased significantly over the last two decades and has even decreased from 83.7 quadrillion Btu in 2002 to 78.6 quadrillion Btu in 2016. However, the US consumption associated with renewable energy has increased steadily from 6.0 in 1982 to 10.2 quadrillion Btu in 2016. It should be noted that hydroelectric power is the main source for the renewable energy contribution. The contribution of nuclear power did not change since 2002 with about 8.0 quadrillion Btu. Indeed, only one nuclear reactor entered service since 1996, the Watts Bar Unit 2 that began operation in 2016 with a rated power

TABLE 4.1 Annual US Energy Consumption by Primary Energy Sources in Quadrillion Btu

Primary energy source	1982	1992	2002	2012	2016
Fossil fuels	63.888	73.396	83.700	77.487	78.569
Nuclear power	3.131	6.479	8.145	8.062	8.422
Renewable energy	5.980	5.821	5.726	8.706	10.161
Total	73.099	85.782	97.643	94.416	97.394

Source: EIA, 2017. Annual Energy Review. Department of Energy, Energy Information Administration. http://www.doe.eia.gov/.

of 1150 MW. The average age of the 100 operating US commercial nuclear reactors is estimated to be 36 years with a total power generation capacity of 100,350 MW (EIA, 2017).

Fig. 4.1 shows the annual total energy use by sector since 1950 (EIA, 2017). It is clear that the energy consumption for the industrial sector has been decreasing for the last two decades since reaching its peak in late 1990s. The energy consumption for the building sector has stabilize since 2005.

Table 4.2 provides the average energy prices for each primary fuel type. Over the years, coal remains the cheapest energy source. The cost of oil has decreased since reaching high values during 2008–14 period. As illustrated in Table 4.2, the prices of all energy sources have remained stable or decreased since 2012. In particular, the cost of natural gas has actually decreased over the last few years.

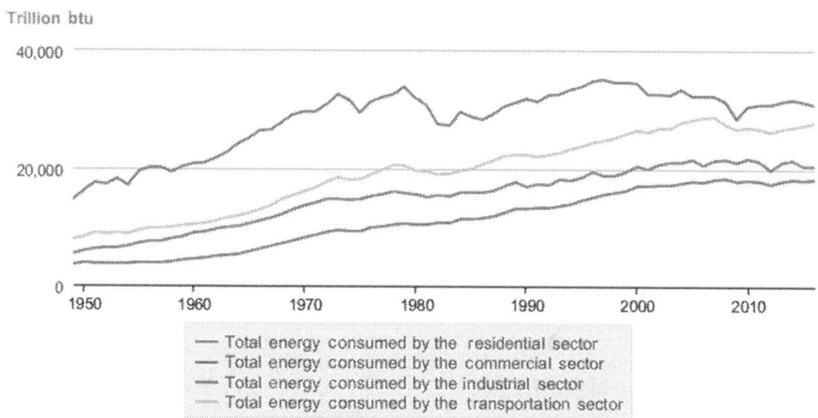

FIGURE 4.1 Annual sectoral US energy consumption since 1950. *(EIA, 2017. Annual Energy Review. Department of Energy, Energy Information Administration. http://www.doe.eia.gov/).*

TABLE 4.2 Consumer Price Estimates for Energy in Real Values in US $/Million Btu

Energy source	1982	1992	2002	2012	2016
Coal	2.2	1.3	0.9	1.6	1.5[a]
Oil	10.6	6.8	6.5	13.4	7.6
Natural gas	5.2	4.1	4.3	4.5	4.0
Electricity	21.0	17.2	13.8	15.2	15.3

[a] *This price was estimated by Quandl (2017) since EIA does not provide coal prices after 2012.*
Source: EIA, 2017. Annual Energy Review. Department of Energy, Energy Information Administration. http://www.doe.eia.gov/.

4.2.1 Electricity

4.2.1.1 Overall Consumption and Price

In the US, the use of coal to power electrical power plants is steadily decreasing as shown in Table 4.3. However, gas-fired power plants are becoming more common in generating electricity due to more efficient, cleaner, and reliable combustion turbines.

The electricity sold by US utilities has remained mostly stable for the residential and commercial sectors as indicated by the data summarized in

TABLE 4.3 Annual US Electric Energy Generated by Utilities by Primary Energy Sources in Billion kWh

Primary energy source	1972	1982	1992	2002	2012	2016
Coal	771	1192	1576	1933	1514	1240
Natural gas	376	305	264	691	1226	1380
Petroleum products	274	147	89	95	23	24
Nuclear power	54	283	619	780	769	805
Renewable energy	274	314	254	343	516	630
Total	1749	2241	2797	3,859	4048	4079

Source: EIA, 2017. Annual Energy Review. Department of Energy, Energy Information Administration. http://www.doe.eia.gov/.

TABLE 4.4 Annual US Electric Energy Sold by Utilities by Sector in Billion kWh

End-use sector	1972	1982	1992	2002	2012	2016
Residential	539	730	936	1265	1375	1407
Commercial	359	526	850	1205	1327	1360
Industrial	641	745	973	990	986	936

Source: EIA, 2017. Annual Energy Review. Department of Energy, Energy Information Administration. http://www.doe.eia.gov/.

Table 4.4. The electricity consumption could be higher without the various energy conservation programs implemented by the Federal or State governments and utilities. For instance, it is estimated that the demand-side-management (DSM) programs provided by utilities have saved about 54 billion kWh in electrical energy use during 2002 and over 69 billion kWh in 2007 (EIA, 2009).

While the contribution of renewable energy systems remain small in US electricity generation and consumption, it has been growing at a respectable rate as shown in Fig. 4.2 especially for geothermal, wind, and solar since early 2000s.

The prices of electricity for all end-use sectors have remained stable between 1992 and 2008 after a recovery period from the 1973 energy crisis as illustrated in Table 4.5. As expected, industrial customers enjoyed the lowest electricity price over the years. Meanwhile, the cost of electricity for residential customers remained the highest.

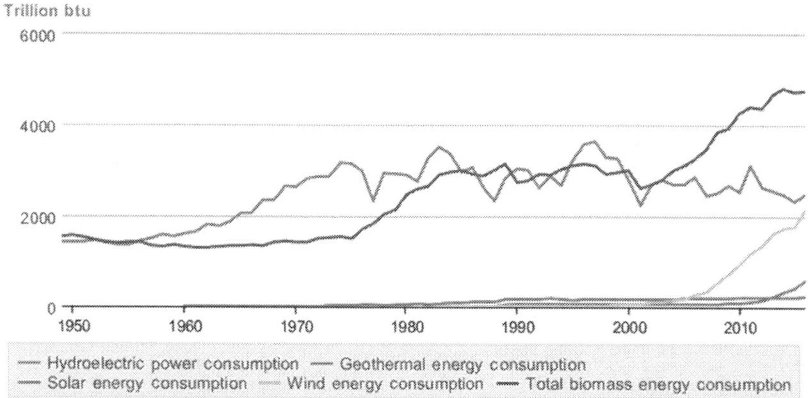

FIGURE 4.2 Annual energy consumption attributed to various renewable energy sources. *(EIA, 2017. Annual Energy Review. Department of Energy, Energy Information Administration. http://www.doe.eia.gov/).*

TABLE 4.5 Average Retail Prices of Electric Energy Sold by US Utilities by Sector in Cents per kWh

End-use sector	1972	1982	1992	2002	2012	2016
Residential	7.9	11.0	9.5	8.1	11.88	12.50
Commercial	7.5	11.0	8.9	7.6	10.09	10.37
Industrial	4.1	8.0	5.6	4.7	6.67	6.75

Source: EIA, 2017. Annual Energy Review. Department of Energy, Energy Information Administration. http://www.doe.eia.gov/.

4.2.1.2 Future of US Electricity Generation

Currently, the electricity market is in the midst of a restructuring period and is becoming increasingly competitive. Several innovative technologies are being considered and tested to generate electricity. A relatively recent approach to produce electricity using small and modular generators is the distributed generation concept. The small generators with capacities in the range of 1 kW–10 MW can be assembled and relocated in strategic locations (typically near customer sites) to improve power quality and reliability, and flexibility to meet a wide range of customer and distribution system needs.

A number of technologies have emerged in the last decade that allows generation of electricity with reduced waste, cost, and environmental impact. It is expected that these emerging technologies will improve the viability of distribution generation in a competitive deregulated market. Among these technologies are fuel cells, microturbines, combustion turbines, gas engines, and diesel engines. Chapter 7 discusses some of the emerging technologies in electricity generation.

4.2.1.3 Utility Deregulation Impact

Following the acceptance of the Energy Policy Act of 1992, which requires open access to utility transmission lines, the US Federal Energy Regulatory Commission issued orders to allow the establishment of a wholesale power market with independent system operators. As a result of significant increase in the quantities of bulk power sale, delivery of energy to users has become increasingly difficult especially through the existing transmission and distribution networks. The frequent power outages, experienced in the last few years especially in western US, illustrates the precarious stability of the transmission system.

Moreover, several States have started to implement retail access, which allows customers to choose among several electric service providers based on a competitive market that may offer a variety of customized services such as premium power quality. Unfortunately, existing distribution networks are not designed to support multiple suppliers or to channel value-added services. Ironically, utilities have built fewer than half the transmission capacity between

1990 and 1995 they built in the previous 5 years (1985–1990). This reduction of investment in the transmission grid is largely due to the uncertainty about ongoing electric utility deregulation and restructuring.

In addition to adding new transmission capacity, it is believed that existing transmission and distribution networks and their control have to be upgraded using advanced and new technologies to ensure high reliability and safety of power delivery systems. Among the technologies that are being considered to upgrade the power delivery system are the following:

- Super-conducting transmission cables: the discovery of high-temperature superconducting (HTS) materials using ceramic oxides in 1986 has lowered the cost of superconducting transmission cables to a reasonable level. It is estimated that an HTS cable could carry 500 MW of electric power at voltages as low as 50 kV.
- High-voltage electronic flexible AC transmission system (FACTS) controllers are now used by several utilities to increase the capacity of transmission lines and improve overall delivery system reliability. Unlike conventional electromechanical controllers, FACTS controllers are sufficiently fast to reduce bottlenecks and transient disturbances in power flow and thus reduce transmission system congestion and improve overall delivery reliability.
- Cost-effective distributed generation and storage technologies that offer flexibility to meet a wide variety of customer needs. Among distribution generation systems under development and testing are microturbines with capacities ranging from 10 to 250 kW, and fuel cells which offer clean, efficient, compact, and modular generation units.
- Diversified and integrated utility services to meet the divergent needs of various market segments. For instance, innovative rate structures such as RTP rates are being offered to customers that are demanding lower rates. Moreover, some utilities are integrating electricity with other services such as Internet access, telecommunications, and cable television using fiber-optic networks. However, the move to integrate utility functions requires new hardware and software technologies. For instance, low-cost electronic meters with two-way communications are needed to provide real-time pricing and billing options for multiple utility services.

4.2.2 Natural Gas

As indicated in Table 4.6 and Fig. 4.3, the total US consumption of natural gas has actually declined between 1972 and 1990. However, after this decline, natural gas consumption has dramatically increased since early 2000 due to the electric power sector as illustrated in Fig. 4.3. Moreover, the industrial sector has started to use more natural gas since 2010. As indicated in Table 4.7, the prices of natural gas remained generally stable in the last two decades. During the 2010s, the prices of natural gas have actually decreased since the gas supplies became more certain.

TABLE 4.6 Annual US Consumption of Natural Gas by Sector in Trillion Cubic Feet

End-use sector	1972	1982	1992	2002	2012	2016
Residential	5.13	4.63	4.69	4.90	4.15	4.39
Commercial	2.61	2.61	2.80	3.14	2.90	3.12
Industrial	9.62	6.94	8.70	8.62	8.62	9.28

Source: EIA, 2017. Annual Energy Review. Department of Energy, Energy Information Administration. http://www.doe.eia.gov/.

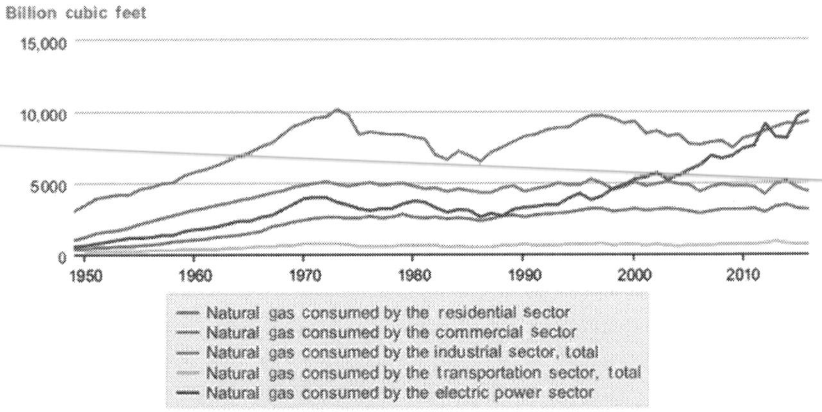

FIGURE 4.3 Annual sector US natural gas use. *(EIA, 2017. Annual Energy Review. Department of Energy, Energy Information Administration. http://www.doe.eia.gov/).*

TABLE 4.7 Average Retail Prices of Natural Gas by Sector in Dollars per 1000 Cubic Feet

End-use sector	1972	1982	1992	2002	2012	2016
Residential	4.01	8.24	6.82	7.57	10.65	10.06
Commercial	2.92	7.68	5.65	6.36	8.10	7.25
Industrial	1.49	6.17	3.29	3.86	3.88	3.51

Source: EIA, 2017. Annual Energy Review. Department of Energy, Energy Information Administration. http://www.doe.eia.gov/.

In the future, it is expected that the natural gas market will continue to expand and its price to be competitive. In particular, the future for natural gas as a primary energy source for electricity generation is considered to be promising as a cleaner alternative to coal and oil products. Indeed, gas-fired power plants are competitive because of their high efficiencies (approaching 50%) and are environmentally attractive since they produce significantly lower carbon and sulfur emissions than plants powered by coal or oil.

4.2.3 Petroleum Products

Overall, the US consumption of fuel oil and other petroleum products have remained stable between 1972 and 2016 as indicated in Table 4.8. However, the oil price fluctuated significantly over the last three decades after the 1973 energy crisis. In the building sector (i.e., residential and commercial applications), the US consumption level of petroleum products has remained small and stable over the years. However, the use of petroleum fuels has steadily increased in the transportation sector.

Table 4.9 clearly indicates that after a drastic increase (of almost four folds) between 1972 and 1982, the crude oil prices have decreased in 1998 to levels even lower than those experienced in 1972. However, another significant in-

TABLE 4.8 US Consumption of Petroleum Products by Sector in Million Barrels per Day

End-use sector	1972	1982	1992	2002	2012	2016
Residential/ Commercial	2.25	1.24	1.19	1.20	0.82	1.04
Industrial	4.19	4.06	4.52	4.93	4.54	4.61
Transportation	8.57	9.31	10.88	13.21	13.03	13.66
Electric utilities	1.36	0.69	0.43	0.43	.10	.11
Total	16.37	15.30	17.03	19.77	18.76	19.42

Source: EIA, 2017. Annual Energy Review. Department of Energy, Energy Information Administration. http://www.doe.eia.gov/.

TABLE 4.9 Average Crude Oil Price in the US in 2000 Dollars per Barrel

Year	1972	1982	1992	2002	2012	2016
Price	11.24	45.47	18.51	21.61	94.52	38.29

Source: EIA, 2017. Annual Energy Review. Department of Energy, Energy Information Administration. http://www.doe.eia.gov/.

crease in oil prices occurs between 2006 and 2014. Since 2015, the oil prices have significantly lowered and are fluctuating around $40/barrel.

4.2.4 Coal

In the US, coal is primarily used as an energy source for power generation by electric utilities as shown in Table 4.10. Indeed, the total US consumption of coal has increased between 1972 and 2008 due primarily to the growth in coal use by electric utilities. However, coal consumption in the US has been steadily decreasing since 2010 in all sectors (i.e., buildings, industrial, and power). This decrease is expected to be maintained in the near future for all sectors. In fact, the share of electricity generation attributed to coal is reduced due to the reliance on other generation technologies and resources (i.e., natural gas) as discussed in Section 4.2.1.

The abundant coal reserve base and the lingering excess production capacity have helped maintain low coal prices especially during the last decade as indicated by Table 4.11. In the future, however, coal price is expected to rise slowly due to reserve depletion and slow growth in labor productivity. The higher coal prices coupled with the environmental concerns may cause a future decline of coal consumption in the US. Fig. 4.4 shows the annual US coal consumption

TABLE 4.10 Annual US Consumption of Coal by Sector in Millions of Short Tons

End-use sector	1972	1982	1992	2002	2012	2016
Residential/ commercial	11.7	8.2	6.2	4.4	2.1	1.2
Industrial	160.1	103.0	106.4	84.4	63.59	51.15
Electric utilities	351.8	593.7	795.1	977.5	823.5	677.3
Total	524.3	706.9	907.7	1066.4	889.2	727.6

Source: EIA, 2009. Annual Energy Review. Department of Energy, Energy Information Administration. http://www.doe.eia.gov/.

TABLE 4.11 Average Coal Price in the US in per Short Ton

Year	1972	1982	1992	2002	2012[a]	2016[a]
Price	25.59	43.44	24.34	17.26	71.0	49.0

[a] *The prices for 2012 and 2016 were estimated from Quandl (2017) since EIA does not provide coal prices since 2011.*
Source: EIA, 2017. Annual Energy Review. Department of Energy, Energy Information Administration. http://www.doe.eia.gov/.

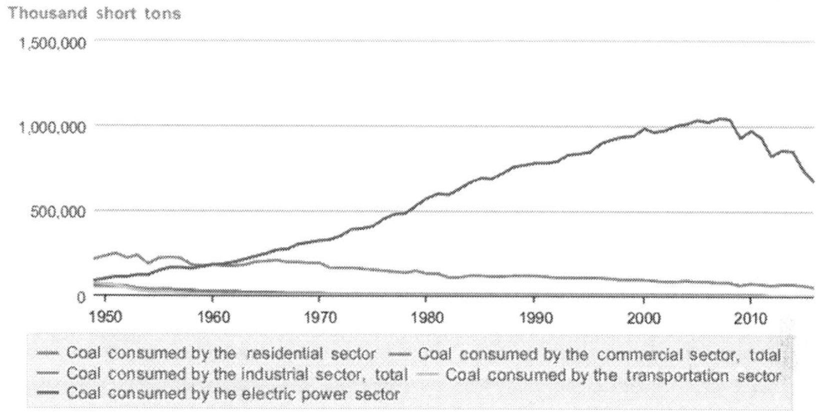

FIGURE 4.4 Annual sectoral US coal consumption. *(EIA, 2017. Annual Energy Review. Department of Energy, Energy Information Administration. http://www.doe.eia.gov/).*

for various sectors. It is clear that coal use in the US has been declining significantly since 2007 especially for generating electrical power.

4.3 ELECTRICITY RATES

To generate electricity, utilities have to consider several operating costs to determine their rates. Typically, an electrical utility is faced with the following cost items:

- *Generation Plant:* The cost of operating a power plant to generate electricity represents typically the highest cost category. Indeed, power generation plants have to meet several regulations and safety requirements. For instance, several plants are required to meet strict pollution standards especially if they operate in highly populated areas. Meeting these regulations can increase significantly the generating costs.
- *Transmission/Distribution Systems:* To deliver electricity from the generation plant where it is produced to areas where it is utilized, transmissions lines, substations, and distribution networks have to be used. The cost of the transmission/distribution systems depends on the distances to be covered as well as transformers, capacitors, and meters to be used. Moreover, the delivery energy losses can be a significant part of the transmission/distribution costs.
- *Fuel Costs:* Electricity is generated using a primary fuel source depending on the power plant. The fuel cost can be small as in case of hydroelectric plants or significant as in conventional fuel oil or coal power plants. The cost of fuel may fluctuate depending on the world markets.
- *Administrative Costs:* The salaries of management, technical, and office staff as well as insurance and maintenance costs for the power plant equipment are part of the administrative costs. These administrative costs can be large especially for nuclear power plants.

Other factors that affect the cost of electricity include the generating capacity of the utility and the demand/supply conditions at a given time (i.e., on-peak and off-peak periods).

Utilities allocate the cost of electricity differently depending on the type of the customers by offering various rate schedules. Three customer types are generally considered by utilities: (1) residential, (2) commercial, and (3) industrial customers. Each utility may offer several rate structures for each customer type. It is therefore important that the auditor knows the various rate structures that can be offered to the audited facility.

Utilities can tailor their rates to customer needs for electricity using several methods. Some of the common rate structures used by US utilities are as follows:

- block pricing rates
- seasonal pricing rates
- innovative rates

In addition to these rates, utilities provide some riders and discounts to their customers. A rider may modify the structure of a rate based on specific qualifications of the customer. For instance, utilities can change summer energy charge of their residential customers using an air-conditioning rider. Moreover, utilities can reduce energy and/or demand charges using discount provisions. For instance, several utilities offer voltage discount when the customer is willing to receive voltages higher than the standard voltage level. To benefit from this discount, the customer may have to install and maintain a properly sized transformer.

More detailed discussion of each of the commonly available rates is provided in the following sections with some examples to illustrate how an energy auditor can apply the utility rate provisions to check the calculation of utility bills. First, common features of all electric utility rates are discussed in details.

4.3.1 Common Features of Utility Rates

There are several utility rate features and concepts that the auditor should be familiar with to be able to interpret and analyze correctly the utility billing procedure. Some of these concepts are described in the following sections.

4.3.1.1 Billing Demand

The demand that is billed by the utility is referred to as the billing demand. The billing demand is often determined from the peak demand obtained for one month (or any billing cycle). The peak demand, also known as actual demand, is defined as the maximum demand or maximum average measured demand in any 15-min period in the billing cycle. To better understand the concept of billing demand, consider two different monthly load profiles: a rugged profile A and a flat profile B as illustrated in Fig. 4.5. It is further assumed that the average demand for profile A coincides with that of profile B (and thus the total energy use in kWh for both profiles A and B is the same). It is not equitable for the util-

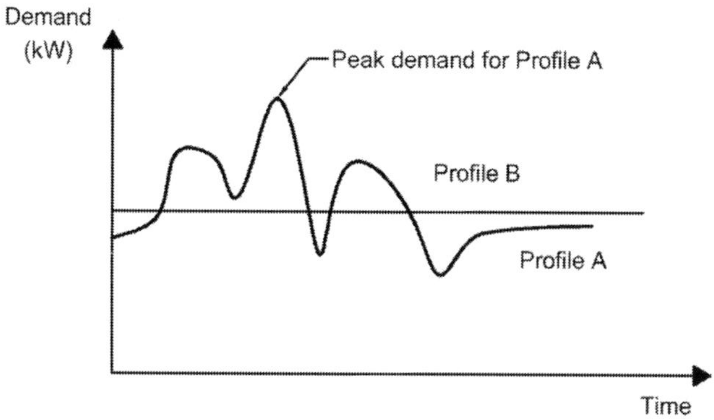

FIGURE 4.5 Peak demand for two different electric load profiles.

ity to bill the same charges for the two profiles. Indeed, profile A requires that the utility supplies higher demand and thus increases its generation capacity for only a short period of time. Meanwhile, profile B is ideal for the utility since it does not change over time. Therefore, some utilities charge their customers for the peak demand incurred during the billing period. This demand charge may serve as an incentive to the customers to reduce or "shave" their peak demand.

In some rates, the billing demand is determined based on the utility specifications and may be different from the maximum demand actually measured during a billing period. For instance, ratchet and power factor clauses can change the determination of the billing demand. Moreover, a minimum billing demand can be specified in a contractual agreement between the utility and the customer. This minimum demand is often called the contract demand.

4.3.1.2 Power Factor Clause

The power factor is defined as the ratio of actual power used by the consumer (expressed in kW) to the total power supplied by the utility (expressed in kVA). Fig. 4.6 shows the power triangle to illustrate the concept of power factor.

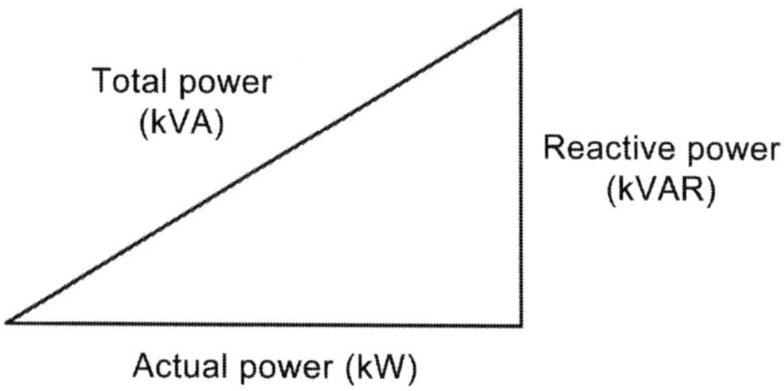

FIGURE 4.6 Power triangle diagram.

For the same actual power consumed by two customers but with different power factor values, the utility has to supply higher total power to the customer with the lower power factor. To penalize for low power factors (generally lower than 0.85), some utilities use the power factor clause to change the billing demand or to impose new charges (on total power demand or reactive power demand) depending on the power factor. For instance, the billing demand can be increased whenever the measured power factor of the customer is below a reference value or base power factor:

$$\text{Billed demand} = \text{Actual demand} * \left(\frac{\text{pf}_{\text{base}}}{\text{pf}_{\text{actual}}} \right) \tag{4.1}$$

$$\text{Power factor (pf)} = \frac{\text{Actual power}}{\text{Total power}}$$

Example 4.1 illustrates how the power factor clause may affect the utility bill.

Example 4.1

A utility has the following monthly billing structure for their industrial customers:

- Customer charge: $450/month
- Demand charge: $20/kW
- Energy charge: $0.03/kWh

A power factor clause states that the billing demand, upon which the demand charge is based, shall be the actual demand for the month corrected for the power factor by multiplying the measured actual demand by 80 and divide the product by the actual average power factor expressed in percent. The power factor clause applies whenever the average monthly power factor is less than 80%.

1. Calculate the utility bill for an industrial facility with the following energy use characteristics during a specific month:
   - Actual demand: 300 kW
   - Energy consumption: 50,000 kWh
   - Average monthly power factor: 60%
2. Determine the cost savings achieved for the month if the facility power factor is improved to be always above 80%.

Solution

1. First, the billing demand for the month is determined using the power factor clause since the average power factor is below 80%:

$$\text{Billed demand} = \text{Actual demand} * 80 / \text{pf} = 300 * 80 / 60 = 400 \text{ kW}$$

Then, the monthly bill can be calculated using the utility rate structure:

(a) Customer charge	$450.00
(b) Demand charge = 400 kW*$20/kW	$8000.00
(c) Energy charge = 50,000 kWh*$0.03/Wh	$1500.00
Total monthly charges = (a) + (b) + (c)	$9950.00

The actual bill may include other components such as fuel adjustment cost and sales taxes.

2. The cost savings from the improved power factor are due to a reduction in the billing demand. Indeed, the billing demand for the considered month will be the actual demand (i.e., 300 kW). Thus, a reduction of 100 kW in billing demand results in a reduction of demand charges of

$$\text{Cost savings} = 100\ \text{kW} * \$20/\text{kW} = \$2000/\text{month}$$

The reader is referred to Chapter 5 to determine the cost-effectiveness for power factor improvement measures.

4.3.1.3 Ratchet Clause

Typically, the utility charges are billed monthly. In particular, the demand charges are based on monthly peak demand. However, when the peak demand for one month is significantly higher than for the other months (such as the case for buildings with high cooling loads in the summer months), the utility has to supply the required peak demand and thus operate additional generators only for 1 or 2 months. For the rest of the year, the utility will have to maintain these additional generators. To recover some of this maintenance cost and to encourage demand shaving, some utilities use a ratchet clause in the determination of the billed demand. For instance, the billed demand for any given month is a fraction of the highest maximum demand of the previous 6 months (or 12 months) or the actual demand incurred in the month. Example 4.2 illustrates the calculation procedure of the utility bill with a ratchet clause.

Example 4.2

The utility has added a ratchet clause to the rate structure described in Example 4.1. Specifically, the ratchet clause states that no billing demand shall be considered as less than 70% of the highest on-peak season maximum demand corrected for the power factor previously determined during the 12 months ending with the current month.

Calculate the utility bill for the industrial facility considered in Example 4.1 taking into account the ratchet clause. Assume that the previous highest demand (during the last 12 months) is 700 kW.

Solution

First, the minimum billing demand determined by the ratchet clause is calculated:

Minimum billing demand (ratchet clause) = 700 kW*0.70 = 490 kW

Since the billing demand based on the power factor clause was found to be 400 kW (see Example 4.1), the billing demand for the month is

$$\text{Billed demand} = \max(490\ \text{kW}, 400\ \text{kW}) = 490\ \text{kW}$$

Then, the monthly bill can be calculated using the utility rate structure:

(a) Customer charge	$450.00
(b) Demand charge = 490 kW*$20/kW	$9800.00
(c) Energy charge = 50,000 kWh*$0.03/Wh	$1500.00
Total monthly charges = (a) + (b) + (c)	$11,750.00

Thus, the ratchet clause increases the utility bill for the month.

4.3.1.4 *Fuel Cost Adjustment*

Most utilities have to purchase primary energy sources (fuel oil, natural gas, and coal) to generate electricity. Since the cost of these commodities change over time, the utilities impose an adjustment to their energy charges to account for any cost variation of their primary energy sources. Generally, utilities provide in the description of their rate structure a formula that they use to calculate the fuel cost adjustment. In addition to fuel cost adjustment, utilities may levy taxes and surcharges to recover imposts required from them by federal or state governments or agencies. In some cases, the fuel cost adjustment can be a significant proportion of the utility bill. Example 4.3 illustrates the effect of both the fuel cost adjustment and sales tax on a monthly utility bill.

Example 4.3

The utility has added a ratchet clause to the rate structure described in Example 4.2. Specifically, the ratchet clause states that no billing demand shall be considered as less than 70% of the highest on-peak season maximum demand corrected for the power factor previously determined during the 12 months ending with the current month.

Calculate the utility bill for the industrial facility considered in Example 4.2 taking into account both a fuel cost adjustment of 0.015/kWh and a sales tax of 7%.

Solution

The fuel cost adjustment should be applied to the energy use, while the sales tax should be applied to the total cost. Thus, the monthly bill for the industrial facility considered in Example 4.2 becomes

(a) Customer charge	$450.00
(b) Demand charge = 490 kW*$20/kW	$9800.00
(c) Energy charge = 50,000 kWh*[$0.03/kWh +$0.015/kW]	$2250.00
Total monthly charges (before sales tax) = (a) + (b) + (c)	$12,500.00
Total monthly charges (after sales tax) = $12,500.00*1.07	$13,375.00

The cumulative effect of both fuel cost adjustment and sales tax increased the utility bill for the month by [$13,375 − $11,750 = 1625].

4.3.1.5 Service Level

Utilities typically offer several rate structures for a given customer depending on the type of service. For instance, utilities may have different rates depending on the voltage level provided to the customers. The higher the delivery voltage level, the cheaper is the energy rate. In particular, utilities offer reduced rate for demand and/or energy charges to customers that own their service transformers This type of rate is generally described under various clauses of a utility bill including customer-owned transformer, voltage level, or service type. The owner has to determine the cost-effectiveness of owning a transformer and thus receive a higher voltage level by comparing the utility bill savings against the cost of purchasing, leasing, and maintaining the transformer. Typically, transformer energy losses are charged to the owner. In addition, the owner should have a standby transformer or make another arrangement (such as standby utility service) in case of breakdown of the service transformer. Example 4.4 presents an estimation of savings in utility bills due to owning a service transformer.

Example 4.4

The industrial facility of Example 4.1 has an option of owning and operating its own service transformer with the advantage of reduced rate structure as described here:

- Customer charge: \$650/month
- Demand charge: \$15/kW
- Energy charge: \$0.025/kWh

Calculate the reduction in the utility bill during the month for which the energy use characteristics are given in Example 4.1. In the savings calculation, do not consider the power factor clause, ratchet clause, sales tax, and fuel cost adjustment.

Solution

First, the utility bill is determined in the case the facility owns its service transformer:

(a) Customer charge	\$650.00
(b) Demand charge = 400 kW*\$15/kW	\$6000.00
(c) Energy charge = 50,000 kWh*\$0.025/Wh	\$1250.00
Total monthly charges = (a) + (b) + (c)	\$7900.00

Thus, the savings in the utility bill for the month is

$$\text{Utility bill savings} = \$9950 - \$7900 = \$2050$$

The reader is referred to Chapter 5 to determine the cost-effectiveness of replacing or owning transformers.

4.3.2 Block Pricing Rates

In these rates, the energy price depends on the rate of electricity consumption using either inverted blocks or descending blocks. An inverted block pricing rate structure increases the energy price as the consumption increases. On the other hand, a descending (also referred to as a declining) block rate structure reduces the price as the energy consumption increases. Typically, the rate is referred to as a "flat" rate when the energy price does not vary with the consumption level. Tables 4.12 and 4.13 show examples of block pricing rates for energy and demand charges. Examples 4.5 and 4.6 illustrate the calculation details of utility bills.

Based on a comprehensive survey of existing rate structures offered by US utilities (GRI, 1993), block pricing rate structures are commonly used. Indeed, almost 60% of the surveyed electric utilities offer descending or inverted block pricing structure for the energy charges. For the residential customers, a combination of descending, flat, and inverted rate structures are used throughout the US. However, descending energy rate structures are used almost exclusively for commercial and industrial customers. Some electric utilities offer block pricing rates for demand as well as energy charges.

Example 4.5

Using utility rate structure A, calculate the utility bill for a residence during a summer month when the energy use is 800 kWh.

Solution

Considering all the charges imposed by utility rate A (see Table 4.12), the monthly bill can be calculated as follows:

(a) Customer charge	\$7.50
(b) Energy charge = 400 kWh*\$0.0874/kWh + 400 kWh*\$0.1209/kWh	\$83.32
(c) Fuel cost adjustment	\$0.00
(d) Taxes = 6.544% [(a) + (b) + (c)] =	\$5.94

Total monthly charges = (a) + (b) + (c) + (d) = \$96.76

Thus, the average cost of electricity for the month is \$96.76/800 kWh = \$0.12095/kWh.

4.3.3 Seasonal Pricing Rates

Some electric utilities offer seasonal rate structures to reflect the monthly variations in their generation capacity and energy cost differences. Generally, the utilities that provide seasonal rate structures use different energy and/or demand

TABLE 4.12 An Electric Utility Rate With No-demand Charges for Residential General Service (Utility Rate A)

Billing item	Winter (November–April)	Summer (May–October)
Customer charge	$7.50	$7.50
Minimum charge	$7.50	$7.50
Fuel cost adjustment	0.00	0.00
Tax rate	6.544%	6.544%
No. of energy blocks	1	3
Block 1 energy size (kWh)	0	400
Block 2 energy size (kWh)	0	400
Block 3 energy size (kWh)	0	800
Block 1 energy charge ($/kWh)	0.088	0.0874
Block 2 energy charge ($/kWh)	0.000	0.1209
Block 3 energy charge ($/kWh)	0.000	0.1406

charges during winter and summer months. The summer charges are typically higher than winter charges for most electric utilities due to higher energy consumption attributed to cooling of buildings. Tables 4.14 and 4.15 show examples of seasonal pricing rates while Examples 4.5 and 4.6 illustrate the calculation procedure for monthly utility bills.

Based on a survey conducted by GRI (1993), more than 55% of US electric utilities offer seasonal pricing rates for residential customers. Only 5% of electric utilities had residential rates where the winter rate is actually higher than the summer rate. These utilities are located in the northeast and the west regions of the US. The same survey reveals that over 42% of US utilities use seasonal pricing rates for commercial and industrial customers. Only 7% of the utilities surveyed offer rates with higher winter prices for their commercial and industrial customers.

Example 4.6

Using utility rate structure B, calculate the utility bill for a commercial facility during a winter month when the energy use is 70,000 kWh, the billing demand is 400 kW, and the average reactive demand is 150 kVAR.

Solution

Accounting for all the charges considered by utility rate B (see Table 4.13), the electric energy bill for the winter month can be calculated as follows:

(a) Customer charge	\$0.00
(b) Energy charge = $40,000\,kWh * \$0.059/kWh + 30,000\,kWh * \$0.042/kWh$	\$3620.00
(c) Demand charge = $50\,kW * \$12.39/kW + 350\,kW * \$11.29/kW$	\$4571.00
(d) Fuel cost adjustments = 70,000 kWh*\$0.01605	\$1123.50
(e) Reactive demand charge = 150 kVAR*\$0.20/kVAR	\$30.00
(f) Taxes	\$0.00
Total monthly charges = (a) + (b) + (c) + (d) + (e) + (f)	\$9344.50

The average cost of electricity for the month is then \$9,344.50/70,000 kWh or \$.1335/kWh.

TABLE 4.13 An Electric Utility Rate With Demand Charges for Commercial General Service (Utility Rate B)

Billing item	Winter	Summer
	(October–May)	(June–September)
Customer charge	0.00	0.00
Minimum charge	\$25.00	\$25.0
Fuel cost adjustment	0.01605	0.01377
Tax rate	0.00	0.00
No. of energy blocks	3	3
Block 1 energy size (kWh)	40000	40000
Block 2 energy size (kWh)	60000	60000
Block 3 energy size (kWh)	>100000	>100000
Block 1 energy charge (\$/kWh)	0.059	0.065
Block 2 energy charge (\$/kWh)	0.042	0.047
Block 3 energy charge (\$/kWh)	0.039	0.042
No. of demand blocks	2	2
Block 1 demand size (kW)	50	50
Block 2 demand size (kW)	>50	>50
Block 1 demand charge (\$/kW)	12.39	13.71
Block 2 demand charge (\$/kW)	11.29	12.52
Reactive demand charge (\$/kVAR)	0.20	0.20

4.3.4 Innovative Rates

Due to the increased focused on integrated resource planning, DSM, and competitive energy market, several innovative rates have been implemented by utilities. These innovative rates have the main objective to profitably meet customer needs. Moreover, some utilities foster new technologies through the use of innovative rates to retain their customers.

There are several categories of rates that can be considered to be innovative rates. In the US, innovative rates can be classified into seven categories (GRI, 1999).

4.3.4.1 Time-of-Use Rates

The time-of-use (TOU) rates are time-differentiated rates with the cost of electricity varying during specific times of the day and/or the year. The TOU rates, which appeared first in the 1940s, set "on-peak" and "off-peak" periods with different energy and/or demand charges. Generally, the on-peak periods occur during daytime hours and have higher cost of energy and demand than the off-peak periods, which occur during the night time. Table 4.14 presents a TOU rate structure for residential customers. Example 4.7 illustrates the calculation approach of a utility bill for the TOU rate presented in Table 4.14.

Example 4.7

Using utility rate structure C, calculate the utility bill for a residence during a summer month when the energy use is 2000 kWh with 55% of the energy consumption occurring during the on-peak period.

TABLE 4.14 An Electric Time-of-Use Rate for Residential Service (Utility Rate C)

Billing Item	Winter	Summer
	(October–May)	(June–September)
Customer charge	$9.85	$9.85
Minimum charge	$9.85	$9.85
Tax rate	3.00%	3.00%
Energy charge ($/kWh)		
On-peak hours:	0.1412	0.1500
Off-peak hours:	0.0335	0.0335

On-peak hours are defined as: 6:00 a.m. to 1:00 p.m. and 4:00–9:00 p.m. (M–F) during October–March; 10:00 a.m. to 9:00 p.m. (M–F) during April–September. Off-peak hours are all the remaining hours including holidays.

Solution

With all the charges considered by utility rate C (see Table 4.14), the electric energy bill for the winter month can be calculated as follows:

(a) Customer charge	$9.85
(b) Energy charge $= 2000\,\text{kWh} * [0.55 * \$0.1500/\text{kWh} + 0.45 * \$0.0335/\text{kWh}]$	$195.15
(c) Taxes = 3%*[(a) + (b)]	$6.15
Total monthly charges = (a) + (b) + (c)	$211.15

The average cost of electricity for the month is then $211.15/2000 kWh or $0.1056/kWh.

4.3.4.2 *The Real-Time-Pricing Rates*

The RTP rates were first offered by Pacific Gas & Electric in 1985 (Krarti, 2010). Like TOU rates, RTP rates are time-differentiated rates but the cost of electricity varies on an hourly basis. Typically, the utilities inform their customers of the hourly electricity prices only a few hours before they take effect. A more detailed description of the RTP rate structures available in the US is provided later in Section 4.3.4.

4.3.4.3 *The End-Use Rates*

To encourage customers to install and operate specific energy-consuming equipment, some US utilities offer end-use rates. With these rates, the utilities can impose operation periods and/or efficiency standards for selected and predefined equipment. For instance, the air conditioning rate allows electric utilities to interrupt service or cycle off the air conditioning equipment during specific times. Typically, the end-use rates require separate metering of the equipment.

4.3.4.4 *Specialty Rates*

The specialty rates are provided by utilities for specific purposes such as energy conservation and dispatchable customer generation. Energy conservation rates are offered by a limited number of US utilities to foster the use of energy efficient equipment and/or high standards of building materials. Dispatchable customer generation rates are provided to customers that have standby generators on their premises. In exchange for a reduced rate or a credit, the customers are requested to operate the generators whenever the utility needs additional generating capacity.

4.3.4.5 *Financial Incentive Rates*

Financial incentive rates encompass economic development rates, displacement rates, and surplus power rates. The economic development rates are typically offered to encourage new customers to locate or existing customers to expand in specific areas that need to be economically revitalized. The displacement rates are offered to customers that are capable of generating electricity to entice them to use utility-provided electricity. Finally, the surplus power rates are highly reduced energy rates that are offered to large commercial and industrial customers when the utility has an excess in the electric capacity.

4.3.4.6 Nonfirm Rates

The nonfirm rates include interruptible rates, stand-by rates, and load management rates. Interruptible rates are offered to customers that can reduce or even eliminate (interrupt) their electricity needs from the utility. The electricity pricing rates depend on several factors such as the capacity that can be interrupted, the length of interruption, and the notification before interruption.

Stand-by rates are intended for customers that require utility-provided electricity on an intermittent basis since they are capable of generating most of their electricity needs (by using, for instance, cogeneration systems as discussed in Chapter 13). The stand-by rates can be offered using three options:

1. Maintenance rates when the customers need electricity from the utility during predefined down-time periods of the generating equipment to perform routine maintenance.
2. Supplementary rates for customers that use regularly more electricity than they are capable of generating.
3. Back-up rates intended for customers that prefer to have back-up power from the utility in case of unexpected events (i.e., outages in the customers' generators).

Load-management rates are offered by utilities to control the usage of specific equipment such as space conditioning systems during peak periods. Generally, the load management rates are offered in combination with TOU rates.

4.3.4.7 Energy Purchase Rates

The energy purchase rates, also known as buy-back rates, are offered by utilities that want to purchase specific levels of energy or generating capacity from customers. The customers are nonutility electricity generators that qualify under the requirements of Public Utility Regulatory Policies Act such as cogeneration facilities and independent power producers.

4.3.5 Real-Time Pricing Rates

Based on a survey, most of the existing RTP rate structures in the US are rather experimental and are restricted to a selected number of customers with large power demands (varying from 250 kW to 10 MW depending on the utility). The available RTP rate structures can be classified into four categories in the survey as described briefly next (Krarti, 2010):

4.3.5.1 Category 1: Base Bill and Incremental Energy Charge Rates

Under these rates, the customers receive firm hourly electricity prices for the next day before a warning time. Some utilities provide updated prices daily [day-ahead-pricing (DAP)] while other utilities provide hourly prices only once a week (week-ahead-pricing). Typically, the category 1 RTP rates include the following charges:

- Base bill charge so that the utility recovers its revenue requirements. This charge depends on the customer baseline load (CBL), which defines the customer's typical energy use (kWh) for each hour of the year and demand values (kW) for each month.
- Incremental energy charge (or credit) reflects the cost of the energy used by the customer above (or below) its CBL profile.

In this category of RTP, an adequate estimation of the CBL profile is an important factor to determine the final energy bill. Therefore, the customer should negotiate for the most favorable CBL values. In general, the customer would benefit from smaller CBL values since the hourly prices of the RTP rate typically fall below those of the standard rate. Moreover, flexible customers can change their electrical load profile to avoid high costs and even obtain a credit for usage below their CBL.

4.3.5.2 Category 2: Total Energy Charge Rates

The hourly electricity prices for this category are applied to the total energy consumption. In addition, the utility imposes an additional charge to recover its revenue requirements. Some utilities set a monthly fixed charge based on the customer's average apparent demand (kVA) and actual demand (kW) determined using the last 12 months load profile. Other utilities charge for actual demand usually ratcheted for the last 12-month period.

The category 2 RTP rates are not suitable for inflexible customers that cannot change their load profile since they cannot protect themselves against higher electricity prices.

4.3.5.3 Category 3: Day-Type Rates

These rates are similar to the category 2 RTP rates but with several predefined day types (such as off-peak day and normal day) associated with different hourly firm electricity prices. These hourly prices are based on the total energy consumption with possibly a charge for actual demand. The utility informs the customer of the applicable day-type rate one day in advance. Again these rates are only suitable for flexible customers that can vary their energy use profile to accommodate the changes in the electricity price.

4.3.5.4 Category 4: Index-Type Rates

The electricity prices for these rates are determined based on the financial market electricity indexes (in Dow Jones) or the trading prices for electricity future contracts (in New York Mercantile Exchange). Generally, the customers have to forecast the electricity prices (using various tools and sources) since the utility does not provide the prices in advance.

4.3.5.5 Case Study of RTP rates

In this section, a detailed description is presented to calculate a monthly utility bill using RTP rate structures. The RTP rates used in this section are based on the DAP pilot rates proposed by Oklahoma Gas and Electric (OG&E) utility.

In the DAP rates offered by OG&E, there are two types of rates offered under either curtailment or noncurtailment services. Customers with curtailment service are allowed to buy electricity (referred to as "buy-through power" or "emergency power") above the contracted CBL profile during selected hours (or a curtailment period). The total cost of the buy-through power during a curtailment period is known as "buy-through period." Typically, the utility notifies in advance the availability and the hourly prices of emergency power. Table 4.15 shows the hourly prices for DAP rates offered by OG&E both noncurtailment and curtailment days (Krarti, 2010). The curtailment period is set between 3:00 p.m. and 8:00 p.m. with a contracted demand of 500 kW. To illustrate the calculation procedure of utility bills using the RTP rates compared to those based on the conventional rate structures, Example 4.8 is presented using the OG&E DAP rates.

Cost savings in the utility bills based on RTP rate structures can be achieved by shifting the load during the curtailment hours to noncurtailment hours. Example 4.9 illustrates the magnitude of savings achieved by a customer under OG&E DAP rates.

Example 4.8

Calculate the utility bills for an industrial customer serviced by OG&E under two rates:

1. Conventional rate with the following charges:
   - Customer charge: $150.00
   - Energy charge: $0.0264/kWh
   - Demand charge: $13.1/kW
2. RTP rate structure with the hourly prices given in Table 4.15. In addition, a customer charge (referred to as an administrative charge) of $300 is incurred monthly by the customer under the RTP rate.

   The hourly energy use and the contracted CBL profiles (for the RTP rate) are summarized in Table 4.16 for typical week and weekend days.

   Assume that the month has 21 weekdays (including two curtailment days) and 9 weekend days. In the utility bill calculations, do not account for any other charges such as taxes and credits.

TABLE 4.15 Day-Ahead-Price Rates From OG&E Utility for Typical Curtailment and Noncurtailment Days

Hour	Noncurtailment rate price ($/kWh)	Curtailment rate price ($/kWh)
0:00–1:00	0.01673	0.01673
1:00–2:00	0.01683	0.01683
2:00–3:00	0.01730	0.01730
3:00–4:00	0.01780	0.01780
4:00–5:00	0.01900	0.01900
5:00–6:00	0.02130	0.02130
6:00–7:00	0.02730	0.02730
7:00–8:00	0.02890	0.02890
8:00–9:00	0.02932	0.02932
9:00–10:00	0.03213	0.03213
10:00–11:00	0.03575	0.03575
11:00–12:00	0.03423	0.03423
12:00–13:00	0.03456	0.03456
13:00–14:00	0.03677	0.03677
14:00–15:00	0.03977	0.03977
15:00–16:00	0.04254	1.06350
16:00–17:00	0.04354	1.08950
17:00–18:00	0.03595	0.89875
18:00–19:00	0.02587	0.64675
19:00–20:00	0.01950	0.48750
20:00–21:00	0.01523	0.01523
21:00–22:00	0.01532	0.01532
22:00–23:00	0.01723	0.01723
23:00–24:00	0.01722	0.01722

Solution

(1) First, the utility bill is determined using the conventional rate structure bases on the actual energy use (see Table 4.16):

(a) Customer charge	\$150.00
(b) Energy charge = \$0.0264 / kWh * [21 * 42,461 kWh +9 * 10,510 kWh] = \$26,037.55	\$26,037.55
(c) Demand charge = \$13.1/kW*3015 kW	\$39,496.50
Total monthly charges = (a) + (b) + (c)	\$65,684.0

TABLE 4.16 Actual and CBL Hourly Energy Use Profiles for the Industrial Facility Assumed in Example 4.8

Hour	Typical weekday actual energy use	Typical week-end day actual energy use	Noncurtailment typical day CBL energy use	Curtailment typical day CBL energy use
	(kWh)	(kWh)	(kWh)	(kWh)
0:00–1:00	345	250	220	220
1:00–2:00	455	270	230	230
2:00–3:00	676	305	325	325
3:00–4:00	785	320	360	360
4:00–5:00	980	360	455	455
5:00–6:00	1200	405	645	645
6:00–7:00	1800	460	1715	1715
7:00–8:00	2905	510	2450	2450
8:00–9:00	2825	525	2545	2545
9:00–10:00	2800	520	2480	2480
10:00 –11:00	2845	535	2560	2560
11:00–12:00	3015	550	2765	2765
12:00–13:00	2500	565	2745	2745
13:00–14:00	2765	560	2730	2730
14:00–15:00	2890	535	2650	2650
15:00–16:00	2815	525	2590	500
16:00–17:00	2825	530	2520	500
17:00–18:00	2780	510	2370	500
18:00–19:00	1755	495	2175	500
19:00–20:00	1100	480	2080	500
20:00–21:00	845	395	1945	1945
21:00–22:00	635	345	1680	1680
22:00–23:00	535	290	1230	1230
23:00–24:00	385	270	575	575
Total	42,461	10,510	42,040	32,805

The average cost of electricity for the month based on the conventional rate structure is then $65,684.05/[21*42,461 kWh + 9*10,510 kWh] or $0.06660/kWh.

(2) The utility bill based on the RTP rate is described as follows:

(a) Customer charges	$300.00
(b) Energy charge for week days (noncurtailment days) (see Table 4.17)	$19*55.99
(c) Energy charge for week days (curtailment days) (see Table 4.18)	$2*8196.79
(d) Standard charge based on the CBL profile (see note below)	
i. Customer charge	$150.00
ii. Energy charge	$0.0264/kWh*958,960 (see note) $26,037.55
iii. Demand charge	$13.1/kW*2765 kW $36,221.50

$$\text{Total monthly charge} = \text{i.} + \text{ii.} + \text{iii.} = \$62,409.05$$

$$\text{Total monthly charge} = (\text{a}) + (\text{b}) + (\text{c}) + (\text{d}) = \$80,166.44$$

The average cost of electricity for the month based on the RTP rate structure is then $80,166.44/[21*42,461 kWh + 9*10,510 kWh] = $0.08128/kWh.

Therefore, the conventional rate structure is more cost-effective for the industrial facility considered in this example (at least during the month for which the utility bills were calculated as just shown).

Note: The total monthly energy use based on the CBL profiles is calculated as follows:

kWh(CBL) = 19 days*42,040 kWh/day + 2 days*32,805 kWh/day + 9 days*10,510 kWh/day

Or,

kWh (CBL) = 968,960 kWh

Example 4.9

Calculate the cost savings in electric utility bills for an industrial customer serviced by OG&E RTP rate used in Example 4.8 if the excess electrical load experienced during the curtailment hours (between 3 p.m. and 8 p.m.) is shifted uniformly to the next 4 h (between 8 p.m. and midnight). With this load shifting measure, the demand during the curtailment hours would be 500 kW and the total daily energy use for typical days would remain the same as shown in Table 4.18.

Solution

(a) The excess load during the curtailment hours is first determined as follows:

$$\text{Excess Load} = 2815\text{ kWh} + 2825\text{ kWh} + 1755\text{ kWh} + 1100\text{ kWh} - 4 * 500\text{ kWh} = 6495\text{ kWh}$$

TABLE 4.17 RTP Hourly Energy Cost During a Typical Noncurtailment Day for the Industrial Facility of Example 4.8

Hour	Typical weekday actual energy use	Noncurtailment typical day CBL energy use	Noncurtailment typical day hourly energy price	Noncurtailment typical day hourly energy cost
	(kWh)	(kWh)	($/kWh)	($)
0:00–1:00	345	220	0.01673	2.09
1:00–2:00	455	230	0.01683	3.79
2:00–3:00	676	325	0.01730	6.07
3:00–4:00	785	360	0.01780	7.57
4:00–5:00	980	455	0.01900	9.98
5:00–6:00	1200	645	0.02130	11.82
6:00–7:00	1800	1715	0.02730	2.32
7:00–8:00	2905	2450	0.02890	13.15
8:00–9:00	2825	2545	0.02932	8.21
9:00–10:00	2800	2480	0.03213	10.28
10:00 –11:00	2845	2560	0.03575	10.19
11:00–12:00	3015	2765	0.03423	8.56
12:00–13:00	2500	2745	0.03456	−8.47
13:00–14:00	2765	2730	0.03677	1.29
14:00–15:00	2890	2650	0.03977	9.54
15:00–16:00	2815	2590	0.04254	9.57
16:00–17:00	2825	2520	0.04354	13.28
17:00–18:00	2780	2370	0.03595	14.74
18:00–19:00	1755	2175	0.02587	−10.87
19:00–20:00	1100	2080	0.01950	−19.11
20:00–21:00	845	1945	0.01523	−16.75
21:00–22:00	635	1680	0.01532	−16.01
22:00–23:00	535	1230	0.01723	−11.97
23:00–24:00	385	575	0.01722	−3.27
Total	42,461	42,040		55.99

(b) For each hour during the next 4 h (i.e., between 8 p.m. and midnight), the additional load will be 6495 kWh/4 = 1623.75 kWh. The additional energy cost during the non-curtailment hours associated to this shift is then:

$$\text{Energy Cost Increase} = 1623.75\,\text{kWh} * [(0.01523 + 0.01532 + 0.01723 + 0.01722)\$/\text{kWh}] = \$105.54$$

TABLE 4.18 RTP Hourly Energy Cost During a Typical Curtailment Day for the Industrial Facility of Example 4.8

Hour	Typical weekday actual energy use	Curtailment typical day CBL energy use	Curtail-ment typical day hourly energy price	Curtailment typical day hourly energy cost
	(kWh)	(kWh)	($/kWh)	($)
0:00–1:00	345	220	0.01673	2.09
1:00–2:00	455	230	0.01683	3.79
2:00–3:00	676	325	0.01730	6.07
3:00–4:00	785	360	0.01780	7.57
4:00–5:00	980	455	0.01900	9.98
5:00–6:00	1200	645	0.02130	11.82
6:00–7:00	1800	1715	0.02730	2.32
7:00–8:00	2905	2450	0.02890	13.15
8:00–9:00	2825	2545	0.02932	8.21
9:00–10:00	2800	2480	0.03213	10.28
10:00–11:00	2845	2560	0.03575	10.19
11:00–12:00	3015	2765	0.03423	8.56
12:00–13:00	2500	2745	0.03456	−8.47
13:00–14:00	2765	2730	0.03677	1.29
14:00–15:00	2890	2650	0.03977	9.54
15:00–16:00	2815	500	1.06350	2462.00
16:00–17:00	2825	500	1.08950	2533.09
17:00–18:00	2780	500	0.89875	2049.15
18:00–19:00	1755	500	0.64675	811.67
19:00–20:00	1100	500	0.48750	292.50
20:00–21:00	845	1945	0.01523	−16.75
21:00–22:00	635	1680	0.01532	−16.01
22:00–23:00	535	1230	0.01723	−11.97
23:00–24:00	385	575	0.01722	−3.27
Total	42,461	32,805		8196.79

(c) The shift of electrical load during the curtailment hours has the effect of eliminating any hourly energy cost during the curtailment hours from 3 p.m. to 7 p.m. as shown in Table 4.18.

(That is, $2533.09 + $2049.15 + $811.67 + $292.50 = $5686.41).

(d) Since all the other charges remain unchanged, the reduction in cost savings is simply the energy charge avoided during the curtailment hours minus any energy cost increase during noncurtailment hours, or

Utility cost savings = [$5686.41/ day − 105.54 / day] *2 days = $11,161.74 for the month.

4.4 NATURAL GAS RATES

The rate structures for natural gas are similar than those described for electricity. However, the rates are generally easy to understand and apply. For instance, natural gas utilities rarely charge for peak demands. However, energy charges using block rates or seasonal rates are commonly offered. Examples of natural gas utility rates are illustrated in Table 4.19 for energy block pricing rates and Table 4.20 for energy and demand block pricing rates (GRI, 1999). Calculation details of monthly utility bills for natural gas are shown in Examples 4.10 and 4.11.

Example 4.10

Using utility rate structure D, calculate the gas utility bill for a residence during a month when the energy use is 10.4 MMBtu.

Solution

Considering all the charges imposed by utility rate D (see Table 4.19), the monthly gas bill can be calculated as shown here:

(a) Customer charge	\$4.50
(b) Energy charge = 2.5 MMBtu * \$5.145 / MMBtu +7.9 MMBtu * \$4.033 / MMBtu	\$44.72
(c) Fuel cost adjustment	\$0.00
(d) Taxes = 4.00% [(a) + (b) + (c)]	\$8.05
Total monthly charges = (a) + (b) + (c) + (d)	\$57.27

Thus, the average gas cost for the month is \$57.27/10.4 MMBtu or 5.51/MMBtu.

TABLE 4.19 A Gas Utility Rate for Residential General Service (Utility Rate D)

Billing item	Any month (January–December)
Customer charge	\$4.50
Minimum charge	\$4.50
Fuel cost adjustment	0.00
Tax rate	4.00%
No. of energy blocks	2
Block 1 energy size (MMBtu)	2.5
Block 2 energy size (MMBtu)	>2.5
Block 1 energy charge (\$/MMBtu)	5.145
Block 2 energy charge (\$/MMBtu)	4.033

Example 4.11

Using utility rate structure E, calculate the gas utility bill for a commercial facility during a month when the energy use is 15,000 MMBtu and the billing demand is 500 MMBtu/day.

Solution

Accounting for all the charges considered by utility rate E (see Table 4.20), the electric energy bill for the winter month can be calculated as follows:

(a) Customer charge	$0.00
(b) Energy charge $= 10{,}000\ \text{MMBtu} * \$3.482/\text{MMBt} + 5{,}000\ \text{MMBtu} * \$3.412/\text{MMBt}$ (b)	$51,880.00
(c) Demand charge = 10 * $5.50 + 490 * $4.50	$2,260.00
(d) Taxes = 4.00%*[(a) + (b) + (c)]	$2,165.60
Total monthly charges = (a) + (b) + (c) + (d)	$56,305.60

The average cost of electricity for the month is then $56,305.60/15,000 MMBtu or $3.75/MMBtu.

TABLE 4.20 A Gas Utility Rate With Demand Charges for Commercial General Service (Utility Rate E)

Billing item	Any month (January–December)
Customer charge	0.00
Minimum charge	$300.00
Fuel cost adjustment	0.00
Tax rate	4.00%
No. of energy blocks	3
Block 1 energy size (MMBtu)	10000
Block 2 energy size (MMBtu)	10000
Block 3 energy size (MMBtu)	>20000
Block 1 energy charge ($/MMBtu)	3.482
Block 2 energy charge ($/MMBtu)	3.412
Block 3 energy charge ($/MMBtu)	3.385
No. of demand blocks	2
Block 1 demand size (MMBtu/day)	10
Block 2 demand size (MMBtu/day)	>10
Block 1 demand charge ($/MMBtu/day)	5.50
Block 2 demand charge ($/MMBtu/day)	4.50

In addition to energy and demand charges, the price of natural gas is determined based on the interruptible priority class selected by the customer. A customer with a low priority has a cheaper rate but can be curtailed whenever a shortage in the gas supply is experienced by the utility. However, some small quantities of gas are generally supplied to prevent the pipes from freezing and to keep the pilot lights burning.

4.5 UTILITY RATES FOR OTHER ENERGY SOURCES

The utility rate structures for energy sources other than electricity and natural gas are generally based on a flat rate. For instance, crude oil is typically charged per gallon while coal is priced on a per ton basis. The prices of oil products and coal are set by the market conditions but may vary within geographical areas depending on local surcharge and tax rates. Moreover, fuel oil or coal can be classified in a number of grades. The grades of fuel oil depend on the distillation process. For instance, No. 1 oil is a distillate and is used a domestic heat oil. On the other hand, No. 6 oil has a very heavy residue left after the other oils have been refined. The grades of coal depend on the sulfur content and percent moisture. Chapter 8 describes in more detail some of the properties of fuel oils.

In some applications, it may be possible and desirable to purchase steam or chilled water to condition buildings rather than using primary fuel to operate boilers and chillers. Steam can be available from large cogeneration plants. Chilled water and steam may also be produced based on the economies of scale in district heating/cooling systems. Generally, steam and chilled water are both charged based on either a flat rate or a block rate structure for energy and demand. The steam is charged based on pound per hour (for demand charges) or thousands of pounds (for energy charges). Meanwhile, chilled water is charged on the basis of tons (for demand charges) or ton-hours (for energy charges).

4.6 OVERVIEW OF MICROGRIDS

4.6.1 Introduction

Microgrids are localized power networks that allow the integration of distributed generation resources to meet electrical loads of part or all the facilities present within the networks. The microgrids can operate as standalone systems such as in remote locations or as grid-connected networks to meet critical loads and mitigate grid disturbances. Even when connected to the main or utility grid (or macrogrid), microgrids should be able to operate autonomously in island mode to provide higher reliability and resilience. Several distributed generation resources can be considered for microgrids including fossil-fuel generation and cogeneration systems as well as renewable resources such solar and wind technologies. Other technologies can also be included in microgrids such as fuel cells, energy storage, and load management systems.

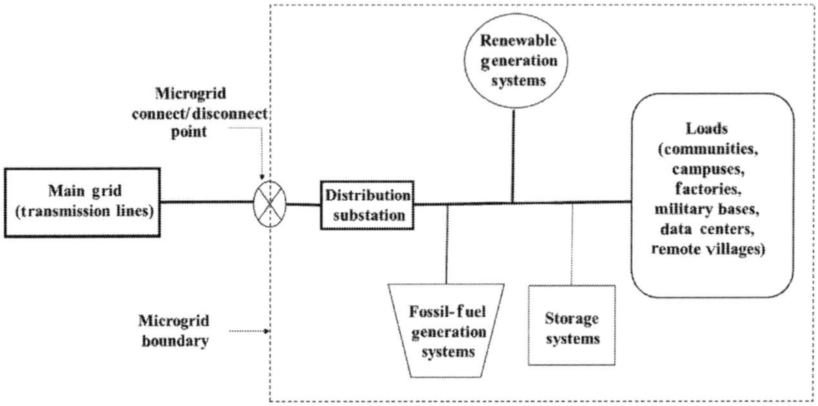

FIGURE 4.7 Typical components and structure of a microgrid.

Fig. 4.7 shows the basic architecture of a microgrid including its main components, boundary, and connection point with the main grid. Microgrids can meet electrical loads for several applications including college campuses, military bases, factories, data centers, and entire communities.

4.6.2 Benefits of Microgrids

The main benefits of microgrids include enhanced reliability and improved energy efficiency. Indeed, one of the main driver for the development of microgrids is to reduce disturbances and vulnerabilities of power supply associated with natural disasters and terrorist attacks (Campbell, 2012; NRC, 2012). It is estimated that sustained power interruptions from the main grid system (lasting over 5 min) could cost the US over $26 billion per year (Cagle, 2012). The US grid reliability issues is expected to increase since the power infrastructure is aging while the demand continues to grow. Microgrids operating in island-mode offer back-power and help reduce and even eliminate blackouts and power disturbances. For instance and during hurricane Sandy that hit the US east coast in 2012, several organizations including businesses and institutions were able to weather the sustained blackouts associated with the damages incurred by the main grid infrastructure. Indeed, on-site power generation systems including back-up diesel generators allowed colleges, data centers, stores, and homes to partially or fully operate during grid outages (LaMonica, 2012). For instance, Princeton University was able to serve power to part of the campus with its local 11-MW generation system during hurricane Sandy. Similarly, a 40-MW combined heat and power plant in the Bronx was used to provide electricity and heat to a large housing complex.

In addition to increasing power quality and reliability, microgrids can achieve significant reductions in fuel use, carbon emissions, and ultimately operating costs. Indeed, microgrids can be optimally designed and operated to serve power required by specific consumers including communities, campuses, or remote villages. Through smart load management strategies and distributed generation technologies, consumers can have notable economic and environmental benefits by reducing their reliance on fossil fuels, lowering their carbon emissions, and decreasing their operating costs (Cagle, 2012).

While it has some clear benefits, the deployment of microgrids remain slow and incremental with specific targets and applications. Indeed, most of the existing microgrids are small and largely limited to demonstration projects. As of 2016, a Navigant study has identified about 1700 microgrid projects operating, under construction, or proposed with a total capacity of 16,500 MW (Navigant, 2017). As is the case for smart grids, the main challenges for deployment of microgrids are related to physical and cyber security issues. Indeed, microgrids have to be designed and operated to ensure that equipment components such as circuit breakers and control sequences are protected against any physical tampering and cyber-attacks. Moreover, there is a lack of well-established business models for microgrid deployment resulting in significant financial risk.

4.6.3 Types of Microgrids

There are several types of microgrids that have been deployed and operated worldwide. The market segments and the applications for which they are best suited typically categorize microgrids. Specifically, five types of microgrids are commonly considered:

1. *Remote Microgrids:* These types are designed to be standalone and to operate always off-grid. These microgrids are suitable for serving remote villages or isolated islands that are not connected to the main grid.
2. *Military Base Microgrids:* These local power systems are deployed with the main objective of enhancing physical and cyber security for military facilities. These types include mobile microgrids suitable for forward operating bases in remote locations.
3. *Campus Microgrids:* They consist of on-site generation for college campuses and research parks to meet multiple loads associated to electrical and thermal loads. Typical capacities of these microgrid types range from 5 to 50 MW.
4. *Industrial and Commercial Microgrids:* These types are considered to serve industrial facilities and/or commercial buildings. Microgrids can be used to serve critical loads such as data centers and hospitals.
5. *Utility and Community Microgrids:* These types are integrated within the existing utility networks to serve multiple customers within communities to provide more reliable, resilient, and sustainable power.

4.6.4 Interconnection With the Main Grid

The main advantages to interconnect any power source including a microgrid to the main grid include the following:

- *Availability:* The grid acts as an additional power source in case of maintenance or repair needs for the microgrid.

- *Stability:* The grid can facilitate the stable operation of the microgrid. Moreover, the grid can act as an energy storage for renewable energy sources.
- *Economics:* The extra power generated by the microgrid can be sold to the grid to achieve some economic benefits. Moreover, power can be acquired from the grid during off-peak periods when electricity prices are low.

While the microgrid can be connected to the main grid, it can operate in different modes as illustrated in Fig. 4.8. In particular, three modes of operation are generally considered:

1. *Isolated Operation:* The microgrid can meet at least the critical loads when there is a power outage. In this case, the microgrid is designed to provide emergency and backup power. Another option of isolated operation consists of a standalone microgrid that is not connected to the main grid as shown in Fig. 4.4A. In an isolated operation, the microgrid does not directly interact with the main grid distribution system since it is connected only when the main power plant is offline.
2. *Roll-Over Operation*: In this operation, the loads can be served by either the main grid or the microgrid but not both. When required, a transfer switch can connect the loads to one of the two power sources as noted in Fig. 4.4B. The transfer can be automatic or manual depending on the design. However, energy storage systems may be needed since there is always some period when the loads are not fed especially when the transfer is made automatically. In this operation, the microgrid is operating in parallel with the main grid. Thus, the microgrid is not injecting any energy (i.e., islanding) to the main grid even in case of outages thus avoiding hazardous conditions to people and damage to the power distribution system.
3. *Parallel Operation:* In this mode, the main grid and microgrid operate simultaneously to serve a fraction of the loads as illustrated in Fig. 4.4C. When one power source fails, the other source would exclusively serve the loads without any interruption. In this mode of operation, the microgrid has to be in synchronism with the main grid requiring more complex and sophisticated monitoring, protection, and controls compared to the other two operation modes. This operation is typically considered when the excess power generated by the microgrid has to be fed or sold to the main grid. Several design configurations can be considered when parallel operation is needed.

Three methods are typically used to connect microgrids to the main grid:

1. *Directly Through Switchgear:* This option is relatively simple and inexpensive but is slow for connection and disconnection (requiring three to six cycles). For this method, the use of circuit breakers require that the electrical characteristics of the microgrid match those of the grid. Moreover, the power flow cannot be effectively controlled.
2. *Static Switches*: This method is more complex and expensive than using the circuit breakers. In addition, bypass switches are used to allow maintenance

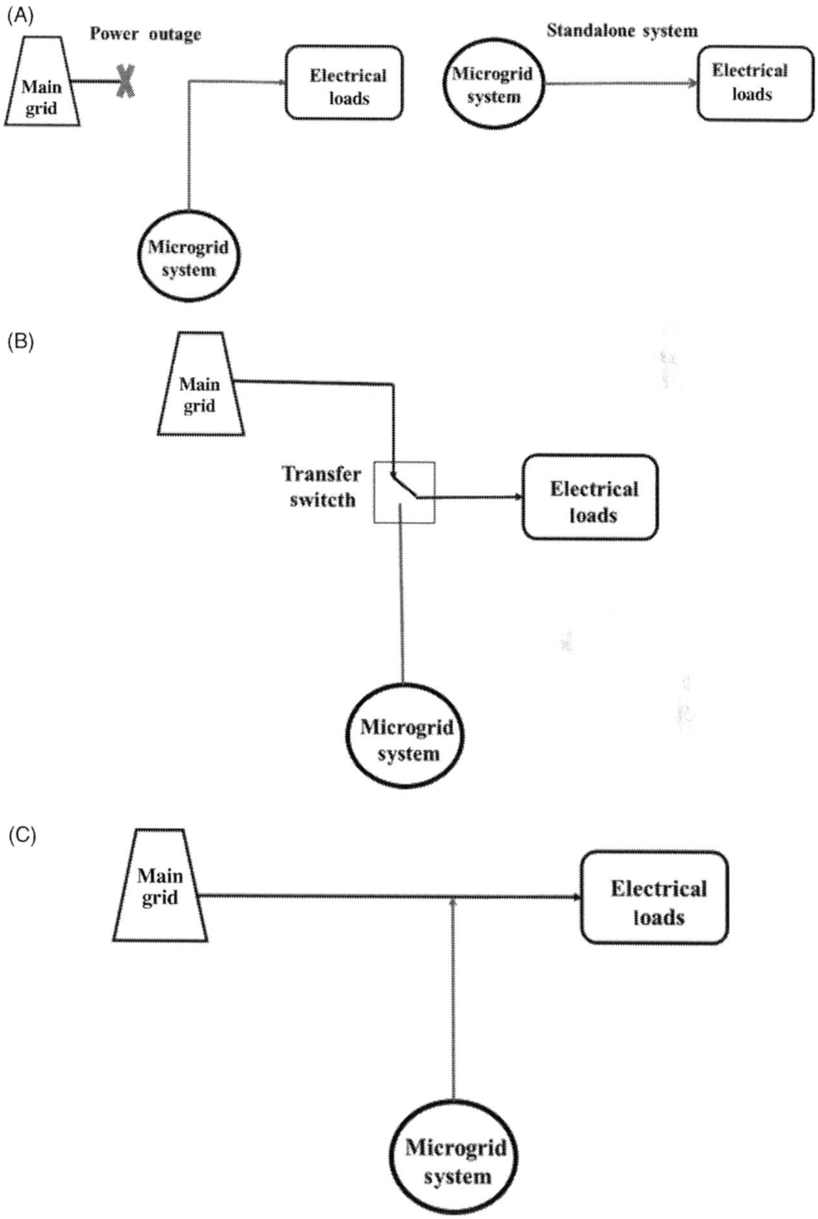

FIGURE 4.8 Operation modes for microgrids. (A) Isolated, (B) roll-over, and (C) parallel configurations.

of the microgrid. Static switches are faster to act than circuit breakers (typically 0.5 to 1 cycle response). However, the power flow cannot be properly controlled with this method.

3. *Power Electronic Interfaces*: this method is the most expensive. It allows for the power distribution characteristics of the microgrid to be different from those of the main grid. Moreover, power flow can be controlled using this connection option. The reaction time is fast and is generally similar to that of static switches (i.e., 0.5 to 1 cycle). Circuit breakers may be required to physically disconnect the microgrid from the grid.

When the microgrid is connected in parallel with the main grid, several challenges have to be considered to ensure seamless operation. Some of these challenges are the following:

- frequency regulation
- voltage regulation
- disconnection and reconnection protocols in the event of grid disturbances
- safe intentional islanding operation
- control of faults when in grid-connected mode
- harmonization of technical standards among and within countries
- protection coordination

There are several standards specifying the different aspects of grid interconnection of a local power generation source. The most important standard is IEEE 1547, which includes several sections. The main sections are highlighted:

- IEEE Standard 1547.1 "IEEE Standard Conformance Test Procedures for Equipment Interconnecting Distributed Resources with Electric Power Systems."
- EEE Standard 1547.2 "IEEE Application Guide for IEEE Std 1547, IEEE Standard for Interconnecting Distributed Resources with Electric Power Systems."
- IEEE Standard 1547.3 "IEEE Guide for Monitoring, Information Exchange, and Control of Distributed Resources Interconnected with Electric Power Systems."
- IEEE Standard 1547.4 "IEEE Guide for Design, Operation, and Integration of Distributed Resource Island Systems with Electric Power Systems."
- IEEE Standard 1547.5 to address issues when interconnecting electric power sources of more than 10 MVA to the power grid.
- IEEE Standard 1547.6 "IEEE Recommended Practice for Interconnecting Distributed Resources with Electric Power Systems Distribution Secondary Networks."
- IEEE Standard 1547.8 to provide supplemental support for implementation methods for expanded use of the previous standards, for example when addressing issues with high penetration of residential PV systems.

4.7 POWER QUALITY

When connecting microgrids or other power sources to the main grid, disturbances in power supply can significantly affect the reliability, safety, and integrity of not only the electrical distribution system components including the connected loads. This section outlines common power quality disturbances and the means to mitigate them to minimize significant damages to electrical equipment and maintain reliable performance for the power generation and distribution systems. Voltage sags, voltage swells, interruptions, transients, voltage unbalances, harmonic distortions, and electrical noises are some of the power quality disturbances common in power distribution systems and are presented in the following sections.

4.7.1 Types of Disturbances

4.7.1.1 Voltage Sags

A voltage sag occurs when the RMS voltage value decreases over a limited time period before it settles and returns to its average value. There are several types of sags including the following (Caramia et al., 2009):

- Instantaneous sags lasting typically from 0.5 cycles to 30 cycles,
- Momentary sags that can last 30 cycles to 3 s, and
- Temporary sags that last 3 s to 1 min.

If the sag lasts for more than a minute, it is then classified as an undervoltage. Voltage sag is generally caused by a short circuit fault or by a sudden change in the characteristics of a power source or a load such as the case during a motor starting. If the voltage does sag, some equipment may not be supplied by the amount of voltage they require and their performance may be affected (Kusko and Thompson, 2007). Fig. 4.9 provides a typical time variation of voltage during a sag event.

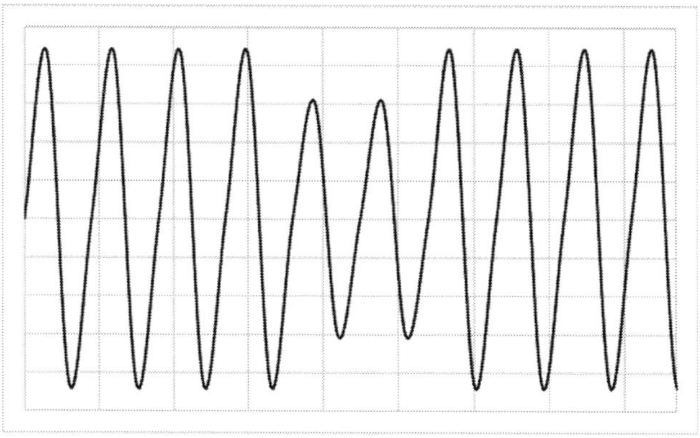

FIGURE 4.9 Variation of voltage during a sag event.

4.7.1.2 *Voltage Swells*

A voltage swell refers to the case when the RMS voltage increases during a time period before it comes back to its average value. Similar to the sag events, several swell types have been identified:

- Instantaneous swells lasting from 0.5 cycles to 30 cycles,
- Momentary swells that can last 30 cycles to 3 s, and
- Temporary swells that last 3 s to 1 min.

Again, if the swell lasts for more than a minute it is then classified as an overvoltage event. Voltage swells occur mostly due to faults in the electrical distribution systems. They do not occur as often as voltage sags, but they can cause more damage to select devices that are sensitive to higher voltages than their rating levels (Kusko and Thompson, 2007). Fig. 4.10 illustrates an example of time variation of voltage in case of a swell event.

4.7.1.3 *Interruptions*

An interruption is an event when the supply voltage decreases to less than 0.1 pu (i.e., 0.1% of its normal value) for a limited period of time. Several types of interruptions can occur:

- Instantaneous interruptions lasting from 0.5 cycles to 30 cycles,
- Momentary interruptions that typically last 30 cycles to 3 s, and
- Temporary interruptions that last 3 s to 1 min.

Like swells, interruptions are events that occur due to faults that may trip protection devices. Fig. 4.11 shows an example of time variation of voltage in case of an interruption event.

4.7.1.4 *Transients*

Transients in electrical power systems can take two forms: (1) impulsive transients and (2) oscillatory transients. Impulsive transients refer to cases when

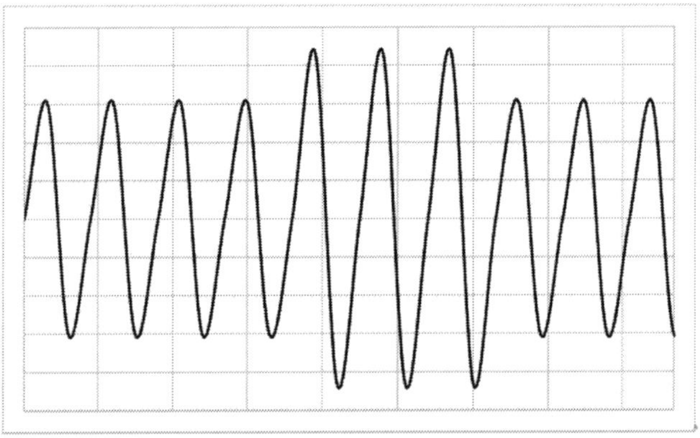

FIGURE 4.10 Variation of voltage during a swell event.

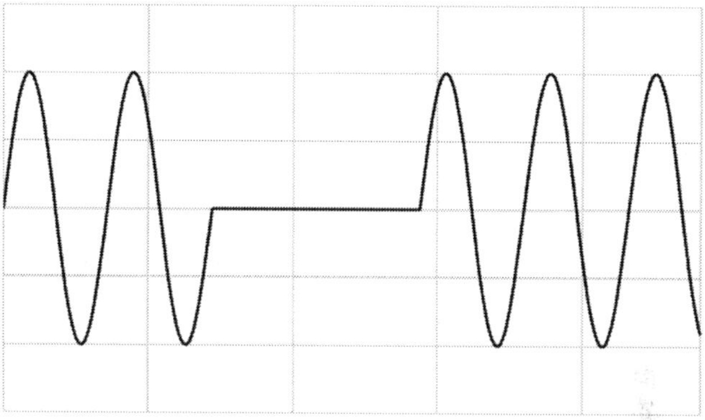

FIGURE 4.11 **Variation of voltage during an interpretation event.**

there are nearly instantaneous changes in the steady-state conditions of voltages or currents. In particular, the voltages or currents suddenly spike and deviate by 30%–90% from their typical values. The most common cause of impulsive transients is a lightning strike. An example of an impulsive transient is illustrated in Fig. 4.12.

When oscillatory transients occur, voltages and/or currents oscillate positively and negatively relative to their typical values. These oscillations are classified into low, medium, or high frequencies. Low frequency transients are usually associated with switching from utility power source to capacitor banks resulting in frequencies ranging from 300 to 900 Hz. Medium frequency transients can be caused by back-to-back capacitor switching and can result in frequencies of 5 to 500 kHz. High frequency transients are typically caused by impulsive transients and result in frequencies of more than 500 kHz. The damages due to transients can be significant and can result in immediate failure of some electrical equipment. An example of an oscillatory transient is shown in Fig. 4.13.

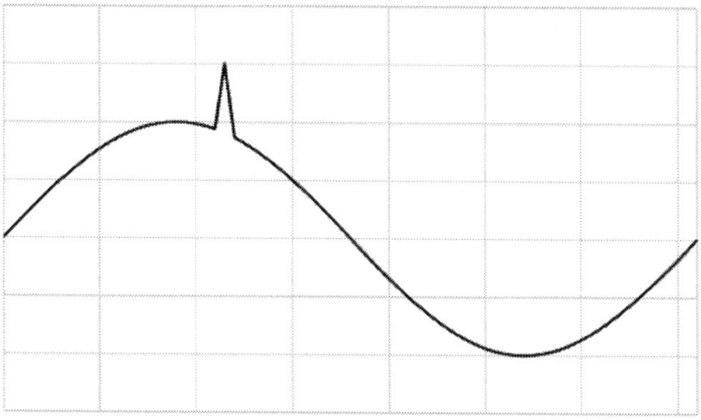

FIGURE 4.12 **Variation of voltage during an impulsive transient.**

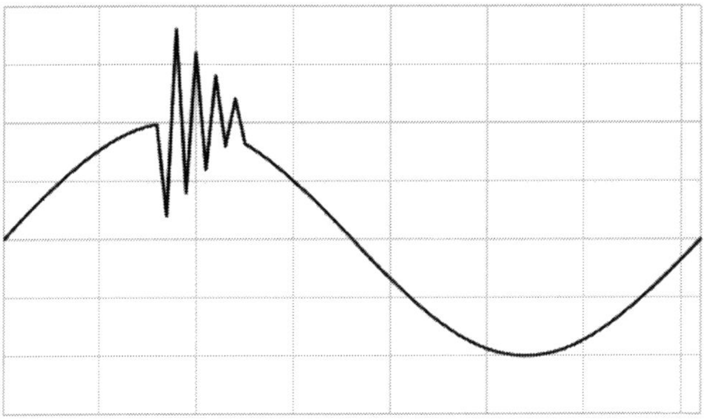

FIGURE 4.13 Variation of voltage during an oscillatory transient.

4.7.1.5 *Voltage Unbalances*

Voltage unbalances are associated to variations in voltages and currents for three-phase systems when the RMS voltage values or the phase angles become not equal between the three phases (Bollen, 2000). Voltage unbalances can occur when there is a load imbalance due to an uneven distribution of single phase loads over three-phase power systems. They can also occur due to a large single-phase loading such as when one of the fuses become not operational for three-phase motors. Voltage unbalances can generate overheating within the windings of induction and synchronous machines.

4.7.1.6 *Harmonic Distortions*

Harmonic distortions are common voltage and current variations due to changes in frequencies within the electrical distribution systems. In particular, there are deviations from the typical sinusoidal variations in voltages or currents. Harmonics are mainly caused by nonlinear loads such as those associated with power electronic converters including variable frequency drives (VFDs) installed for fans and pumps serving building air conditioning systems (Vedam and Sarma, 2009). Specifically, these converters draw a nonsinusoidal current/voltage and introduce nonlinear loads throughout the electrical distribution systems. The problems due to harmonic distortions are nowadays common in buildings since computers and other power electronics are heavily utilized. In particular, harmonic distortions can cause several disturbances and damages including wire overheating and power loss. Fig. 4.14 shows a typical voltage variation due to harmonic distortion.

4.7.1.7 *Voltage Fluctuations*

Voltage fluctuations present different types of voltage and current variations that can cause damages to the electrical power distribution systems. These

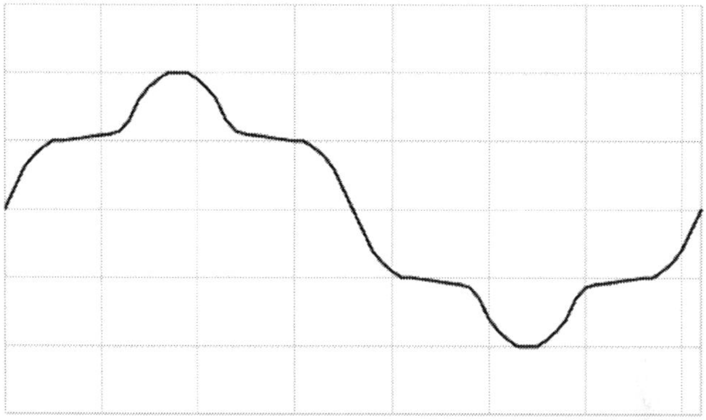

FIGURE 4.14 **Variation of voltage during a harmonic distortion.**

fluctuations, often called flickers, are the result of the application of large loads or power surges that can distort the voltages. Common loads that cause voltage fluctuations are arc furnaces used in steel industrial facilities. Fig. 4.15 shows an example of voltage fluctuations.

4.7.1.8 Electrical Noises

Electrical noise is associated to an unwanted electrical signal with a frequency generally lower than 200 kHz (Caramia et al., 2009). It should be noted that noise disturbances are different from harmonic distortions and transients since they do not act like waveforms. Sources of noise include power electronic devices, control circuits, arcing equipment, and switching power supplies (Vedam and Sarma, 2009). It has been noted that the effects due to noises can be made worse when grounding is introduced to the electrical power distribution system. An example of an electrical noise is outlined in Fig. 4.16.

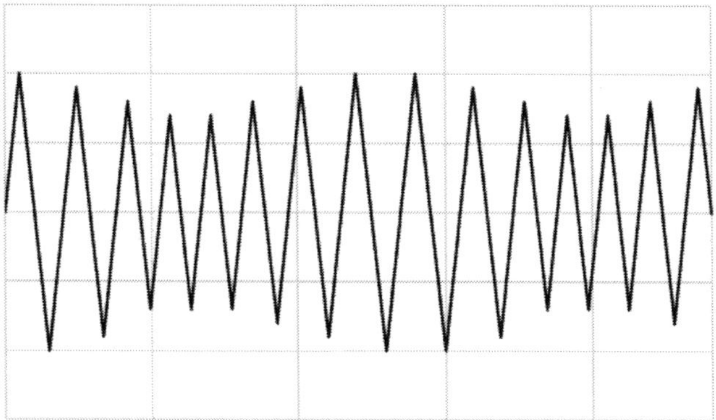

FIGURE 4.15 **Variation of voltage during a voltage fluctuation condition.**

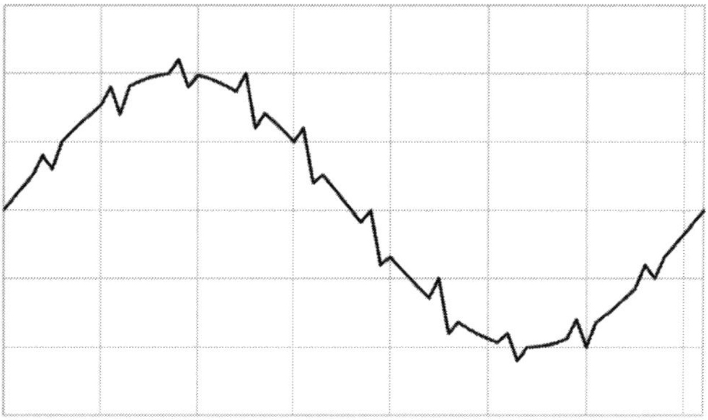

FIGURE 4.16 Variation of voltage during a noise event.

4.7.2 Mitigation Options

Power quality disturbances are becoming common occurrences in power distribution systems especially those with interconnections with multiple power sources. Moreover, the wide range of electrical systems and devices present within modern buildings has increased unwanted and uncontrolled variations of electrical voltages and currents. To eliminate or reduce these disturbances, several strategies and techniques are currently available and can be employed to ensure acceptable and reliable power quality to vital electrical power systems. First, it is important to properly design and select high-quality devices and equipment for the electrical power systems to ensure good power quality. When disturbances occur, methods of correcting power quality can be considered depending on the main causes. In particular and to control voltages, power quality mitigation approaches include the addition of power harmonic filters and the use of voltage compensators. Moreover, capacitor banks can be installed to reduce the reactive power generated by some loads. Special transformers with phase shifting connections can also diminish harmonic currents. Finally, the integration of uninterruptable power supply (UPS) systems can minimize power quality issues within parts of the electrical distribution systems.

In summary, two main criteria should be considered when designing and selecting equipment for electrical power systems. First, it is important to install high-quality electrical equipment with low possibilities of causing power quality issues. Second, it is crucial to use electrical systems with low sensitivity to voltage and current fluctuations. When harmonic distortions do occur, several mitigation systems can be considered as briefly discussed in the following sections.

4.7.2.1 Multipulse Rectifiers

The use of VFDs, also known as variable speed drives, is becoming common in operating motors used in heating, ventilating, and air conditioning (HVAC) systems within buildings. For three-phase motors, the VFDs are typically made up of six-pulse rectifiers (i.e., sets of diodes) that allow the AC currents to be drawn in pulses and converted to DC currents. These pulses generate harmonic current distortions with frequencies h that are multiples of the fundamental frequency with

$$h = np \pm 1 \tag{4.1}$$

with n being any integer number (i.e., 1, 2, 3, etc.) and p being the pulse number.

Thus, for the conventional six-pulse rectifiers ($p = 6$), the significant harmonics are 5, 7, 11, 13, 17, 19, and so on. Typically, the magnitude of each harmonic is inversely proportional to the harmonic number so that 5th is more significant than the 7th, which is more significant that 11th.

To reduce the generated harmonics, higher pulse rectifiers are utilized including 12-pulse, 18-pulse, or 24-pulse. For 12-pulse rectifiers ($p = 12$), the significant harmonics can be determined using Eq. (4.1) and consist of 11th, 13th, 23rd, 25th, and so on. Thus, and relative to the six-pulse rectifiers, the 5th, 7th, 17th, 19th, and so on are cancelled by the use of 12-pulse rectifiers. While they cost more than the six-pulse rectifiers, the high-pulse rectifiers when connected to the VFDs represent an effective mitigation solution for harmonic distortions. Typically, multiple pulse rectifiers are coupled with transformers, which are referred to as multipulse transformers, to reduce the magnitude of harmonic distortions (Kusko and Thompson, 2007).

4.7.2.2 Harmonic Filters

Power harmonic filters are mainly used to either block harmonic currents from flowing through electrical power distribution systems or locally isolate and cancel them (Fuchs and Masoum, 2008). Specifically, harmonic filters are designed to have low impedances to ensure that the harmonic currents are flowing between the loads and the filters, rather than reaching the power source and the other components of the electrical power generation and distribution systems. There are three main classifications of filters: (1) passive, (2) active, and (3) hybrid filters:

1. Passive filters are made up of reactive components like capacitors, inductors, and resistors. They are tuned to provide a low impedance path for specific currents with undesirable harmonic frequencies. Some potential issues can occur with passive systems if not properly selected resulting in resonance circuits parallel with power source leading to worse power quality conditions.
2. Active filters rely on active power conditioning electronic devices to compensate for unfavorable harmonic currents. Specifically, active filters correct for the distorted sinusoidal current waves to effectively eliminate harmonics. Active filters have the advantage to respond to dynamic changes in the electrical distribution systems (Fuchs and Masoum, 2008).

Therefore, active filters can be considered for more complicated and critical systems because they have a better capability to act in response to drastic disturbances. They are, however, more expensive to install than passive filters.

3. Hybrid filters are combinations of active and passive filters. They are used in specific applications to take advantage of the benefits of both filters.

4.7.2.3 Voltage Compensators

Voltage compensators can be used to improve power quality by placing voltage sources between power sources and loads. For instance, voltage sags can be eliminated by inserting voltage components into the voltage waveforms to adjust their profiles. Voltage compensators usually can store only limited amounts of energy so that they can only provide specific voltage compensation range. This range is typically limited to 2 s for 50% dips in voltage and 12 cycles for 10% dips (Kusko and Thompson, 2007). Voltage compensators are relatively inexpensive and are easy to install since they are not bulky.

4.7.2.4 Capacitor Banks

The installation of capacitor banks is a common mitigation strategy to correct power quality problems. Indeed, capacitors when optimally sized and placed to compensate heavily inductive loads can vastly reduce the reactive power drawn by the electrical power system. This reduction in reactive power has several benefits including increase of the overall power factor, improvement of the voltage regulation, lower harmonic distortion, and reduction of energy loss (Fuchs and Masoum, 2008). One of the challenges for utilizing capacitor banks for power quality improvements is determining the optimum location, size, and number of capacitors for a specific electrical distribution system. Indeed, several factors need to be taken into account to control the overall power quality throughout the system.

4.7.2.5 Harmonic Mitigation Transformers

Transformers are crucial elements of electrical distribution systems to transfer power from its source to the loads. In the US, the voltage from a power plant ranges from 13.8 kV to 480 V. While buildings normally require 480 V three-phase and/or 277 V or 120 V single-phase. To achieve the necessary drop in voltage, multiple step-down transformers are used. Specially designed transformers can also reduce some harmonic current distortions including the triplens (i.e., 3rd, 6th, 9th, etc.). These transformers are set up to fit delta–delta connections to wye–delta connections so that the secondary voltages have a 30° phase-shift (Kusko and Thompson, 2007). Other types of transformers can also help in mitigating voltage sags. For instance, constant voltage transformers use secondary transformers and capacitors to achieve a constant output voltage.

4.7.2.6 Uninterruptible Power Supply

For critical loads, the most common effective protection option is a UPS system. Typically, UPS systems consist of battery-powered devices that are charged from the utility power source using inverters and rectifiers. When the source voltage level is reduced due to any disturbances, a bypass switch is activated and power is supplied from the battery. The battery continues to provide steady power to the critical loads until the main power source returns to its normal operating conditions unless the battery is totally discharged. Battery-powered UPS systems come in a variety of sizes from 100 W to 500 kW. It should be noted that these standard sizes of batteries can only maintain the necessary voltage level for short periods ranging from 3 to 10 min. For large critical building loads such as data centers or operating rooms in hospitals, UPS systems include, instead of batteries, standby engine generators. The generators are connected to the critical load through transfer switches and can provide power for extended periods of time.

4.7.3 Codes and Standards

In the US, there are several standards related to power quality of electrical power systems. These standards are mostly developed by IEEE (Institute of Electrical and Electronic Engineering), ANSI (American National Standards Institute), and NEMA (National Electrical Manufacturing Association). It should be noted that NEC (National Electrical Code), developed by the NFPA (National Fire Protection Agency), does not generally provide any specific guidelines on methods and strategies to reduce or eliminated power quality issues. Indeed, the NEC emphasis is to define minimum safety requirements rather than to ensure reliable performance of electrical systems. Although several power quality problems are disruptive to normal operation of electrical distribution systems, they do not necessarily lead to safety issues. For instance, one of the common power quality issues is an excessive voltage drop, which can affect significantly the operation of electrical loads such as lighting systems and motors but is not considered as a safety issue. NEC provides some specifications to systems that can assist in eliminating power quality programs such as surge arrestors and grounding systems. Otherwise, NEC has no special rules about power conditioners, filters, special-purpose transformers, or other mitigation devices.

As one of the main guidelines to improve power quality, IEEE standard 519 provides specifications for controlling harmonic currents. Specifically, the standard sets requirements for power quality from utilities and limits the harmonic currents caused by the end-users. Table 4.21 summarizes the voltage distortion and *THD* limits as outlined by the IEEE standard 519. Table 4.22 provides the allowable harmonic current limits for the end-users. These limits are based on the ratio of the maximum available short circuit current to the maximum demand load current at the point of common coupling (PCC) between the building and the utility. As this ratio increases, the allowable total harmonic distortion for current increases.

TABLE 4.21 Limits for Voltage Distortion Set by IEEE Standard 519

Bus voltage at point of common coupling (PCC)	Individual voltage distortion (%)	Voltage total harmonic distortion, THD (%)
69 kV and below	3.0	5.0
69.001 kV through 161 kV	1.5	2.5
161.001 kV or above	1.0	1.5

TABLE 4.22 Limits for Harmonic Currents for 120 V to 69 kV Distribution Systems Set by IEEE Standard 519

I_{sc}/I_L	$h < 11$	$11 \leq h < 17$	$17 \leq h < 23$	$23 \leq h < 35$	$35 \leq h$	THD
<20	4.0	2.0	1.5	0.6	0.3	5.0
20–50	7.0	3.5	2.5	1.0	0.5	8.0
50–100	10.0	4.5	4.0	1.5	0.7	12.0
100–1000	12.0	5.5	5.0	2.0	1.0	15.0
>1000	15.0	7.0	6.0	2.5	1.4	20.0

h, individual harmonic number; I_L, maximum fundamental frequency demand load at the point of common coupling or PCC; I_{sc}, Maximum short circuit current; *THD*, total harmonic distortion for currents.

Some other standards are available and can be considered for power quality specifications when designing electrical power systems:

- IEEE Standard 1159: Monitoring and Definition of Electric Power Quality
- IEEE Standard 1250: Guide on Equipment Sensitive to Momentary Voltage Disturbances
- IEEE Standard 1346: Guide on Compatibility for ASDs and Process Controllers
- ANSI Standard C62: Guides and Standards on Surge Protection
- ANSI Standard C84.1: Voltage Ratings for Power Systems & Equipment
- ANSI Standard C 57.100: Transformer Derating for Supplying Nonlinear Loads
- NEMA-UPS: Uninterruptible Power Supply Specifications

4.8 DEMAND-SIDE MANAGEMENT

In this section, some programs that electrical utilities typically encourage to reduce their operating costs in providing power to their customers are discussed. First, typical load electrical profiles imposed by buildings are discussed. Then, an overview of DSM programs are discussed including some of the incentives utilities provide their customers to entice them to participate in these programs.

4.8.1 Electric Load Profiles

Electrical load demand for buildings vary throughout the day and the season depending on the occupancy patterns and weather conditions. For office buildings, the loads are typically during the daytime and in most climates during the summer due to the air conditioning requirements to maintain indoor thermal comfort. The load profile of a building, a set of buildings, or a utility is characterized by their load factor, *LF*, defined as the fraction of the average power demand, P_{avg}, over the peak power demand, P_{max}:

$$LF = \frac{P_{avg}}{P_{max}} \tag{4.2}$$

The load factor in buildings can vary from 0.15 for high variability of power demand to 0.90 for uniform and steady power need. Table 4.23 lists typical load factors for different building types.

Electrical loads for buildings and the grids are categorized into three types depending on the frequency of their occurrence during the day and the year:

1. *Baseload* is a load that occurs constantly throughout the day and the year. For instance, refrigerator load in a household is considered a baseload.
2. *Intermediate load* is a load that occurs part of the day or a season. For instance, air conditioning for an office building is typically only during the day and during the summer to ensure indoor thermal comfort for occupants.
3. *Peak load* is a load that occurs rarely and only for a short period. There are daily, monthly, and annual peak loads. Peak load occurrence depends on the building types, operation patterns, and climatic conditions.

TABLE 4.23 Typical Load Factors for Buildings

Building type	Range of load factor
Office buildings	0.45–0.55
Apartments	0.50–0.65
Schools	0.75–0.80
Grocery stores	0.50–0.60
Health care facilities	0.55–0.65
Hotels and motels	0.45–0.60
Restaurants	0.50–0.55
Manufacturing facilities	0.35–0.50

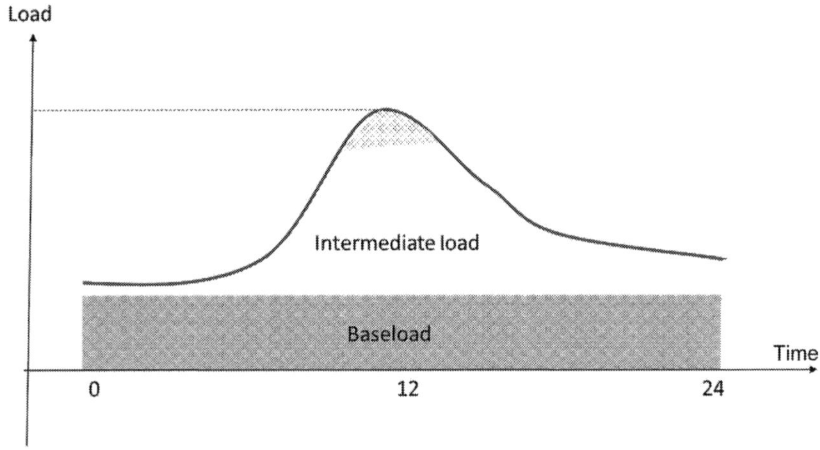

FIGURE 4.17 Typical electrical load profile with a baseload, an intermediate load, and a peak load.

Fig. 4.17 illustrates a daily load profile and identifies the baseload, intermediate load, and the peak load. Depending on the type of loads, the electrical utility operates different power generation plants to optimize operation costs. For baseloads, plants that cannot easily be stopped and then restarted and that have the lowest operating costs are typically used such as those using hydro, nuclear, and large steam engines. For intermediate loads, older large power plants using fuels are generally operated. Finally, peak loads are met with small power plants that are gas-fired. In addition, most utilities have some additional power plants to provide extra power generation beyond the peak demand to cover occasional contingency conditions and allow regular maintenance needs of the regular plans. The additional generation capacity is referred to as the reserve margin and typically varies from 12% to 15% of the total utility electrical peak load to be considered adequate. When the additional plants are operating continuously, they are referred to as spinning reserve. If these pants are idle but can be started within 10 min, they are considered nonspinning reserve. Otherwise, they are considered standby reserve but they have to be available within a short notice of any contingency occurrence.

4.8.2 Types of Demand Side Management Programs

There are several options to change the electrical load profile of a building to match it to the supply side. Fig. 4.18 shows some of the common actions typically considered to vary the building's load through demand or load management incentives:

- Peak clipping to reduce the peak demand as noted in Fig. 4.18A through demand response or peak shaving controls. Typically, time-based rates

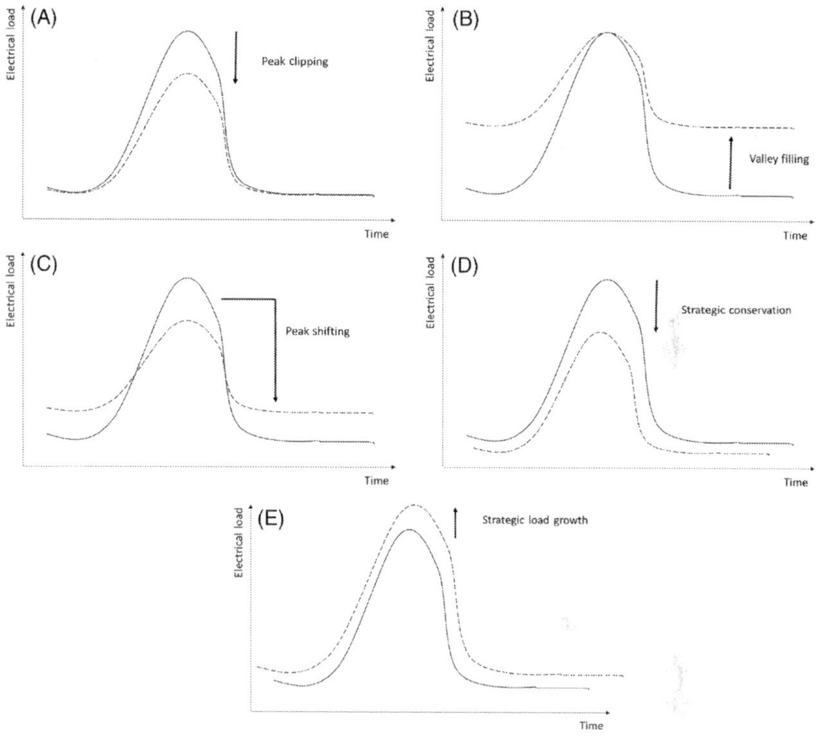

FIGURE 4.18 Typical load management options to vary the electrical load profile for buildings.

(i.e., TOU rates) or incentive-based strategies are used by utilities to encourage peak clipping by building owners and operators. Several options can be considered for peak clipping including delaying the use of some equipment such as appliances or turning off air-conditioning systems for only a few hours to ensure that indoor thermal comfort is not significantly affected.

- Valley filling to make the load profile more uniform through the day as shown in Fig. 4.18B. This load management option is not only specific to electricity usage but can be applicable to other energy sources. The main purpose of the valley filling management is to increase off-peak loads and ultimately boost the load factor. The use of electricity to meet end-uses served by other energy sources is a common strategy for valley clipping such as domestic hot water heating or space heating.
- Load shifting to displace electricity use from on-peak to off-peak periods as indicated in Fig. 4.18C. The use of storage systems for heating and cooling is the most common load shifting strategy used for buildings. The storage systems can be passive (i.e., use of building thermal mass) or active (i.e., use of thermal energy storage to shift on-peak to off-peak cooling).

- Strategic conservation to ensure that the energy end-use is lowered throughout the day as presented in Fig. 4.18D. This load management option is typically implemented by energy conservation measures such those associated with improved energy efficiency of the energy systems of the building including lighting, appliances, and air-conditioning systems.
- Strategic load growth to increase the energy end-use is increased beyond the levels outlined for valley filling as indicated in Fig. 4.18E. The use of electric vehicles or process heating are examples of strategic load growth strategies aimed at reducing the use of fossil fuels to meet some regulations or to improve the energy productivity.

4.8.3 DSM Incentive Programs

To encourage their clients to adhere to the offered DSM programs, utilities offer either price-based incentives or incentive-based programs as outlined in the following sections.

4.8.3.1 Price-Based Incentives

There are three common energy rate structures that encourage building operators to adopt peak clipping, load shifting, and demand response strategies:

1. TOU where the price of electricity is set depending on the time of the day. Examples of TOU rates are provided in Example 4.7.
2. RTP where the price of electricity varies continuously in response to wholesale market prices. Examples 4.8 and 4.9 provide utility bill analysis using RTP rates.
3. Critical peak pricing is typically the rate set by the utility for a few hours to charge high prices during critical periods to reduce the demand.

4.8.3.2 Incentive-Based Programs

The utilities can encourage the building owners to induce demand reduction through the following programs:

- Direct load controls where customers receive some payments for allowing the utilities to control the operation of some equipment (typically appliances and air-conditioning systems) during specific hours of the day and the season.
- Demand bidding programs where customers can offer bids to the utilities to curtail their loads during periods where wholesale market prices are high.
- Demand buyback programs where customers offer to curtail some portions of their loads based on posted prices.
- Emergency demand response programs where customers can receive payments for voluntarily reducing their loads during emergency periods to increase the grid reliability.
- Capacity market programs where customers receive payments or discounts for reducing their loads to lower the capacity requirements during critical periods of high demands.

- Interruptible/curtailable programs where customers receive discounts or credits for allowing the utilities to reduce their loads when requested. Penalties are typically enforced if the customers cannot deliver the desired curtailments.
- Ancillary services market programs where customers receive payments for ancillary services provided requiring them to adjust their loads quickly.

4.9 SUMMARY

In this chapter, various energy sources used to operate and power buildings in the US have been discussed. In addition, the energy price rate structures proposed by US utilities are outlined with a special emphasis on electricity pricing features. Several calculation examples are presented to illustrate the effects of various components of rate structures on the monthly utility bills. The chapter presents some of common utility rate structures and demand-side management programs that utilities offer to their customers. Moreover, the chapter outlines some of the options and associated challenges for microgrids to interconnect with the utility grid.

PROBLEMS

4.1 In this problem, the utility bill for a facility subject to power factor clause is to be estimated for one specific month under two scenarios.

1. Calculate the utility bill for a facility during one month characterized by the following parameters:
   - Actual demand: 450 kW
   - Energy consumption: 85,000 kWh
   - Average power factor: 70%

   The facility is subject to the following rate structure:
   - Customer charge: $150/month
   - Billed demand charge: $10/kW
   - Energy charge: $0.025/kWh

   The rate structure includes a power factor clause that states that the billing demand is determined as the actual demand multiplied by the ratio of 85 and the actual average power factor expressed in percent.
2. Determine the cost savings achieved for the month if the facility power factor is improved to be at least 85%.

4.2 A Company consumed about 355,000 kWh of electrical energy during the month of April. The billed demand is estimated to be 600 kW for April. If the sales tax is 7%, calculate the electric utility bill for the company.

Estimate the reduction in April electric utility bill for the company during following a peak shaving measure that lowered April billed peak demand to 450 kW.

The company is subject to the following electric rate structure:

Rate-1:

(i) Customer charge: $145.00/month
(ii) Energy charge: $0.0325/kWh/month

(iii) Demand charge applicable to the billing demand:
On-peak season: \$500 for the first 100 kW
plus \$5.50 / kW for any additional kW
Off-peak season: \$300 for the first 100 kW
plus \$3.50 / kW for any additional kW

The on-peak season extends from May to October and the off-peak season spans from November to April.

4.3 A manufacturing company consumes on average 250,000 kWh per month and has a billed demand of 1200 kW during each month. The company can select between two rates: Rate-1 (described in Problem 4.2) and Rate-2 (described here). Determine the best rate suitable for the company.

Rate-2:

(iv) Customer charge: \$250.00/month
(v) Energy charge: \$0.0225/kWh/month
(vi) Demand charge applicable to the billing demand:
On-peak season: \$400 for the first 100 kW
plus \$6.00 / kW for any additional kW
Off-peak season: \$250 for the first 100 kW
plus \$4.00 / kW for any additional kW

The on-peak season extends from June to September and the off-peak season spans from October to May.

4.4 The monthly energy consumption and peak demand history of a commercial building is shown in the table for 1 year.

Month	Energy use (kWh)	Peak demand (kW)
January	125,000	455
February	137,500	505
March	155,500	610
April	176,000	675
May	185,500	755
June	201,000	920
July	235,500	1150
August	240,000	1200
September	197,500	895
October	184,500	650
November	164,000	605
December	141,000	550

The commercial building is on rate structure defined in Table 4.13 (utility rate B).

(i) Estimate the annual electric utility bill for the commercial building assuming that the average power factor is 70% during each month.
(ii) Determine the reduction in the annual electric utility bill if the power factor is improved to 90%.

4.5 A utility offers the following rates to its commercial customers:

Service Level 1

Customer charge: $500/bill/month
Energy charge: $0.025 applicable to all kWh/month
Demand charge applicable to all kW/month of billing demand:
On-peak season: $350.00 for first 75 kW or less, and
$3.75 / kW for all additional kW
Off-peak season: $210.00 for first 75 kW or less, and
$2.50 / kW for all additional kW

Service Level 2

Customer charge: $200/bill/month
Energy charge: $0.027 applicable to all kWh/month
Demand charge applicable to all kW/month of billing demand:
On-peak season: $400.00 for first 75 kW or less, and
$4.50 / kW for all additional kW
Off-peak season: $250.00 for first 75 kW or less, and
$2.90 / kW for all additional kW

Service Level 3

Customer charge: $110/bill/month
Energy charge: $0.028 applicable to all kWh/month
Demand charge applicable to all kW/month of billing demand:
On-peak season: $450.00 for first 75 kW or less, and
$4.90 / kW for all additional kW
Off-peak season: $290.00 for first 75 kW or less, and
$3.30 / kW for all additional kW

The on-peak season includes the months of June through October. The off-season includes the remainder of the months of any year.

The billing demand is defined as the maximum demand as determined above the corrected for the power factor (refer to the power factor clause defined below), provided that no billing demand shall be considered as less than a percent of the highest on-peak season maximum demand corrected for the power factor previously determined during the 12 months ending with the current month.

The following power factor clause applies to all the utility rates structures:

Power factor clause: The consumer shall at all times use power so that the power factor shall be as close to 100% as possible, but when the average power factor as determined by continuous measurement of lagging reactive kilovolt-ampere hours is less than 80%, the billing demand shall be determined by multiplying the maximum demand, shown by the demand meter for the billing period, by 80 and dividing the product thus obtained by the actual average power factor expressed in percent.

(i) A large manufacturing company is on the rate schedule of level 3. Their peak demand for the last year is shown in the following table. They found a way to reduce their demand in the off-peak season by 100 kW, but the peak season demand will be the same (i.e., the demand in each month of November through May would be reduced by 100 kW). Assuming they are on the 65%

ratchet clause specified in the rate schedule, what is their dollar savings? Assume the high month was July of the previous year at 1150 kW. If the demand reduction of 100 kW occurred in the peak season, what would be the dollar savings (i.e., the demand in June through October would be reduced by 100 kW).

Month	Demand (kW)	Month	Demand (kW)
January	495	Jul	1100
February	550	Aug	1000
March	580	Sep	900
April	600	Oct	600
May	610	Nov	500
June	900	Dec	515

(ii) Determine among the three available rate structures, the rate level that minimizes annual utility bill for the manufacturing company before any demand reduction.

4.6 A company is on the utility rate structure defined as level 1 in Problem 4.5. In auditing a company, you find they have averaged 65% power factor over the past year. They are on the rate schedule given in the class handout and have averaged 1000 kW/month. Neglecting any ratchet clause and assuming their demand and power factor is constant each month, calculate the savings for correcting to 80% power factor. How much capacitance would be necessary to obtain this correction? (Assume level 1).

4.7 A company has contacted you regarding their rate schedule. The company is on the rate schedule defined in Problem 4.5 as service level 3 (secondary service), but are near transmission lines and so can accept service at a higher level (service level 1) if they buy their own transformers. Assuming they consume 300,000 kWh/month and are billed for 1000 kW each month, how much would they save by owning their own transformers. Ignore any charges other than demand and energy.

REFERENCES

Bollen, M.H., 2000. Understanding Power Quality Problems. Institute of Electrical and Electronics Engineers, Piscataway, NJ.

Cagle, S., 2012. Microgrids could bring big green changes to power systems.

Campbell, R.J., 2012. Weather-Related Power Outages and Electric System Resiliency. Report for Congress. Congressional Research Service, Washington, DC.

Caramia, P., Carpinelli, G., Verde, P., 2009. Power Quality Indices in Liberalized Markets. John Wiley and Sons Ltd, Chichester.

EIA, 2009. Annual Energy Review. Department of Energy, Energy Information Administration. http://www.doe.eia.gov/.

EIA, 2017. Annual Energy Review. Department of Energy, Energy Information Administration. http://www.doe.eia.gov/.

Fuchs, E.F., Masoum, M.A., 2008. Power Quality in Power Systems and Electrical Machines. Elsevier, Boston.

GRI, 1993. Electric and Gas Rates for the Residential, Commercial, and Industrial Sectors, Vols. 1 and 2, by Whitem, L.J., McVicker, C.M, Stiles, E. Gas Technology Institute, formerly Gas Research Institute, Des Plaines, IL.

GRI, 1999. Electric and Gas Rates for the Residential, Commercial, and Industrial Sectors, Volumes 1 and 2, by Whitem L.J., McVicker, C.M, and Stiles, E., Gas Technology Institute, formerly Gas Research Institute.

Krarti, M., 2010. Energy Audit of Building Systems: An Engineering Approach, 2nd ed. Francis and Taylor Publishing, CRC Press.

Kusko, A., Thompson, M.T., 2007. Power Quality in Electrical Systems. McGraw-Hill, New York, NY.

LaMonica, M., 2012. Microgrids Keep Power Flowing Through Sandy Outages. MIT Technology Review. http://www.technologyreview.com/view/507106/microgrids-keep-powerflowing-through-sandy-outages/.

Navigant, 2017. Microgrid Deployment Tracker. Navigant Research, Navigant Consultant Inc. http://www.navigantresearch.com/research/microgrid-deployment-tracker-2q16.

NRC, 2012. Terrorism and the Electric Power Delivery System. Report by National Research Council of the National Academies, Washington, DC.

Quandl, 2017. US Coal Prices by Regions, accessed on August 10, 2017. https://www.quandl.com/.

Vedam, R.S., Sarma, M.S., 2009. Power Quality: VAR Compensation in Power Systems. Taylor & Francis Group, Boca Raton, FL.

Chapter 5

Life-Cycle Cost and Energy Productivity Analyses

Chapter Outline

Optimal Design and Retrofit of Energy Efficient Buildings, Communities, and Urban Centers.
http://dx.doi.org/10.1016/B978-0-12-849869-9.00005-3

5.1 INTRODUCTION

In most applications, initial investments are required to implement energy conservation measures. These initial costs must be generally justified in terms of a reduction in the operating costs (due to energy cost savings). Therefore, most improvements in the efficiency of energy systems have a delayed reward, that is, expenses come at the beginning of a retrofit project while the benefits are incurred later. For an energy retrofit project to be economically worthwhile, the initial expenses have to be lower than the sum of the savings obtained by the reduction in the operating costs over the lifetime of the project.

The lifetime of an energy system retrofit project spans typically over several years. Therefore, it is important to properly compare savings and expenditures of various amounts of money over the lifetime of a project. Indeed, an amount of money at the beginning of a year is worth less at the end of the year and has an even less buying power at the end of the second year. Consequently, the amounts of money due to expenditures or savings incurred at different times of a retrofit project cannot be simply added.

In engineering economics, savings and expenditures of amounts of money during a project are typically called *cash flows*. To compare various cash flows over the lifetime of a project, a life-cycle cost (*LCC*) analysis is typically used. In this chapter, basic concepts of engineering economics are described. First, some of the fundamental principles and parameters of economic analysis are presented. In addition, data are provided to help the reader estimate relevant economic parameters. Then, the general procedure of an economic evaluation of a retrofit project is described. Finally, some of the advantages and disadvantages of the various economic analysis methods are discussed.

5.2 BASIC CONCEPTS

There are several economic parameters that affect a decision between various investment alternatives. To perform a sound economic analysis for energy retrofits, it is important that the auditor be (1) familiar with the most important economic parameters, and (2) aware of the basic economics concepts. The parameters and the concepts that significantly affect the economic decision making include the following:

- The time value of money and interest rates including simple and compounded interest.
- Inflation rate and composite interest rate.
- Taxes including sales, local, state, and federal tax charges.
- Depreciation rate and salvage value.

In the following sections, the listed parameters are briefly described to help the reader better understand the *LCC* analysis procedure to be discussed later on in this chapter. A more detailed description of these parameters and common engineering economic analysis approaches is provided by Krarti (2010).

5.2.1 Interest Rate

When money is borrowed to cover part or all of the initial cost of a retrofit project, a fee is charged for the use of this borrowed money. This fee is called *interest* (I) and the amount of money borrowed is called *principal* (P). The amount of the fee depends on the value of the principal and the length of time over which the money is borrowed. The interest charges are typically normalized to be expressed as a percentage of the total amount of money borrowed. This percentage is called *the interest rate* (i):

$$i = \frac{I}{P} \tag{5.1}$$

It is clear that an economy with low interest rates encourage money borrowing (for investment in projects or simply for buying goods) while an economy with high interest rates encourages money saving. Therefore, if money has to be borrowed for a retrofit project, the interest rate is a good indicator of whether or not the project can be cost-effective.

For the economic analysis of an energy efficiency project, the interest rate is typically assumed to be constant throughout the lifetime of the project. Therefore, it is common to use an average interest rate when an economic analysis is performed for a retrofit project. Table 5.1 provides historical data for long-term interest rates for selected countries. Example 5.1 illustrates the difference between simple and compounded interest rates.

TABLE 5.1 Average-Long Term Interest Rates for Selected Countries

Period/year	France	Germany	Japan	United States
Period				
1961–73	6.9	7.2	7.0	5.3
1974–80	11.2	8.1	8.0	8.6
1981–90	12.0	7.8	6.5	10.3
1990–95	8.5	7.5	5.1	7.2
Year				
1995	7.5	6.9	3.4	6.6
2000	5.4	5.3	1.7	6.0
2005	3.4	3.4	1.4	4.8
2010	4.1	4.0	2.0	4.4
2015	0.8	0.5	0.4	2.1

Source: OECD, 2017. Organization for Economic Cooperation and Development, Economic Statistics. Available at: http://www.ocde.org.

Example 5.1

A homeowner has $2500 available and has the option to invest this money in either (1) a bank that has an annual interest rate of 5% or (2) buying a new furnace for his home. If he decides to invest all the money in the bank, how much will the homeowner have after 15 years? Compare this amount if simple interest rather than compounded interest had been paid.

Solution

If the interest is compounded, with P = \$2500; N = 15; and i = 0.05; the investment will accumulate to the total amount F:

$$F = \$2500 * (1+0.05)^{15} = \$5197$$

Thus, the homeowner's original investment will have doubled over the 15-year period.

If simple interest had been paid, the total amount that would have accumulated is slightly less:

$$F = \$2500 * (1+0.05*15) = \$4375$$

5.2.2 Inflation Rate

Inflation occurs when the cost of goods and services increases from one period to the next. While the interest rate, i, defines the cost of money, the inflation rate, λ, measures the increase in the cost of goods and services. Therefore, the future cost of a commodity, FC, is higher than the present cost, PC, of the same commodity:

$$FC = PC(1+\lambda) \quad (5.2)$$

Table 5.2 presents historical data for inflation rates for selected countries.

TABLE 5.2 Average Inflation Rates for Selected Countries

Period/year	France	Germany	Japan	United States
Period				
1971–80	9.8	5.0	8.8	7.1
1981–90	6.2	2.5	2.1	4.7
1990–95	1.9	2.8	1.0	2.6
Year				
1995	1.8	1.2	−0.1	2.8
2000	1.8	1.4	−0.5	3.4
2005	1.9	1.9	−0.6	3.4
2010	0.7	0.4	−1.4	1.0
2015	0.1	0.2	0.8	0.1

Source: OECD, 2017. Organization for Economic Cooperation and Development, Economic Statistics. Available at: http://www.ocde.org.

If the interest and inflation rates are compounded during the same period, a composite interest rate, θ, can be defined to account for the fact that inflation decreases the buying power of money due to increases in the cost of commodities:

$$\theta = \frac{i - \lambda}{1 + \lambda} \tag{5.3}$$

It should be mentioned that theoretically the composite interest rate can be negative. In this case, the value of money is reduced with time.

Example 5.2

Determine the actual value of the \$2500 investment for the homeowner of Example 5.1 if the economy experiences an annual inflation rate of 3%.

Solution

The composite interest rate can be determined using Eq. (5.3):

$$\theta = \frac{i - \lambda}{1 + \lambda} = \frac{0.05 - 0.03}{1 + 0.03} = 0.01942$$

With P = \$2500 and N = 15; the investment will accumulate to the total amount F:

$$F = \$2500 * (1 + 0.01942)^{15} = \$3336$$

5.2.3 Tax Rate

In most economies, the interest incurred from an investment is subject to taxation. If this taxation has a rate, t, over a period that coincides with the interest period, then the amount of taxes, T, to be collected from an investment, P, with an interest rate, i, is determined as follows:

$$T = tiP \tag{5.4}$$

Therefore, the composite interest rate defined by Eq. (5.3) can be generalized to account for both inflation and tax rates related to present and future values:

$$\theta = \frac{(1 - t)i - \lambda}{1 + \lambda} \tag{5.5}$$

Example 5.3

If the homeowner is in the 28% tax bracket, determine the actual value of his \$2500 investment considered in Example 2.1 if the economy experiences an annual inflation rate of 3%.

Solution

The composite interest rate can be determined using Eq. (3.15):

$$\theta = \frac{(1-t)i - \lambda}{1+\lambda} = \frac{(1-0.28) * 0.05 - 0.03}{1+0.03} = 0.00583$$

With P = \$2500 and N = 15; the investment will accumulate to the total amount F:

$$F = \$2500 * (1+0.00583)^{15} = \$2728$$

5.2.4 Cash Flows

In evaluating energy efficiency projects, it is important to account for the total cash receipts and disbursements due to an implementation of an energy conservation measure (such as the installation of a new furnace) during the entire lifetime of the project. The difference between the total cash receipts (inflows) and total cash disbursements (outflows) for a given period is called a cash flow.

Over the project lifetime, an accurate accounting of all the cash flows should be performed. For energy efficiency improvement projects, the cash flow accounting can be outlined in a tabular format as illustrated in Table 5.3 for the cash flows attributed to the installation of a new furnace. In particular, Table 5.3 accounts for the cost related to the initial cost of a new furnace installation (counted as a

TABLE 5.3 Cash Flows for an Installation of a New Boiler Over a Lifetime of 10 Years

End of year	Total cash receipts	Total cash disbursements	Total cashflows	Comments
0	\$ 0	\$ 4,000	−\$ 4,000	Installation cost of a new furnace
1	\$ 400	\$ 0	+\$ 400	Net cost savings
2	\$ 380	\$ 0	+\$ 380	" "
3	\$ 360	\$ 0	+\$ 360	
4	\$ 340	\$ 0	+\$ 340	
5	\$ 330	\$ 0	+\$ 330	
6	\$ 320	\$ 0	+\$ 320	
7	\$ 310	\$ 0	+ \$ 310	
8	\$ 305	\$ 0	+ \$ 305	
9	\$ 300	\$ 0	+ \$ 300	
10	\$ 295	\$ 0	+ \$ 295	

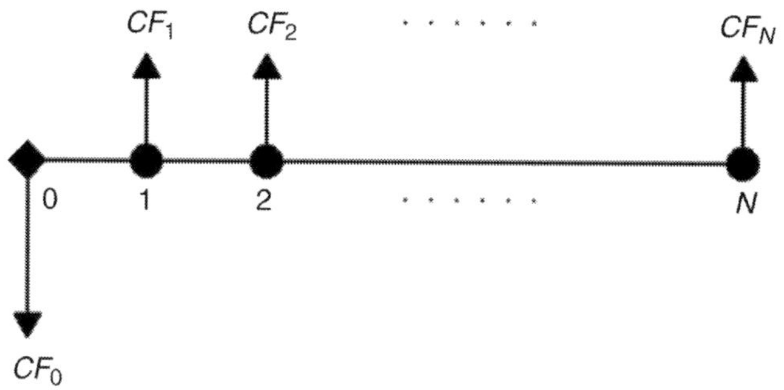

FIGURE 5.1 **Typical cash flow diagram.**

disbursement for Year 0) and the cost savings due to the higher energy efficiency of the new boiler (counted as receipts in Year 1 through Year 10). The reduction in the annual receipts is attributed to the aging of the equipment.

Note that the cash flows are positive when they represent inflows (i.e., receipts) and are negative when they are outflows (i.e., disbursements). To better visualize the evolution over time of cash flows, a cash flow diagram can be used as depicted in Fig. 5.1. Note that in Fig. 5.1, the initial cash flow, $C_0 = -\$4000$ (disbursement), is represented by a downward-pointing arrow. Meanwhile, the cash flows occurring later, C_1 through C_N with $N = 10$, are receipts and are represented by upward-pointing arrows.

It should be noted again that cash flows cannot be simply added since the value of money changes from one period to the next. In the next section, various factors are defined to correlate cash flows occurring at different periods.

5.3 COMPOUNDING FACTORS

Two types of payment factors are considered in this section. These payment factors are useful in the economical evaluation of various retrofit projects. Without loss of generality, the interest period is assumed to be 1 year. Moreover, a nominal discount rate, d, is used for the analysis described in this chapter. The discount rate is an effective interest rate that includes the effects of several parameters such as inflation and taxation as shown by Eq. (5.5).

5.3.1 Single Payments

In this case, an initial payment is made to implement a project by borrowing an amount of money, P. If this sum of money earns interest at a discount rate, d, then the value of the payment, P, after N years is a future amount, F. The ratio F/P is often called single payment compound-amount factor (*SPCA*). The *SPCA* factor is a function of i and N and is defined as

$$SPCA(d,N) = F / P = (1+d)^N \tag{5.6}$$

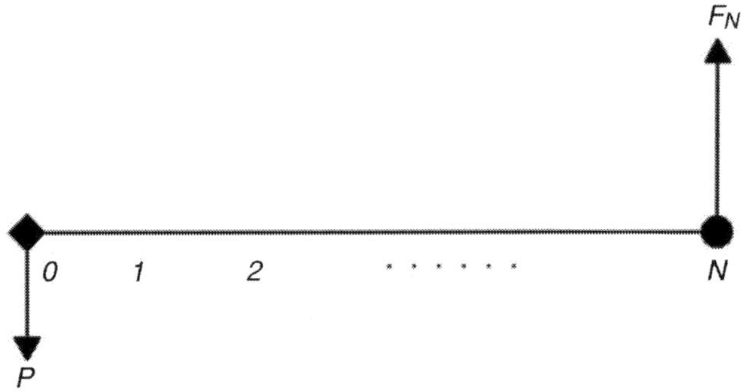

FIGURE 5.2 Cash flow diagram for single payment.

Using the cash flow diagram of Fig. 5.1, the single payment represents the case where $C_0 = P$, $C_1 = \ldots = C_{N-1} = 0$, and $C_N = F$ as illustrated in Fig. 5.2.

The inverse ratio *P/F* allows to calculate the value of the cash flow, *P*, needed to attain a given amount of cash flow *F* after *N* years. The ratio *P/F* is called single payment present worth (*SPPW*) factor and is equal to:

$$SPPW(d,N) = P / F = (1+d)^{-N} \tag{5.7}$$

Table 5.4 provides the *SPPW* values for various discount rate, *d*, and life-cycle, *N*.

5.3.2 Uniform-Series Payments

In most energy retrofit projects, energy cost savings are estimated annually after an initial investment to implement one package of energy efficiency measures. It is hoped that during the lifetime of the project, the sum of all the annual energy cost savings can surpass the initial investment.

Consider then an amount of money, *P*, that represents the initial investment, and a receipt of an amount, *A*, that is made each year and represents the cost savings due to the retrofit project. To simplify the analysis, the amount *A* is assumed to be constant for all the years during the lifetime of the project. Therefore, the cash flows can be set to $C_0 = P$, $C_1 = \ldots = C_N = A$ as depicted in Fig. 5.3.

It can be shown that the ratio *A/P*, called uniform-series capital recovery factor (*USCR*), can be expressed as a function of both *d* and *N*:

$$USCR(d,N) = A / P = \frac{d(1+d)^N}{(1+d)^N - 1} = \frac{d}{1-(1+d)^{-N}} \tag{5.8}$$

TABLE 5.4 *SPPW* Values for Selected Discount Rates, *d*, and Number of Years, *N*

SPPW	Discount rate, *d* (%)										
N (years)	0	1	2	3	4	5	6	7	8	9	10
1	1.000	0.990	0.980	0.971	0.962	0.952	0.943	0.935	0.926	0.917	0.909
2	1.000	0.980	0.961	0.943	0.925	0.907	0.890	0.873	0.857	0.842	0.826
3	1.000	0.971	0.942	0.915	0.889	0.864	0.840	0.816	0.794	0.772	0.751
4	1.000	0.961	0.924	0.888	0.855	0.823	0.792	0.763	0.735	0.708	0.683
5	1.000	0.951	0.906	0.863	0.822	0.784	0.747	0.713	0.681	0.650	0.621
6	1.000	0.942	0.888	0.837	0.790	0.746	0.705	0.666	0.630	0.596	0.564
7	1.000	0.933	0.871	0.813	0.760	0.711	0.665	0.623	0.583	0.547	0.513
8	1.000	0.923	0.853	0.789	0.731	0.677	0.627	0.582	0.540	0.502	0.467
9	1.000	0.914	0.837	0.766	0.703	0.645	0.592	0.544	0.500	0.460	0.424
10	1.000	0.905	0.820	0.744	0.676	0.614	0.558	0.508	0.463	0.422	0.386
11	1.000	0.896	0.804	0.722	0.650	0.585	0.527	0.475	0.429	0.388	0.350
12	1.000	0.887	0.788	0.701	0.625	0.557	0.497	0.444	0.397	0.356	0.319
13	1.000	0.879	0.773	0.681	0.601	0.530	0.469	0.415	0.368	0.326	0.290
14	1.000	0.870	0.758	0.661	0.577	0.505	0.442	0.388	0.340	0.299	0.263
15	1.000	0.861	0.743	0.642	0.555	0.481	0.417	0.362	0.315	0.275	0.239
16	1.000	0.853	0.728	0.623	0.534	0.458	0.394	0.339	0.292	0.252	0.218

(Continued)

TABLE 5.4 *SPPW* Values for Selected Discount Rates, *d*, and Number of Years, *N* (*cont.*)

SPPW	Discount rate, *d* (%)										
N (years)	0	1	2	3	4	5	6	7	8	9	10
17	1.000	0.844	0.714	0.605	0.513	0.436	0.371	0.317	0.270	0.231	0.198
18	1.000	0.836	0.700	0.587	0.494	0.416	0.350	0.296	0.250	0.212	0.180
19	1.000	0.828	0.686	0.570	0.475	0.396	0.331	0.277	0.232	0.194	0.164
20	1.000	0.820	0.673	0.554	0.456	0.377	0.312	0.258	0.215	0.178	0.149
21	1.000	0.811	0.660	0.538	0.439	0.359	0.294	0.242	0.199	0.164	0.135
22	1.000	0.803	0.647	0.522	0.422	0.342	0.278	0.226	0.184	0.150	0.123
23	1.000	0.795	0.634	0.507	0.406	0.326	0.262	0.211	0.170	0.138	0.112
24	1.000	0.788	0.622	0.492	0.390	0.310	0.247	0.197	0.158	0.126	0.102
25	1.000	0.780	0.610	0.478	0.375	0.295	0.233	0.184	0.146	0.116	0.092
26	1.000	0.772	0.598	0.464	0.361	0.281	0.220	0.172	0.135	0.106	0.084
27	1.000	0.764	0.586	0.450	0.347	0.268	0.207	0.161	0.125	0.098	0.076
28	1.000	0.757	0.574	0.437	0.333	0.255	0.196	0.150	0.116	0.090	0.069
29	1.000	0.749	0.563	0.424	0.321	0.243	0.185	0.141	0.107	0.082	0.063
30	1.000	0.742	0.552	0.412	0.308	0.231	0.174	0.131	0.099	0.075	0.057

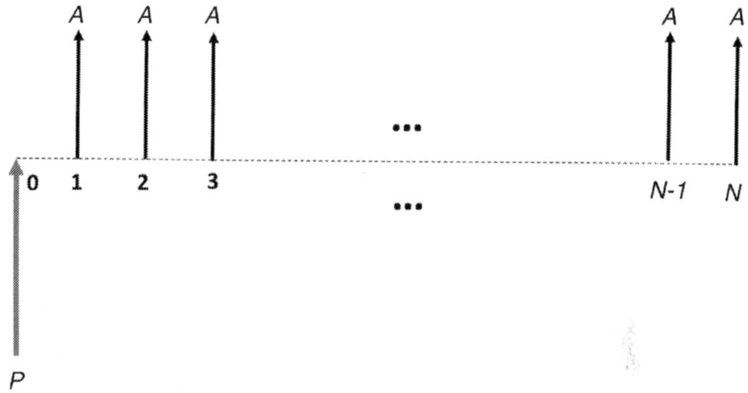

FIGURE 5.3 Cash flow diagram for uniform-series payment.

The uniform-series present worth factor (*USPW*), needed to calculate the value of *P* knowing the amount *A*, is the ratio *P/A* and can be expressed as follows:

$$USPW(d,N) = P/A = \frac{(1+d)^N - 1}{d(1+d)^N} = \frac{1-(1+d)^{-N}}{d} \tag{5.9}$$

Table 5.5 provides the *USPW* values for various discount rate, *d*, and life-cycle, *N*.

5.3.3 Uniform Gradient Series Payments

In some projects, energy use and cost savings can increase uniformly during their life-cycle. It is the case of incremental implementation of large-scale energy efficiency programs. For instance, when 10% of the building stock is retrofitted on an annual basis, the energy use savings increase uniformly by about 10% every year during the 10-year implementation period. The energy cost savings increase also incrementally during the same implementation period. The total cost savings achieved from the project may require the sum of a uniform gradient series as outlined next.

Consider then a gradient increase of an initial payment *G* during *N* years as shown in Fig. 5.4. The present values of these payments can be estimated using the sum of a gradient series for an economy with a discount rate, *d*. In particular, it can be shown that the ratio *P/G*, called uniform-gradient present worth (*UGPW*) factor can be expressed as a function of *d* and *N*:

$$UGPW(d,N) = P/G = \frac{1-(1+d)^{-N}}{d^2} - \frac{N(1+d)^{-N}}{d} \tag{5.10}$$

Table 5.6 provides the *UGPW* values for various discount rate, *d*, and life-cycle, *N*.

Similarly, it can be shown that the future value, *F*, of all the payments can be estimated using the ratio *F/G*, called uniform-gradient future worth (*UGPW*)

TABLE 5.5 *USPW* Values for Selected Discount Rates, *d*, and Number of Years, *N*

USPW	Discount rate, *d* (%)										
N (years)	0	1	2	3	4	5	6	7	8	9	10
1	1.00	0.99	0.98	0.97	0.96	0.95	0.94	0.93	0.93	0.92	0.91
2	2.00	1.97	1.94	1.91	1.89	1.86	1.83	1.81	1.78	1.76	1.74
3	3.00	2.94	2.88	2.83	2.78	2.72	2.67	2.62	2.58	2.53	2.49
4	4.00	3.90	3.81	3.72	3.63	3.55	3.47	3.39	3.31	3.24	3.17
5	5.00	4.85	4.71	4.58	4.45	4.33	4.21	4.10	3.99	3.89	3.79
6	6.00	5.80	5.60	5.42	5.24	5.08	4.92	4.77	4.62	4.49	4.36
7	7.00	6.73	6.47	6.23	6.00	5.79	5.58	5.39	5.21	5.03	4.87
8	8.00	7.65	7.33	7.02	6.73	6.46	6.21	5.97	5.75	5.53	5.33
9	9.00	8.57	8.16	7.79	7.44	7.11	6.80	6.52	6.25	6.00	5.76
10	10.00	9.47	8.98	8.53	8.11	7.72	7.36	7.02	6.71	6.42	6.14
11	11.00	10.37	9.79	9.25	8.76	8.31	7.89	7.50	7.14	6.81	6.50
12	12.00	11.26	10.58	9.95	9.39	8.86	8.38	7.94	7.54	7.16	6.81
13	13.00	12.13	11.35	10.63	9.99	9.39	8.85	8.36	7.90	7.49	7.10
14	14.00	13.00	12.11	11.30	10.56	9.90	9.29	8.75	8.24	7.79	7.37
15	15.00	13.87	12.85	11.94	11.12	10.38	9.71	9.11	8.56	8.06	7.61
16	16.00	14.72	13.58	12.56	11.65	10.84	10.11	9.45	8.85	8.31	7.82
17	17.00	15.56	14.29	13.17	12.17	11.27	10.48	9.76	9.12	8.54	8.02
18	18.00	16.40	14.99	13.75	12.66	11.69	10.83	10.06	9.37	8.76	8.20

19	19.00	17.23	15.68	14.32	13.13	12.09	11.16	10.34	9.60	8.95	8.36
20	20.00	18.05	16.35	14.88	13.59	12.46	11.47	10.59	9.82	9.13	8.51
21	21.00	18.86	17.01	15.42	14.03	12.82	11.76	10.84	10.02	9.29	8.65
22	22.00	19.66	17.66	15.94	14.45	13.16	12.04	11.06	10.20	9.44	8.77
23	23.00	20.46	18.29	16.44	14.86	13.49	12.30	11.27	10.37	9.58	8.88
24	24.00	21.24	18.91	16.94	15.25	13.80	12.55	11.47	10.53	9.71	8.98
25	25.00	22.02	19.52	17.41	15.62	14.09	12.78	11.65	10.67	9.82	9.08
26	26.00	22.80	20.12	17.88	15.98	14.38	13.00	11.83	10.81	9.93	9.16
27	27.00	23.56	20.71	18.33	16.33	14.64	13.21	11.99	10.94	10.03	9.24
28	28.00	24.32	21.28	18.76	16.66	14.90	13.41	12.14	11.05	10.12	9.31
29	29.00	25.07	21.84	19.19	16.98	15.14	13.59	12.28	11.16	10.20	9.37
30	30.00	25.81	22.40	19.60	17.29	15.37	13.76	12.41	11.26	10.27	9.43

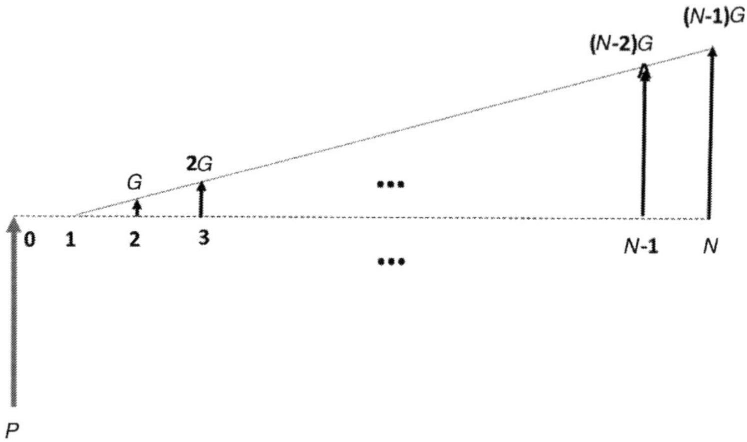

FIGURE 5.4 Present value for uniform gradient series payments.

factor as illustrated in Fig. 5.5. The factor *UGPW* can be calculated using Eqs. (5.6) and (5.10) and expressed as a function of both *d* and *N*:

$$UGFW(d,N) = F/G = (F/P)*(P/G) = \frac{(1+d)^N - 1}{d^2} - \frac{N}{d} \tag{5.11}$$

Finally, the uniform-gradient series can be converted to a uniform-series, *A*, using the ratio (*A/G*), also called uniform gradient uniform series (*UGUS*) factor. The expression of the *UGUS* factor can be determined using both Eqs. (5.8) and (5.10):

$$USUS(d,N) = A/G = (A/P)*(P/G) = \frac{1}{d} - \frac{N(1+d)^{-N}}{1-(1+d)^{-N}} \tag{5.12}$$

Example 5.4 estimates the compounding factors for a given set of a discount rate, *d*, and a life-cycle, *N*.

Example 5.4

Find the various compounding factors for $N = 10$ years and $d = 5\%$.

Solution

The values of the compounding factors for $d = 0.05$ and $N = 10$ years are summarized below:

Compounding factor	Equation used	Value
SPCA	Eq. (5.6)	1.629
SPPW	Eq. (5.7)	0.613
USCR	Eq. (5.8)	0.129
USPW	Eq. (5.9)	7.740
UGPW	Eq. (5.10)	31.65
UGFW	Eq. (5.11)	51.56
UGUS	Eq. (5.12)	4.100

TABLE 5.6 *UGPW* Values for Selected Discount Rates, *d*, and Number of Years, *N*

UGPW	Discount rate, *d* (%)									
N (years)	1	2	3	4	5	6	7	8	9	10
1	0.00	0.00	0.00	0.00	0.00	0.00	0.00	0.00	0.00	0.00
2	0.98	0.96	0.94	0.92	0.91	0.89	0.87	0.86	0.84	0.83
3	2.92	2.85	2.77	2.70	2.63	2.57	2.51	2.45	2.39	2.33
4	5.80	5.62	5.44	5.27	5:10	4.95	4.79	4.65	4.51	4.38
5	9.61	9.24	8.89	8.55	8.24	7.93	7.65	7.37	7.11	6.86
6	14.32	13.68	13.08	12.51	11.97	11.46	10.98	10.52	10.09	9.68
7	19.92	18.90	17.95	17.07	16.23	15.45	14.71	14.02	13.37	12.76
8	26.38	24.88	23.48	22.18	20.97	19.84	18.79	17.81	16.89	16.03
9	33.70	31.57	29.61	27.80	26.13	24.58	23.14	21.81	20.57	19.42
10	41.84	38.96	36.31	33.88	31.65	29.60	27.72	25.98	24.37	22.89
11	50.81	47.00	43.53	40.38	37.50	34.87	32.47	30.27	28.25	26.40
12	60.57	55.67	51.25	47.25	43.62	40.34	37.35	34.63	32.16	29.90
13	71.11	64.95	59.42	54.45	49.99	45.96	42.33	39.05	36.07	33.38
14	82.42	74.80	68.01	61.96	56.55	51.71	47.37	43.47	39.96	36.80
15	94.48	85.20	77.00	69.74	63.29	57.55	52.45	47.89	43.81	40.15
16	107.27	96.13	86.35	77.74	70.16	63.46	57.53	52.26	47.58	43.42

(Continued)

TABLE 5.6 *UGPW* Values for Selected Discount Rates, *d*, and Number of Years, *N* (*cont.*)

UGPW	Discount rate, *d* (%)									
N (years)	1	2	3	4	5	6	7	8	9	10
17	120.78	107.56	96.03	85.96	77.14	69.40	62.59	56.59	51.28	46.58
18	135.00	119.46	106.01	94.35	84.20	75.36	67.62	60.84	54.89	49.64
19	149.90	131.81	116.28	102.89	91.33	81.31	72.60	65.01	58.39	52.58
20	165.47	144.60	126.80	111.56	98.49	87.23	77.51	69.09	61.78	55.41
21	181.69	157.80	137.55	120.34	105.67	93.11	82.34	73.06	65.05	58.11
22	198.57	171.38	148.51	129.20	112.85	98.94	87.08	76.93	68.20	60.69
23	216.07	185.33	159.66	138.13	120.01	104.70	91.72	80.67	71.24	63.15
24	234.18	199.63	170.97	147.10	127.14	110.38	96.25	84.30	74.14	65.48
25	252.89	214.26	182.43	156.10	134.23	115.97	100.68	87.80	76.93	67.70
26	272.20	229.20	194.03	165.12	141.26	121.47	104.98	91.18	79.59	69.79
27	292.07	244.43	205.73	174.14	148.22	126.86	109.17	94.44	82.12	71.78
28	312.50	259.94	217.53	183.14	155.11	132.14	113.23	97.57	84.54	73.65
29	333.49	275.71	229.41	192.12	161.91	137.31	117.16	100.57	86.84	75.41
30	355.00	291.72	241.36	201.06	168.62	142.36	120.97	103.46	89.03	77.08

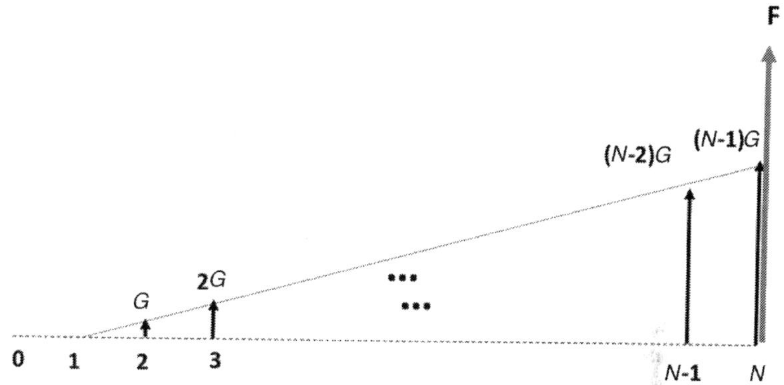

FIGURE 5.5 **Future value for uniform gradient series payments.**

5.4 ECONOMIC EVALUATION METHODS

To evaluate the cost-effectiveness of energy retrofit projects, several evaluation tools can be considered. The basic concept of all these tools is to compare among the retrofit alternatives the net cash flow that can be obtained during the lifetime of the project. As discussed earlier, a simple addition of all the cash flows such as those represented in Fig. 5.1 is not appropriate. However, by using the compound factors discussed in Section 5.3, "conversion" of the cash flows from one period to another is feasible. This section provides a brief description of the common economic evaluation methods used in engineering projects.

5.4.1 Net Present Worth

The basic principle of this method is to evaluate the present worth of the cash flows that occur during the lifetime of the project. Referring to the cash flow diagram of Fig. 5.1, the sum of all the present worth of the cash flows can be obtained by using the *SPPW* factor defined in Eq. (5.7):

$$NPW = -CF_0 + \sum_{k=1}^{N} CF_k * SPPW(d,k) \tag{5.13}$$

Note that the initial cash flow is negative (a capital cost for the project) while the cash flows for the other years are generally positive (revenues).

In the particular common case of a project with constant annual revenues (due to energy operating cost savings), $CF_k = A$, the net present worth is reduced to

$$NPW = -CF_0 + A * USPW(d,N) \tag{5.14}$$

For the project to be economically viable, the net present worth has to be positive or at worst zero ($NPW \geq 0$). Obviously, the higher is the value of net present worth (*NPW*), the more economically sound is the project. The net present worth value method is often called the net savings method since the revenues are often due to the energy cost savings from implementing the project.

5.4.2 Rate of Return

In this method, the first step is to determine the specific value of the discount rate, d', that reduces the net present worth to zero. This specific discount rate is called the rate of return (*ROR*). Depending on the case, the expression of *NPW* provided in Eq. (5.13) or Eq. (5.14) can be used. For instance, in the general case of Eq. (5.13), the *ROR*, d', is solution of the following equation:

$$-CF_0 + \sum_{k=1}^{N} CF_k * SPPW(d',k) = 0 \tag{5.15}$$

To solve accurately this equation, any numerical method can be used. However, an approximate value of d' can be obtained by trial and error. This approximate value can be determined by finding the two d values for which the *NPW* is slightly negative and slightly positive, and then interpolate linearly between the two values. It is important to remember that a solution for *ROR* may not exist.

Once the *ROR* is obtained for a given alternative of the project, the actual market discount rate or the minimum acceptable *ROR* is compared to the *ROR* value. If the value of *ROR* is larger ($d' > d$), the project is cost-effective.

5.4.3 Benefit-Cost Ratio

The benefit-cost ratio (*BCR*) method is also called the savings-to-investment ratio (SIR) and provides a measure of the net benefits (or savings) of the project relative to its net cost. The net values of both benefits (B_k) and costs (C_k) are computed relative to a baseline case. The present worth of all the cash flows is typically used in this method. Therefore, the *BCR* is computed as follows:

$$BCR = \frac{\sum_{k=0}^{N} B_k * SPPW(d,k)}{\sum_{k=0}^{N} C_k * SPPW(d,k)} \tag{5.16}$$

The alternative option for the project is considered economically viable relative to the base case when the *BCR* is greater than 1.0 ($BCR > 1.0$).

5.4.4 Discounted and Simple Payback Periods

In this evaluation method, the period Y (expressed typically in years) required to recover an initial investment is determined. Using the cash flow diagram of Fig. 5.1, the value of Y is solution of the following equation:

$$CF_0 = \sum_{k=1}^{Y} CF_k * SPPW(d,k) \tag{5.17}$$

If the payback period Y is less than the lifetime of the project N ($Y < N$), then the project is economically viable. The value of Y obtained using Eq. (5.17) is typically called discounted payback period (*DPP*) since it includes the value of money.

In the vast majority of applications, the time value of money is neglected in the payback period method. In this case, Y is called simple payback period (*SPP*) and is the solution of the following equation:

$$CF_0 = \sum_{k=1}^{Y} CF_k \tag{5.18}$$

In the case where the annual net savings are constant ($CF_k = A$), the *SPP* can be easily calculated as the ratio of the initial investment over the annual net savings:

$$Y = \frac{CF_0}{A} \tag{5.19}$$

The values for the *SPP* are shorter than for the *DPP*s since the undiscounted net savings are greater than their discounted counterparts. Therefore, acceptable values for *SPPs* are typically significantly shorter than the lifetime of the project.

5.4.5 Cost of Energy

For power-generating projects, the cost of energy is often used to determine their cost effectiveness compared to the baseline options such as buying power from the utility grid system. Specifically, the cost of energy is the cost of generating 1 kWh by the power-generating facility or technology accounting for its capital costs, financing costs, fuel costs, and fixed and variable operating and maintenance (O&M) costs over an assumed lifetime. Typically, the term levelized cost of energy (*LCOE*) is utilized when the present worth of all costs are considered. The specific calculations for *LCOE* depend on the generating technology but a general method is provided by Eq. (5.20):

$$LCOE = \frac{\sum_{k=0}^{N} C_k * SPPW(d,k)}{\sum_{k=0}^{N} E_k} \tag{5.20}$$

where C_k are the total costs incurred in year k and include investment expenditures as well as O&M and fuel costs, while E_k is the total electrical energy produced at year k. For a generating project or technology to be economically competitive, *LCOE* should be lower than the baseline (typically from the grid) price, COE_b.

5.4.6 Summary of Economic Analysis Methods

Table 5.7 summarizes the basic characteristics of the economic analysis methods used to evaluate single alternatives of an energy retrofit project. Example 5.5 uses the presented economic analysis methods to assess the cost-effnness of a boiler retrofit.

It is important to note that the economic evaluation methods just described provide an indication of whether or not a single alternative of a retrofit project is cost-effective. However, these methods cannot be used or relied on to compare and rank various alternatives for a given retrofit project. The *LCC* analysis method is more appropriate.

TABLE 5.7 Summary of the Basic Criteria for the Various Economic Analysis Methods

Evaluation method	Equation	Criterion
Net present worth (*NPW*)	$NPW = -CF_0 + \sum_{k=1}^{N} CF_k * SPPW(d,k)$	$NPW > 0$
Rate of return (*ROR*)	$-CF_0 + \sum_{k=1}^{N} CF_k * SPPW(d',k) = 0$	$d' > d$
Benefit-cost ratio (*BCR*)	$BCR = \frac{\sum_{k=0}^{N} B_k * SPPW(d,k)}{\sum_{k=0}^{N} C_k * SPPW(d,k)}$	$BCR > 1$
Discounted payback period (*DPP*)	$CF_0 = \sum_{k=1}^{Y} CF_k * SPPW(d,k)$	$Y < N$
Simple payback period (*SPP*)	$Y = \frac{CF_0}{A}$	$Y \ll N$
Levelized cost of energy (*LCOE*)	$LCOE = \frac{\sum_{k=0}^{N} C_k * SPPW(d,k)}{\sum_{k=0}^{N} E_k}$	$LCOE < COE_b$

Example 5.5

After finding that the old boiler has an efficiency of only 60% while a new boiler would have an efficiency of 85%, a building owner of Example 5.1 has decided to invest $10,000 in getting a new boiler. Determine whether this investment is cost-effective if the lifetime of the boiler is 10 years and the discount rate is 5%. The boiler consumes 5000 gallons per year at a cost of $1.20 per gallon. An annual maintenance fee of $150 is required for the boiler (independently of its age). Use the first five methods summarized in Table 5.7 to perform the economic analysis.

Solution

The baseline case for the economic analysis presented in this example is the case where the boiler is not replaced. Moreover, the salvage value of the boiler is assumed insignificant after 10 years. Therefore, the only annual cash flows (A) after the initial investment on a new boiler are the net savings due to higher boiler efficiency as calculated here:

$$A = \text{Fuel} - \text{Use}_{\text{before}} * \left(1 - \frac{\eta_{\text{before}}}{\eta_{\text{after}}}\right) * \text{Fuel} - \text{cost/gallon}$$

Thus,

$$A = 5000 * \left(1 - \frac{0.60}{0.85}\right) * \$1.20 = \$1765$$

The cost-effectiveness of replacing the boiler is evaluated as indicated below:

(1) Net present worth: For this method CF_0 = $10,000 and $CF_1 = \ldots = CF_{10} = A$, $d = 0.05$, and $N = 10$ years. Using Eq. (5.11) with $USPW = 7.740$ (see Example 5.4):
$NPW = \$3682$
Therefore, the investment in purchasing a new boiler is cost-effective.

(2) Rate of return: For this method also CF_0 = $10,000 and $CF_1 = \ldots = CF_{10} = A$, while $SPPW(d',k)$ is provided by Eq. (5.7). By trial and error, it can be shown that the solution for d' is
$d' = 12.5\%$
Since $d' > d = 5\%$, the investment in replacing the boiler is cost-effective.

(3) Benefit-cost ratio: In this case, $B_0 = 0$ and $B_1 = \ldots = B_{10} = A$ while C_0 = $10,000 and $C_1 = \ldots = C_{10} = 0$.
Note that since the maintenance fee is applicable to both old and new boilers, this cost is not accounted for in this evaluation method (only the benefits and costs relative to the baseline case are considered).
Using Eq. (5.16):

$$BCR = 1.368$$

Thus, the benefit–cost ratio is greater than unity ($BCR > 1$) and the project of getting a new boiler is economically feasible.

(4) Discounted payback period: For this method, CF_0 = $10,000 and $CF_1 = \ldots = CF_{10} = A$. Using Eq. (5.18), Y can be solved: $Y = 6.9$ year.

Thus, the discounted payback period is shorter than the lifetime of the project ($Y > N = 10$ years) and therefore replacing the boiler is cost-effective.

(5) Simple payback period: For this method, $CF_0 = \$10,000$ and $A = \$1765$. Using Eq. (5.19), Y can be easily determined: $Y = 5.7$ years.
Thus, the simple payback period method indicates that the boiler retrofit project can be cost-effective.

5.5 LIFE-CYCLE COST ANALYSIS METHOD

5.5.1 General Analysis Approach

The *LCC* analysis method is the most commonly accepted method to assess the economic benefits of energy efficiency projects over their lifetime. Typically, the method is used to evaluate at least two alternatives of a given project (for instance, evaluate two alternatives for the same retrofit project: replace a furnace or consider a ground source heat pump to condition a house). Only one alternative will be selected for implementation based on the economic analysis.

The basic procedure of the *LCC* method is relatively simple since it seeks to determine the relative cost effectiveness of various alternatives. For each alternative including the baseline case, the total cost is estimated over the project lifetime. The cost is commonly determined using one of two approaches: (1) the present worth or (2) the annualized cost estimate. Then, the alternative with the lowest total cost (or *LCC*) is typically selected.

Using the cash flow diagram of Fig. 5.1, the *LCC* amount for each alternative can be computed by projecting all the costs (including costs of acquisition, installation, maintenance, and operating of the energy systems related to the energy-conservation project) to

(1) One single present value amount that can be computed as follows:

$$LCC = \sum_{k=0}^{N} CF_k * SPPW(d,k) \tag{5.21}$$

This is the most commonly used approach in calculating *LCC* in energy retrofit projects.

Or

(2) Multiple annualized costs over the lifetime of the project:

$$LCC_a = USCR(d,N) * \left[\sum_{k=1}^{N} CF_k * SPPW(d,k) \right] \tag{5.22}$$

Note that the two approaches for calculating the *LCC* values are equivalent.

In several energy efficiency projects, the annual cash flow remains the same after the initial investment. In these cases, *LCC* can be estimated based on the initial cost, *IC*, and the annual cost, *AC*, as follows:

$$LCC = IC + USPW(d, N) * AC \tag{5.23}$$

Example 5.6 provides an economic analysis for three projects for retrofitting a burner–boiler system using the *LCC* approach.

Example 5.6

The building owner of Example 5.5 has three options to invest his money as briefly described here:

(A) Replace the entire older boiler (including burner) with more efficient heating system. The old boiler/burner system has an efficiency of only 60% while a new boiler/burner system has an efficiency of 85%. The cost of this replacement is $10,000.

(B) Replace only the burner of the old boiler. This action can increase the efficiency of the boiler/burner system to 66%. The cost of the burner replacement is $2000.

(C) Do nothing and not replace neither the boiler nor the burner.

Determine the best economical option for the building owner. Assume that the lifetime of the retrofit project is 10 years and the discount rate is 5%. The boiler consumes 5000 gallons per year at a cost of $1.20 per gallon. An annual maintenance fee of $150 is required for the boiler (independently of its age). Use all the life-cycle cost analysis method to determine the best option.

Solution

The total cost of operating the boiler/burner system is considered for the three options. In this analysis, the salvage value of the boiler or burner is neglected. Therefore, the only annual cash flows (A) after the initial investment on a new boiler are the maintenance fee and the net savings due to higher boiler efficiency. To present the calculations for *LCC* analysis, it is recommended to present the results in a tabular format and proceed as shown below:

Cost item	Option A	Option B	Option C
Initial investment			
(a) Replacement cost ($)	10,000	2,000	0
Annual operating costs:			
(b) Fuel use (gallons)	3,530	4,545	5,000
(c) Fuel cost ($) [$1.2*(b)]	4,236	5,454	1,000
(d) Maintenance fee ($)	150	150	150
(e) Total operating cost ($)[(c) + (d)]	4,386	5,594	6,150
USPW factor [$d = 5\%$, $N = 10$, Eq. (3.22)]	7.740	7.740	7.740
LCC ($) [(a) + *USPW**(f)]	43,948	45,298	47,601

Therefore, the life-cycle cost for option A is the lowest. Thus, it is recommended for the building owner to replace the entire boiler/burner system.

This conclusion is different from that obtained by using the simple payback analysis [Indeed, the payback period for option A—relative to the base case C—is SPP(A) = ($10,000)/($1765) = 5.66 years; while for option B, SPP(B) = ($2000)/($546) = 3.66 years].

Note: If the discount rate was d = 10% (which is unusually high for most markets), the *USPW* would be equal to *USPW* = 6.145 and the life-cycle cost for each option will be

$$LCC(\text{A}) = \$36{,}952 \quad LCC(\text{B}) = \$36{,}375 \quad LCC(\text{C}) = \$37{,}791$$

Therefore, option B will become the most effective economically and will be the recommended option to the building owner.It should be noted that the *LCC* analysis can be applied to determine the cost effectiveness of large-scale energy efficiency programs such as the implementation of new technology or retrofitting existing buildings. Case Study No. 1 outlines the cost-effectiveness analysis of retrofitting roofs of US residential and commercial buildings to have high albedo based on a detailed analysis performed by Akbari et al. (1999).

5.5.2 Application: Cost-Effectiveness of Cool Roof Coatings

Akbari et al. (1999) have conducted a comprehensive simulation-based analysis of cool roofs to estimate the potential energy use savings from applying high albedo coatings to roofs for residential and commercial buildings in 11 cities throughout the US including Atlanta, Chicago, Los Angeles, Dallas/Fort Worth, Houston, Miami/Fort Lauderdale, New Orleans, New York City, Philadelphia, Phoenix, and DC/Baltimore. The building prototypes modeled in each city consisted of a new/old residential building, new/old office building, new/old retail building, primary and secondary school, hospital, nursing home, and a grocery store. All building prototypes were single-story, since the energy use savings associated with cool roofs are assumed to only affect the top floors. For all old vintage building prototypes, schools, health care buildings, and grocery stores, moderate insulation levels of R-11 for roofs and R-7 for walls were considered. For all new vintage building prototypes, R-19 roofs and R-11 walls were used. Whole-building energy simulations were completed for each building type and location for two cases: (1) standard roof reflectance and (2) high roof reflectance. A standard roof reflectance of 0.25 was used for residential and commercial prototypes, and a high roof reflectance of 0.55 and 0.70 were used for residential and commercial prototypes, respectively. The roofs were modeled using a single-layered construction with a thermal emittance of 0.91.

The study results indicated that cool roofs achieved annual cooling electricity savings per square meter of flat roof area ranging from 1.1 to 5.2 kWh (7%–17%) for old residential buildings, 0.4–2.8 kWh (6%–15%) for new residences, 1.8–6.0 kWh (5%–9%) for old offices, 1.1–3.3 kWh (3%–7%) for new offices, 2.6–7.8 kWh (5%–10%) for old retail stores, and 1.4–4.5 kWh (3%–7%) for new retail stores. However, the cool roofs resulted also in annual heating energy use penalties per square meter of flat roof area ranging from 0.002 to 0.17 therm (2%–5%) for old residences, 0–0.13 therm (0%–5%) for new residences, 0–0.19 therm (0%–14%) for old offices, 0–0.11 therm (0%–12%) for new offices, 0–0.10 therm (0%–12%) for old retail stores, and 0–0.04 therm (0%–7%) for new retail stores. The relative energy use savings were found to be most significant for shell-dominated residential building prototypes. Moreover, buildings with longer hours of occupancy or operation (such as hospitals and nursing homes) showed larger savings per roof area than those with shortened occupancy schedules. In addition, old residential, office, and retail store prototypes provided larger electricity savings and gas penalty than those of the new vintage. This result is associated with lower roof *R*-values specific to the old vintage buildings. Additionally, prototypes with gas heating systems provided larger net cost savings than prototypes with electric heat pumps because of the higher cost of electric heat compared to natural gas (Akbari et al., 1999).

Moreover, building location was found to have a significant effect on energy use savings and penalties associated with cool roofs. For example, new offices with electric heat pumps in Chicago and Philadelphia, new residences with electric heat pumps in New York City, and new residences with gas furnaces in Philadelphia all displayed limited impact of cool roofs with net energy deficits or savings very close to zero. The simulation results confirmed that cooling energy use savings and heating energy use penalties were found to vary as linear functions of changes in roof albedo for each building type and location. Thus, to estimate heating energy use penalties and cooling energy use savings from other differences in roof solar reflectance values not considered in the study, the results of the analysis can be adjusted by the ratio of the reflectance differences. Using the results obtained for individual building prototypes, energy use savings associated with cool roofs were scaled up using the building stock statistics for 11 US metropolitan areas. The total impact of using cool roofs for all 11 metropolitan areas were estimated to include electricity savings of 2597 GWh; natural gas deficit of 6884 GBtu, annual energy bill net savings of $194 M, and a peak electrical demand savings of 1741 MW.

Since the incremental costs for applying high albedo coating vary depending on the type of roofing systems, two scenarios are considered to assess the cost-effectiveness using *NPW* analysis of retrofitting all US buildings considered by the Akbari et al. (1999) analysis. Specifically, the labor and material costs for scenarios 1 and 2 are $1.5/m^2 and 3.0/m^2, respectively. Using a discount rate of 5% and 30-year analysis period (i.e., $USPW(d, N) = 15.37$), *NPW* values can be estimated for both scenarios using Eq. (5.11) and the net annual savings from the retrofit program of $194/year as well as the total roof area of all buildings of 1200 Mm^2 (i.e., millions of square meters) documented by Akbari et al. (1999).

Scenario 1 (Cool roof coating cost = $1.5/m^2):

$$NPW = -(1200 * 1.5) + 194 * 15.37 = \$1182\text{M}$$

Scenario 2 (Cool roof coating cost = \$3.0/m^2):

$$NPW = -(1200 * 3.0) + 194 * 15.37 = -\$618M$$

It is clear that the program can be cost-effective under scenario 1 but may not be under scenario 2 depending on the actual costs of applying the high albedo coatings on the existing roofs. It should be noted that the analysis ignores other benefits of the program including reduction in capacities of air conditioning equipment, potential reductions in building new power plants, and even potential increase of thermal comfort within buildings. The energy productivity (*EP*) analysis introduced in the following sections can account for these additional potential benefits as long as their monetary values can be estimated.

5.6 GENERAL PROCEDURE FOR AN ECONOMIC EVALUATION

It is important to remember that the recommendations for energy conservation projects should be based on an economically sound analysis. In particular, the auditor should ask several questions before making the final recommendations:

- Will project savings exceed costs?
- Which design solution will be most cost-effective?
- What project size will minimize overall building costs?
- Which combination of interrelated project options will maximize net savings?
- What priority should be given to various project alternatives if the owner has a limited investment level?

As discussed earlier, the best suitable economic assessment method is the *LCC* method described in Section 5.5. Before the application of the *LCC*, several data are needed to perform an appropriate and meaningful economic analysis. To help the auditor collect the required information and apply the *LCC* method, the following systematic approach in any economic evaluation is proposed:

- Step 1: Define the problem that the proposed retrofit project is attempting to address and state the main objective of the project. For instance, a house has an old furnace that does not provide enough heat to condition the entire house. The project is to replace the boiler with the main objective of heating all the conditioned spaces within the building.
- Step 2: Identify the constraints related to the implementation of the project. These constraints can vary in nature and include financial limitations or space requirements. For instance, the new boiler cannot be gas-fired since there is no supply of natural gas near the house.
- Step 3: Identify technically sound strategies and alternatives to meet the objective of the project. For instance, three alternatives can be considered for

the old boiler replacement: (1) a new boiler with the burner of the old boiler, (2) a new boiler/burner system, and (3) a new boiler/burner system with an automatic air-fuel adjustment control.

- Step 4: Select a method of economic evaluation. When several alternatives including the base case (which may consist of the alternative of "doing nothing"), the *LCC* method is preferred for energy projects. When a preliminary economic analysis is considered, the *SPP* method can be used. As mentioned earlier, the payback period method is not accurate and should be used with care.
- Step 5: Compile data and establish assumptions. The data includes the discount rates, the energy costs, the installation costs, operating costs, and maintenance costs. Some of these data are difficult to acquire and some assumptions or estimations are needed. For instance, an average discount rate over the life-cycle of the project may be assumed based on a historical data.
- Step 6: Calculate indicators of economic performance. These indicators depend on the economic evaluation method selected. The indicators include *SPP*, *NPW*, and *LCC*.
- Step 7: Evaluate the alternatives. This evaluation can be performed by simply comparing the values of *LCC* obtained for various alternatives.
- Step 8: Perform sensitivity analysis. Since the economic evaluation performed in step 6 are typically based on some assumed values (for instance the annual discount rate), it is important to determine whether or not the results of the evaluation performed in step 7 depend on these assumptions. For this purpose, the economic evaluation is repeated for all alternatives using different but plausible assumptions.
- Step 9: Take into account unqualified effects. Some of the alternatives may have some effects that cannot be accounted properly in the economic analysis but may be a determining factor in the decision making. For instance, the environmental impacts (emission of pollutants) can be important to disqualify an otherwise economically sound alternative.
- Step 10: Make recommendations. The final selection will be based on the findings of the three previous steps (i.e., steps 7–9). Typically, the alternative with the lowest *LCC* value will be recommended.

Once the project for energy retrofit is selected based on an economic analysis, it is important to decide on the financing options to actually carry out the project and implement the measures that allow a reduction in energy cost of operating the facility.

5.7 ENERGY PRODUCTIVITY ANALYSIS

In the last decade, various benefits and impacts of building energy efficiency programs on sustainable economic growth have been discussed and assessed. In this section, a new analysis framework is described to evaluate the benefits of building energy efficiency actions. The analysis framework considers both energy and nonenergy benefits of building energy efficiency at the macro- and the microeconomic scales. Specifically, the approach applies the concept of *EP* to the building sector and accounts for value added and energy savings of energy efficiency measures. Moreover, the proposed analysis accounts for all quantifiable benefits of energy efficiency actions including economic, environmental, and social impacts. One of the main advantages of the proposed analysis approach is to inform decision policy makers on how energy efficiency actions in the building sector can contribute to an economy's overall *EP*. In this section, the *EP* analysis is applied to evaluate large-scale energy efficiency programs for existing and new buildings. First, the concepts of *EP* and nonenergy benefits of energy efficiency actions are briefly presented.

5.7.1 Overview of Energy Productivity

To assess the effectiveness of using energy by an economy or one of its sectors, *EP* has been introduced and increasingly utilized as a metric to assess how energy resources can be allocated to optimize economic growth. Generally, the term *EP* is simply defined as the value of services and goods that can be produced by one unit of energy. For an economy, the *EP* is typically estimated as the ratio of the gross domestic product (*GDP*) per total primary energy supply (*TPES*). The higher this ratio, the more effective and productive the economy is to extract value to generate goods and services from the energy it consumes. It is clear that the service sector has inherently higher *EP* values than other energy intensive sectors such industry and transport sectors. Thus, countries with a service-driven economy have generally higher *EP*. Table 5.8 lists the basic economic indicators of select countries including their population, *GDP*, *TPES*, *EP*, and *GDP* per capita using data obtained from IEA (2016). Fig. 5.6 compares the *EP* indices of GCC to G-7 countries based on data reported by IEA (2015). United Kingdom ranks among the top in *EP* since most of all its economy is based on services (Blok et al., 2015). It is clear that GCC countries have low *EP* compared to most G-7 countries. In fact, the annual growth of *EP* in the GCC region remains low and in some cases has decreased during the last decade.

Two main factors affect an economy-wide *EP*: (1) enhancements in energy efficiency due primarily to the adoption of new technologies and behavioral changes for various sectors and (2) structural economic changes associated with a shift toward less energy intensive sectors such as financial services (KAPSARC, 2015). To better understand the best approaches to improve *EP* of an economy, it is important to evaluate how its various sectors utilize energy resources.

EP has been generally defined and utilized as a macroeconomic indicator to indicate the level of energy efficiency of a country to develop its economy. Recently, *EP* for various sectors such as transport and industry has been defined and evaluated. However, no specific analysis has been reported on sectorial *EP*

TABLE 5.8 Economic Indicators for Select Countries Based on 2014 Data

Countries	Population (millions)	*GDP* ($2005 billions)	*TPES* (Mteps)	*EP* ($/Tep)	*GDP*/capita (1000$/per)
Australia	23.51	853.52	127.25	6.71	36.31
Bahrain	1.36	52.61	14.76	3.56	38.68
Canada	35.55	1349.37	256.61	5.26	37.96
China	1364.00	14944.43	3073.20	4.86	10.96
France	66.25	2051.96	243.49	8.43	30.97
Germany	80.28	2979.38	301.29	9.89	37.11
India	1259.70	6272.04	831.66	7.54	4.98
Italy	60.56	1625.66	146.46	11.10	26.85
Japan	127.10	4066.41	443.10	9.18	31.99
Kuwait	3.50	242.20	36.25	6.68	69.11
Oman	3.96	141.62	24.08	5.88	35.76
Qatar	2.33	260.10	46.03	5.65	111.65
Saudi Arabia	29.37	1382.80	205.26	6.74	47.08
United Arab Emirates	9.67	510.27	73.78	6.92	52.77
United Kingdom	64.56	2296.11	177.94	12.90	35.57
United States	318.43	14801.18	2216.79	6.68	46.48

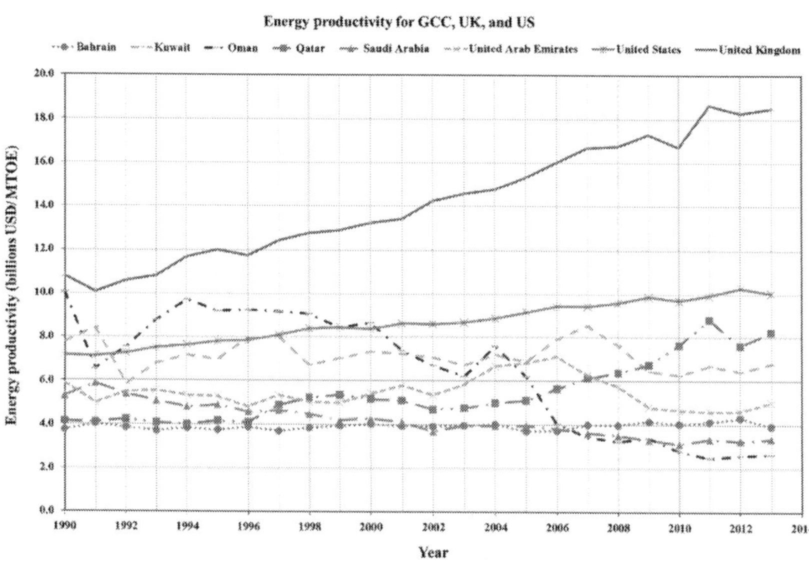

FIGURE 5.6 Annual energy productivity of GCC countries and select other countries.

indicators and their impact of overall economy *EP*. In particular, no analysis has been found on the *EP* of the building sector. This paper provides some guidelines of the impact of energy efficiency actions on the building sector *EP*. As applications, large-scale building energy retrofit programs for the GCC countries are considered to outline the proposed *EP* analysis approaches to assess the macroeconomic benefits and impacts of improving energy efficiency for buildings. The *EP* analysis is also applied to the microeconomic level to assess the benefits of improving energy efficiency for individual buildings to both owners and occupants.

5.7.2 Benefits of Building Energy Efficiency

In addition to the well documented energy use and cost savings (Bernstein and Russo, 2011; Kats, 2010; WGBC, 2013; Yu et al., 2011), energy efficiency improvements of buildings have other benefits. In particular, several nonenergy benefits (NEBs) have been observed and discussed. However, these NEBs are often ignored when assessing the cost–benefits of energy efficiency interventions since evaluating their economic impacts has been challenging and difficult to estimate. Some NEBs include the following:

- comfort, health, and safety enhancements for occupants;
- productivity enhancements for businesses; and
- operation system cost reductions for electric utilities.

While their monetary values are generally difficult to quantify, the impacts of NEBs have started to be recognized and included in measuring and assessing the cost-effectiveness of large-scale energy efficiency programs (Russell et al., 2015).

In particular, the impact of indoor thermal comfort on work productivity has been evaluated by several studies using specific case studies and surveys of subjects (Fisk and Rosenfeld, 1997; Kosonena and Tan, 2004; Romm and Browning, 2004; Wyon, 1996). In particular, Fisk and Rosenfeld (1997) estimated that improving indoor environment in US office buildings would result in a direct increase in productivity of 0.5%–5%, worth \$12 billion to \$125 billion annually. Similarly, Kosonena and Tan (2004) revealed that a mere 2°C change between indoor air temperature and neutral comfort temperature (typically 24°C) can result in more than 10% productivity loss. Using typical salaries in US office buildings, it is estimated the productivity reduction approximately represents an annual loss of \$200/m^2.

Several studies and analyses have shown that asset values of sustainable buildings are higher than conventional structures. Using available data throughout several countries, including mostly LEED rated office buildings in the US, it is found that prices of certified green buildings are higher by up to 30% of those of noncertified buildings (WBGC, 2013). In addition, it is reported that LEED and Green Star–rated buildings typically command rental premiums up to 17% (WBGC, 2013). Similarly, increases ranging from 15% to 40% in sales per floor area have been reported in various US retail stores due to energy efficient features such as daylighting (Colliers International, 2012).

In addition, it is reported that energy efficient and sustainable buildings typically provide better indoor quality and healthy environment. For instance, increased attendance by 3 days per year, 5%–14% improvement in test scores, and 20%–26% faster learning rates have been reported in schools with optimal daylight (Montgomery et al., 1994). It has been well documented that better ventilated buildings with outdoor air can maintain healthy indoors and reduce cases of Sick Building Syndrome. For instance, it has been reported that adequate ventilation to keep indoor carbon dioxide levels similar to outdoor levels can reduce Sick Building Syndrome symptoms by 70%–85% (Apte et al., 2000).

Several studies have discussed approaches to define and estimate the monetary values of NEBs related to a wide range of energy efficiency programs (Amann, 2006; LeBaron, 2011; Russell et al., 2015; Skumatz, 2014). Some of the main NEBs and their value estimates for various stakeholders are summarized in Table 5.9 based on reported studies and results of surveys conducted on US building energy efficiency programs (Russell et al., 2015; WBGC, 2013).

TABLE 5.9 Select NEBs for Energy Efficiency Programs for Various Stakeholders

Owners/occupants of residential buildings		Operators/occupants of commercial buildings		Power generators and utility companies	
Impact/ benefits	**Value (%)[a]**	**Impact/ benefit**	**Value**	**Impact/ benefit**	**Value**
Lighting maintenance reduction	28%	Improved productivity	Up to 10% increase in worker productivity	Avoided costs of T&D capacity	\$0–\$200 per kW-year
Increase durability	10%	Reduced maintenance costs	7% of energy use savings	Avoided costs of generating capacity	\$22–\$434 per kW-year
Increase of marketability of rental units	8%	Increased sales	Up to 17% relative to conventional buildings	Avoided costs of energy	\$0.02–\$0.19 per kWh
Increased safety, comfort, and respect from community	18%	Enhanced public image	Increased attendance 3 days/year	Demand reduction induced price effects	\$0-\$0.024 per kWh and \$0.62-\$34 per kW-year

[a]*Percent of the energy cost savings.*

5.8 BASIC ENERGY PRODUCTIVITY ANALYSIS APPROACH

5.8.1 Macroeconomic Analysis

The macroeconomic *EP*, is generally estimated using the *GDP* and the total primary energy consumption (*TPEC*) reported for an economy as expressed by Eq. (5.24):

$$EP = \frac{GDP}{TPEC} \tag{5.24}$$

This global economy *EP* can be decomposed using its sectorial *EP* values as indicated in Eq (5.25):

$$EP = \sum_{i=1}^{S}\left(\frac{GDP_i}{TPEC_i}\right)\left(\frac{TPEC_i}{TPEC}\right) = \sum_{i=1}^{S} f_i EP_i \tag{5.25}$$

where f_i is the contribution in percent of sector, *i*, in *TPEC*. Fig. 5.7 indicates the share of various sectors in the *TPEC* of GCC countries, US, and UK. EP_i is the *EP* for sector, *i*.

Eq. (5.25) clearly indicates that a global economy *EP* is heavily affected by the *EP* values of energy-intensive sectors. Thus, and to increase the global *EP*, two main options can be considered: (1) improve *EP* of energy-intensive sectors through energy efficiency actions (i.e., by increasing their EP_i values), or (2) shift the economy to less energy-intensive sectors (i.e., by increasing their f_i values).

As noted in Fig. 5.6, the share of the building sector in *TPEC* are rather low for most GCC countries compared to that of the UK and to lesser extent the US. Industry is dominant in all GCC countries (except Oman) representing up to 70% of the total primary national energy consumption. Using the decomposition of the *EP* of Eq. (5.25) and the results shown in Fig. 5.7 for 2013,

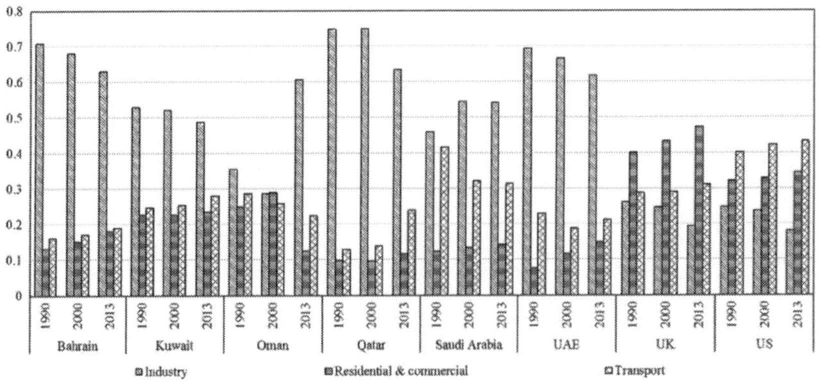

FIGURE 5.7 Share of building, industry, and transport sectors in primary energy consumption of GCC countries and select other countries.

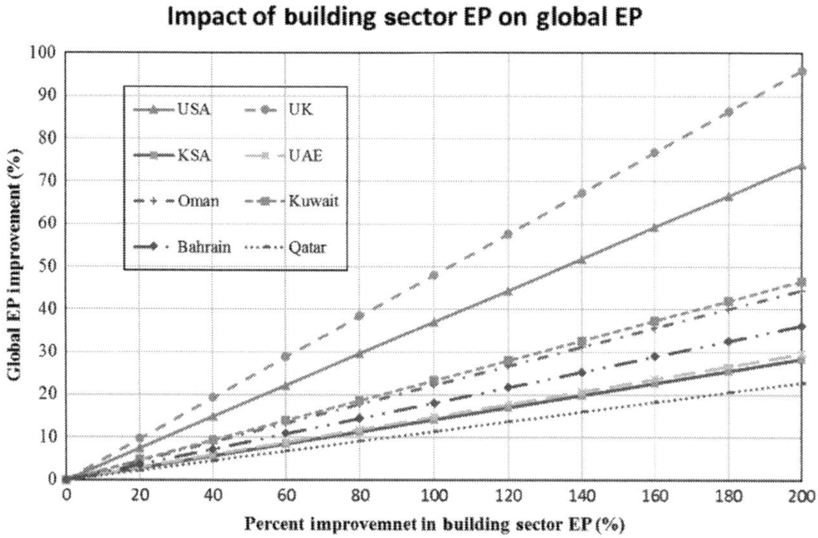

FIGURE 5.8 **Impact of building sector *EP* improvements on global country *EP* changes.**

Fig. 5.8 illustrates the potential impact of improving *EP* for the building sector on the global economy *EP* for GCC countries and US. When the building *EP* is doubled (i.e., 100% improvement), the global *EP* would increase by 14% for the Kingdom of Saudi Arabia (KSA) and by 35% for the US.

Typically, the *EP* for any sector including the building sector, EP_B, is defined using the value added, VA_B, and the total final energy consumption, TFC_B, attributed to the sector as outlined in Eq. (5.26):

$$EP_B = \frac{VA_B}{TFC_B} \tag{5.26}$$

Fig. 5.9 shows the annual variation of *EP* values for three sectors representing buildings, transport, and industry for three GCC countries. As noted in Fig. 5.9, the *EP* for the building sector in KSA is higher than that of the other two more energy-intensive sectors. However, KSA *EP* indicators for all sectors have been steadily decreasing since the 1990s. A slight improvement of the building sector *EP* can be noted from 2005 through 2013.

The energy efficiency of buildings is generally estimated using the energy use intensity (*EUI*) or index, EUI_B, defined as ratio of the total final energy consumption of the buildings, TFC_B, and the total usable or useful area, $TUFA_B$ as noted in Eq. (5.27):

$$EUI_B = \frac{TFC_B}{TUFA_B} \tag{5.27}$$

This indicator can be expanded to the entire building sector and used to determine its *EP* by reformulating Eq. (5.26):

$$EP_B = \left(\frac{VA_B}{TUFA_B}\right)\left(\frac{TUFA_B}{TFC_B}\right) = \left(\frac{VA_B}{TUFA_B}\right)\frac{1}{EUI_B} \tag{5.28}$$

Eq. (5.28) indicates that the building *EP* is a function of two indicators:

- Economic productivity indicator: the ratio ($VA_B/TUFA_B$) represents the economic output per a building unit area. Unlike the case for the industry and transport sectors, the *VA* for the building sector is rather difficult to estimate. Different added value estimates can be considered for buildings including their asset value, their rental value, work productivity value (office buildings), or sale value (retail stores). For macroeconomic analysis, the value

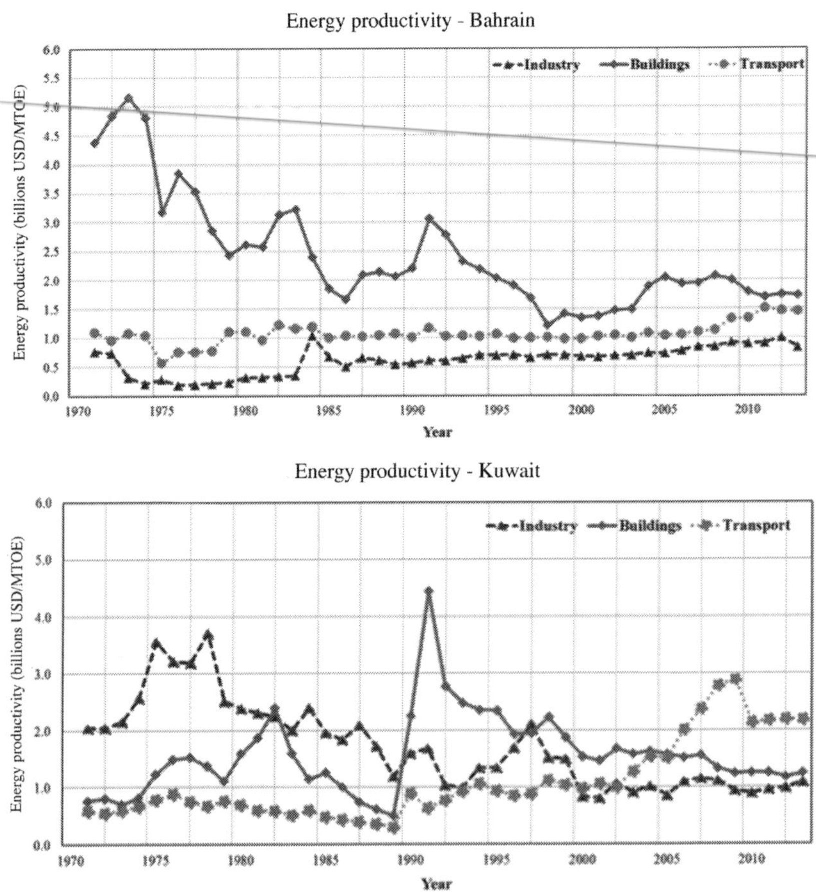

FIGURE 5.9 **Annual variation of *EP* for three sectors for select GCC countries.**

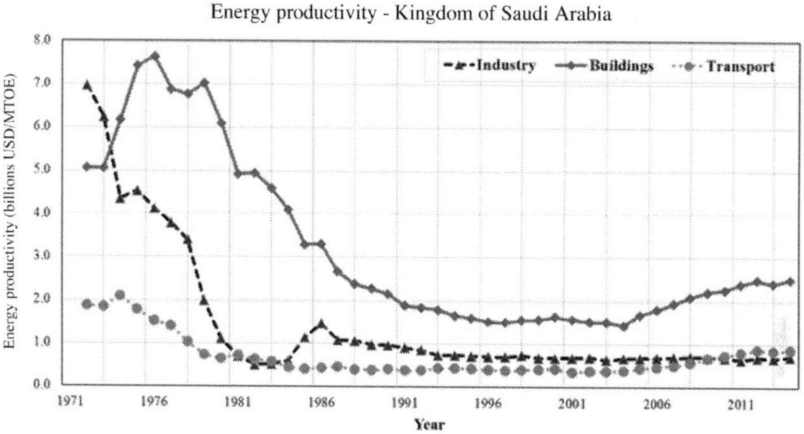

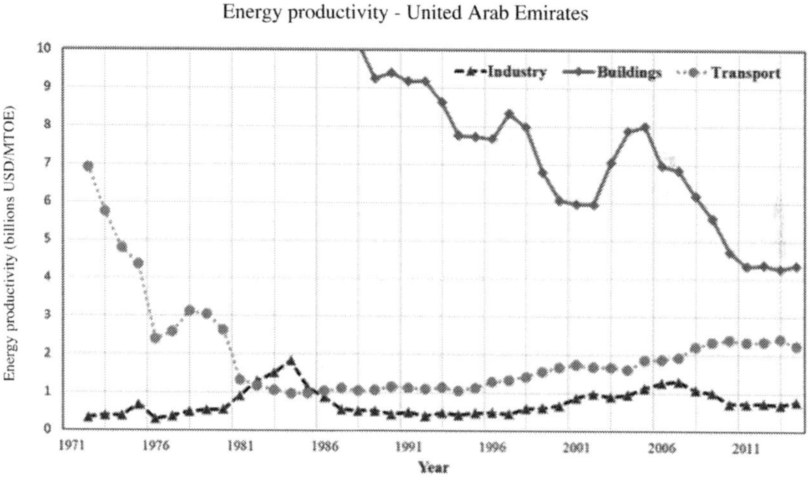

FIGURE 5.9 ***(cont.)***

added for the building sector can be estimated from the reported sectorial data.

- Energy efficiency indicator, the reciprocal of the *EUI*, $1/EUI_B$, and represents the building useful area that can be served by one unit of energy. This indicator is sometimes referred to as the energy affordability index and can be used to assess the energy performance of individual buildings.

Fig. 5.10 illustrates the concept for the building sector *EP* considering the two indicators outlined by Eq. (5.28): energy efficiency and economic productivity attributed to the building sector. It should be noted that *EP* analysis of the building sector encompasses all building types and a wide range of benefits including NEBs.

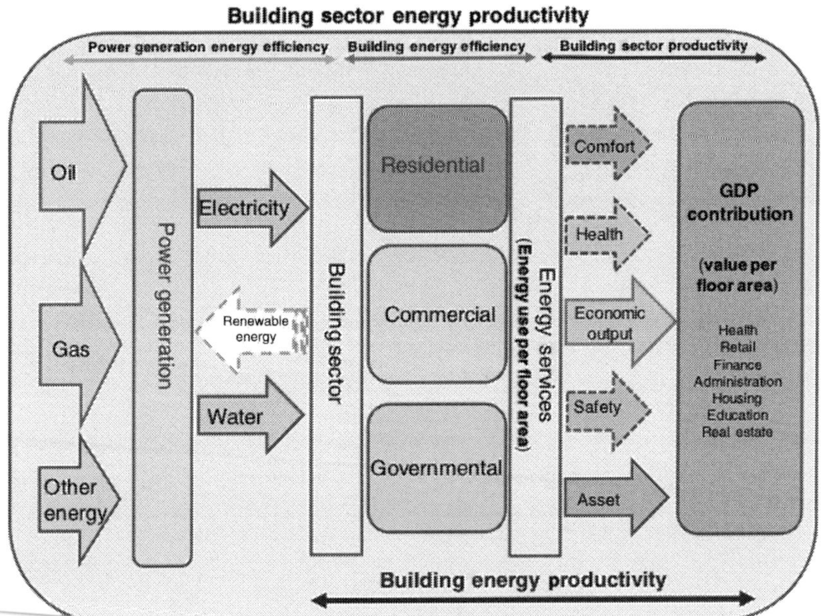

FIGURE 5.10 **Basic concept for the building sector energy productivity.**

To improve the building sector *EP*, EP_B, provided by Eq. (5.28) two main actions can be considered:

- The economic output per unit floor area needs to be increased either by fostering productive economic activities and/or by enhancing effective use of building spaces.
- The energy use index, EUI_B, is reduced through either implementation of energy efficiency measures or integration of renewable energy technologies within buildings.

Since the building sector includes several building types such as housing, offices, hotels, banks, retail stores, and schools, the *EP* for the building sector can be further decomposed to determine the contribution of each building type. One decomposition option is through the energy use index, EUI_B, which can be expressed as follows:

$$\frac{1}{EUI_B} = \frac{TUFA_B}{TSEC_B} = \sum_{k=1}^{N} \left(\frac{TUFA_{B,k}}{TSEC_{B,k}} \right) \left(\frac{TSEC_{B,k}}{TSEC_B} \right) = \sum_{k=1}^{N} r_k \frac{1}{EUI_k} \tag{5.29}$$

where r_k is the fraction of the building type k relative of the entire building stock in terms of usable total floor area and EUI_k is the energy use index specific to each building type k.

Using the available data for energy consumption and floor area, the fraction r_k can be estimated. In the GCC region, floor area for housing is significant compared to other building types. Therefore, it is important to reduce the energy use index for the residential buildings to affect the overall energy efficiency of the building sector.

In this study, large-scale implementation of energy efficiency measures and/or renewable energy systems within buildings are considered and their impact on the building sector *EP* is evaluated for all GCC countries. First, a general analysis approach is proposed for assessing the impact of any energy efficiency measure on the building sector *EP*.

5.8.2 Evaluation of Impact of Energy Efficiency Interventions

The change in *EP*, ΔEP_B, associated to any energy efficiency measure in the building sector can be estimated as noted in Eq. (5.30):

$$\Delta EP_B = EP_{B,n} - EP_{B,e} = \frac{VA_{B,n}}{TFC_{B,n}} - \frac{VA_{B,e}}{TFC_{B,e}} \tag{5.30}$$

where $EP_{B,n}$ and $EP_{B,e}$ are the new and existing building sector *EP* values respectively, $VA_{B,n}$ and $VA_{B,e}$ are the new and existing building sector value added values, respectively, $EP_{B,n}$ and $EP_{B,e}$ are the new and existing building sector total final energy consumption values, respectively.

The energy efficiency measure may change the value added and the total final energy used depending on the quantifiable resulting benefits, ΔVA_{EE}, and energy savings, ΔTFC_{EE}. As discussed in the background section, the benefits can encompass a wide range of impacts including the avoided costs associated with electricity and fuel use, avoided costs to energy generation and distribution (from lower electricity power demands), reductions in carbon emissions, and other benefits such as increased worker productivity (due to improved indoor air quality and thermal comfort). Based on the estimation of the monetary value of the benefits incurred from the energy efficiency action, ΔVA_{EE}, the new value added can be determined:

$$VA_{B,n} = VA_{B,e} + \Delta VA_{EE} \tag{5.31}$$

The new total final consumption can be determined from the energy use savings, ΔTFC_{EE}:

$$TFC_{B,n} = TFC_{B,e} - \Delta TFC_{EE} \tag{5.32}$$

Thus, the change in *EP* can be expressed as

$$\Delta EP_B = \frac{VA_{B,e} + \Delta VA_{EE}}{TFC_{B,e} - \Delta TFC_{EE}} - \frac{VA_{B,e}}{TFC_{B,e}} \tag{5.33}$$

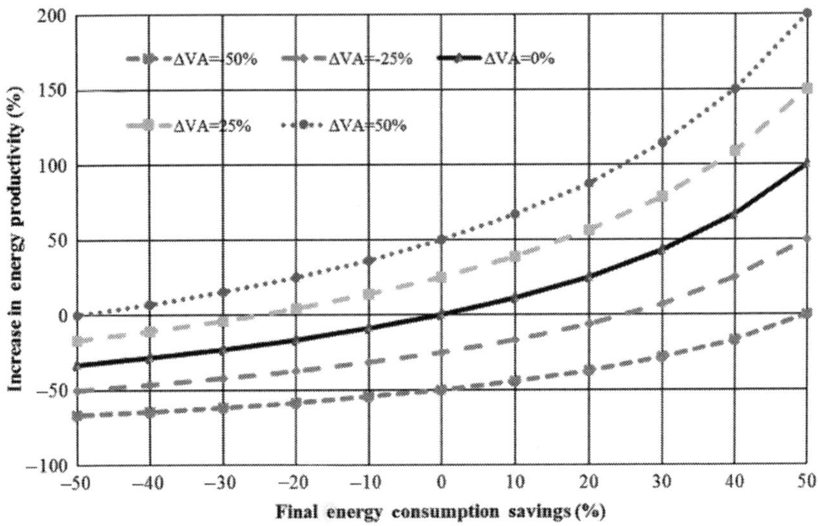

FIGURE 5.11 **Variation of energy productivity as a function of energy use savings and change in added value for the building sector.**

The percent increase in the building sector *EP* can then be determined simply as a function of percent changes in value added and energy consumption:

$$\Delta EP_B(\%) = \frac{1+\Delta VA_{EE}(\%)}{1-\Delta TFC_{EE}(\%)} - 1 = \frac{\Delta VA_{EE}(\%)+\Delta TFC_{EE}(\%)}{1-\Delta TFC_{EE}(\%)} \quad (5.34)$$

Fig. 5.11 illustrates the percent increase of *EP* as a function of how any intervention stimulates relative changes in value added and energy consumption. Based on the profiles shown in Fig. 5.11 and the expression in Eq. (5.34), two basic principles can be formulated to assess the impact of any policy or program on the building sector's *EP*:

1. Any intervention that saves energy consumption and increases the value added raises *EP*. In other terms, any cost-effective energy efficiency measure increases *EP*.
2. Any intervention that reduces the value added, even if it reduces energy consumption, may lower *EP*. Such as the case of an EEM that is not cost-effective. The threshold of value-added reduction, after which *EP* starts to decrease, due to any EEM can be estimated using Eq. (5.34):

$$\Delta VA_{EE}(\%) < -\Delta TFC_{EE}(\%) \quad (5.35)$$

3. Any intervention that increases the value added but increases energy consumption can increase the *EP* such as the case for an investment to add a building system to enhance a service including thermal comfort. The threshold of energy consumption increase, after which *EP* starts to increase,

due to an action that increases the value added can be estimated using Eq. (5.34):

$$\Delta TFC_{EE}(\%) > -\Delta VA_{EE}(\%) \tag{5.36}$$

The impact of energy efficiency intervention on sectoral *EP* and overall *EP* can be compared using Eq. (5.34) when the change of value added is not significant (i.e., $\Delta VA_{EE} = 0$). Indeed, Eq. (5.34) provides the relative change for both sectoral and overall *EP* due to reduction in *TFC* value:

$$\Delta EP(\%) = \frac{\Delta TFC_{EE}(\%)}{1 - \Delta TFC_{EE}(\%)} \tag{5.37}$$

Thus, the relative change in building sectorial *EP* can be estimated as follows from an energy efficiency:

$$\Delta EP_B(\%) = \frac{\Delta TFC_{EE} / TFC_B}{1 - \Delta TFC_{EE} / TFC_B} \tag{5.38}$$

The relative change in overall economy *EP* for the same energy efficiency intervention can be determined using the sectoral energy consumption share $f_B = TFC_B/TFC$ introduced in Eq. (5.37):

$$\Delta EP(\%) = \frac{f_B\, \Delta TFC_{EE} / TFC_B}{1 - f_B\, \Delta TFC_{EE} / TFC_B} \tag{5.39}$$

The impacts of energy efficiency interventions on the relative changes of the building sector and the total economy *EP* values are shown in Fig. 5.12 for select countries using Eqs. (5.38) and (5.39). For all the countries, the building sector *EP* doubles with a program that achieves 50% savings in the sector's energy consumption. However, the same program leads to an increase in the total economy *EP* of 32% in the UK, 23% in the US, and only 7% in KSA. A program that achieves 50% savings in the industrial sector's energy consumption would result in almost 40% increase in KSA's total economy *EP* in the conservative scenario when no value added is considered.

5.8.3 Microeconomic Analysis

A microeconomic *EP*, EP_b, for an individual building or an energy system within a building (such as lighting, an AC, or an appliance) can be defined using a similar definition given by Eq. (5.26):

$$EP_b = \frac{VA_b}{EU_b} \tag{5.40}$$

where VA_b is the average annual value provided by the building including one of the combinations of any values added such as asset value (housing), rental value

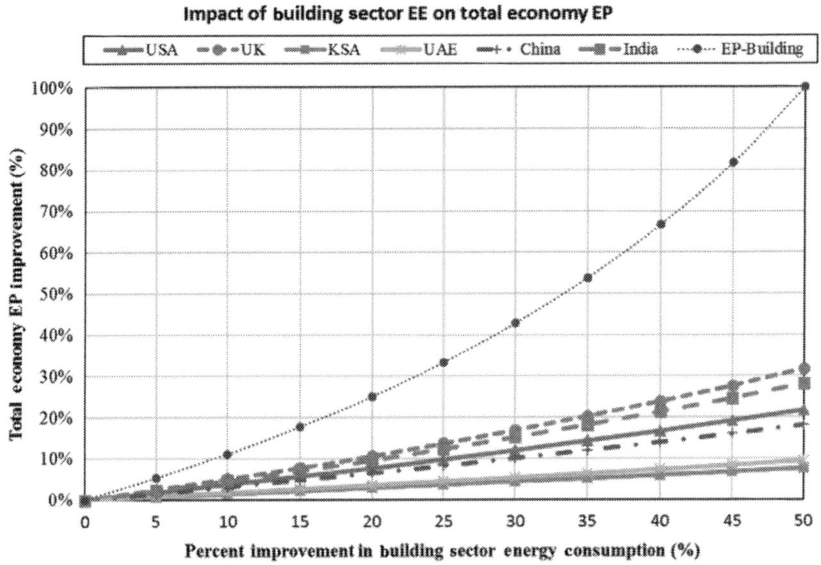

FIGURE 5.12 Variation of total economy and sectorial energy productivity as a function of energy use savings achieved in the building sector for select countries.

(office building), and sale value (retail); and EU_b is the annual energy used by the building or building energy system.

A decomposition of Eq. (5.40) can be introduced to estimate the individual building EP as a function of its EUI value, EUI_b:

$$EP_b = \left(\frac{VA_b}{FA_b}\right)\left(\frac{FA_b}{EU_b}\right) = \left(\frac{VA_b}{FA_b}\right)\frac{1}{EUI_b} \tag{5.41}$$

As discussed earlier, the term $1/EUI_b$, expressing the floor area served by a unit of energy is often considered as a measure of energy affordability, EA_b:

$$EA_b = \frac{1}{EUI_b} = \frac{FA_b}{EU_b} \tag{5.42}$$

Fig. 5.13 shows the energy affordability, EA_b, expressed in terms of the potential building floor area served by per 10,000 kWh/year to maintain acceptable indoor environment quality for a villa in Riyadh and Abha. The impacts on EA_b of various energy efficiency measures including addition of wall insulation, installing low-e glazing, using LED lighting, and high efficiency air conditioning system are presented in Fig. 5.13 based on the results from analyses carried out by Alaidroos and Krarti (2015) and Krarti and Dubey (2017). As expected, measures that increase energy efficiency enhance energy affordability and thus energy productivity.

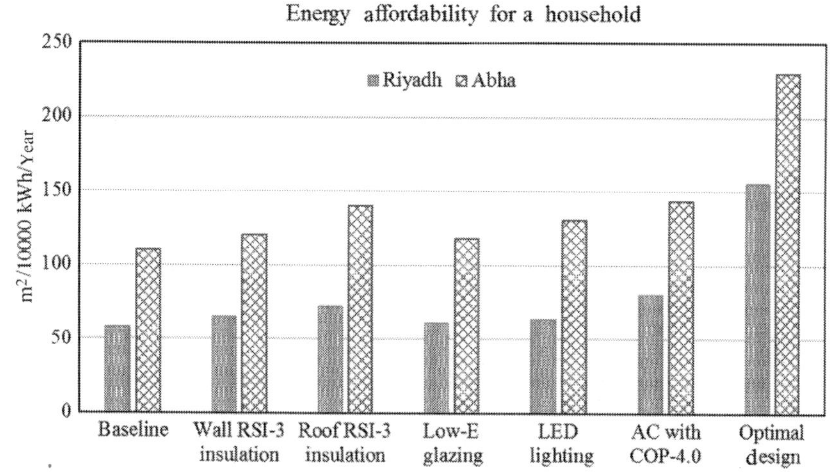

FIGURE 5.13 **Household energy productivity expressed in floor area covered by 10,000 kWh/year electricity consumption in Riyadh and Abha, KSA.**

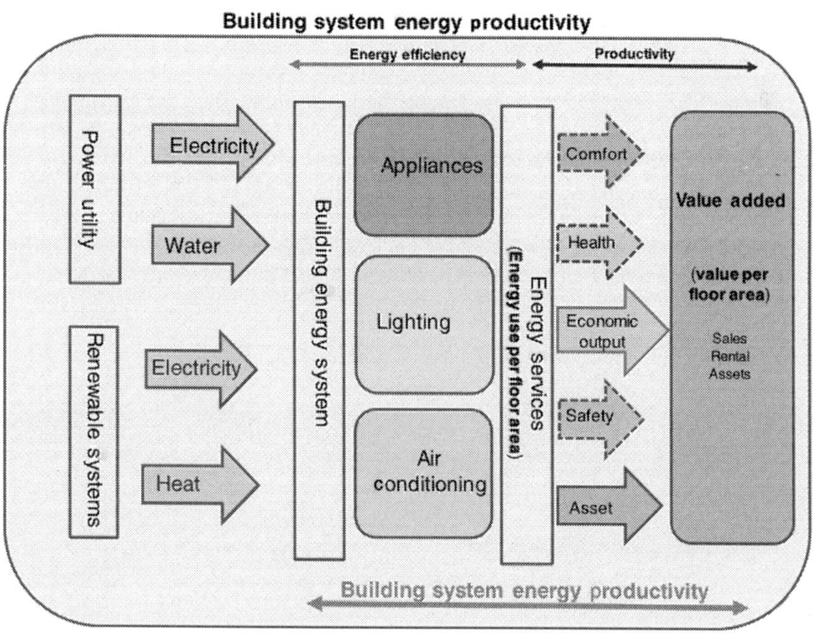

FIGURE 5.14 **Basic concept of individual building system energy productivity.**

Fig. 5.14 illustrates the *EP* concept proposed for an individual building or for a building subsystem incorporating two indicators: (1) the energy efficiency and (2) the value-added productivity. The value added can account for quantifiable and measurable indicators for NEBs such as the sale levels and rental rates.

To illustrate, at the microeconomic level, the distinction between *EP* and energy efficiency, examples 5.7 and 5.8 consider for a small retail store the benefits of investing in an energy-efficient air conditioning system compared to, respectively, no cooling system and a standard efficiency cooling system.

Example 5.7

Consider 500-m^2 retail store with energy use intensity (*EUI*) of 172 kWh/m^2/year and annual sales of \$600/m^2. Perform an *LCC* and *EP* analyses to assess the effectiveness of investing in a more energy-efficient air conditioning system that can reduce the *EUI* to 125 kWh/m^2/year. Assume that (1) the cost of the air conditioning system increases from \$5000 to 6500 when upgrading from standard to energy efficient, (2) electricity cost is \$0.10/kWh, and (3) life of the air conditioning system is 15 years and the discount rate is 5% (so that *USPW* = 10.4 years), (4) sales remain unchanged.

Solution

The results of both *LCC* and *EP* analyses are summarized in Table 5.10 to compare the cost-effectiveness and any change of energy productivity for the retail store when investing in an energy-efficient air conditioning system. Using the cost-effectiveness analysis, the energy-efficient system has lower annual costs (i.e., \$6875/year) than the baseline cost (i.e., \$9081/year), resulting in annual savings of \$2206/year. When evaluating the two options using the energy productivity analysis, the highly efficient air conditioning system provides more benefits with higher *EP* value (\$4.69/kWh) than the baseline system (with an *EP* value of \$3.38/kWh), delivering an additional *EP* value of \$1.31/kWh. Installing the high-efficiency air conditioning system is cost-effective since it has lower annualized costs and boosts productivity of the retail store, using the sales value and added income from energy use cost savings as the measurable economic output.

TABLE 5.10 Comparison of Annualized Cost and *EP* Indicators for the Retail Store With Standard and Energy-Efficient Air Conditioning System.

Indicator	Baseline	Improved Air Conditioner
EUI (kWh/m^2/year)	172	125
Annual energy use (kWh/year)	86,000	62,500
Initial cost (\$)	5,000	6,500
Annualized cost (\$/year)	9,081	6,875
Annual sales (\$)	300,000	300,000
Value added (\$/year)	290,919	293,125
EP (\$/kWh)	3.38	4.69

Example 5.8

For the same 500-m^2 retail store considered in Example 5.7, the baseline case in this example has no air conditioning system with an energy use intensity (*EUI*) of 110 kWh/m^2/year and annual sales of \$594/m^2. Perform an *LCC* and *EP* analyses to assess the effectiveness of investing in an energy-efficient air conditioning system that can improve thermal comfort but increase the *EUI* to 125 kWh/m^2/year. Assume that (1) the cost to install an air conditioning system is \$6500, (2) electricity cost is \$0.10/kWh, and (3) life of the air conditioning system is 15 years and the discount rate is 5% (so that *USPW* = 10.4 years), (4) sales increases from \$500/m^2 to \$600/m^2 due to the improved comfort.

Solution

The results of both *LCC* and *EP* analyses are provided in Table 5.11 for the case outlined in Example 5.8. Using the cost-effectiveness analysis, the baseline with no air conditioning system has lower annualized cost (i.e., \$5500/year) than the case with an energy-efficient air conditioning system (i.e., \$6985/year). When evaluating the two options using the energy productivity analysis, the highly efficient air conditioning system provides more benefits with higher *EP* value (\$4.69/kWh) than the baseline system (with an *EP* value of \$4.45/kWh), delivering an additional *EP* value of \$0.24/kWh due mostly to the increased sales associated with better thermal comfort level.

The energy productivity analysis of select energy efficiency measures for buildings is discussed in the following section. In particular, the analysis provided in the application section defines and compares indicators for *LCC* and *EP* to evaluate the impact of various energy efficiency measures applied to a residential building.

TABLE 5.11 Comparison of Annualized Cost and *EP* Indicators for the Retail Store With no Air-Conditioning Equipment and Energy-Efficient Air Conditioning System

Indicator	Baseline (No AC)	With Air Conditioner
EUI (kWh/m^2/year)	110	125
Annual energy use (kWh/year)	55,000	62,500
Initial cost (\$)	0	6,500
Annualized cost (\$/year)	5,500	6,875
Annual sales (\$)	250,000	300,000
Value added (\$/year)	244,500	293,125
EP (\$/kWh)	4.45	4.69

5.8.4 Application: Evaluation of Improved Designs for Villas

With the adoption of proven energy efficiency and renewable energy technologies, buildings can have low *EUI* and even be net-zero energy. As discussed in Krarti and Dubey (2017), residential buildings in KSA can be designed to consume less than 60% of current consumption levels. Four design options for a 500-m^2 villa are considered including:

- Baseline design with no energy efficiency design features.
- Baseline design incorporating thermal insulation ($RSI = 3.0\ m^2\ K/W$) for both exterior walls and roof. This case represents the current mandatory energy conservation regulations for KSA buildings.
- Low-energy building design incorporating a set of energy efficiency features as outlined in the optimization analysis conducted by Krarti and Dubey (2017).
- Baseline design with a 20-kW PV roof-mounted panel that can generate electricity to offset part of the villa energy needs using either a storage system or through smart metering with the grid.

In particular, Table 5.12 provides the annual electricity consumption for each design as well as the cost of the energy efficiency and PV systems. Moreover, Table 5.13 provides the *LCCs* for each design using a lifetime of 30 years and a discount rate of 5%. The cost of the PV system is considered to be $3000/kW. The *LCC* indicates the costs effectiveness of each design alternative relative to the baseline option to the owner using the current electricity rate structure. As indicated in Table 5.12, the low-energy design is the most cost-effective due to lower energy consumption and the lower energy prices applied to low-energy consumers. The addition of the PV system is not cost-effective relative to the baseline. Table 5.13 provides the results of the *EP* analysis of the four design

TABLE 5.12 Annual Energy Consumption and Life Cycle Costs for Four Villa Designs in Riyadh, KSA

Villa design option	Annual energy use (kWh)	PV size (kW)	Energy cost (USD)	Implementation costs (USD)	PV cost (USD)	Life Cycle cost (USD)
Baseline	153,594	0	7,924	0	0	121,811
Insulation in walls and roof	115,196	0	5,193	3,500	0	83,329
Low energy design	57,444	0	1,653	21,000	0	46,411
Baseline with PV System	118,694	20	5,407	0	60,000	143,119

TABLE 5.13 Estimation of Energy Productivity and Energy Intensity for Various Villa Design in Riyadh, KSA

Villa design option	Net present value (USD)	Value added (USD/year)	Annual energy use (kWh)	Energy productivity (USD/kwh)	Energy intensity (kWh/USD)
Baseline	0	32525.72	153,594	0.211764	4.722232
Insulation	38,482	30022.4	115,196	0.260621	3.836985
Low energy design	75,401	27620.8	57,444	0.480829	2.079743
Baseline with PV	21,308	33911.8	118,694	0.285708	3.500079

options when considering the construction and operating costs of the villa as an indicator of the value added for the baseline design. For the other designs, the annualized costs of various additional energy efficiency and renewable energy features accounted for using the analysis outlined in Section 5.8. These annualized costs are estimated using whole-building analysis described by Alaidroos and Krarti (2015). As shown in Table 5.13, the low-energy design has the highest *EP* value while the baseline has the lowest value. The installation of the PV system, while not cost-effective as noted in Table 5.12, would increase the building *EP* more than the baseline or the addition of thermal insulation.

5.8.5 Estimation of Change in Value Added

The change in value added due to any energy efficiency action applied to the entire building sector or an individual building energy system can be estimated using the *NPV* analysis to account for implementation costs, *IC*, initial monetary benefits, B_0, and annual cash flows from the action:

$$NPV = -(IC - B_0) + \sum_{n=1}^{N} CF_n SPPW(r_d, n) \tag{5.43}$$

where CF_n are the annual cash flows and include typically the potential energy cost savings, ΔEC_n, the operation and maintenance costs, OM_n, NEBs (such as emission reductions, work productivity, and increase of sales), B_n, and other costs (such as replacement and resale costs), O_n:

$$CF_n = \Delta EC_n + B_n + OM_n + O_n \tag{5.44}$$

SPPW is the single present payment worth factor as provided by Eq. (5.7) and depends on the annual average discount rate, *d*, and the lifetime, *N*, expressed by the number of year for the energy efficiency action.

The annual change of value added from the project can be estimated using the annualized costs, *AC*, of the energy efficiency project obtained from the present worth value, *NPW*, and the uniform series present worth factor, *USPW*:

$$\Delta VA = AC = \frac{NPV}{USPW} \tag{5.45}$$

USPW is provided by Eq. (5.9) and depends on the life-cycle period *N* and discount rate, *d*, of the economy. The concept of uniform series is shown in Fig. 5.15A when all the annual cash flows are the same.

In particular, the lifetime avoided energy (i.e., electricity or fuel) consumption, U_{tot}, from an energy efficiency measure that results in uniform annual savings, *A*, can be estimated from Eq. (5.46):

$$U_{tot} = N \cdot A \tag{5.46}$$

However, when a large-scale energy program is implemented incrementally in the initial phase of *M* years, the annual cash flows are not uniform but follow a gradient series as illustrated in Fig. 5.15B. In this case, the gradient series present worth, *GSPW*, is used to convert the annual cash flows to the present using the parameters, *d*, *M*, and *N*:

$$GSPW = \frac{1-(Md+1)(1+d)^{-M}}{d^2} + \frac{(1+d)^{-M} - (1+d)^{-N}}{d} \tag{5.47}$$

Moreover, the lifetime avoided energy consumption, G_{tot}, can be computed from the final annual energy savings, *A*, as follows:

$$G_{tot} = \left(N + \frac{M}{2}\right) \cdot A \tag{5.48}$$

It should be noted that both the net present worth, *NPW*, and the annual cost, *AC*, analysis methods can be used to assess the cost effectiveness of the energy efficiency action.

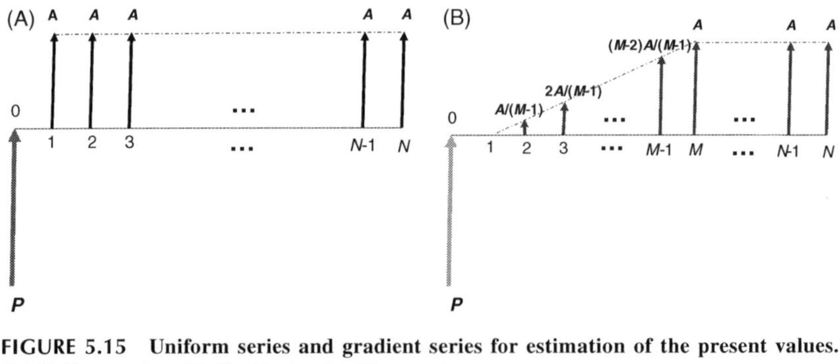

FIGURE 5.15 Uniform series and gradient series for estimation of the present values. (A) Uniform series. (B) Gradient and uniform series.

5.8.6 Uncertainty Analysis of Energy Productivity Changes

Using Eq. (5.34), an uncertainty analysis is carried out to estimate the uncertainty level in the change in building *EP* based on the uncertainties associated with changes of value added and total final energy consumption. Specifically, based on propagation of error analysis, the uncertainty in ΔEP, $U_{\Delta EP}$, can be expressed based on the uncertainty levels of ΔVA and ΔTFC, $U_{\Delta VA}$, and $U_{\Delta TFC}$ as follows:

$$U_{\Delta EP} = \sqrt{A^2 U_{\Delta TFC}^2 + B^2 U_{\Delta VA}^2} \tag{5.49}$$

with the coefficients A and B defined as

$$A = \frac{\partial U_{\Delta EP}}{\partial U_{\Delta TFC}} = \frac{1}{(1-\Delta TFC)^2} \tag{5.50}$$

$$B = \frac{\partial U_{\Delta EP}}{\partial U_{\Delta VA}} = \frac{1}{(1-\Delta TFC)} \tag{5.51}$$

Using Eqs. (5.49) through (5.51), the uncertainty $U_{\Delta EP}$ can be determined for various uncertainty levels $U_{\Delta VA}$ and $U_{\Delta TFC}$ as illustrated in Figs. 5.16 and 5.17 for the existing KSA building stock retrofit program level 1 and level 3, respectively (Krarti et al., 2017). Refer to the following section and to Chapter 9 for more details on large-scale analysis of energy retrofit of building stock for a country. As noted in Figs. 5.16 and 5.17, the uncertainty level in change of

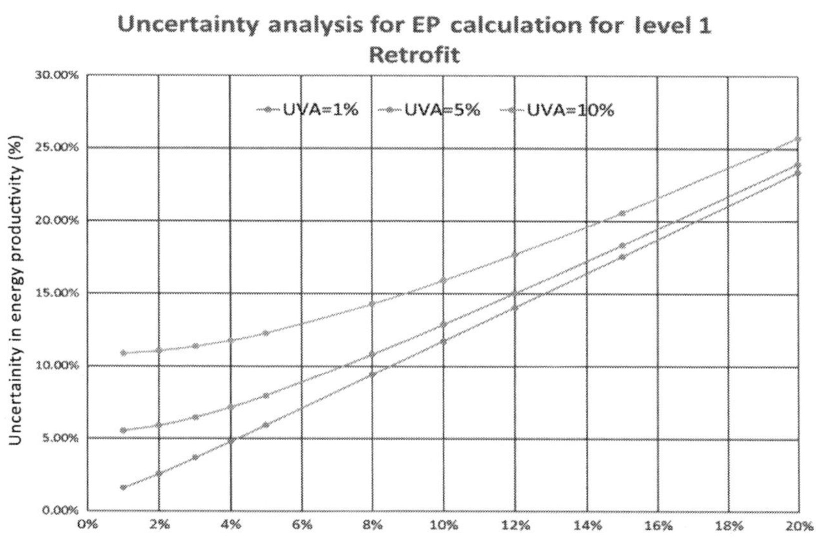

FIGURE 5.16 Variation of uncertainty of ΔEP with the uncertainties ΔTFC and ΔVA for Level 1 building energy retrofit program in KSA.

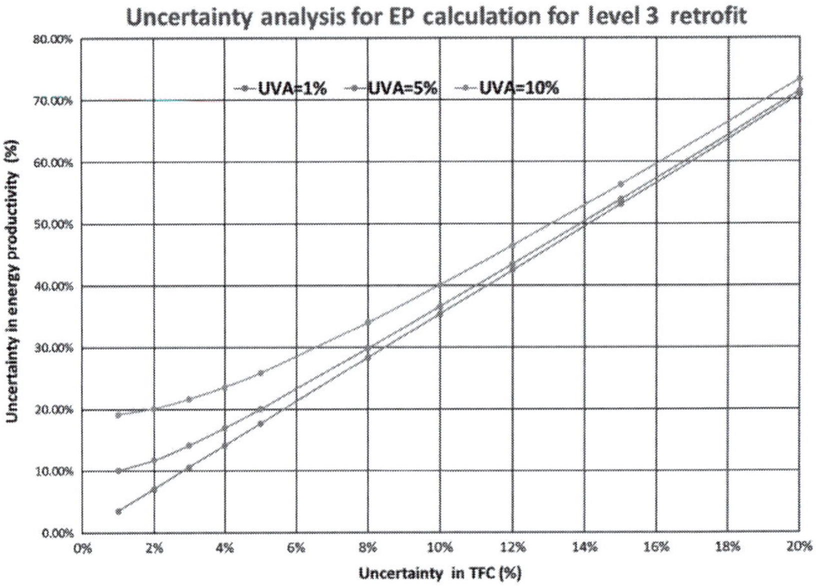

FIGURE 5.17 Variation of uncertainty of ΔEP with the uncertainties ΔTFC and ΔVA for Level 3 building energy retrofit program in KSA.

the building *EP* is highly affected by the uncertainty level for estimating the change in final energy consumption, that is, the actual energy savings associated with the retrofit programs. The uncertainty level for the change in value added is rather limited in estimating the change in building *EP*. Thus, it is important to have a high confidence in estimating the energy savings associated with the building retrofit programs especially when deep retrofits are considered (i.e., level 3 program). For instance, when the uncertainty for determining both ΔVA and ΔTFC is 5%, the uncertainty in estimating ΔEP is 8% for Level 1 and 20% for Level 2 retrofit programs. However, when both ΔVA and ΔTFC are computed with 10% uncertainty, ΔEP can only be estimated at 16% and 40% for respectively, level 1 and level 3 retrofit programs.

5.9 APPLICATION TO LARGE-SCALE BUILDING RETROFIT PROGRAM ANALYSIS

5.9.1 Background

Currently, there are no specific energy efficiency mandatory regulations for buildings in Oman (Krarti and Dubey, 2017). Due to high population growth and the lack of any comprehensive energy efficiency program, electricity consumption and peak demand have significantly increased in Oman especially during the last few years as noted in Fig. 5.18. In particular, the building sector consumes 77% of the total electricity generated in Oman (OPWP, 2014).

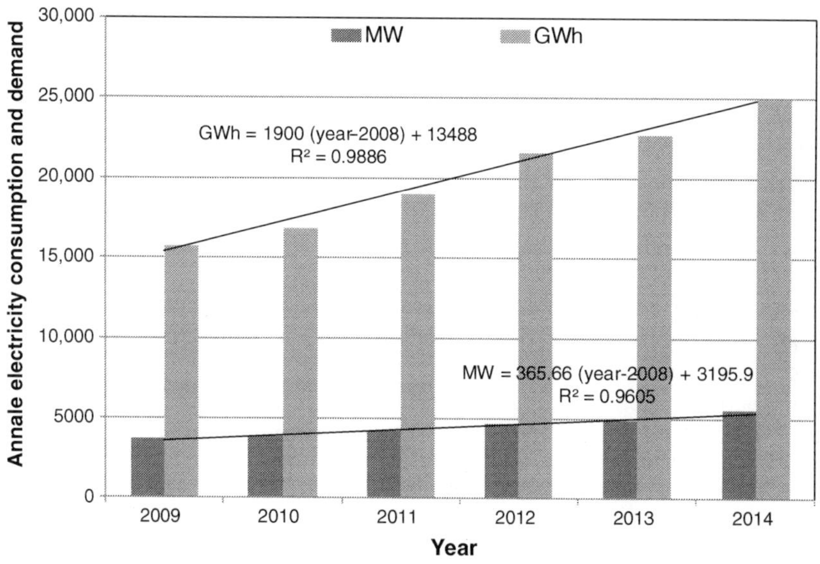

FIGURE 5.18 Variation of annual electrical consumption and peal demand between 2009 and 2014 in Oman.

Specifically, residential, commercial, and governmental buildings accounted for, respectively, 48, 20, and 9% of the total electricity consumed in Oman during 2014. As illustrated in Fig. 5.18, a regression analysis of available historical data indicates that there is a consistent growth of the electrical demand over the last decade with an annual increase of 1900 GWh for the electricity consumption and 365.7 MW for the peak demand. If this growth remains unchanged, the annual electricity consumption and peak demand are predicted to be respectively, 55,288 GWh and 11,240 MW by 2030 (i.e., double the current electricity demands).

To improve energy efficiency in the building sector, the Omani government has initiated some actions and pilot programs to promote sustainability for the built environment (OBG, 2013; TRC, 2014). However, very limited studies and analyses have been published to assess the benefits of improving current practices for designing, constructing, and operating Omani buildings to increase the energy performance of buildings. For instance, a reported analysis has investigated the potential of load management strategies using gas cooling and ice storage systems to reduce electrical peak demand as well as the impacts of introducing time-of-use tariffs (JICA, 1998). Another study has indicated that implementation of energy-efficient air conditioning and lighting systems in commercial and governmental buildings can reduce peak demand up to 596 MW and electricity consumption up to 44,000 GWh by 2024 (Malik, 2007). A simulation-based analysis has shown that simple control strategies such as reducing lighting use by 25% and increasing the cooling temperature from 20°C to

24°C can provide savings of 25.6% in annual energy consumption for an Omani building (Mallela et al., 2010). Similarly, the benefits of clean power generation technologies including the use of renewable energy resources to generate electricity in Oman have been evaluated by few studies. In particular, potential energy savings of 335 GWh have been estimated from a large-scale installation of solar water heaters in the Seeb district (Gastli and Charabi, 2011). In general, it is found that Oman has significant solar and wind energy resources that can be harnessed to generate enough electricity to meet all the national needs (Al-Badi, 2011a,b; Al-Hatmi et al., 2014; Gastli and Charabi, 2010; Malik and Al-Badi, 2009; Solanki et al., 2013). However, the lack of incentives and legal framework are the main barriers of the limited deployment of energy efficiency and renewable energy technologies in Oman. Through a survey of the Omani construction stakeholders, the main challenges to promote energy efficiency technologies and practices in the building sector have been identified and include the following (Saleh and Alalouch, 2015):

1. Economic challenges due to higher costs associated with energy-efficient products including thermal insulation for building envelope systems. Indeed, most of these products are not manufactured locally and have to be imported. Moreover, the high energy subsidies make any investment in energy efficiency not a cost-effective strategy for building owners or developers (Al-Badi et al., 2015; Powmya and Abidin, 2014).
2. Lack of knowledge and expertise on sustainable construction practices has been identified as one of the barriers in promoting high-performance buildings in Oman. In addition to the lack of any codes and standards to promote energy efficiency in the building sector, the Omani construction industry relies on foreign workforce with limited knowledge of green technologies and materials (Al-Badi et al., 2015; Powmya and Abidin, 2014).
3. Social challenges due to resistance from various building sector stakeholders to change the conventional construction methods to follow more sustainable practices. These challenges are made more ingrained due to the lack of any legal and economic incentives to promote high energy performance buildings (Al-Badi et al., 2015; Powmya and Abidin, 2014).

In this section, the impacts of large-scale building energy efficiency programs in Oman are evaluated using the *EP* analysis approach (Krarti and Dubey, 2017). Energy efficiency programs targeting new and existing building are considered in the study. As part of the analysis, the benefits of implementing a wide range of energy efficiency measures are first determined using a whole-building energy simulation analysis applied to energy models representing the existing Omani building stock. Economic and environmental benefits of a wide range of energy efficiency programs are then quantified for the Omani building stock using a bottom–up analysis approach. In particular, the impacts of developing and enforcing new building energy efficiency codes are evaluated. In addition, different retrofit levels of existing buildings are considered. Finally, the implementation costs and the impacts of the energy efficiency programs on the *EP* for the Oman building sector are evaluated.

5.9.2 Analysis of Impacts of Energy Efficiency Programs for New Buildings

As noted earlier, there are no specific mandatory energy conservation requirements for buildings in Oman. If thermal insulation for walls and roofs as well as better glazing for windows are required for new buildings, an estimated reduction of annual energy consumption and electrical peak demand of at least 25% can be achieved (Krarti and Dubey, 2017). Using the bottom–up analysis outlined in Chapter 9 and carried out by Krarti (2015), the economic and environmental benefits of the thermal insulation requirements on new buildings can be estimated as summarized in Table 5.14. The carbon emissions for generating electricity are assumed to be 0.690 $kgCO_2/kWh$ and the cost of a power plant is estimated to be $1700/kW. It is estimated that new buildings represent 6% of the existing building stock estimated for 2014 (Krarti and Dubey, 2017).

A more stringent building energy efficiency code can be developed and adopted in Oman for all new buildings. This stringent code can achieve at least 50% savings in both energy consumption and peak demand associated with the new building stock (Krarti and Dubey, 2017). These savings are set lower than the potential levels that could be achieved for all building prototypes as found by a detailed analysis (Krarti and Dubey, 2017). The conservative values are considered to account for behavioral variations including some rebound effects (Jacobsen and Kotchen, 2013; Majcen et al., 2013). The benefits of a

TABLE 5.14 Economic and Environmental Benefits for Requiring Better-Insulated Building Envelope Systems for All New Buildings in Oman Based on 2014 Building Stock Estimates

Building type	Annual energy use savings (GWh/year)	Peak demand savings (MW)	Annual CO_2 emissions savings (10^3 tons/year)	Annual energy cost savings ($ million/year)	Peak demand savings ($ million)
Residential Buildings	179	40	124	15	68
Commercial Buildings	76	17	37	6	29
Governmental Buildings	55	12	27	5	20
Total	310	69	188	26	117

TABLE 5.15 Economical and Environmental Benefits for Implementing a Stringent Building Energy Efficiency Code for All New Buildings in Oman Based on 2014 Building Stock Estimates

Building type	Annual energy use savings (GWh/year)	Peak demand savings (MW)	Annual CO_2 emissions savings (10^3 tons/year)	Annual energy cost savings ($ million/year)	Peak demand savings ($ million)
Residential buildings	359	80	248	30	136
Commercial buildings	152	34	75	13	58
Governmental buildings	109	24	54	9	41
Total	620	139	377	52	236

comprehensive building energy efficiency code applied to the new buildings are estimated and are summarized as shown in Table 5.15.

5.9.3 Analysis of Impacts of Energy Retrofit Programs for Existing Buildings

Three levels of energy efficiency building retrofits are suggested for the Sultanate of Oman with different economic and environmental benefits (ASHRAE, 2011; Hong et al., 2015; Krarti, 2015; Krarti and Dubey, 2017):

- *Level 1 of energy efficiency retrofit:* Involves low-cost energy efficiency measures such as installation of programmable thermostats, use of CLF or LED lighting fixtures, and weatherization of building shell to reduce air infiltration. As detailed in Table 5.16, the estimated average savings from a level 1 retrofit program are 8% for all building types based on the simulation analysis carried out for this study as well as case studies reported for residential, commercial, and governmental buildings.
- *Level 2 of energy efficiency retrofit:* Includes use of energy-efficient cooling systems as well as temperature and lighting controls. Based on the simulation results obtained in this study as noted in Table 5.16 and confirmed by the reported findings in the existing literature, average savings of 23% can be achieved for level 2 retrofits for all building types (Ameer and Krarti, 2016; Hong et al., 2015).
- *Level 3 of energy efficiency retrofit:* Requires the implementation of capital-intensive measures including addition of roof thermal insulation, cooling system replacement, and installation of daylighting control systems. While deep retrofits are typically costly, they can provide significant energy use

TABLE 5.16 Options for Energy Efficiency Measures Specific to Three Retrofit Levels of Omani Residential Buildings

Recommended options	Retrofit description ([a])	Retrofit level for residential buildings		
		Level-1	Level-2	Level-3
1.	List of EEMs	EEM-1	EEM-1, EEM-2, and EEM-3	EEM-1, EEM-2, EEM-3, and EEM-4
	Energy use savings	12%	28%	54%
	Range of reduction in savings due to behavioral and rebound effects ([b])	0–6%	0–6%	0–6%
2.	List of EEMs	EEM-2	EEM-4	EEM-2, EEM-3, and EEM-6
	Energy use savings	10.0%	29%	51.0%
	Range of reduction in savings due to behavioral and rebound effects ([b])	0–4%	0–4%	0–4%
3.	List of EEMs	EEM-3	EEM-5	EEM-5, and EEM-6
	Energy use savings	10%	28%	52%
	Range of reduction in savings due to behavioral and rebound effects ([b])	0–4%	0–6%	0–6%

[a]*Description of EEMs: EEM-1: Increase the cooling set from 21° to 23°C, from 22° to 24°C, or from 23° to 25°C, depending on the existing operating conditions. EEM-2: Replace existing lighting fixtures by LEDs. EEM-3: Seal air leakage sources around building envelope (i.e., window and door frames so ACH = 0.21). EEM-4: Replace the existing AC unit by high efficiency system (COP = 4.0). EEM-5: Better lighting controls including dimming daylighting and occupancy sensors for commercial buildings. EEM-6: Insulate the roof using RSI-3.*

[b]*The behavioral and rebound effects are estimated based on previous studies. Typically, the effects are higher for measures that relay on temperature and lighting controls (i.e., EEM-1 and EEM-5).*

savings exceeding 50% as noted in the simulation analysis results specific to buildings in Oman (Krarti and Dubey, 2017).

Table 5.16 provides specific packages of simple yet proven energy efficiency measures that can be considered for the three energy retrofit levels for residential buildings in Oman (Krarti and Dubey, 2017). The potential annual energy use savings for each package is provided in Table 5.16. Moreover, ranges for the rebound effects associated with changes in occupant behaviors are estimated for each energy efficiency package and building retrofit level using results from reported analyses (Borenstein, 2015; Jabobsen et al., 2013; Majcen et al., 2013).

The economic benefits of the building energy efficiency retrofit programs are estimated using the actual cost of 32.4 Bz/kWh required for generating and distributing electricity (i.e., \$0.08347/kWh) rather than the average subsidized price of 15.7 Bz/kWh (i.e., \$0.040/kWh) using the reported AER data (AER, 2014). Moreover, the environmental benefits are estimated using an average carbon emission factor of 690 gCO_2/kWh for Oman (Solanski, 2013). Table 5.17 summarizes the annual CO_2 emissions savings (associated with reducing the amount of fuel needed to generate electricity), annual energy cost savings (associated with using less fuel to generate electricity), and the peak demand savings (associated with avoiding the construction of new power plants) for level 1, level 2, and level 3 building energy efficiency retrofit programs in Oman. As clearly shown in Table 5.17, significant economic and environmental benefits can be achieved for all the levels of the building energy retrofit programs. While requiring higher implementation costs, larger savings and more benefits can be achieved for level 2 and level 3 retrofit programs compared to the level 1 retrofit program. The economic and environmental benefits that can be realized for residential buildings are significantly higher than those obtained for commercial or governmental buildings for all energy retrofit levels. Indeed, over 58% of the benefits can be achieved by solely retrofitting residential buildings in Oman as indicated in Table 5.17.

The implementation of the large-scale building energy efficiency retrofit programs has to be gradual requiring several years due to several reasons: (1) the significant investments needed for renovating the entire existing building stock as estimated in the Cost-Effectiveness Analysis section, and (2) the lack of qualified energy efficiency contractors in Oman as discussed in the Job Creation section. However, any of the three energy retrofit programs can result in significant economic and environmental benefits for Oman even when only a small fraction of the existing stock is targeted (Krarti and Dubey, 2017).

5.9.3.1 *Program Implementation Considerations*

Energy prices are highly subsidized in Oman. Table 5.18 shows the current electricity prices for residential, governmental, and commercial customers in Oman. While the commercial customers have a flat rate of \$0.05194/kWh, the residential and governmental customers are charged based on the level of their consumption ranging from a low of \$0.026/kWh to \$0.078/kWh. The Omani Authority for Electricity Regulation has estimated that the average price charged to all its customers is 15.7 Baiza/kWh (\$0.0408/kWh) during 2014 (AER, 2014). Since the minimal economic cost for generating electricity is reported to be 25.3 Baiza/kWh (i.e., \$0.0743/kWh) for 2014, the Omani government provides subsidies of about 9.7 Baiza/kWh (\$0.0252/kWh) or 38% the economic cost of generating electricity. The subsidies level is estimated to be even higher reaching 46% during 2015 (AER, 2014). Overall, the International Monetary Fund has estimated that during 2015 the total energy subsidies amounted to \$7.27 billion or 8.9% of the Omani *GDP* (IMF, 2015).

TABLE 5.17 Economic and Environmental Benefits for Three Levels of Building Energy Retrofit Programs for Oman Based on 2014 Building Stock Estimates

Retrofit program	Level-1	Level-2	Level-3
Annual energy savings (GWh/year)			
Residential buildings	957	2751	5980
Commercial buildings	404	1162	2525
Governmental buildings	291	836	1817
Total existing building stock	1652	4748	10322
Peak demand savings (MW)			
Residential buildings	214	616	1339
Commercial buildings	90	260	565
Governmental buildings	65	187	407
Total existing building stock	370	1063	2311
Annual CO_2 savings (kilo-tons/year)			
Residential buildings	660	1898	4126
Commercial buildings	200	574	1247
Governmental buildings	144	413	898
Total existing building stock	1003	2885	6271
Annual avoided fuel costs ($million/year)			
Residential buildings	80	230	499
Commercial buildings	34	97	211
Governmental buildings	24	70	152
Total existing building stock	138	396	862
Avoided power plants costs ($million)			
Residential buildings	364	1047	2276
Commercial buildings	153	442	961
Governmental buildings	111	318	692
Total existing building stock	628	1807	3929

The cost-effectiveness of implementing large-scale energy efficiency programs for existing buildings can be estimated from the perspective of the Omani government (Krarti and Dubey, 2017). Indeed, investing in energy efficiency measures is generally not cost-effective for building owners due to highly subsidized energy prices. As outlined earlier, the overall economic,

TABLE 5.18 Electricity Prices for Omani Residential, Governmental, and Commercial Buildings

Commercial buildings (flat rate)	20 BZ/kWh ($0.05194/kWh)
Residences/government buildings	
0–3000 kWh	10 Bz/kWh ($0.026/kWh)
3001–5000 kWh	15 Bz/kWh ($0.039/kWh)
5001–7000 kWh	20 Bz/kWh ($0.052/kWh)
7001–10000 kWh	25 Bz/kWh ($0.065/kWh)
10000+ kWh	30 Bz/kWh ($0.078/kWh)

environmental, and social benefits of the energy retrofit programs are significant for Oman and therefore their cost-effectiveness should be evaluated from the perspective of the government of Oman. First, the costs needed to implement the three energy efficiency programs are estimated. Then, the cost-effectiveness analysis of the retrofit programs is carried out for various building types. Finally, the job creation potential required for the implementation of the energy retrofit programs is also determined and discussed in this section.

5.9.3.2 *Cost-Effectiveness Analysis*

The specific implementation costs for the building energy retrofits depend on several factors including the building size and the physical conditions of the building energy systems (Krarti and Dubey, 2017). Based on various sources including cost of labor and materials in the GCC region, the average costs of completing energy retrofit for buildings are estimated (Ameer and Krarti, 2015; Krarti, 2011; Krarti, 2012; Krarti and Ihm, 2014). Table 5.19 summarizes the implementation costs for all three level of energy retrofit programs specific to residential and commercial buildings including costs for performing energy audits as well as for installing any recommended energy efficiency measures (Krarti and Dubey, 2017).

TABLE 5.19 Average Costs Expressed in US$ for Energy Retrofits of Buildings in Oman

Building type	Level-1	Level-2	Level-3
Residential buildings	250	2,500	5,000
Commercial buildings	5,000	25,000	50,000
Governmental buildings	10,000	50,000	100,000

The cost–benefit analysis for the building energy efficiency retrofit programs to upgrade the existing buildings in Oman are estimated using the latest available building stock statistics (NCSI, 2015). The results, summarized in Table 5.20, indicate that the three levels of energy retrofit programs are highly cost-effective when deployed for the residential buildings. In fact, the implementation costs for the level 1 retrofit program would be recovered from the avoided costs of building new power plants to generate electricity needed to meet the energy requirements of households if no energy efficiency actions are taken. The payback periods for all three retrofit levels are found to be long for commercial and governmental buildings most likely due to the high estimated implementation costs of energy efficiency measures for these buildings. It is

TABLE 5.20 Cost-Effectiveness Analysis of Three Energy Efficiency Retrofit Programs in Oman

Retrofit program	Level-1	Level-2	Level-3
Implementation program costs ($million)			
Residential buildings	174	1739	3477
Commercial buildings	581	2906	5881
Governmental buildings	278	1391	2781
Total existing building stock	1033	6035	12070
Annual avoided fuel costs ($million/year)			
Residential buildings	80	230	499
Commercial buildings	34	97	211
Governmental buildings	24	70	152
Total existing building stock	138	396	862
Avoided power plants costs ($million)			
Residential buildings	364	1047	2276
Commercial buildings	153	442	961
Governmental buildings	111	318	692
Total existing building stock	628	1807	3929
Payback period (years)			
Residential buildings	0	3.0	2.4
Commercial buildings	12.6	25.4	23.3
Governmental buildings	7.0	15.3	13.7
Total existing building stock	2.9	10.7	9.4

therefore highly recommended that a retrofit program be developed first by the Omani government to improve the energy efficiency of existing residential building stock as summarized in Table 5.20 (Krarti and Dubey, 2017).

5.9.3.3 *Job Creation and Market Potential of EE Industry*

The other economic impact of the building energy efficiency retrofit programs is the potential to create new jobs in Oman. As outlined by Hyland et al. (2013), Krarti (2015), and Pollin et al. (2009), the direct effects for retrofitting buildings include jobs needed to implement the energy efficiency measures while the indirect effects are associated with jobs needed to produce and supply energy efficiency equipment and materials. Most of the jobs created for building retrofits are within the construction and manufacturing industries with a wide range of pay level and technical specialization including electricians, HVAC technicians, insulation installers, energy auditors, building inspectors, and construction managers.

Using the job creation model considered in the analysis performed by Krarti (2015), \$1.0 million investment in building energy retrofit can create 11.9 job-years with 7.0 direct and 4.9 indirect jobs. The direct jobs are those required to audit and implement energy efficiency measures in the buildings while the indirect jobs are associated with producing and supplying energy efficiency equipment and materials (Pollin et al., 2009). For the case of Oman, up to 41,376 new job-years can be created when renovating the existing residential building stock using the level 3 retrofit option as illustrated in Table 5.21. When the entire building stock is considered, up to 143,633 job-years would be created to implement a level 3 retrofit program. In particular, if a level 3 energy efficiency retrofit program is implemented during a 10-year period, 14,363 jobs per year can be created (Krarti and Dubey, 2017).

5.9.3.4 *Energy Productivity Analysis*

To determine the cumulative impacts of the benefits that can be incurred from building energy efficiency programs, the monetary values of these benefits are estimated and accounted for as part of the *EP* analysis outlined in this chapter. Specifically,

TABLE 5.21 Number of Job Years That Can be Created From a Building Energy Retrofit Program in Oman

Building type	Level-1	Level-2	Level-3
Residential	2,071	20,694	41,376
Commercial	6,914	34,581	69,984
Governmental	3,308	16,553	33,094
Total	12,293	71,817	143,633

three main benefits are considered for the three levels of energy retrofit programs proposed for the existing building stock in Oman (Krarti and Dubey, 2017):

1. Energy consumption reduction resulting in two impacts on the *EP* of the Omani building sector as defined by Eq. (5.26): (1) the reduction in final energy consumption, TFC_B, and (2) the increase in value added, VA_B, associated with the avoided cost of oil required to generate electricity. To estimate the value added of avoided oil consumption, oil price of $45/barrel, oil production cost in Oman of $5.3/barrel (Knoema, 2016), and an average efficiency of 36% for electricity power plants (IEA, 2015) are utilized.
2. Electrical peak demand reduction lowering the number of power plants required to meet future energy needs of the building sector. An average cost of $1700/kW is considered to estimate the increase in added value, VA_B, due to the avoided power plant capacities.
3. Carbon emission reduction due to decreasing the need to combust oil or gas to generate electricity in power plants. A value of $10/ton is used to estimate the increase in added value, VA_B, due to the lower carbon emissions (EPA, 2016; Russell et al., 2015).

Fig. 5.19 shows the effects of the three levels of energy retrofit programs, considered in this study, on the *EP* for the Omani building sector (Krarti and Dubey, 2017). The impacts of the benefits just outlined on the building sector

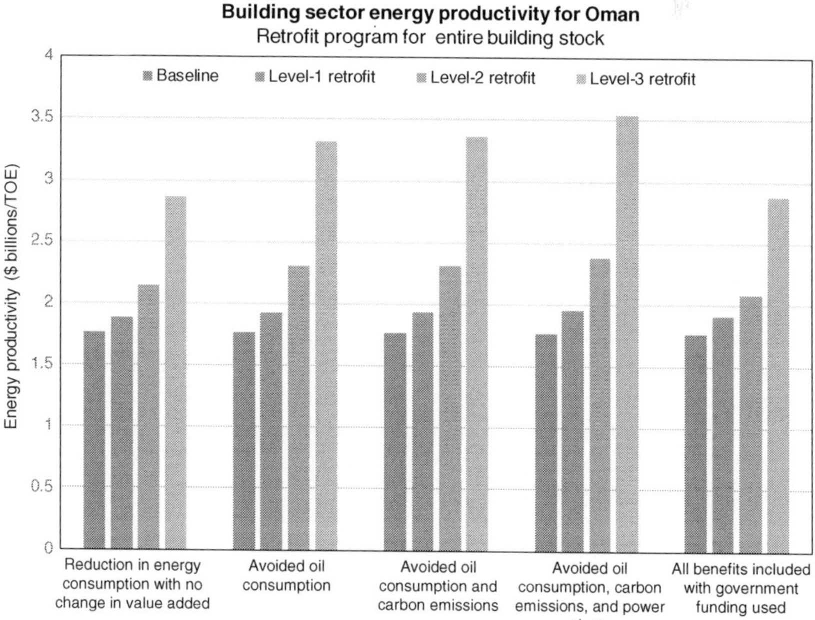

FIGURE 5.19 Impact on Omani building sector energy productivity for three retrofit programs of the entire building stock.

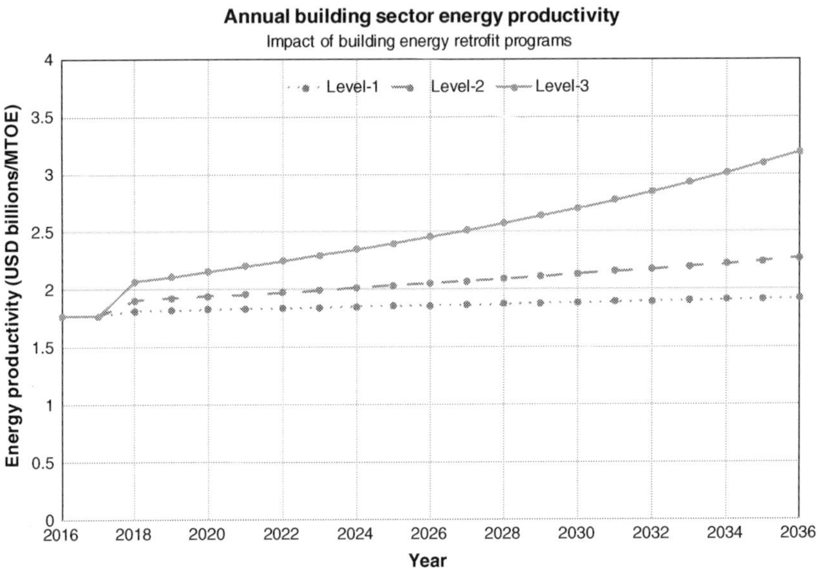

FIGURE 5.20 Annual progression for Omani building sector energy productivity for three retrofit programs of the entire building stock during a 20-year implementation period.

EP are evaluated gradually for the three retrofit programs. When accounting for only the savings in energy consumption, without accounting for the potential monetary added value of avoided oil use, the *EP* can be increased by up to 75% when the level 3 retrofit program is applied to the building stock. However, when considering all the benefits and their added values assuming the private sector can provide the necessary investments needed for the programs' implementation, the *EP* can be doubled when the entire building stock is retrofitted. When the government finances the entire retrofit programs, the *EP* can be increased by up to 160% for the entire building stock. Fig. 5.20 illustrates the progression of the building sector *EP* when all the benefits are considered for the three retrofit levels when the implementation period is set for 20 years (so that each year 5% of the building stock is retrofitted) with a 2-year start-up period for building capacity of energy auditors and planning of the implementation phase. As shown in Fig. 5.20, the level 3 retrofit program provides the best option to increase the *EP* of the Omani building sector. It should be noted that the sole implementation of mandatory building energy efficiency codes for new buildings increases incrementally but slightly the *EP* of the Omani building sector by up to 4% when a comprehensive set of requirements is considered and up to 2% when thermal insulation is only required for walls and roofs (Krarti and Dubey, 2017).

5.9.3.5 Analysis Conclusions

In this section, a general analysis approach is developed and applied to estimate the impacts of implementation of energy retrofit programs on the building sector

EP. The analysis can account for the energy and nonenergy benefits that can be incurred from any energy efficiency improvement action applied at large-scale or at the individual building scale. The main advantage of the proposed *EP* analysis is its ability to account using one metric for both economic and energy performance impacts of energy efficiency actions. The *EP* indicator can assist on deciding on the merit of any action in maximizing economic benefits while minimizing energy consumption.

The analysis has been applied to assess the benefits of large-scale energy retrofits for the existing building stock as well as the implementation of energy efficiency requirements for new buildings in Oman. Over the last decade, Oman, like the GCC region, has experienced significant economic growth and high energy consumption. The analysis for the building sector *EP* has indicated that improving the existing building stock offers an effective option to extract economic value from reducing the national energy consumption. In addition to the reduction in energy use, other benefits from the retrofit programs proposed for the Omani building stock include additional income from avoided fuel used to generate electricity and reduced costs for power generation as well as lower carbon emissions and creations of high-skilled jobs. It has been shown that even when governments provide the entire investments needed to implement the large-scale retrofit programs, the *EP* of the building sector can be doubled.

5.10 SUMMARY

This chapter described a general but systematic approach to perform the economic evaluation of various alternatives of energy projects. In particular, several analysis methods described in this chapter (such as the net present worth, the *BCR*, the *ROR*, and the payback period) are suitable for evaluating single alternatives but may not be used to rank among alternatives. The *LCC* analysis method is more suitable for evaluating several alternatives and should be used to select the most economical option among possible alternatives for an energy project. Specially for large-scale energy efficiency projects and programs, the *EP* analysis framework is presented to account for energy and nonenergy benefits. As outlined in this chapter, the *EP* analysis can provide the decision-makers with better evaluation indicators on the economic merit of energy projects. Chapter 9 provides more examples for *EP*–based analysis of large-scale energy efficiency and renewable energy projects.

PROBLEMS

5.1 A homeowner is considering adding a sunspace to his home. The project is carried out during 4 years and he has to spend \$20,000 the first year, \$15,000 the second year, \$10,000 the third year, and \$5000 the fourth year.

Determine how much money the homeowner needs to set aside to pay for the sunspace addition project:

(a) If the interest rate is 4% per year, compounded annually.

(b) If the interest rate is 8% per year, compounded annually.

5.2 You want to buy a new home at a cost of $250,000. Determine how much money you should pay every year for 20 years to pay for the house based on equal annual payments:

(a) If the interest rate is 5% per year, compounded annually.

(b) If the interest rate is 10% per year, compounded annually.

5.3 An energy audit of a residential heating system reveals that the boiler-burner efficiency is only 60%. In addition, the energy audit showed that an electric water heater is used. The house uses annually 1500 gallons of oil at a cost of $1.40 per gallon. The total electric bill averaged $84.50 per month for an average monthly consumption of 818 kWh. Of this total, about 35% was for domestic water heating. It is suggested to the owner of the house to equip the boiler with a tank-less domestic water heater replacing the existing electric water heater. The existing water heater is 12 years old and has no resale value and little expected life. The owner of the house is expected to spend $300 within the coming year to replace the electric water heater. The cost of the tank-less heater is $400.

To improve heating efficiency, it is suggested to

(i) Replace the existing burner with a new one (this burner replacement will improve efficiency to 65%) costing $520.

(ii) Replace the entire heating plant with higher efficiency (85%) with a cost of $2000.

For two discount rates (5% and 10%), provide the ***LCC*** analysis of the following options and make the appropriate recommendations:

- ☐ Keep the boiler-burner and replace the electric water heater with like system.
- ☐ Replace burner and electric water heater with like systems.
- ☐ Replace boiler-burner with efficient boiler and replace electric heater with like system.
- ☐ Replace existing boiler-burner and electric water heater with new boiler/take-less domestic water heater.

Note: State all the assumptions made in your calculations.

5.4 An energy audit indicates that a home consumption is 20,000 kWh per year. By using Energy Star appliances and CFLs, a savings of 20% can be achieved for the electrical energy use. The cost for the home retrofit is about $15,000. Assuming that the average energy charge is $0.09 per kWh, is the expenditure justified based on a minimum *ROR* of 18% before taxes. Assume a 20-year life-cycle and use the present worth, annual cost, and *ROR* methods.

5.5 An electrical energy audit indicates that lighting consumes 5000 kWh per year. By using more efficient lighting fixtures, electrical energy savings of 15% can be achieved. The additional cost for these energy efficient lighting fixtures is about $900. Assuming that the average energy charge is $0.08/kWh, is the expenditure justified based on a minimum *ROR* (i.e., a discount rate) of 7.0? Assume a 10-year life-cycle and use the present worth, annual cost, and *ROR* methods.

5.6 A residential building is heated with a 78% efficient gas-fired furnace (costing $1600). The annual heating load is estimated at 160 MMBtu. The cost of gas is $6.00/MMBtu. For the discount rate of 5% and 10-year cycle, determine:

(a) The life-cycle of the existing heating system.

(b) If it is worth considering a 90% efficient gas-fired furnace that costs $2800.

(c) The variation of the *LCC* savings between the two systems [the existing system and the proposed in (b)] versus the life of the system for life-cycles ranging from 3 to 20 years.

Conclude.

5.7 An air-conditioner (AC) consumes 7500 kWh annually with an overall efficiency of COP = 3.0. If this AC is replaced by a more energy efficient AC (COP = 3.7) at a cost of $3500

(a) Determine the *SPP* of the chiller replacement.

(b) If the expected life of the old chiller is 15 years. Is it cost-effective to replace the chiller in an economy with a discount rate of 7%?

5.8 Two air conditioner systems are proposed to cool a house. Each AC has a rated capacity of 5 tons and is expected to operate 500 full-load equivalent hours per year. Air conditioner A has a standard efficiency with a COP of 2.8 and costs $1600 while AC B is more efficient with a COP of 3.5 at a cost of $2750. The electricity cost is estimated to be $0.09/kWh. For a lifetime of 20 years and a discount rate of 5%,

(a) Estimate the payback time for using air conditioner B instead of A.

(b) Calculate the *LCC*s for both air conditioners. Conclude.

(c) Estimate the *ROR* for using air conditioner B instead of A.

(d) Determine the highest cost for air conditioner B at which air conditioner A is more competitive.

(e) Determine the cut-off electricity price at which both air conditioners A and B have the same *LCC*.

Assume that the average cost of electricity is $0.08/kWh.

5.9 Consider the macroeconomic data for a country listed in the table including annual value added (*VA*) for the building sector and all the sectors (i.e., *GDP* for the entire economy) as well as associated total final energy consumption (*TFC*) values.

Macro-economic data	2008	2009	2010	2011	2012	2013
VA-buildings ($billions)	164.49	166.26	161.42	167.31	172.20	174.13
VA-all ($billions)	1150.80	1138.64	1145.95	1183.11	1212.95	1231.31
TFC-buildings (MTOE)	63.956	60.48423	59.19308	62.85592	61.48592	59.89367
TFC-all (MTOE)	175.7826	165.2469	167.1425	172.9924	180.4211	180.8411

1. Estimate the energy productivity (*EP*) for all the sectors and the building sector. Comment on the variation over the years.
2. Considering the *EP* value for the building sector for 2013, consider the impact of energy efficiency programs on the *EP* value under the three scenarios for the annual changes in both *VA* and *TFC* (relative to the values in 2013) (1) $\Delta VA = -10\%$ and $\Delta TFC = -5\%$, (2) $\Delta VA = 0\%$ and $\Delta TFC = -5\%$, and (3) $\Delta VA = 10\%$ and $\Delta TFC = -5\%$. Assess under these scenarios if the energy program is effective in improving the energy productivity for the building sector.
3. Under the best scenario, determine the impact of the energy efficiency program on the overall economy energy productivity.

REFERENCES

AER, 2014. Annual Report for 2014, Authority for Electricity Regulation, Oman.

Akbari, H., Konopacki, S., Pomerantz, M., 1999. Cooling energy savings potential of reflective roofs for residential and commercial buildings in the United States. Energy 24, 391–407.

Alaidroos, A., Krarti, M., 2015. Optimal design of residential building envelope systems in the Kingdom of Saudi Arabia. Energy Build. 86, 104–117.

Al-Badi, A.H., 2011a. Wind power potential in Oman. Int. J. Sustainable Energy 30, 110–118.

Al-Badi, A.H., 2011b. Hybrid (solar and wind) energy system for Al Hallaniyat Island electrification. Int. J. Sustainable Energy 30, 212–222.

Al-Badi, S.A., Malik, A., Gastli, A., 2015. Sustainable energy usage in Oman- Opportunities and barriers. Renewable Sustainable Energy Rev. 15 (4), 3780–3788.

Al-Hatmi, Y., Tan, C.S., AlBadi, A., Charabi, Y., 2014. Assessment of the consciousness levels on renewable energy resources in the Sultanate of Oman. Renewable Sustainable Energy Rev. 40, 1081–1089.

Amann, J., 2006. From Valuation of No-energy Benefits to Determine Cost-Effectiveness of Whole-House Retrofits Program: A Literature Review. American Council for an Energy-Efficient Economy, Washington, DC.

Ameer, B., Krarti, M., 2016. Impact of Subsidization on high energy performance designs for Kuwaiti residential buildings. Energy Build 116, 249–262.

Apte, M.G., Fisk, W., Daisey, J.M., 2000. Associations between indoor CO_2 concentration and sick building syndrome symptoms in US office buildings: an analysis of the 1994–1996 base study data. Indoor Air 10 (4), 246–257.

ASHRAE, 2011. Procedures for Commercial Building Energy Audits, second ed., Guide Developed by the American Society for Heating, Refrigerating, and Air Conditioning Engineers, Atlanta, GA.

Blok, K., Hofheinz, P., Kerkhoven, J., 2015. The 2015 Energy Productivity and Economic Prosperity Index: How Efficiency Will Drive Growth, Create Jobs, and Speed Weelbeing Throughout Society, Report by Ecofys, Quintel Intelligence, and the Lisobon Council.

Bernstein, H.M., Russo, M.A., 2011. Business Case for Energy Efficient Building Retrofit and Renovation. McGraw Hill Construction SmartMarket Reports.

Borenstein, S., 2015. A microeconomic framework for evaluating energy efficiency rebound and some implications. Energy J. 36 (1).

Colliers International, 2012, Building Choice Crucial to Attracting Retaining Quality Office Staff. Available from: http://geelong.colliers.com.au/News.

EPA, 2016. The Social Cost of Carbon, Climate Change, http:www3.epa.org.

Gastli, A., Charabi, Y., 2010. Solar electricity prospects in Oman using GIS-based solar radiation maps. Renewable Sustainable Energy Rev 14, 790–797.

Gastli, A., Charabi, Y., 2011. Solar water heating initiative in Oman energy saving and carbon credits. Renewable Sustainable Energy Rev 15 (4), 1851–1866.

Hong, T., Piette, M.A., Chen, Y., Lee, S.H., Taylor-Lange, S.C., Zhang, R., Sun, K., Price, P., 2015. Commercial building energy saver: an energy retrofit analysis toolkit. Appl. Energy 159, 298–309.

Hyland, M., Lyons, R.C., Lyons, S., 2013. The value of domestic building energy efficiency, evidence from Ireland. Energy Econ. 40, 943–952.

IEA, 2016. IEA Statistics: Energy Balance. http://www.iea.org/statistics, Paris, France.

Jacobsen, G.D., Kotchen, M.J., 2013. Are building codes effective at saving energy? Evidence from residential billing data in Florida. Rev. Econ. Stat. 95 (1), 34–49.

KAPSARC, 2015. Global Shift: the energy productivity transformation. KAPSRAC Energy Workshop Series. The King Abdullah Petroleum Studies and Research Center, Riyadh, http://www.kapsarc.org.

Kats, G., 2010. Greening our built world: costs. Benefits and Strategies. Island Press, Washington D.C.

Knoema, 2016. Cost of Oil Production by Country, http://www.knoema.org.

Kosonena, R., Tan, F., 2004. Assessment of productivity loss in air-conditioned buildings using PMV index. Energy Build. 36, 987–993.

Krarti, M., 2012. Weatherization of Residential Buildings Textbook. Taylor and Francis Publishing, New York, NY.

Krarti, M., 2015. Evaluation of large scale building energy efficiency retrofit program in Kuwait. Renewable Sustainable Energy Rev. 50, 1069–1080.

Krarti, M., Dubey, K., 2017. Energy productivity evaluation of large scale building energy efficiency programs for Oman. Sustainable Cities Soc. 29, 12–22.

LeBaron, R., 2011. Getting to Fair Cost-Effectiveness Testing Using the PACT, Best Practices for the TRC and Beyond, National Home Performance Council, September.

Majcen, D., Itard, L., Visscher, H.H., 2013. Actual and theoretical gas consumption in Dutch dwellings: what causes the differences? Energy Policy 61, 460–471.

Malik, A.S., 2007. Impact on power planning due to DSM in commercial and government sectors with rebound effect – a case study of central grid of Oman. Energy 32, 2157–2166.

Malik, A., Al-Badi, A.H., 2009. Economics of wind turbine as an energy fuel saver – a case study for remote application in Oman. Energy 34, 1573–1578.

Mallela, V., Solanki, P., Al-Harasi, Y., 2010. Regulation of energy demands via DSM policies: a case study in Oman. Can. J. Electr. Electron. Eng., 54–59.

Montgomery, J., Heubach, J., Weimer, W., Heerwagen, J., 1994. Impacts on Satisfaction and Productivity in a Laboratory Renovation. Internal Report by Pacific Northwest National Laboratory, PNNL.

NCSI, 2015. 2015 Statistical Year Book. Report by National Centre for Statistics and Information, Sultanate of Oman.

OBG, 2013. Oxford Business Group Report: Oman 2013, Going Green: sustainable building practices should help address housing shortages. Available from: http://www.oxfordbusinessgroup.com.

OPWP, 2014. Annual Report for 2014, Oman Power and Water Procurement Corporation, Member of Nama Group, Muscat, Oman.

Powmya, A., Abidin, N.Z., 2014. The challenges of green construction in Oman. Int. J. Sustainable Constr. Eng. 5 (1), 33–41.

Romm, J.J., Browning, W.D., 2004. Greening the Building and the Bottom Line. A report by Rocky Mountain Institute, Boulder, CO.

Russell, C., Baatz, B., Cluett, R., Amann, J., 2015. Recognizing the Value of Energy Efficiency's Multiple Benefits. American Council for an Energy-Efficient Economy, Washington, DC.

Saleh, M.S., Alalouch, C., 2015. Towards sustainable construction in Oman: challenges and opportunities. Procedia Eng. 118, 177–184.

Skumatz, L.A., 2014. Non-Energy Benefits/Non-Energy Impacts (NEBs/NEIs) and Their Role and Values in Cost-Effectiveness Tests: State of Maryland, Final Report. Skumatz Economic Research Associates, Inc., Superior, CO.

Solanki, P.S., Mallela, V.S., Zhou, C., 2013. Estimation and diminution of CO_2 emissions by clean development mechanism option at power sector in Oman. Int. J. Energy Environ. 4 (4), 641–652.

TRC, 2014. The Research Council: Oman Eco-House Design Competition. Available from: https://home.trc.gov.om.

WGBC, 2013. The Business Case for Green Buildings: A Review of the Costs and Benefits for Developers, Investors and Occupants, a report by World Green Building Council. http://www.worldgbc.org.

Wyon, D.P., 1996. Indoor environmental effects on productivity. IAQ 96 Paths to better building environments/Keynote address, Y. Kevin. Atlanta. ASHRAE Trans., 5–15.

Yu, S., Yong, T., Chenxi, L., 2011. Green Retrofitting Costs and Benefits: A New Research Agenda, IRES Working Paper Series. National University of Singapore, Institute of Real Estate Studies, Singapore.

Chapter 6

Integrated Design and Retrofit of Buildings

Chapter outline

6.1 INTRODUCTION

In the United States, homes built before 2000 represent 92% of the US housing stock (DOE, 2005). Similar conditions are encountered by several countries as outlined in Chapter 1 with aging and energy inefficient housing stock. Unfortunately, most of the existing retrofit and weatherization programs are limited

Optimal Design and Retrofit of Energy Efficient Buildings, Communities, and Urban Centers
http://dx.doi.org/10.1016/B978-0-12-849869-9.00006-5

in scope and impact with only few predefined retrofit measures typically considered and even fewer measures are implemented and supported through incentive programs. Common evaluated and implemented energy efficiency measures in residential buildings include attic insulation, programmable thermostat, weather-stripping, and cooling/heating system replacement. To reduce residential energy use significantly, novel approaches to retrofit programs should be considered to combine a comprehensive list of energy efficiency strategies and renewable energy technologies. Even with current technologies, there is a significant potential to improve cost-effectively the energy efficiency of existing buildings. A recent study by the World Council for Sustainable Development (WBCSD, 2009) found that several energy efficiency projects are feasible with today's energy costs. Specifically, the study found that at oil prices of $60 per barrel, investments in existing building energy efficiency technologies can reduce related energy use and carbon footprints by 40% with five discounted payback years.

Comprehensive retrofit programs that encompass several energy efficiency measures require more detailed and sophisticated analysis tools that only recently have been applied to improve the energy performance of existing buildings. For instance, concepts such low energy or zero-energy buildings have been applied traditionally to design new buildings and often utilize detailed simulation analysis tools coupled with optimization techniques. These same concepts are only now being extended to existing buildings to select the most cost-effective energy efficiency measures that minimize initial costs and maximize energy savings. In particular, optimization of energy efficiency measures suitable for existing buildings entails evaluating various retrofit options until a set of measures is identified that achieves minimum cost and/or maximum energy savings. While exhaustive enumeration can be performed, in which every possible combination of possible energy efficiency measures is evaluated, optimization techniques are often utilized to minimize the number of combinations that need to be analyzed.

While the concept of zero net energy (ZNE) buildings has been mostly applied to new construction, it can be considered for retrofit projects including residential buildings. Typically, ZNE is defined either in terms of site energy or resource energy (Torcellini and Pless, 2010). Site energy consists of energy produced and consumed at the building site. Source or primary energy includes site energy as well as energy used to generate, transmit, and distribute this site energy. Therefore, source energy provides a better indicator of the energy use of buildings and their impact on the environment and society and thus is better suited for ZNE building analysis. Indeed, an analysis based on source energy effectively allows different fuel types, such as electricity and natural gas commonly used in buildings, to be encompassed together. It is generally understood that ZNE buildings produce as much energy as they consume on-site annually. These buildings typically include aggressive energy efficiency measures and active solar water heating systems. Moreover, ZNE buildings employ grid-tied, net-metered renewable energy generation technologies, generally photovoltaic (PV) systems, to produce electricity. Effectively, ZNE buildings use the grid as a "battery storage" to reduce required generation system capacity.

It should be noted that other definitions of ZNE has been proposed include the life-cycle net zero energy (LC-NZE). A life-cycle energy analysis accounts not only for operational energy or energy use but also the embodied energy, which includes energy needed for the manufacturing, transportation, construction, replacement, and demolition of building components and systems. As the energy efficiency of buildings is increased and thus the operational energy is reduced, the relative share of embodied energy increases (Berggren et al., 2013; Hernandez and Kenny, 2010). Moreover, ZNE buildings are defined by accounting for other metrics such as carbon emissions, in addition to energy used and generated by buildings using different weighting factors (Noris et al., 2014; Sartori et al., 2012).

Fig. 6.1 illustrates the concept of a ZNE building to account for energy balance between the energy delivered to the building and the energy delivered to the grid associated with the excess of on-site generation compared to the building needs.

Fig. 6.2 shows how the energy balances between demand and supply change depending on the application of energy efficiency (EE) and renewable energy (RE) measures in the cases of (a) the ZNE and (b) positive energy buildings.

Over the last two decades, few tools have been developed for selecting optimal packages of energy efficiency measures for residential buildings. Most of these tools are for designing new homes and are based on the sequential search technique. The sequential search methodology was first applied for optimizing the energy efficiency level of the residential building sector (Meier, 1982). Then, the methodology was implemented to a spreadsheet-based analysis tool (DEG, 1994) and to the EnergyGauge-Pro software (FSEC, 2007). The sequential search methodology was used to find a path to design zero-energy homes using a building

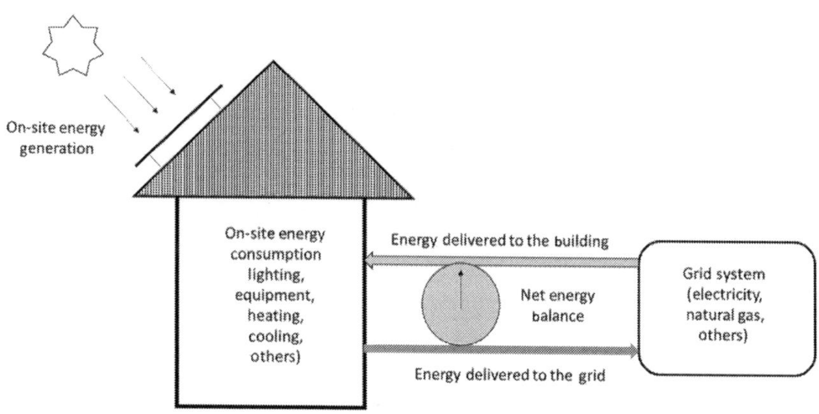

FIGURE 6.1 Concept of zero net energy buildings.

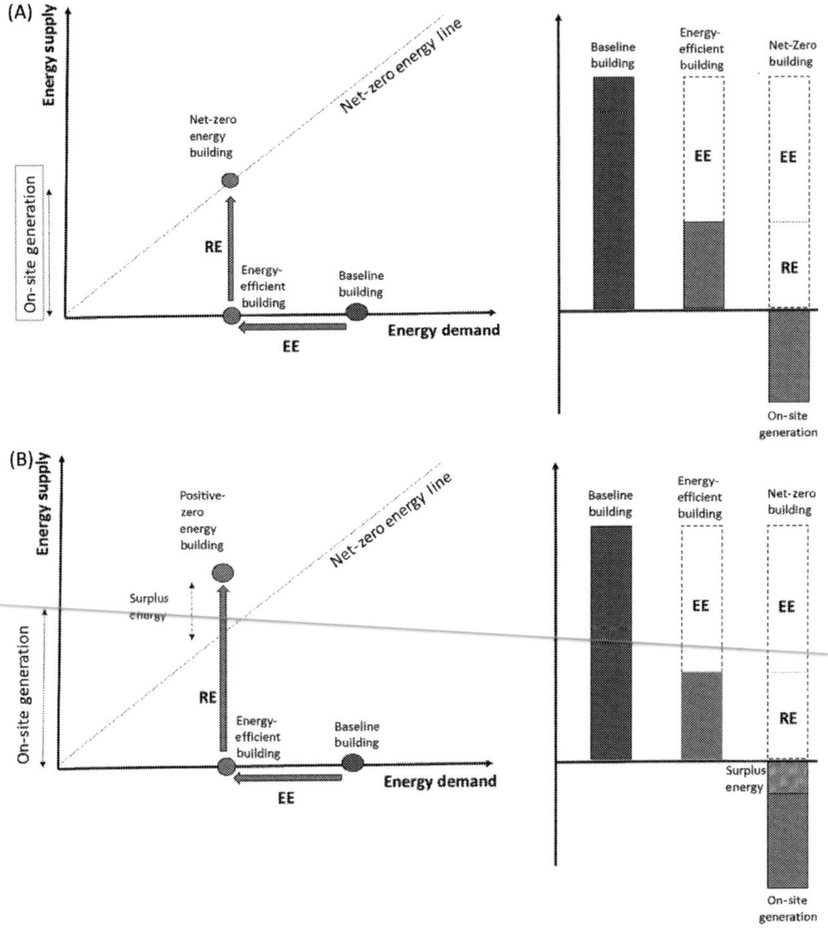

FIGURE 6.2 Balances between energy demand and supply. (A) Zero net energy (ZNE) buildings and (B) positive energy buildings.

energy simulation-based computer program referred to as the building energy optimization or *BE*opt tool (Christensen et al., 2004). Starting from a reference building model, *BE*opt identifies the cost-optimal set of building energy options needed to achieve a variety of energy savings levels including ZNE option (i.e., 100% energy savings). The building energy options are chosen from predefined or custom efficiency and renewable energy measures. Energy savings for each measure are calculated relative to a reference building on a source energy basis.

Originally, *BE*opt was used primarily to optimize design options for new homes. It can now be used for optimizing the energy retrofit options for existing residential buildings even though automatic calibration capabilities are not available in the current version of the tool. A conceptual plot of the cost/energy saving

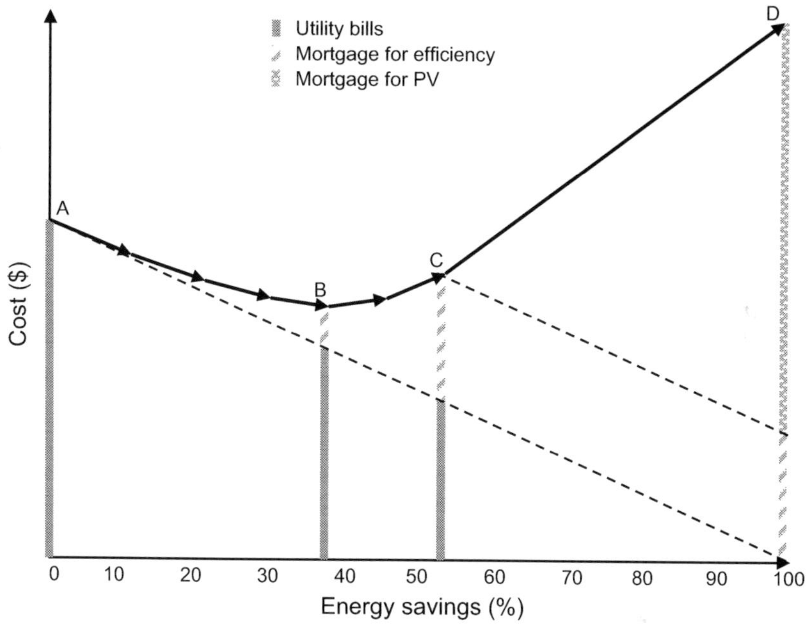

FIGURE 6.3 *BE*opt Pareto graph of the optimal path to ZNE.

Pareto graph from *BE*opt is illustrated in Fig. 6.3. At the starting point of the optimization (point A), only utility bills comprise the building's energy-related cost as no efficiency upgrades have been included to cause an increase in mortgage, which includes the energy use cost and the capital cost for implementing energy efficiency and/or renewable systems. As energy efficiency measures are introduced into the building, incremental mortgage costs increase and utility bills decrease until the marginal cost of saved energy equals the cost of utility power. At this point, the curve reaches a minimum and the global cost-optimum point is reached (point B). Additional efficiency measures with marginal costs more expensive than the cost of fuel are introduced until the marginal cost of saved energy equals the marginal cost of producing PV energy (point C). At this point, PV capacity is added until all source energy use is offset (point D). For roofs where there is a limit to the number of PV panels that can fit, additional energy efficiency measures are employed after the PV system selection until ZNE is reached. This additional energy efficiency measures may have a marginal cost of saved energy that is larger than that of the produced energy from PV. If all energy efficiency measures in the optimization search space are exhausted before ZNE is attained, the optimization stops at its maximum possible energy savings level.

Other optimization techniques and algorithms have been considered to optimize building energy use (Tuhus and Krarti, 2009; Wetter, 2004; Wright et al., 2002). Some of these optimization techniques are briefly presented in the following section.

6.2 BUILDING ENERGY OPTIMIZATION APPROACHES

Inherently, building energy retrofits consider several energy efficiency measures and thus are multivariate. The energy efficiency measures encompass a wide range of options including building envelope insulation, HVAC (heating, ventilation, and air conditioning), equipment, appliances, lighting, water heating, and possibly renewable energy systems. To accurately select the best package for the energy efficiency measures, all the possible combinations of the measures would need to be evaluated. This approach, often referred to as the brute force optimization approach, can be time-consuming and in several cases not needed to select the optimal package. Instead, some optimization techniques and methodologies can be considered. Specifically, several optimization techniques have been used in a wide range of applications to improve the energy performance of buildings. In particular, optimization-based analysis approaches have been widely used to improve the control and operation of buildings (Fong et al., 2006; Huang and Lam, 1997; Wright et al., 2002). Optimization techniques have been proposed to select building envelope and heating and cooling system design features (Bichiou and Krarti, 2011; Caldas and Norford, 2003; Christensen et al., 2004; Tuhus-Dubrow and Krarti, 2010; Wanga et al., 2005). In particular, Tuhus-Dubrow and Krarti (2010) developed a simulation environment using a genetic algorithm optimization technique to select the best combinations of several building envelope features to optimize energy consumption and life-cycle costs.

Several optimization tools are now available and can be applied to both new design retrofit of buildings. For instance, GenOpt is a generic optimization program that can be connected with an external building energy simulation program to minimize a cost function. It includes several optimization methodologies suitable for continuous, discrete, univariate, or multivariate optimization problems (Wetter, 2004). In this section, a set of optimization techniques are briefly presented. These techniques can handle multivariate and discrete optimizations and have been successfully applied to select an optimal package of energy efficiency measures for residential buildings (Tuhus-Dubrow and Krarti, 2009).

6.2.1 Sequential Search

The sequential search technique, used in *BE*opt, is a direct search method that identifies the building energy option that will best decrease the cost function at each iterative step. It begins by simulating a user-defined reference building model. The reference model is typically a calibrated energy simulation model for an audited building. A simulation is performed for each potential energy efficiency measure individually, and the most cost-effective measure is chosen and used in the building model description for the next point along the path. There are a number of discrete options for a given energy efficiency measure category such as glazing type, heating system, and ceiling insulation. The most cost-effective option is defined as the one that gives the largest reduction in annual costs for the smallest reduction in source energy use. Annual costs are a combination of mortgage costs (which increase as more expensive energy-efficient options are included) and utility costs. The process is repeated, ultimately defining a path from the reference building to the minimum cost point, and then to a ZNE building.

Without modifications, this simple algorithm would not reliably find the correct least-cost path, due to the problem of interactive effects between different options. Three special cases have been identified for the sequential search technique as implemented in *BE*opt—invest/divest, large steps, and positive interactions (Christensen et al., 2004). The invest/divest case is a result of negative interactive effects. *BE*opt attempts to account for this by evaluating the removal of options in the current building design, which may result in a more cost-optimal point. For example, a highly efficient HVAC system may have been selected as the most cost-effective option at an early point in the process. Later on, however, the improvement of the building envelope may cause the efficient HVAC option to not be cost-optimal, so it is removed from the building design.

The large steps case is another example of negative interaction among options. There may be a large energy-saving option that is available at a current point, but is less cost-effective than another option that does not save as much energy. The latter option is chosen, and then the most cost-effective option is again chosen at that second point, which results in a third point. However, it is possible that the original large energy-saving option available at the first point is more cost-optimal than the third point. To solve this problem, the sequential search keeps track of points from previous iterations and compares them to the current point. If a previous point is more cost-optimal, it replaces the current point.

A positive interaction occurs if two options are more cost-effective when present together than they would be separately. An example could be the presence of both large south-facing windows and thermal mass for passive solar heating. The sequential search will only find these positive interactions if one of the options is first selected individually. This inability to always identify synergistic options is a potential deficiency with the sequential search method.

6.2.2 Genetic Algorithm

Genetic algorithm (GA) uses the evolutionary concept of natural selection to converge on an optimal solution over many generations. This technique differs from traditional optimization methods in a number of areas. First, rather than working with one potential solution at a time, the technique works with a set of solutions called a population. This ensures a global approach to the optimization and helps the GA avoid getting stuck in local minima, which can be a problem with other methods. Second, the GA works with encodings of the parameters, not the parameters themselves. Parameters are traditionally encoded as binary strings, although other methods are possible. Finally, GAs use probabilistic methods for determining the parameter values in each successive iteration, rather than deterministic rules. This means that each time a GA is run, the path toward convergence is different, and the end result may be different as well.

Each individual in the population represents a different solution to the problem. Every option for each parameter has a corresponding binary representation, and the parameters are concatenated to form the complete binary string. A new generation is formed at the end of each iteration of the algorithm, consisting of a new population, and this process is repeated until satisfactory convergence criteria are reached, or the maximum number of generations is reached. The algorithm uses only three operators to produce a new population for the next generation—selection, crossover, and mutation.

There are a number of different ways to handle selection. One method is to rank the population in ascending order by fitness value (after the cost function is evaluated for each individual), and assign probabilities for selection based on each individual's rank. This is called rank weighting. A virtual roulette wheel is spun (by generating a random number between 0 and 1) to determine the members in the new population selected for reproduction.

Once the population for reproduction is selected, the individuals are paired off and mated using the crossover process. A cross-point is selected at random for each pairing, and two new individuals are created by joining the first part of the first string with the second part of the second string, and vice versa. Mutation is the last step in the formation of the population for the next generation, and involves flipping a bit at random in the population from a 0 to a 1 or vice versa. Mutation is intended to prevent the GA from converging prematurely and helps to maintain a global search. The mutation rate is set at the beginning of the algorithm. Finally, this mutated population becomes the population of the next generation, and the process is repeated until convergence is reached.

6.2.3 Particle Swarm Optimization

Particle swarm optimization (PSO) is one of the simplest techniques to implement that can deal with discrete options. PSO shares many similarities with GAs. Like GAs, the technique works with a set of solutions called a population. Each potential solution is called a *particle*. Instead of using evolutionary methods, however, the PSO is based on the social behavior of bird flocks or fish schools. Each particle is characterized by a velocity with which it explores the cost function. The velocity and position of each particle are updated after each successive iteration of the algorithm. The particle velocity and position are governed by Eqs. (6.1) and (6.2):

$$v^{new} = v^{old} + c_1 r_1 (p^{localbest} - p^{old}) + c_2 r_2 (p^{globalbest} - p^{old}) \tag{6.1}$$

$$p^{new} = p^{old} + v^{new} \tag{6.2}$$

where v is particle velocity, p is particle position, r_1,r_2 is independent uniform random numbers between 0 and 1, c_1 is cognitive acceleration constant, c_2 is social acceleration constant, $p^{localbest}$ is the best local solution (best particle in current population), and $p^{globalbest}$ is the best global solution (best particle so far in all generations).

The two acceleration constants are usually numbers between 0 and 4. The PSO technique has become popular for the same reasons as the GA, in that it is easy to implement with relatively few parameters to adjust.

6.3 BUILDING LOW-ENERGY SYSTEMS

In addition to the energy efficiency technologies discussed in Chapter 2, there are several low-energy systems that can be incorporated as part of the design of new buildings or integrated as part of deep retrofit of existing buildings to reduce their energy use consumption and improve indoor thermal comfort and indoor air quality. Some of these low-energy systems are briefly described in this section.

6.3.1 Natural Ventilation

Natural ventilation is one of the most fundamental techniques to reduce energy usage in buildings. If the cooling capacity of ambient air can be harnessed to increase indoor thermal comfort, then the necessity for mechanical space conditioning lessens. Natural ventilation relies on natural forces: wind from the surrounding environment as well as buoyancy forces that develop due to temperature gradients within the building. Buildings can be designed to take advantage of either driving force or a combination of both. Though natural ventilation is an old technology, it has experienced a recent resurgence of interest, particularly in Europe where several major research initiatives have been conducted (Ghiaus et al., 2006; Op't Veld, 2008).

Using natural ventilation in buildings offers additional benefits besides energy usage reductions. Research has shown that the acceptable thermal comfort range for natural ventilated buildings is significantly larger than for buildings with standard mechanical HVAC systems (De Dear and Brager, 2002). To this end, ASHRAE Standard 55 has been revised to include an adaptive thermal comfort standard specifically for naturally ventilated buildings, which allows for increasing indoor temperatures as outdoor temperature increases. The generally high level of occupant control associated with naturally ventilated buildings is also thought to contribute to the acceptance of warmer indoor temperature. Additionally, natural ventilation systems have been shown to consistently outperform mechanical systems with respect to complaints of Sick Building Syndrome (SBS) and its associated symptoms. Seppanen and Fisk (2002), in a review of 18 different studies on SBS and ventilation systems, found that the prevalence of SBS symptoms was 30%–200% higher in air conditioned buildings as compared to naturally ventilated ones. In general, when building occupants are more satisfied with their working environment, productivity and job satisfaction also increase. Thus, the benefits of using natural ventilation are significant even beyond the energy benefit.

These benefits have led to the increased use of natural ventilation in green building design in areas with mild summers like the UK and Europe (Roth et al., 2006). In more extreme climates, such as the warm and humid areas of the United States, studies have shown that systems using purely natural ventilation are insufficient to maintain acceptable thermal comfort, excluding some coastal areas (Emmerich, 2006). In conditions where natural ventilation forces are deemed insufficient, hybrid ventilation, the use of supplemental mechanical forces along with natural forces, offers a compromise. Hybrid ventilation allows a building to capitalize on the benefits of natural ventilation, with insurance that the building can operate at a desirable thermal comfort level during extreme outdoor conditions. Thus, hybrid ventilation can be implemented in a wider range of climates, where purely natural ventilation is insufficient to provide all the cooling needs of a space. Studies have shown that the use of hybrid ventilation can result in considerable HVAC energy savings in warmer climates, humid subtropical (Axley, 2001), tropical (Haase and Amato, 2009), and arid (Ezzeldin et al., 2009), by using natural forces when they are sufficient but with mechanical assistance when necessary. The use of hybrid ventilation represents an opportunity to make better use of energy consumption for HVAC, by utilizing free natural cooling as much as possible.

A variety of simple analytical models have been developed for different natural ventilation flow configurations. Most are based on applying the mass and energy conservation equations. Some models have been verified experimentally.

When natural ventilation occurs from one opening within one single-zone spaces, Yamanaka et al. (2006) experimentally investigated the validity of two single-sided wind-driven ventilation models: the pulsation theory (Cockroft and Robertson, 1976) and the mixing layer theory (Warren, 1978), which are also given in the ASHRAE handbook. Pulsation theory assumes that air flow is quantified by the fluctuation of pressure across the opening. Yamanaka proved this model was inaccurate for all wind incidence angles except 0°. The mixing layer theory assumes a velocity and pollutant distribution inside the single opening, and ventilation rate is given by

$$Q = CAU_w \tag{6.3}$$

where Q is ventilation flow rate, C is airflow coefficient, A is opening area, and U_w is local velocity (wind velocity along the wall at the outside edge of mixing layer).

Yamanaka's experimentation proves the validity of this model, though the value of C varies depending on geometry and wind incidence direction.

A semianalytical model for single-sided ventilation driven by buoyancy alone has been developed, based on the fact that the volume flow rate is driven by the pressure difference across the opening and using the following assumption, valid for small ΔT (Allocca et al., 2003):

$$\frac{\rho_{out} - \rho_{in}}{\rho_{in}} \cong \frac{T_{out} - T_{in}}{T_{in}} = \frac{\Delta T}{T_{out}} \tag{6.4}$$

The volume flow rate for this configuration, from Bernoulli principles, is

$$Q = \frac{C_d A}{3}\sqrt{gh\frac{\Delta T}{T_{out}}} \tag{6.5}$$

where h is the height of the opening. This equation is also given in the ASHRAE handbook.

The general form of the equation for airflow from stack, wind, and other driving pressure differences is the power law for orifices:

$$Q = C_d A\sqrt{\frac{2\Delta p}{\rho}} \tag{6.6}$$

Phaff et al. (1980) performed a detailed experimental study into the terms and coefficients comprising the Δp in Eq. (6.6). They considered contributions from wind, buoyancy, and also included a turbulence constant. Their empirical equation, based on meteorological wind speed, V_{met}, rather than local wind speed, is

$$Q = \frac{A}{2}\sqrt{0.001V_{met}^2 + 0.0035 \cdot h\Delta T + 0.01} \tag{6.7}$$

For complex natural ventilation configurations with multiple openings and multiple zones, a number of network models and analysis tools have been developed. The most common analysis tools used for modeling natural ventilation are COMIS and CONTAM (Walton and Dols, 2006). The full-building simulation tool ESP-r also has a network model built in. A version of CONTAM has been implemented in the full-building simulation tool EnergyPlus (Walton and Dols, 2006).

6.3.2 Breathing Walls

Breathing wall systems, also referred to as dynamic walls, provide air ventilation through the walls while recovering thermal energy. Essentially using a heat exchanging mechanism, filtrated fresh air is introduced through the walls from outside to inside and is heated or cooled by the thermal energy wasted by conduction and convection with the walls. The thermal load of air ventilation (or air infiltration) is rather difficult to assess. It has been traditionally thought that the sensible thermal load, Q_{OA}, due to fresh air introduced to a building, is simply calculated as follows:

$$Q_{OA} = \rho.c_{p,a}.\dot{V}_{OA}.(T_i - T_o) = \dot{m}_{OA}.c_{p,a}.(T_i - T_o) \tag{6.8}$$

Eq. (6.8) assumes that the fresh air entering the condition space (kept at temperature T_i) has the same temperature as the outdoor temperature, T_o. However, various recent studies showed that actually infiltrating air exchanges heat with the building envelope before entering a condition space. This heat exchange occurs especially when air flows through the walls in a diffuse manner (through long airflow channels inside the building envelope). As a consequence, the actual thermal load due to outdoor air is lower than that determined by Eq. (6.8) by a fraction that depends on the heat exchange rate between outdoor air flow and heat conduction through the building envelope. In particular, several analytical studies have indicated that the outdoor air thermal load can be reduced by a heat recovery factor, *f*, as shown in Eq. (6.9):

$$Q_{OA} = (1-f).\dot{m}_{OA}.c_{p,a}.(T_i - T_o) \tag{6.9}$$

When the building envelope is not airtight, it can be assumed that the outdoor airflow occurs through direct and short paths from outdoors to the indoors without significant heat recovery and thus Eq. (6.9) can be used to estimate the thermal load due to outdoor air intake through either ventilation or infiltration. However, when the outdoor flow is diffuse with long paths, a significant heat recovery can occur. In the case of diffuse airflow through the walls as shown in the simplified house model of Fig. 6.4, analytical solutions have been developed to estimate the value of the heat recovery factor, *f*. For instance, Krarti (1994) estimated that the heat recovery factor, *f*, can be calculated as follows:

$$f = 1 - \left(\frac{PeBi_i e^{\frac{Pe}{2}}}{[2Bi_i Bi_o + Pe(Bi_i - Bi_o)]\sinh\left(\frac{Pe}{2}\right) + Pe(Bi_i + Bi_o)\cosh\left(\frac{Pe}{2}\right)} \right) \tag{6.10}$$

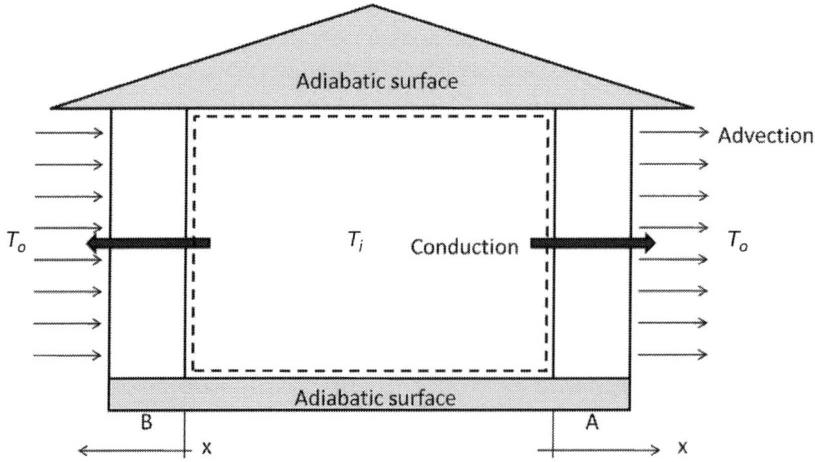

FIGURE 6.4 Simplified house model using breathing walls.

where Pe is the Peclet number and is defined as $Pe = \frac{\dot{m}c_p}{UA}$; Bi_i is the Biot number at the inside surface: $Bi_i = \frac{h_i}{U}$; and Bi_o is the Biot number at the outside surface: $Bi_o = \frac{h_o}{U}$.

Note that when the Biot numbers, Bi_o and Bi_i, are large (i.e., the surface temperatures are the same as the air temperatures), then Eq. (6.10) can be reduced to the solution reported by Anderlind (1985):

$$f = \frac{2*\left(e^{Pe} - Pe - 1\right)}{Pe*(e^{Pe} - 1)} \tag{6.11}$$

Fig. 6.5 illustrates the variation of the heat recovery factor as a function of the convection heat transfer and ventilation rate expressed in ACH.

As a result of the coupling between conduction and air ventilation or air infiltration within the building envelope, the heating and cooling thermal loads may be reduced. The reduction of the building thermal load can be expressed as the ratio of the building thermal load when coupling occurs over the thermal load without coupling (Solupe and Krarti, 2014):

$$\frac{Q_{IHR}}{Q_{no\,IHR}} = 1 - \frac{2 + \left(1 + \frac{1}{\mathrm{Bi_i}} + \frac{1}{\mathrm{Bi_o}}\right)\left(1 - \mathrm{Bi_i Bi_o} + \frac{\mathrm{Pe}(\mathrm{Bi_o}(1+e^{2Pe}) + \mathrm{Pe}(e^{2Pe}-1)) + \mathrm{Bi_i}(\mathrm{Bi_o}(e^{2Pe}-1) + \mathrm{Pe}(e^{2Pe}+1))}{(\mathrm{Bi_o} e^{Pe}\mathrm{Pe} + \mathrm{Bi^i}(\mathrm{Bi_o}(e^{Pe}-1)+\mathrm{Pe}))(\mathrm{Bi_o Pe} + \mathrm{Bi_i}(\mathrm{Bi_o}(e^{Pe}-1) + e^{Pe}\mathrm{Pe}))}\right)\mathrm{Pe}}{1 + \left(1 + \frac{1}{\mathrm{Bi_i}} + \frac{1}{\mathrm{Bi_o}}\right)\mathrm{Pe}} \tag{6.12}$$

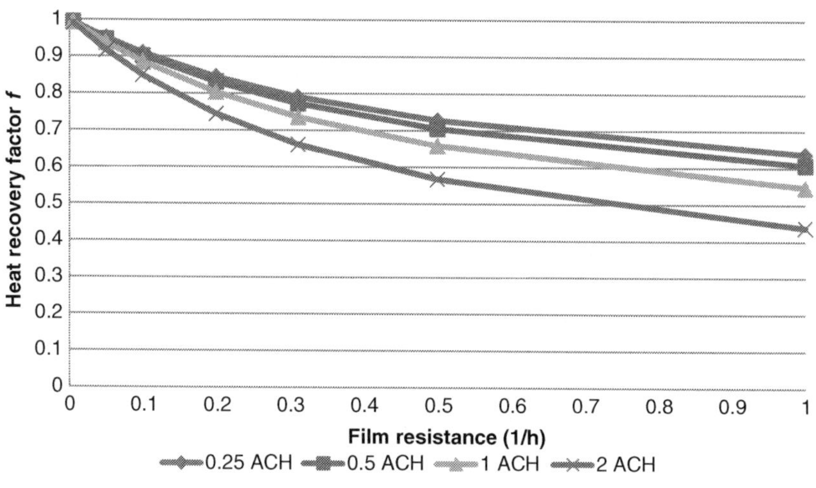

FIGURE 6.5 Variation of the heat recovery factor, *f*, with air infiltration rate and convection heat transfer coefficients.

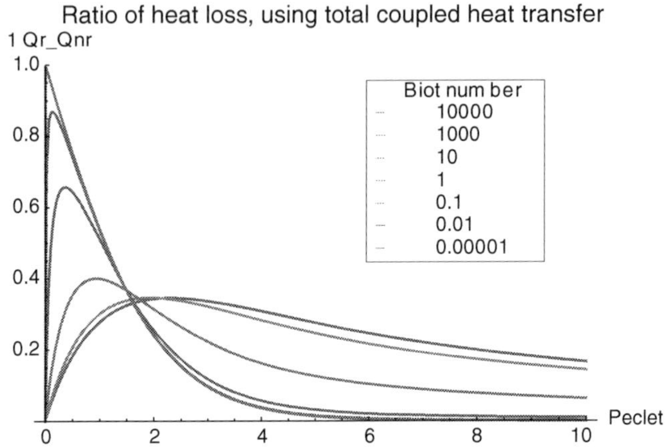

FIGURE 6.6 Variation of the heat recovery efficiency, η_{IHR}, with the outdoor air flowrate (Peclet number) and convection heat transfer coefficients (Biot number).

An efficiency of the heat recovery can be then defined:

$$\eta_{IHR} = 1 - \frac{Q_{IHR}}{Q_{no\,IHR}} \tag{6.13}$$

Fig. 6.6 illustrates the variation of the reduction in thermal load due to heat recovery between infiltrating air and walls as a function of infiltration rate expressed by the Peclet number for various Biot values (Krarti, 1994). The concept of heat recovery is being used for designing dynamic walls (also referred to as breathing walls) to provide air ventilation through the building envelope components and therefore reduce the energy penalty for outdoor air intake for residential and commercial buildings. Note that for residential buildings, the Peclet number varies between 0 and 2 depending on their leakage characteristics while Biot numbers range between 2 and 20. Thus, the efficiency of the heat recovery is in the range of 0%–35%.

When breathing walls are integrated with building envelope components, their performance is affected by various building energy systems. Fig. 6.7 illustrates one simplified building model with breathings walls equipped with a ventilator to exhaust air and enhance outdoor fresh air intake (Krarti, 1994). There are two main options for fresh air to enter a building: either through the breathing walls (i.e., ventilation air) or through cracks in the building envelope (i.e., air infiltration). It can be shown that the thermal efficiency of breathing walls depends on the ventilation airflow and the *UA* value, UA_{walls}, of the breathing walls. The term UA_{walls} can be normalized using the total building load coefficient, *BLC*, as noted in Eq. (6.14):

$$C_{cond} = \frac{UA_{walls}}{BLC} \tag{6.14}$$

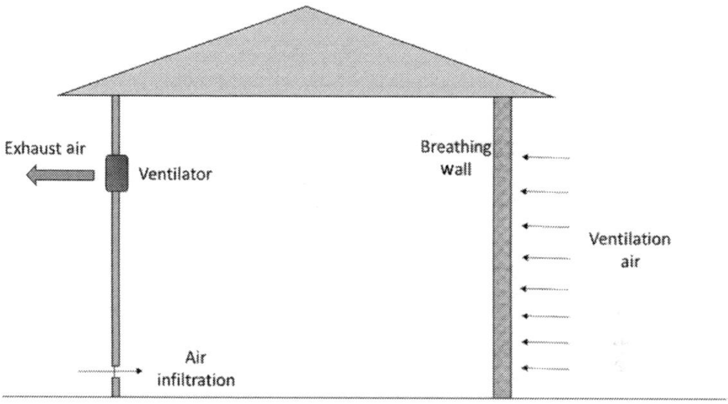

FIGURE 6.7 Simplified building model with breathing walls.

The efficiency of the breathing walls, η_{BW}, is defined as the reduction in building thermal loads as noted by Eq. (6.15):

$$\eta_{BW} = 1 - \frac{BLC_{BW}}{BLC} \tag{6.15}$$

where BLC is the reference building load coefficient without breathing walls and BLC_{BW} is the building load coefficient when breathing walls are considered.

It should be noted that the airflow through the breathing walls cannot exceeds the air exhausted from the building through the ventilator. It can be shown that the Peclet number of the airflow through the breathing walls has to be limited to a threshold that depends on the coefficient C_{cond} (Krarti, 1994):

$$Pe = \frac{\dot{m}_{BW} c_p}{UA_{walls}} \le \frac{1 - C_{cond}}{C_{cond}} \tag{6.16}$$

where $\dot{m}_{BW}$ is the mass flowrate of outside air circulating within the breathing walls.

Fig. 6.8 shows the variation of the steady-state thermal efficiency of breathing walls as a function of the Peclet number (i.e., ventilation airflow rate) and C_{cond} (i.e., UA-value of walls) for typical Biot numbers for the walls are $Bi_e = 100$ and $Bi_i = 25$. The dashed line in Fig. 6.8 represents the limit of the breathing walls' efficiency taking into account the condition of Eq. (6.16). The breathing walls can achieve a steady-state thermal efficiency of up to 23%. The highest efficiency can be achieved when $C_{cond} = 0.5$.

Examples 6.1 and 6.2 illustrate the calculation procedure to estimate the efficiency for breathing walls in homes.

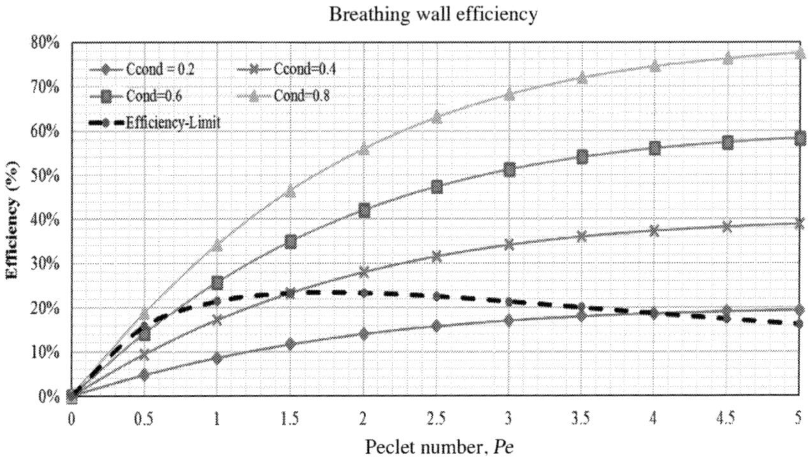

FIGURE 6.8 Variation of the efficiency, η_{BW}, for breathing walls.

Example 6.1:

Problem: Estimate the thermal efficiency for breathing walls considered for a house with R-19 walls with 0.3 ACH air ventilation flowing through the breathing walls. The total opaque wall area is 1124 ft^2 (104 m^2). The volume of the house, located at sea level, is 15,000 ft^3 (425 m^3). The total *BLC* of the house including the ventilation and air filtration is 300 Btu/h°F (162 W/°C).

Solution

First, the *U*-value of the wall is estimated. Taking into account the frame factor and the *R*-value of the insulation: *U*-value = 0.05 Btu/h ft^2°F (0.29 W/m^2°C) so the UA_{walls} = 56.2 Btu/h°F (30.3 W/°C) and C_{cond} = 56.2/300 = 0.19. Using the volume of the house and 0.3 ACH flow rate, the mass flowrate for the ventilation can be estimated to be $\dot{m}_{OA}c_p = 81.0$ Btu/h ft^2°F (43.7 W/°C). Using *UA*-value of the walls, the Peclet can be computed:

$$Pe = \dot{m}_{OA}c_p \,/\, UA_{walls} = 1.44$$

Using Fig. 6.8, the efficiency of the breathing walls is estimated to be about η_{BW} = 12% for C_{cond} = 0.19. Note that this value is lower than the limit efficiency of 22% at the *Pe* value of 1.4 since using Eq. (6.16), $Pe = 1.4 < (1 - C_{cond})/C_{cond} = (1 - 0.19)/0.19 = 4.2$. Thus, the breathing walls have the potential to save 12% of heating and cooling of the house.

Example 6.2:

Problem: Consider a house in Denver, CO has the following building characteristics:

- Exterior walls with a total *R*-value of 19 (i.e., *R* = 19 h°F ft^2/Btu = 3.35°C m^2/W) insulation and total area of 1124 ft^2 (104 m^2).

- A roof with a total R-value of 30 (i.e., $R = 30$ h°F ft^2/Btu = 5.28°C m^2/W) insulation and total area of 1540 ft^2 (143 m^2).
- Double pane clear windows with R-value of 2 (i.e., $R = 2$ h°F ft^2/Btu = 0.35°C m^2/W) insulation and total area of 185 ft^2 (17.2 m^2).
- A ventilation rate of 0.30 air-changes per hour (ACH) with a total volume of 13,860 ft^3 or 392 m^3.

Determine the maximum breathing wall efficiency for this house and estimate the annual potential heating energy savings using 6000°F-days for Denver and a furnace annual average efficiency of 0.85. Assume the house is heated 24 h/day.

Solution

The next table summarizes the calculation of the *BLC* of the house using the *UA* values of the various components. Thus, $C_{cond} = 59.2/329.0 = 0.18$ and $Pe = 126/59.2 = 2.13$. From Fig. 6.8, the steady-state thermal efficiency for the breathing walls is $\eta_{BW} = 14\%$.

Components	*U*-value Btu/h ft²°F (W/m²°C)	Area ft² (m²)	*UA* Btu/h°F (W/°C)
Exterior walls	0.053 (0.30)	1124 (104)	59.2 (31.2)
Roof	0.033 (0.19)	1540 (143)	51.3 (27.1)
Windows	0.500 (2.89)	185 (17.2)	92.5 (48.8)
Ventilation/ infiltration	–	–	126.0 (66.6)
Total s			329.0 (173.7)

Using the degree-day method, the potential heating energy use savings for the house associated with the breathing walls can be estimated using the following equation:

$$\Delta EU_H = \eta_{BW} \cdot 24 \cdot f_H \cdot BLC \cdot DD_H(T_b) / \eta_H$$

Thus, the heating fuel savings expressed in therms can be computed:

$$\Delta EU_H = (0.14) \cdot (24) \cdot (1) \cdot (329) \cdot (6000) / (0.85) / 100{,}000 = 78 \text{therms}$$

6.3.3 Evaporative Coolers

Evaporative coolers, also called swamp coolers, provide energy-efficient alternatives to cool houses in dry regions. There are several types of evaporative coolers including the following:

Direct evaporative cooler: This type of cooler often called a swamp cooler uses a blower to force air through a permeable water-soaked pad as illustrated in Fig. 6.9. As the air passes through the pad, it is filtered, cooled, and humidified. Fig. 6.10 illustrates an indirect cooling process in a psychrometric chart.

Indirect evaporative cooler: A secondary heat exchanger is used to prevent humidity from being added to the airstream that enters the conditioned space as shown in Fig. 6.11. The psychrometric process of an indirect evaporative cooler is presented in Fig. 6.12.

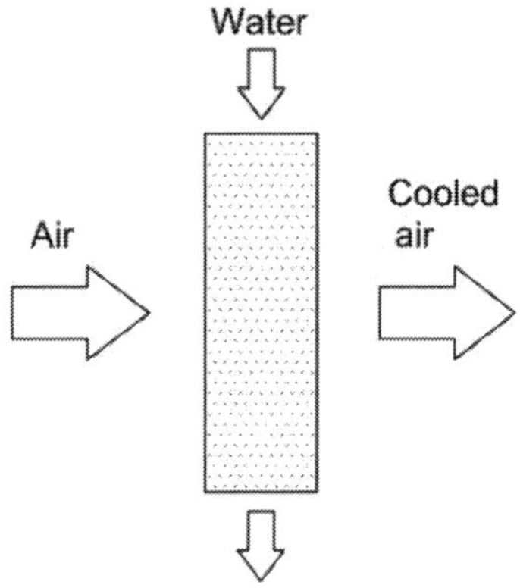

FIGURE 6.9 Basic direct evaporative cooler.

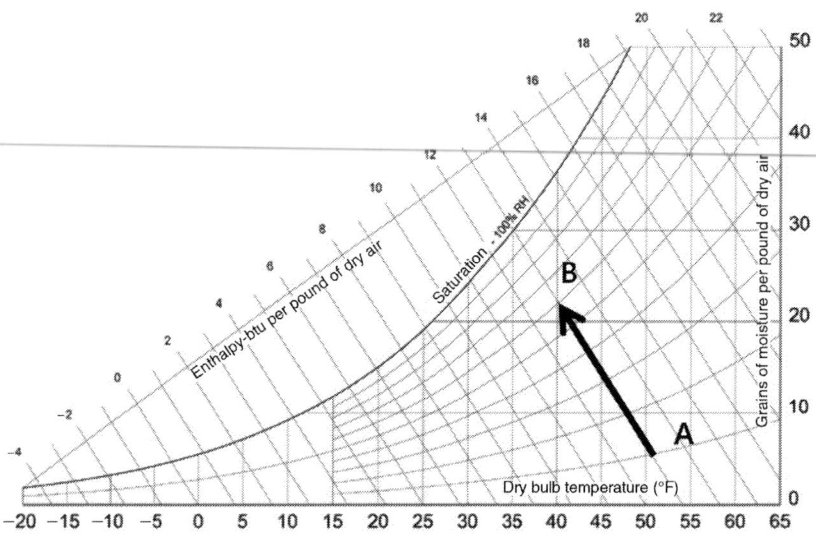

FIGURE 6.10 Direct cooling process in a psychrometric chart.

Indirect/direct evaporative cooler: This combines indirect and direct evaporative cooling. In addition to the water-soaked pad, a secondary heat exchanger is used to prevent humidity from being added to the airstream that enters the conditioned space as shown in Fig. 6.13. The psychrometric process of an indirect evaporative cooler is presented in Fig. 6.14.

Ultracoolers: These coolers are able to cool air below the thermodynamic wet-bulb temperature associated with the dry-bulb temperature of the outside ambient air. Ultracoolers typically use direct and indirect processes operating in parallel and in stages to achieve cooler air than a direct or indirect system alone could achieve.

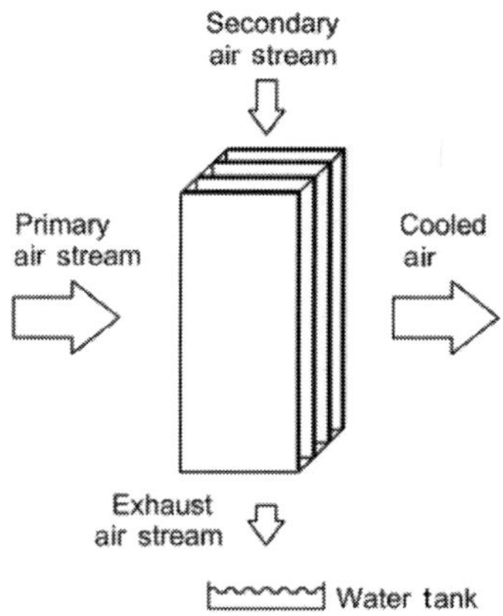

FIGURE 6.11 Basic indirect evaporative cooler.

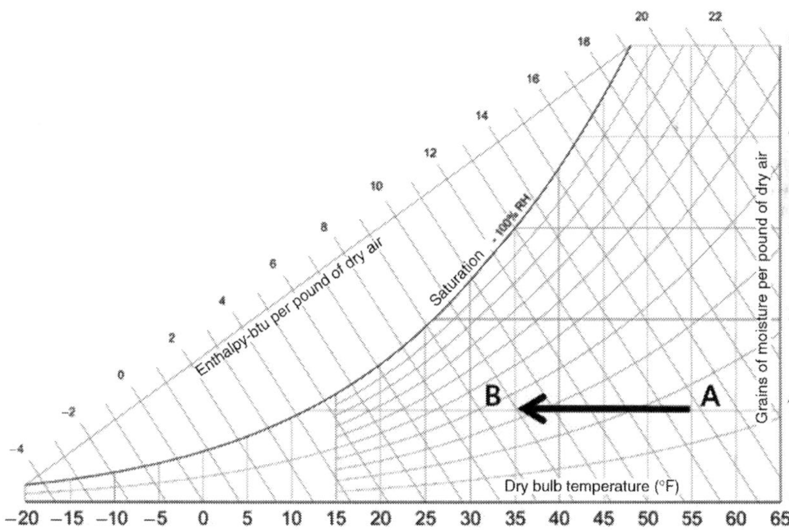

FIGURE 6.12 Indirect cooling process in a psychrometric chart.

To assess the effectiveness of the evaporative cooling systems, wet-bulb efficiency is used to determine how effectively the system cooled the air relative to its wet-bulb temperature:

$$\varepsilon_{wb} = \frac{T_{db,in} - T_{db,out}}{T_{db,in} - T_{wb,in}} \tag{6.17}$$

Table 6.1 illustrates the wet-bulb efficiency for various evaporative cooling systems including direct, indirect, and indirect/direct to deliver the same dry-bulb temperature to the conditioned space.

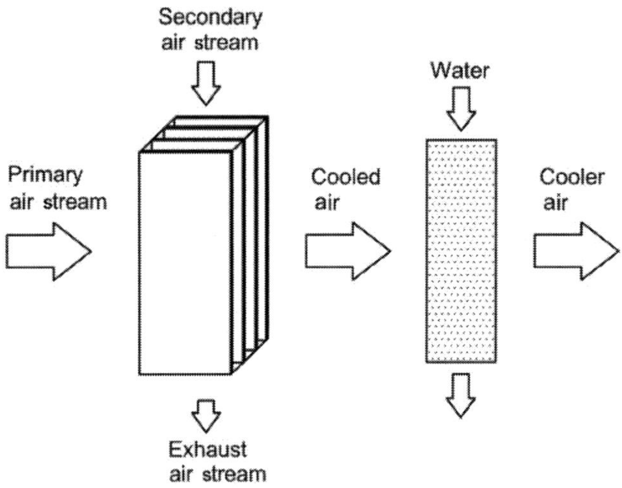

FIGURE 6.13 Basic indirect/direct evaporative cooler.

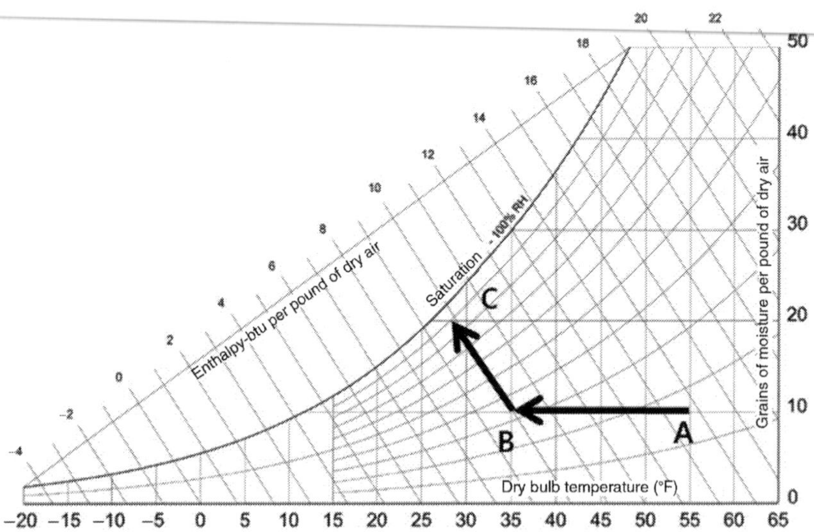

FIGURE 6.14 Indirect/direct cooling process in a psychrometric chart.

It should be noted that evaporative coolers can be used as a standalone cooling system in a home or in combination with a refrigeration-based air conditioning system. However, conventional air conditioners should not be operated simultaneously with direct evaporative coolers, because air conditioners dehumidify while evaporative coolers humidify the indoor air.

TABLE 6.1 Wet-Bulb Efficiency for Different Types of Evaporative Coolers

Evaporative cooler type	$T_{db,in}$	$T_{wb,in}$	$T_{db,out}$	ε_{wb} (%)
Direct	110°F (43.3°C)	72°F (22.2°C)	78°F (25.5°C)	84
Indirect	100°F (37.8°C)	65°F (18.3°C)	78°F (25.5°C)	63
Indirect/direct	113°F (45°C)	70°F (21.1°C)	78°F (25.5°C)	81

6.3.4 Earth Tubes

Away from any building, soil daily temperature at any depth can be estimated based on the ground surface temperature as expressed by Eq. (6.18):

$$T(z,t) = T_m + T_a e^{-z/d} \sin\left[\omega t - \theta - \frac{z}{d} - \frac{\pi}{2}\right] \tag{6.18}$$

where $\omega = \frac{2\pi}{8760}$ is annual angular frequency; $d = \sqrt{\frac{2\alpha_s}{\omega}}\sqrt{\frac{2\alpha_s}{\omega}}$ is the dampening depth; t is time expressed in hours starting from January 1st; T_m is the annual average soil surface temperature; T_a is the annual amplitude of soil surface temperature variation; and Θ is the phase lag between ambient air and soil surface temperatures.

The soil surface temperature variation over time can be expressed typically as a sine function of time:

$$T(0,t) = T_m + T_a \sin\left[\omega t - \theta - \frac{\pi}{2}\right] \tag{6.19}$$

Table 6.2 provides the values of T_m, T_a, and Θ for select US sites using the data provided by Kusuda and Achenback (1965). There are some simplified models that can predict the annual soil surface temperature variation including the three parameters T_m, T_a, and Θ using ambient air temperature (Krarti et al., 1995). It is very common to assume that the annual average soil temperature, T_m, is close to the deep ground water temperature.

Fig. 6.15A illustrates the annual soil temperatures for Fort Collins, CO at various depths from the ground surface. Fig. 6.15B shows the soil temperature variation with depth for 4 days representing four seasons. As shown in Fig. 6.15, the soil temperature becomes constant at about 10 m or 30 ft.

An analytical model for the thermal performance of earth tube systems has been developed (Krarti and Kreider, 1996). Fig. 6.16 shows that the outlet temperature depends significantly upon the pipe diameter, D_h, as well as the airflow velocity, V_f. As illustrated in Fig. 6.16A, an increase in the diameter results in a higher outlet air temperature. This result can be explained by the fact that

TABLE 6.2 Soil Surface Temperatures for Selected US Sites (Refer to Eq. 6.19)

		IP Unit			SI Unit		
		Avg. temp	Amplitude	Phase angle	Avg. temp	Amplitude	Phase angle
Site	State	°F	°F	Radian	°C	°	Radian
Auburn	Alabama	65	17	0.49	18.3	−8.3	0.49
Tempe	Arizona	70	20	0.47	21.1	−6.7	0.47
Davis	California	66	19	0.63	18.9	−7.2	0.63
Fort Collins	Colorado	50	24	0.54	10.0	−4.4	0.54
Moscow	Idaho	47	18	0.73	8.3	−7.8	0.73
Argonne	Illinois	51	23	0.7	10.6	−5.0	0.7
Burlington	Iowa	54	30	0.57	12.2	−1.1	0.57
Manhattan	Kansas	55	26	0.61	12.8	−3.3	0.61
Lexington	Kentucky	55	23	0.6	12.8	−5.0	0.6
Upper Marlboro	Maryland	56	25	0.56	13.3	−3.9	0.56
East Lansing	Michigan	50	24	0.6	10.0	−4.4	0.6
St. Paul	Minnesota	48	25	0.65	8.9	−3.9	0.65
State Univ.	Mississippi	67	21	0.58	19.4	−6.1	0.58
Kansas City	Missouri	54	22	0.56	12.2	−5.6	0.56
Bozeman	Montana	44	21	0.68	6.7	−6.1	0.68
Lincoln	Nebraska	54	28	0.52	12.2	−2.2	0.52
New Brunswick	New Jersey	53	21	0.69	11.7	−6.1	0.69
Ithaca	New York	49	19	0.69	9.4	−7.2	0.69
Columbus	Ohio	53	22	0.65	11.7	−5.6	0.65
Lake Hefner	Oklahoma	64	23	0.63	17.8	−5.0	0.63
Ottawa	Ontario	47	21	0.64	8.3	−6.1	0.64
Corvallis	Oregon	56	18	0.53	13.3	−7.8	0.53
Calhoun	South Carolina	64	22	0.49	17.8	−5.6	0.49
Madison	South Dakota	47	26	0.59	8.3	−3.3	0.59
Jackson	Tennessee	60	20	0.44	15.6	−6.7	0.44
Temple	Texas	70	21	0.58	21.1	−6.1	0.58
Salt Lake City	Utah	51	21	0.48	10.6	−6.1	0.48
Seattle	Washington	53	15	0.64	11.7	−9.4	0.64

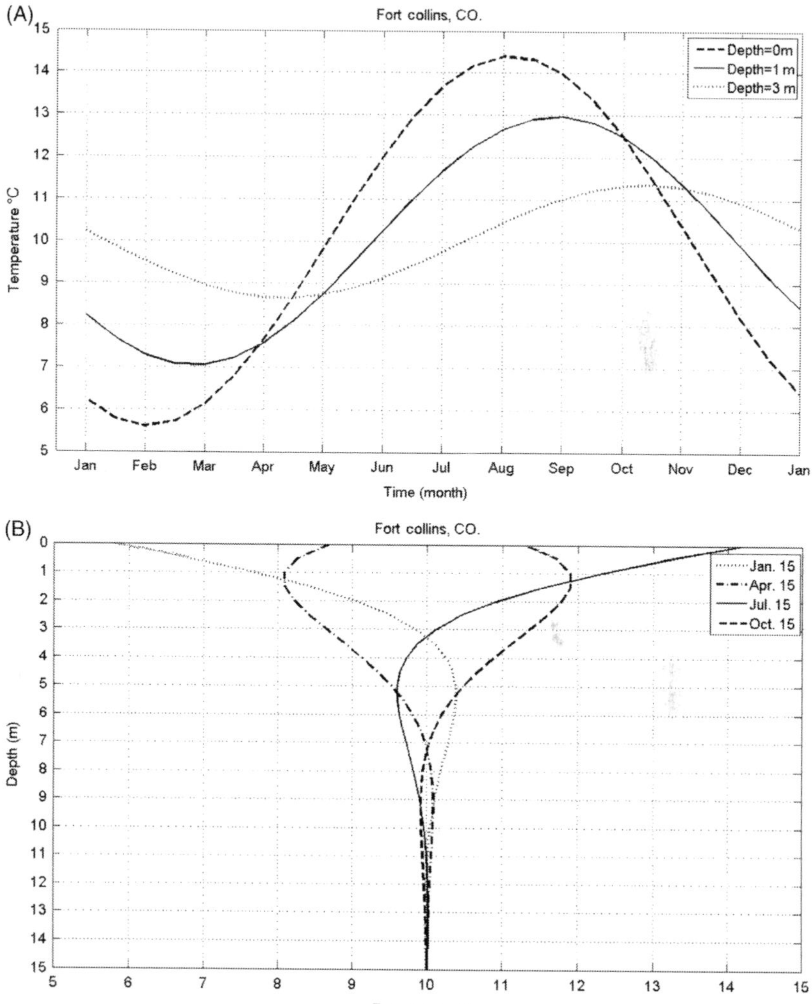

FIGURE 6.15 Soil temperature variation (A) with time (B) with depth for four seasons.

the volume of air flowing with the tubes is proportional to D_h^2 while the heat exchanged between the air and the soil is proportional to D_h. Therefore, as the pipe diameter, D_h, increases, less heat is exchanged between the air within the tubes and the surrounding soil resulting in a smaller decrease of the outlet air temperature. From Fig. 6.16B, it can be seen that an increase in air velocity, that is, an increase of the air mass flow rate through the pipe, results in an increase of the outlet temperature since the air has a shorter residence time in contact with the soil.

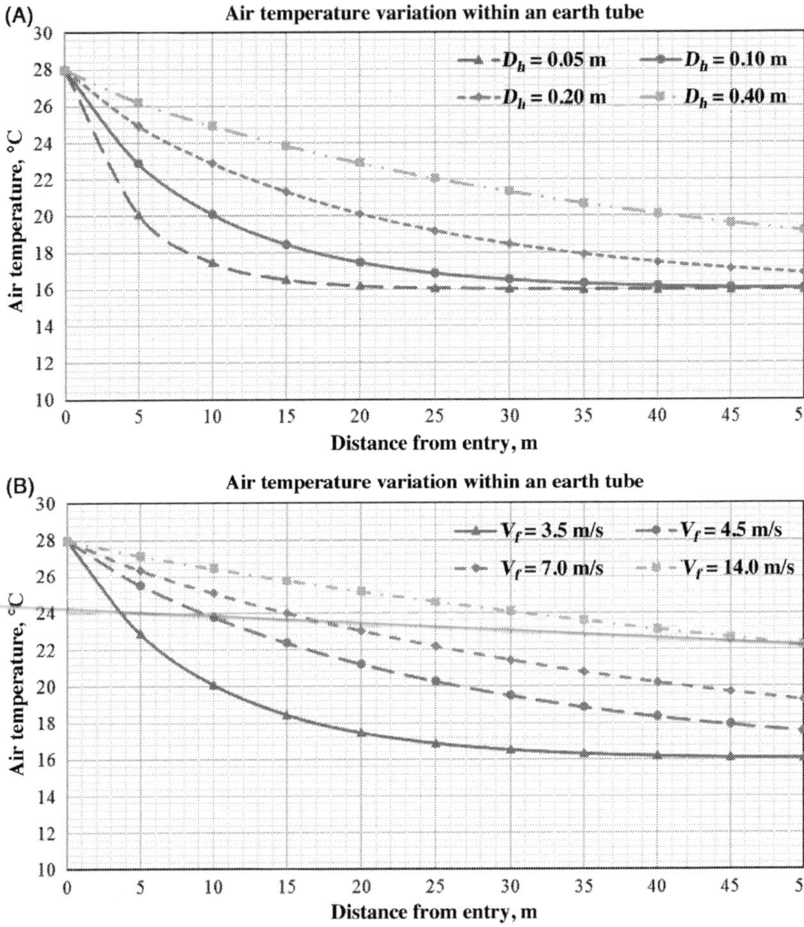

FIGURE 6.16 Variation of the air temperature within the earth-tube as a function of (A) the tube diameter and (B) air velocity.

Fig. 6.17 presents the variation of cooling rate as a function of pipe diameter, D_h, and air velocity, V_f. Increasing the diameter improves significantly the cooling rate as shown in Fig. 6.17A. This result can be explained by the fact that an increase in pipe diameter implies an increase in air mass flowing through the pipe that makes up for the outlet temperature drop. Fig. 6.17B illustrates that the effect of the air velocity on the cooling rate is similar to that of the pipe diameter. However, the increase in air velocity does not improve the cooling rate as significantly as an increase in the diameter of the pipe.

6.3.5 Radiant Heating Systems

Radiant floor heating systems consist typically of embedded hot water coils in floor slabs of residential and commercial buildings to provide space heating as

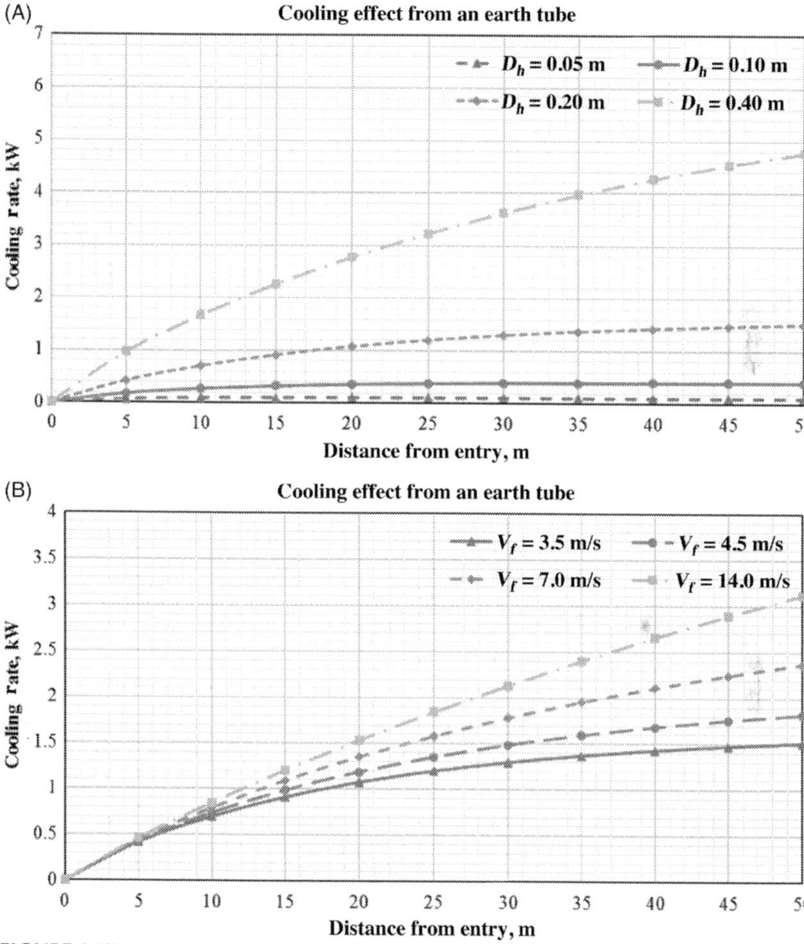

FIGURE 6.17 Variation of the cooling effect by the earth-tube as a function of (A) the tube diameter and (B) air velocity.

illustrated in Fig. 6.18. Recently, radiant systems have received renewed interest in the US due to their inherent advantages compared to conventional all-air heating systems including low noise, potential energy savings, uniform temperature distribution within spaces, and superior thermal comfort (ASHRAE, 2008). Experimental studies have demonstrated that radiant heating can achieve 33%–52% savings when compared to air-to-air heat pump and electric baseboard heating systems, respectively, in residential houses (CADDET, 1999).

The thermal mass of a heated floor acting as heat storage medium absorbs heat from the embedded heating sources. Heat is then conducted from the heating sources to the panel surfaces. By radiation, the surfaces directly heat objects without heat-transferring media (such as air). However, the room air will be also heated to some degree due to convection heat transfer from the panel surfaces. Therefore, radiant heating systems may require lower energy use due to lower

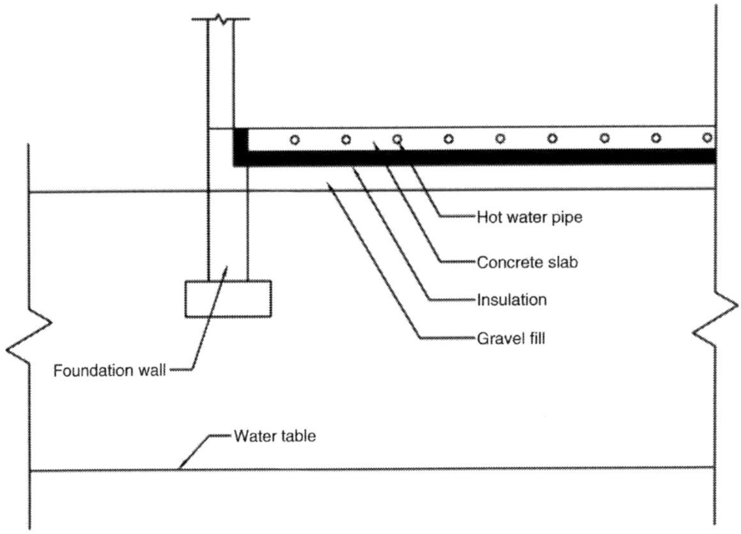

FIGURE 6.18 Cross-section of radiant floor heating.

room setpoint temperature when compared to conventional forced air systems. Similar indoor comfort levels can be accomplished with 3–4°C (5.5–7°F) lower space temperatures. Additionally, infiltration as a part of heating loads is reduced for radiant floor systems due to relatively smaller temperature difference between indoor and outdoor air temperature than air-heating systems. Radiant heating systems also provide better thermal comfort to people since the warm panels reduce heat losses through radiation from the human body.

The ASHRAE Handbook of HVAC Systems and Equipment (2016) provides useful information to analyze, design, and control a radiant floor. Similarly, the Radiant Heating and Cooling Handbook (Watson and Chapman, 2002) provides a detailed overview of radiant heating systems including radiant heat transfer, building thermal comfort, design guidelines, and typical control strategies. As an example, the design guidelines for a residential building include the following recommendations:

- Room temperature: 65–72°F (18–22°C)
- Supplied hot water temperature: 95–140°F (35–60°C)
- Floor surface temperature: 75–85°F (24–30°C)
- Temperature drop of water temperature: 15–20°F (8–11°C)
- Maximum length of loop: 200 ft (60 m) in 3/8-inch tube size, 300 ft (90 m) in 1/2-inch tube size
- Tube size: 3/8 inch
- Tube spacing: 4–9v inch (10–23 cm).

Controlling the performance of radiant systems can be challenging due to the high thermal mass of the floor and ground medium. Control strategies for radiant floor heating systems use typically temperature or heat flux modulation techniques. Temperature-modulation control sets the supply water temperature to be proportional to either outdoor temperature or to the difference between a desired set-point and room air temperatures. Flux-modulation control attempts to ensure that a slab delivers heat to the space at a rate per unit area proportional to the difference between the room air temperature and the slab surface temperature. The heat flux provided to the slab is proportional to the difference between the supply and return water temperatures.

In addition to the mean air temperature (MAT), mean radiant temperature (MRT) or operative temperature (OT) is used for control radiant heating systems. Fig. 6.19 illustrates the impact of temperature settings on total heating energy use and the number of hours when thermal comfort levels exceed the acceptable range (i.e. $-0.5 \leq PMV \leq +0.5$) for a typical house located in Denver, CO (Ihm and Krarti, 2005). As shown in Fig. 6.19, total heating energy use is generally proportional to the increasing setpoint temperature for all three control strategies. Depending on the control strategy, lower setpoint temperature can save some heating energy use at the expense of a deterioration in the indoor thermal comfort. To further evaluate the performance of control strategies, the setpoint temperatures are selected based on achieving low energy use and satisfying building thermal comfort. Typical setpoint temperatures of 20°C (68°F) for MAT-based control, 24°C (75.2°F) for MRT-based control, and 22°C (71.6°F) for OT-based control are recommended to operate radiant heating systems.

To better operate radiant heating systems, optimal control schemes have been proposed. For instance, Simmonds (1993) used optimal controls to demonstrate that additional energy savings could be achieved if the control is based on maintaining the Fanger's predicted mean vote (*PMV*) index rather than mean air temperature within a thermal comfort range. A *PMV* value within ±0.5 is acceptable for thermal comfort according to the ASHRAE comfort standard (2001). Ihm and Krarti (2005) have evaluated optimal control strategies to operate radiant floor heating systems in three locations: Denver, CO, Minneapolis, MN, and Albany, NY. They found that compared to the conventional controls using fixed mean air temperature or MAT set-point to maintain thermal comfort during occupied periods, the optimal control can reduce the total building energy use by 26.9% for Denver, CO, 11.3% for Minneapolis, MN, and 22.2% for Albany, NY. These results indicated that optimal controls have the potential to save significant energy use relative to conventional MAT-based controls for radiant heating systems especially for milder climates.

6.3.6 Radiant Cooling Systems

Radiant cooling systems utilize chilled water pipes to distribute cooling energy to various conditioned spaces as alternatives to conventional DX systems that rely on cooled air supplied through ducts. Radiant cooling systems provide direct cooling of occupants by radiative heat transfer from typically cooling

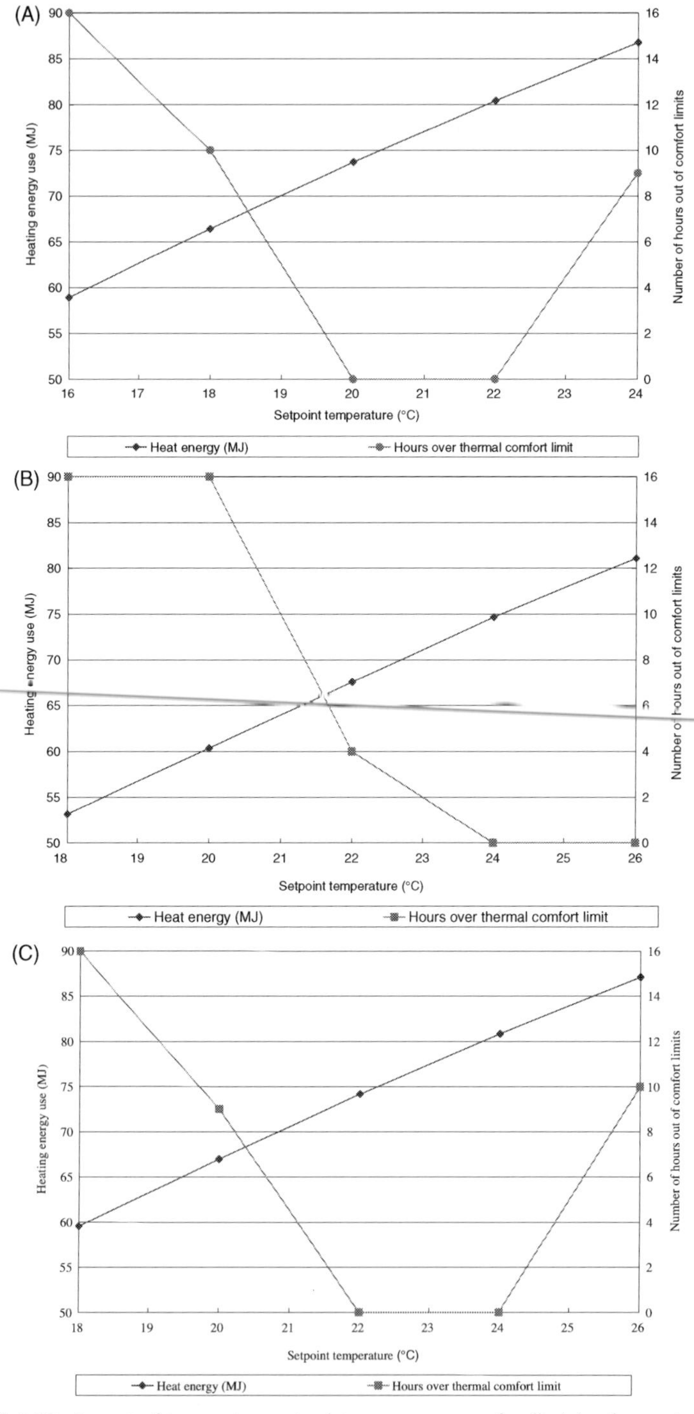

FIGURE 6.19 Impact of temperature set-point on energy use of radiant heating system for a house in Denver, CO. (A) Performance of mean air temperature-based proportional control. (B) Performance of mean radiant temperature-based proportional control. (C) Performance of operative temperature-based proportional control.

ceiling surfaces maintained at temperatures of about 65°F (18.3°C). Heat generated by occupants is absorbed by the radiant cooling surfaces.

There are several systems to provide chilled water in radiant cooling surfaces. Panel systems are the most common of these systems. Aluminum panels that carry tubing can be surface-mounted or embedded on floors, walls, or ceilings. Capillary tube systems are also used to provide chilled water through mats of small, closely spaced tubes that are embedded in plastic, gypsum, or plaster on walls and ceilings. Moreover, similar to hydronic radiant heating systems described earlier, concrete layers with embedded tubes can provide the conduit and thermal storage capacity for cooling systems. Fig. 6.20 illustrates construction details for a radiative wall that can be used for cooling and heating homes. In most radiant cooling systems, water is mixed with glycol and cooled by an air-to-water heat pump, a cooling tower, a ground-source heat pump, or even well water.

6.3.7 Analysis of Low-Energy Systems

In this section, a comparative analysis of several alternative systems to provide heating and cooling for a residential building is presented. Passive and active heating and cooling options are considered. The analysis is based on detailed simulation modeling analysis for an apartment building located in Denver, CO (Seo and Krarti, 2010) as depicted in Fig. 6.21. The systems are evaluated to determine their performance in terms of energy efficiency and thermal comfort. A brief description of the modeling approach for each heating and cooling system is provided next:

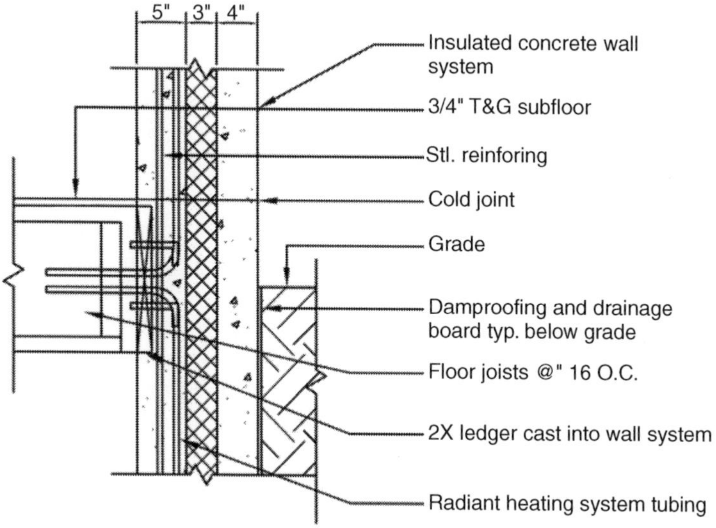

FIGURE 6.20 Cross-section of a radiative wall.

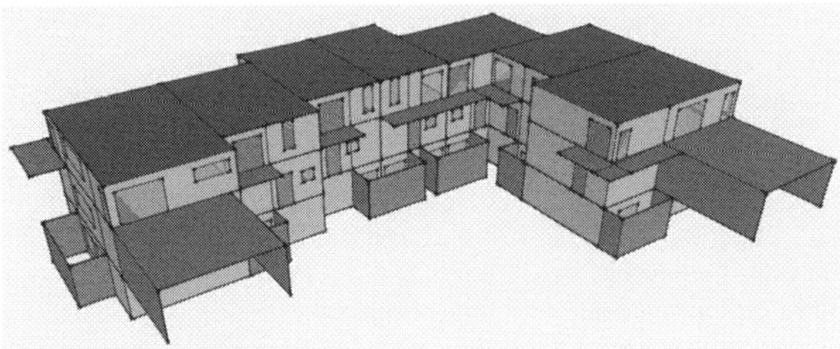

FIGURE 6.21 Model rendering of apartment building in Denver, CO.

System 1—No HVAC System: The building has no heating or cooling system. Energy is used only for systems such as lighting, appliances, and DHW. This case was considered as a reference case to assess the level of thermal discomfort within the building if no heating and cooling systems are installed.
System 2—Natural Ventilation: With this option, the building is cooled using natural ventilation. No mechanical system is used for heating or cooling. For this building, natural ventilation is considered by opening windows only when outside air temperature is lower than 23°C and indoor and outdoor temperature difference is over 1°C than indoor temperature.
System 3—Earth Tube: This option utilizes the heating/cooling energy stored within the ground medium. In this building, an earth tube is used to conditioned air supplied to each apartment unit in the building. A design flow rate of 0.1 m^3/s (212 cfm) is used for each unit with the following earth tube fan control options:

- Cooled/heated air is provided to each unit using fans
- No fan operation for the period ranging from October 1 to May 30 due to the very small heat extraction capabilities of the earth tube (the outlet air temperature from the tube is lower than indoor air temperature).

System 4—Radiative Walls: Radiative walls, as depicted in Fig. 6.20, are used to heat and cool the apartment unit. Three thermostats per unit are used to avoid significant temperature stratification. It is assumed that the heating season starts October 29 and ends April 17 and that the cooling season settings are set for the remainder of the days. To operate of the system, operative temperature control is used. Simulation analysis to compare various control strategies has indicated that controls based on operative temperature provide more savings in heating/cooling energy use than controls based on mean-air temperature. Chilled water in this system is supplied through a district cooling system. The pumping energy and site energy for generating the chilled water are accounted for in evaluating this system.
System 5—DX Air Conditioner and Gas Furnace: This system, most commonly used in the US to condition residential buildings, consists of a

gas-furnace and an electric air conditioner (AC). A COP of 3.0 for the AC and an efficiency of 80% are assumed for this system.

System 6—Radiative Wall with Cooling Tower: This system is a variation of system 4 with an open-type cooling tower (condenser) loop directly connected to the radiative wall system loop to reduce cooling energy consumption.

System 7—Radiative Wall with Evaporative Cooler: This system is also a variation of system 4 with an evaporative cooler used to cool the water supplied to the radiant walls.

System 8—Radiative Floors: Instead of the radiative walls, radiative floors are utilized to provide heating and cooling throughout each apartment unit of the building. A radiative floor per floor level for each unit is considered.

System 9—Heating with Radiative Floor and Cooling with Radiative Wall: This system used two separate water loops: (1) a hot water loop to provide heating through the radiative floors as in system 8, and (2) a chilled water loop to provide cooling through radiative walls as in system 4.

Fig. 6.22 summarizes the energy performance of all the heating and cooling systems just described and the potential energy savings relative to the conventional air conditioning system (i.e., system 5 with gas furnace and DX cooling). Fig. 6.23 compares the thermal comfort performance of all the systems expressed in terms of the number of hours where the *PMV* is within the −0.5 to 0.5 range for all the apartment units in the building as recommended by ASHRAE Standard 55 for indoor thermal comfort.

The results provided in Figs. 6.22 and 6.23 indicate the following:

- Without any heating and cooling system, thermal comfort can be achieved 65% of the time. Specifically, the total comfort hours (*PMV* index within −0.5 to 0.5) are 17,000 for system 1 compared to 26,000 achieved by

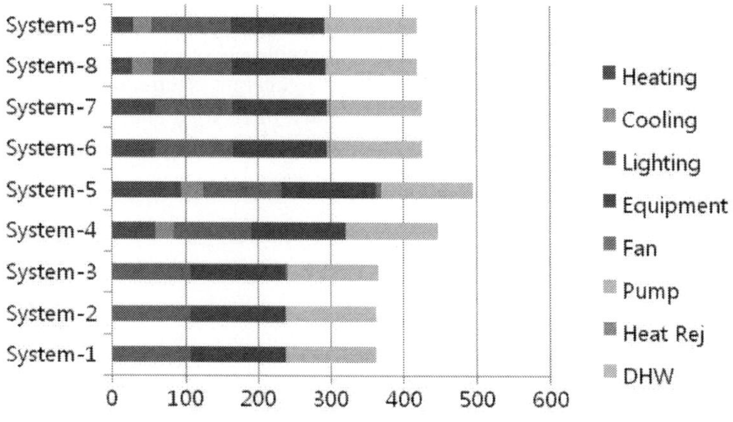

FIGURE 6.22 Summary of annual energy end-use for various heating and cooling systems considered for the apartment building.

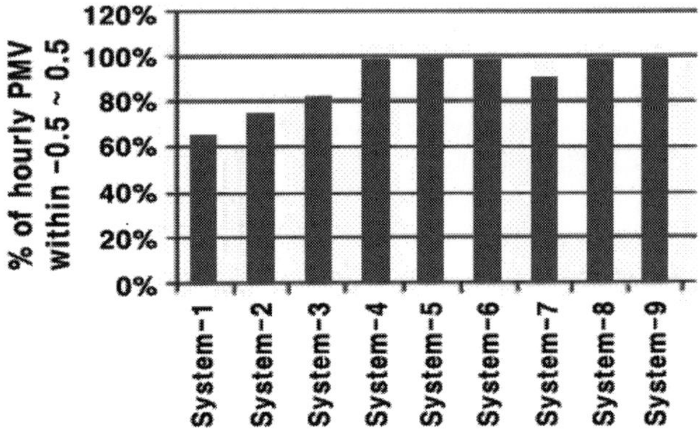

FIGURE 6.23 Summary of annual thermal comfort level for various heating and cooling systems considered for the apartment building.

systems with heating and cooling capabilities including system 5 through system 9) out of all 26,280 possible hours (i.e., sum of three floors per unit, i.e., 3 * 7760 = 26,280).

- System 3 is only effective during the summer since room temperatures in the winter are typically higher than the air temperatures supplied by the earth tube. A detailed estimation for the change-over period (from heating to cooling) is needed to better improve the performance of this system.
- The radiative floor system (system 8) saves almost 50% in heating energy use when compared to the radiative wall system (system 4).
- Evaporative cooling with a cooling tower or an evaporative cooler is effective and is highly recommended for cooling.

6.4 RENEWABLE ENERGY SYSTEMS

Throughout this book, various energy efficiency measures have been presented and evaluated. In this section, selected renewable energy systems are briefly described to meet thermal loads as well as to generate electrical energy needs for retrofitted residential buildings. Specifically, three solar energy systems are discussed: (1) passive solar heating systems, (2) hot water systems, and (3) photovoltaic (PV) systems.

6.4.1 Passive Solar Heating Systems

Using existing building envelope components, passive solar heating strategies can be considered when retrofitting residential buildings. Indeed, a passive solar heating system typically requires windows for solar energy aperture and walls and floors to capture and store solar energy. Two passive solar systems can be suitable for retrofit applications including direct gain and sunspace configurations as illustrated in Fig. 6.24. As shown in Fig. 6.24A, by increasing the size

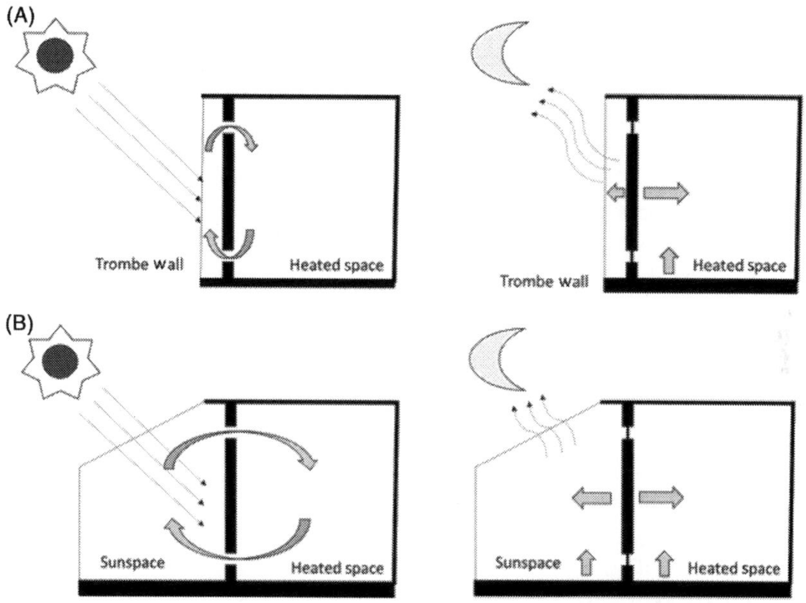

FIGURE 6.24 Passive solar heating systems suitable for retrofit applications. (A) Trombe wall. (B) Sunspace.

of the window and/or adding more thermal mass, solar radiation heat can be effectively stored in the building and distributed uniformly to all spaces. Similarly, heat can be captured and released with some time delay to the house when a sunspace or greenhouse space is present or added as illustrated in Fig. 6.24B.

Several analysis methods are available to assess the potential energy savings associated with passive solar heating systems. In addition, a set of rules of thumb and simplified evaluation methods can be used such as the load collector ratio or *LCR* method (Balcomb et al., 1980). Moreover, most detailed building energy simulation programs presented in Chapter 3 can be used to model the performance of the solar passive heating systems.

6.4.2 Solar Thermal Collectors

For residential buildings, solar thermal collectors can be used for domestic water heating, for space heating, or a combination domestic water and space heating. Solar water heaters are considered among the oldest high-efficiency water heating technologies available in the US. Indeed, the first commercially available solar water heater, the Climax, was made available in 1891 and was primarily sold in Southern California. Solar water heater sales in Southern California peaked at 1000 units sold in 1920, but sales rapidly declined when cheap natural gas became widely available (Butti and Perlin, 1980). Solar water heater sales then moved to Florida in 1923, and average sales were between 4000 and

10,000 units in Miami alone between 1935 and 1941. Solar water heater manufacturing stopped during World War II, and competition from electric water heaters after the war drastically reduced solar water heater sales. Sales of solar water heaters picked up again after the first oil embargo of 1973, peaking at over 20 million square feet of solar thermal collectors shipped in 1980. A combination of the repeal of tax credits for solar water heaters and a bad reputation coming from some poorly manufactured solar water heaters drastically reduced solar water heater sales in the 1980s (Ramlow and Nusz, 2006). Solar water heater sales today have increased compared to sales in the 1980s due to the return of federal tax credits as well as many state incentives for solar water heaters.

Solar water heaters offer an opportunity to significantly reduce natural gas or electricity consumption required for hot water. A solar water heater uses the sun to heat water for use in domestic hot water or hydronic space heating applications. Solar water heaters typically provide over half of the energy required by a household for water heating, while a backup water heating system provides the remaining energy (DOE, 2009a). The backup heating system could be any residential water heater, including a condensing or heat pump water heater. There are currently over 100 different models of solar water heaters on the market in the US (DOE, 2009b). However, solar water heaters currently make up less than 1% of the market in the US.

There are several types of collectors that can be used for solar water heating. The two most common options are a flat plate collector and an evacuated tube collector, but there are also systems with the water storage integrated into the collector (these are commonly referred to as ICS, or integrated collector storage systems). Flat plate collectors consist of a collector surface that absorbs the solar radiation, a glazing to prevent the absorber from reradiating the solar energy, a heat transfer medium (for domestic hot water applications, this is almost always water or a propylene glycol base heat transfer fluid, depending on the climate and system type), and insulation on the sides and back to prevent losses by conduction and convection (Goswami et al., 2000). Evacuated tube collectors consist of several absorber surfaces with heat transfer fluid flowing through them encased in a vacuum. The vacuum around the collector minimizes the losses to the environment from the collector since in it drastically reduces the conduction and convection losses from the system. However, they are typically much more complicated, and therefore more expensive, than flat plate collectors. ICS systems typically consist of a flat plate collector with a large enough volume to also act as a storage tank. Since they are especially susceptible to freezing, they are typically only used in climates where the outdoor air temperature rarely gets below freezing.

The decision of which collector type is used typically depends on the inlet fluid parameter, which itself is dependent on the ambient temperature, the inlet fluid temperature, and the incident solar radiation intensity (Siegenthaler, 2008). For the temperatures used in most residential applications, flat plate collectors are more efficient, although for higher temperature applications an evacuated tube collector may be more efficient. Flat plate collectors also make up the 76% of the collector sales for domestic water heating applications in the US (Weiss et al., 2006), and are therefore the only solar water heating technology considered here. Evacuated tube solar water heaters are better suited for commercial or light industrial applications.

Solar collectors have been in use for a relatively long time compared to other high-efficiency water heating technologies and because of this a rating procedure for collector efficiency is well established. Solar collectors in the US are rated by the Solar Rating and Certification Corporation (SRCC), which tests different collector designs under standardized conditions to characterize the thermal performance of the collector (SRCC, 1994). The SRCC procedure takes data from standardized tests (either ASHRAE 93-77 or 96-1980) and produces predictions for the amount of energy captured by the panel per day depending on the weather and application. The SRCC test also produces an efficiency equation for the collector that provides the efficiency as a function of the inlet fluid parameter. This data are used for projection by the SRCC of energy savings for each solar water heater in a variety of locations. The SRCC also makes these data available so that they can be used for the modeling of solar water heaters.

Fig. 6.25 illustrates typical efficiency curves for three solar collectors: (1) unglazed, (2) flat-plate, and (3) evacuated-tube systems.

In the following sections, two applications of solar collectors including solar domestic hot water (SDHW) systems as well as combi-systems are briefly described.

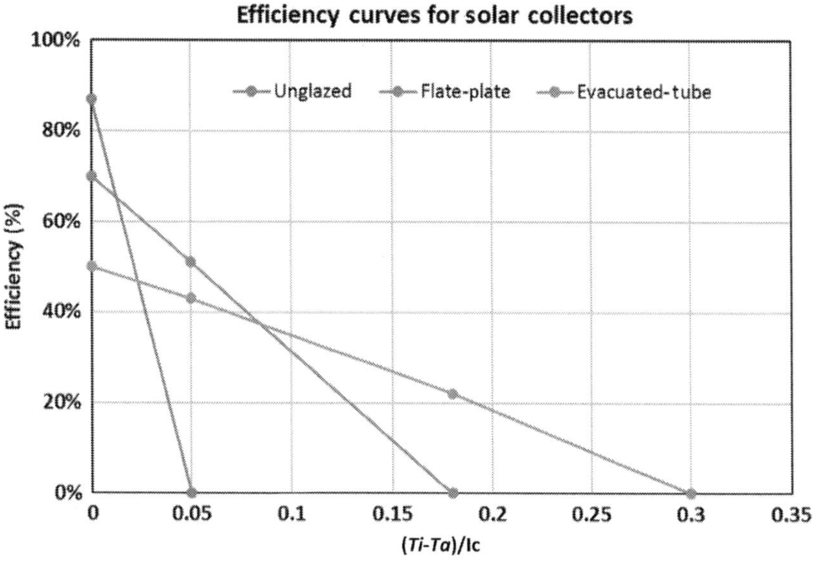

FIGURE 6.25 Typical thermal efficiency curves for three types of solar collectors.

6.4.3 Solar Domestic Hot Water Systems

SDHW systems are often cost-effective systems to consider for residential buildings. There are several types of SDHW systems and are divided into two primary categories: direct and indirect. In direct or open loop systems, water is directly heated through the solar collectors. However, heat exchangers are used to heat water for indirect or closed loop systems. Moreover, SDHW systems can be passive (i.e., natural convection) or active (i.e., pump) systems to circulate water from the collector to the tank. Thermosiphon and integrated collector systems or ICS are examples of direct and passive SDHW systems. Drainback systems are examples of indirect and active SDHW. Fig. 6.26 shows a typical drainback SDHW system suitable for residential buildings located in cold climates. The main feature of a drainback system is the fail-safe set-up used to ensure that the collector loop system including the collector and the pipes would not freeze by removing water from the loop when the system is not collecting solar heat.

The thermal efficiency of an SDHW system is closely related to the thermal efficiency of its collector efficiency, typically expressed as a function of the inlet fluid (i.e., water for direct systems) temperature, ambient air temperature, and solar radiation falling on the collector:

$$\eta_C = F_r(\tau\alpha)_n - F_rU_l * (T_{inlet} - T_{ambient}) / G_{net} \quad (6.20)$$

where η_c is collector efficiency, $F_r(\tau\alpha)_n$ is collector optical gain coefficient, F_rU_l is collector heat loss coefficient, T_{inlet} is fluid temperature at the inlet of the collector, $T_{ambient}$ is ambient air temperature, G_{net} is net incident radiation (includes short-wave solar and net sky infrared radiation).

Table 6.3 provides typical parameter values for selected SDHW collector types including a selective and nonselective metal-glass collector as well as a number of polymer collectors: a glazed, with selective coating; a glazed standard; and an unglazed.

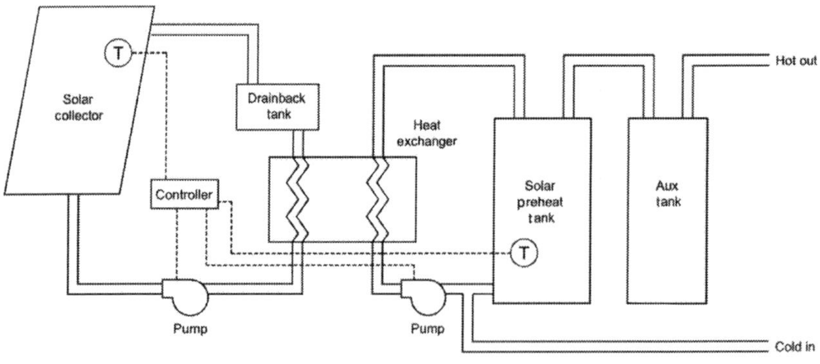

FIGURE 6.26 Typical components of a drainback solar domestic hot water system.

TABLE 6.3 Typical Collector Parameters

Collector type	$F_r(\tau\alpha)_n$	F_rU_l (W/m^2)
Selective: metal–glass	0.779	4.77
Nonselective: metal–glass	0.768	7.245
Polymer: selective	0.779	4.77
Polymer: nonselective	0.739	8.216
Polymer: unglazed	$0.88 - 0.029 * v_{wind}$	$10.24 - 4.69 * v_{wind}$

v_{wind}, wind refers to the wind speed expressed in m/s.

6.4.4 Solar Combi-Systems

Solar combi-systems utilize solar collectors to meet thermal loads associated with space heating and domestic hot water. These systems are typically considered for detached single-family homes, groups of family homes, and multifamily buildings. As compared to conventional SDHW systems, combi-systems have the advantage of their high solar collector's utilization independent of occupant hot water use. However, these systems require a relatively large capital investment and their design can be intricate. The market share of solar collectors installed in solar combi-system applications as compared to SDWH applications is about 50% for several northern European countries including Austria, Switzerland, Denmark, and Norway (Weiss, 2003). Solar combi-systems first appeared in the US and Europe in the mid-1970s. In the US, federal tax credits allowed the technology to be cost-effective. During the 1980s, the solar water heater/space heat market in the US began to fade due to the end of tax credits. Beginning in the 1990s, European installers and solar companies began offering simpler and cheaper solar combi-systems but these systems were often custom-built systems (Weiss, 2003). Today, limited commercialized systems exist in the US and Europe. As a result, most combi-systems are still custom-designed and built.

Solar combi-systems consist of five components: (1) a solar collector loop, (2) a storage tank, (3) a control system, (4) an auxiliary heater, and (5) a heat distribution system. Combi-systems typically use a stratified sensible heat storage tank ranging from 100 to 1000 gallons and are coupled to up to 400 ft^2 of solar collectors. The heat distribution system consists of a hydronic heating loop serving radiant floors, baseboard heaters, or air handler heating coils that are directly or indirectly heated by the storage tank. Solar combi-systems typically reduce auxiliary space heating energy usage by 10%–60% depending on the size and type of system installed. Even higher solar fractions can be achieved with large seasonal thermal storage. Seasonal storage systems store summer solar heat in a large stratified tank for later use in the winter. These systems typically have a storage volume between 5000 and 20,000 gallons (Weiss, 2003).

Detailed simulation programs such as TRNSYS can be used to model the performance of combi-systems.

6.4.5 PV Systems

Most PV systems installed for residential buildings are used to generate electricity that can be either used directly in the house (standalone systems) or sold to the grid (grid-connected systems). There is an interest in using hybrid PV systems to generate electricity and heat through photovoltaic/thermal (PV/T) collectors. A brief overview of grid-connected PV systems and PV/T collectors is provided in this section.

6.4.5.1 Grid-Connected PV Systems

Fig. 6.27 illustrates the basic design for a grid-connected PV system suitable for a residential building. A PV system typically includes a PV panel made up of several arrays to generate electricity from solar radiation, an inverter to convert direct current generated by the PV panel to alternating current commonly provided to the house by the utility company, and a set of regulators and controllers and protection devices.

Flat-plate PV panels are commonly installed in residential buildings with fixed mounting systems. However, other PV panels exist such as 1- or 2-axis tracking systems and concentrating systems. More recently, flexible shingle-type PV systems have been used as both PV panels and roofing shingle materials. The efficiency of a PV panel is typically by its *I–V* curve as presented in Fig. 6.28 with the highlighted maximum power point (MPP) where the generated electrical power is maximized as indicated in Fig. 6.29. It should be noted that the two characteristic points are often given as part of the manufacturers' specifications of PV panels including the short-circuit current or I_{sc} (obtained when $V = 0$) and the open circuit voltage or V_{oc} (obtained when $I = 0$). Table 6.4

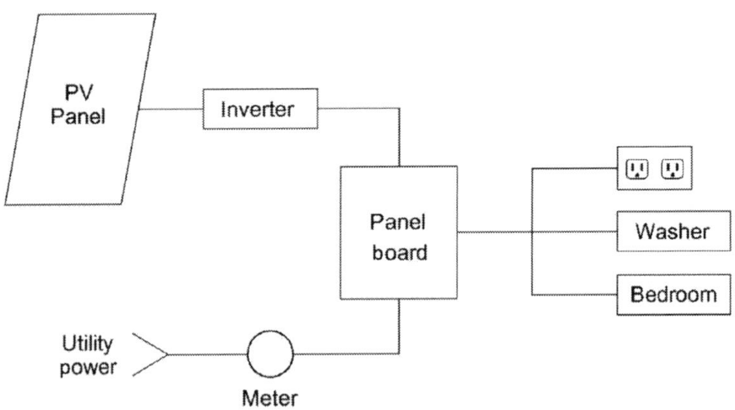

FIGURE 6.27 Typical grid-connected photovoltaic (PV) system for a house.

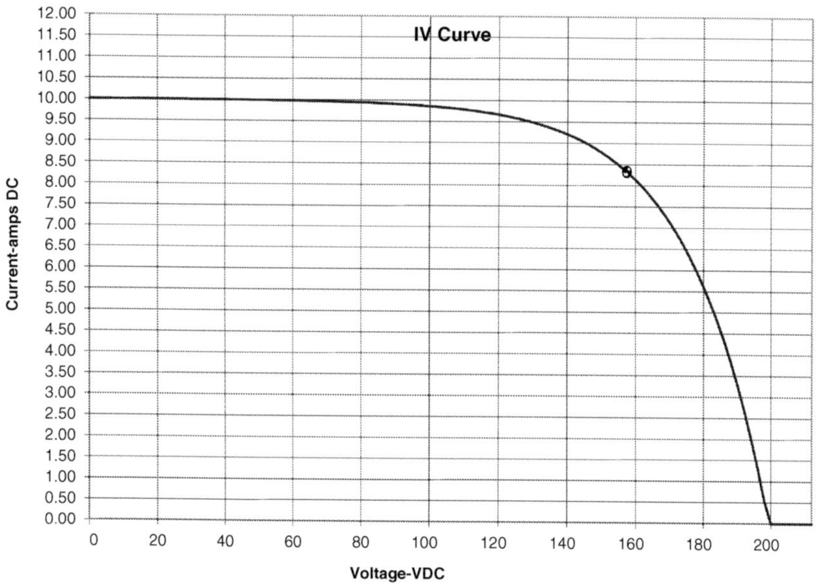

FIGURE 6.28 Typical *I–V* curve for a PV panel.

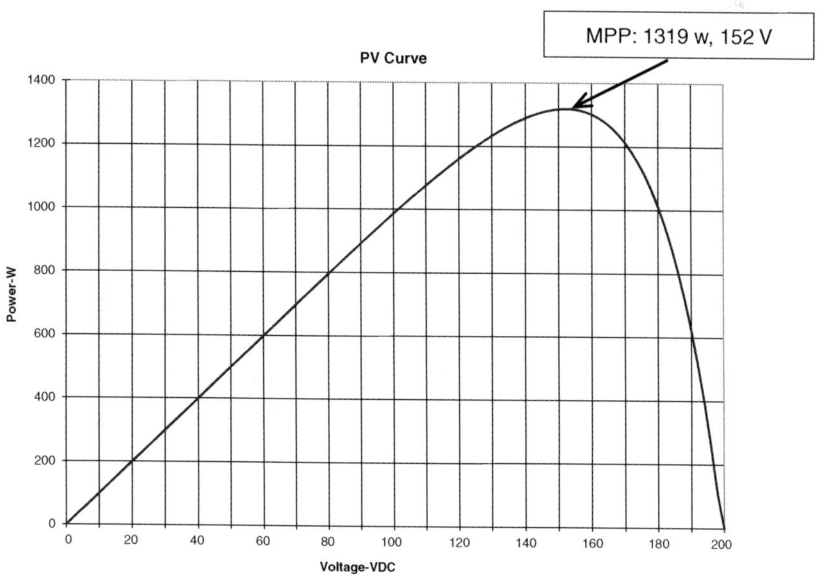

FIGURE 6.29 Variation of electrical power generated by a PV panel.

TABLE 6.4 Typical PV Module Characteristics

PV cell type	Rated power (W)	Short circuit current, I^{SC} (A)	Open circuit voltage, V_{oc} (V)	Efficiency (%)
Single crystal	175	5.55	44.4	13.5
Multi crystal	75	4.75	21.8	11.6
Thin film	40	2.68	23.3	9.4

PV, Photovoltaic.

presents typical PV module characteristics including rated power, short-circuit current, open circuit voltage, and efficiency.

As noted in Table 6.4, the conversion efficiency of commercially available PV modules is in the 10%–13% range under rated conditions. It should be noted that the thin-film PV modules are being used for building integrated PV applications (such as shingle-PV systems) and their efficiency is improving. A simple analysis of PV systems is to utilize the efficiencies for all of its components. In particular, for a typical PV system with a PV panel, an inverter, and MPP tracker, the overall system efficiency can be estimated as follows:

$$\eta_{PV,system} = \eta_{PV,Module} * \eta_{INV} * \eta_{MPPT} * \eta_{VR} \tag{6.21}$$

where $\eta_{PV,Module}$ is PV panel efficiency (refer to Table 6.4 for typical efficiency values), η_{INV} is inverter efficiency (typically over 95%), η_{MPPT} is maximum power point tracker efficiency (typically over 95%), and η_{VR} is voltage regulator efficiency (typically over 95%).

Then the electrical energy generated, E_{gen}, by the PV system can be estimated on an hourly basis:

$$E_{gen} = \eta_{PV,system} * A_{PV} * G_{net} \tag{6.22}$$

where A_{PV} is area of the PV panel and G_{net} is net incident radiation hitting the PV panel and depends on the tilt angle and tracking system used for the PV system.

Several tools are available to estimate the electrical energy generated by PV systems such PVWatts, TRNSYS, and PV-Chart.

6.4.5.2 Photovoltaic/Thermal Collectors

Hybrid PV/T collectors convert solar energy into electricity and heat. A typical residential PV/T collector is analogous to a flat plate solar thermal collector except that a PV panel is attached on the top of the metallic absorber plate. The PV cells are composed of semiconductor materials that convert high-energy photons of incident solar radiation into electricity. The lower energy photons are absorbed by the PV panel and generate heat within the cells. This heat is rejected to the ambient conditions while also being removed from behind the PV cells by a heat transfer fluid. If extracted, heat from the PV panels can offer multiple benefits. The heat can be used to meet various thermal loads while improving the PV electrical efficiency due to its lower operating temperature. Both air and fluids are used to recover heat from the PV systems. The amount of thermal energy and its quality are important factors when considering the value of the output heat. The performance of a PV/T system is characterized by both the amount of electrical energy and useful heat produced. Currently, there are only very limited PV/T systems commercially available.

6.5 NEAR-OPTIMAL ANALYSIS METHODOLOGY

6.5.1 General Methodology Description

Instead of using an optimization technique to find the best package energy and renewable energy measures to retrofit an existing building to zero-net energy, a sequential search analysis procedure similar to the *BE*opt approach can be utilized. The basic sequential search process involves the following steps:

Step 1: Start by the evaluation of different types of energy efficiency measures (e.g., wall insulation addition, window replacement, and HVAC equipment efficiency improvement) to determine the most cost-effective measure in terms of marginal cost. The baseline building model for this first iteration is a calibrated model using either utility data or monitored data as described in Chapter 10. Using the Pareto graph similar to Fig. 6.30, the life-cycle cost is shown in the *y*-axis and the percent savings of source energy is provided in the *x*-axis for each measure. The measure with the steepest negative slope (i.e., ratio cost over percent energy savings) is selected as shown in Fig. 6.30.

Step 2: The chosen measure in step 1 is then removed from the potential energy efficiency measures for the second iteration analysis. In this step,

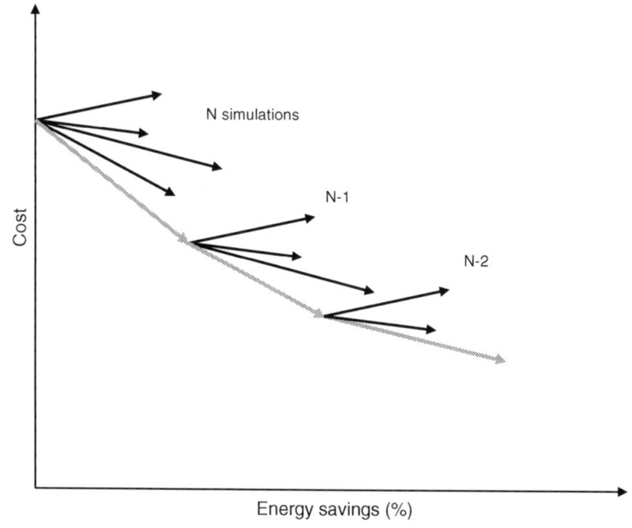

FIGURE 6.30 Sequential analysis procedure for selecting energy efficiency measures.

the new baseline building model consists of the calibrated model modified to include the selected measure after the first iteration outlined in step 1. The remaining energy efficiency measures are then simulated using the new baseline building model. Again, the measure with the most-effective marginal cost is selected after this iteration as outlined in Fig. 6.30.

Step 3: Repeat the same procedure described for step 2 until an optimal point is reached and the slope of the life-cycle cost (*LCC*) over percent energy savings become positive as illustrated in Fig. 6.31. The positive slope indicates that a reduction in building energy use results in increasing the *LCC* due to significant implementation costs. It should be noted that now the measure with the lowest positive slope is selected.

Step 4: After reaching an optimal point, the most cost-effective efficiency measure's marginal cost is compared to the cost of PV energy. At the point where further improving the building has a higher marginal cost, PV is employed until zero net source energy is achieved, as illustrated in Fig. 6.30.

Step 5: A final check is needed to ensure that the size of the required PV system can be installed based on the constraints of the available spaces (for instance the house roof area). If no sufficient space is available to install the required PV system, then additional energy efficiency measures need to be selected until 100% source energy savings are achieved. In this case, the selection process is similar to that described for step 4 without the need to compare with the marginal cost of the PV system.

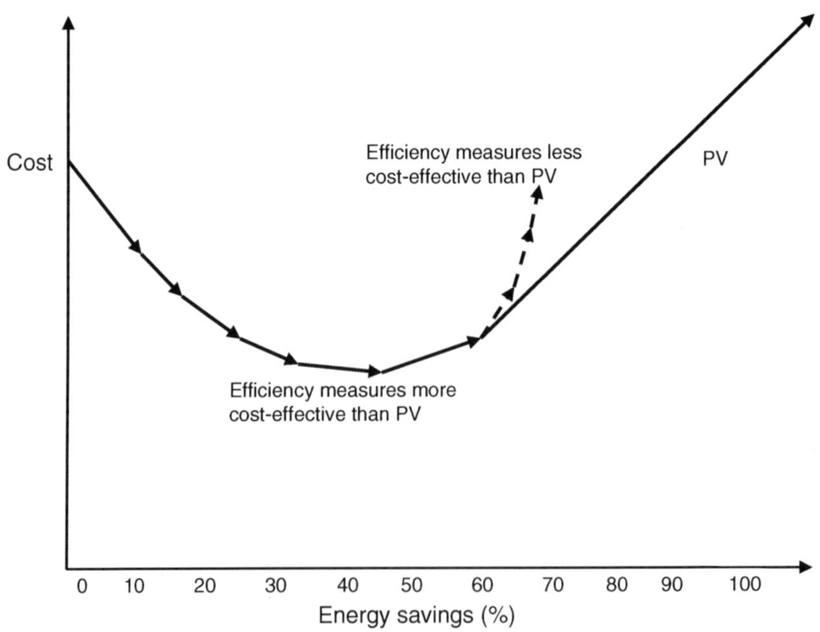

FIGURE 6.31 Sequential selection procedure for the path to ZNE.

6.5.2 Validation of Optimization Results

The results obtained from the sequential search optimization were compared with a "brute-force" search approach using the full combination options of energy efficiency measures to find the optimum design package for a prototypical single-family home by Ihm and Krarti (2012). The computational efforts required using the brute force search to find optimal design values for 11 energy efficiency measures (EEMs) (i.e., about 11.1 million possible combinations of building design options) are significant and may take several months to complete using the current state-of-the art computing processors. Instead, three analysis cases were considered by Ihm and Krarti (2012) to validate the results of the sequential search optimization approach. The three cases considered by Ihm and Krarti (2012) consisted of different combinations of design options:

- 4-EEM package: WWR, glazing type, lighting level, infiltration rate;
- 6-EEM package: exterior wall insulation, roof insulation, WWR, glazing type, lighting level, infiltration rate; and
- EEM package: indoor and exterior wall insulation, roof insulation, WWR, glazing type, lighting level, infiltration rate, and cooling set point.

Figs. 6.32–6.34 validate the sequential search optimization results obtained for the three EEM-packages for a villa against those obtained with the full brute-force approach (i.e., all combinations of options are considered) against

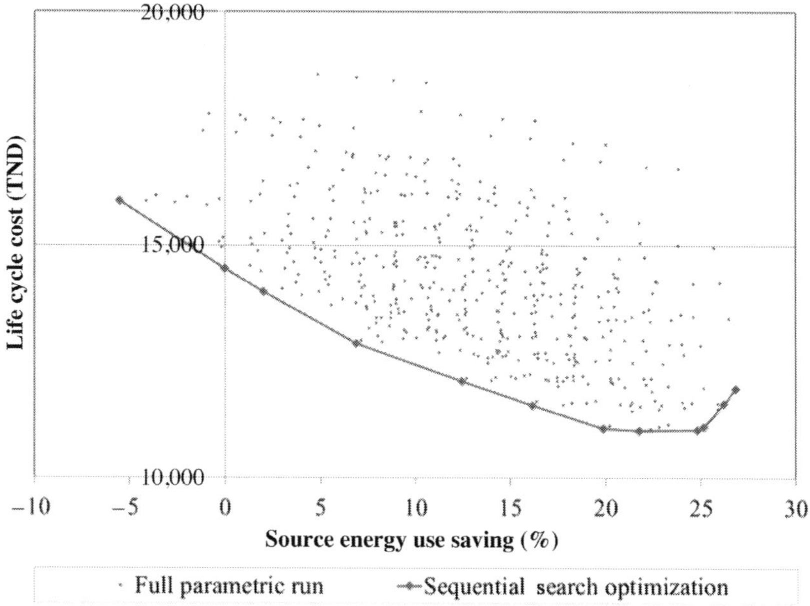

FIGURE 6.32 Comparison of optimal results obtained by the brute-force analysis and the sequential search optimization for 4-energy efficiency measure (EEM) package for a villa.

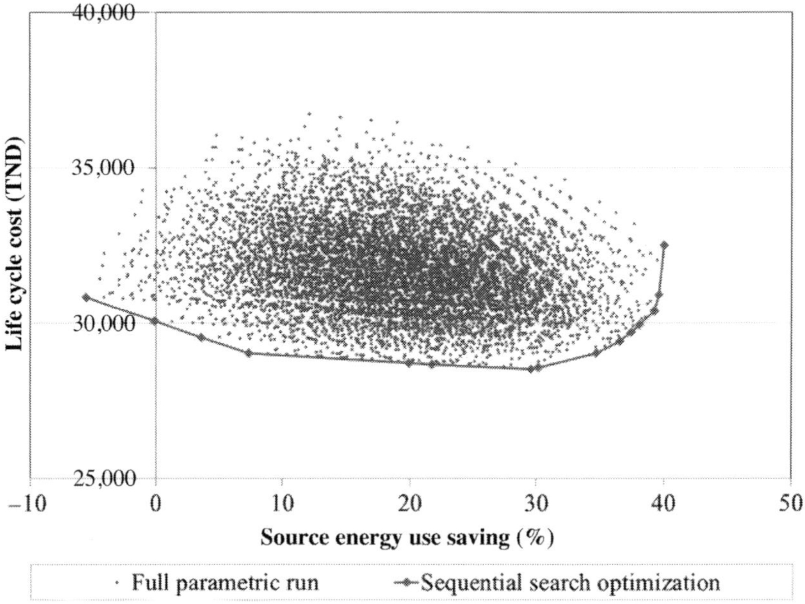

FIGURE 6.33 Comparison of optimal results obtained by the brute-force analysis approach and the sequential search optimization for the 6-EEM package for a villa.

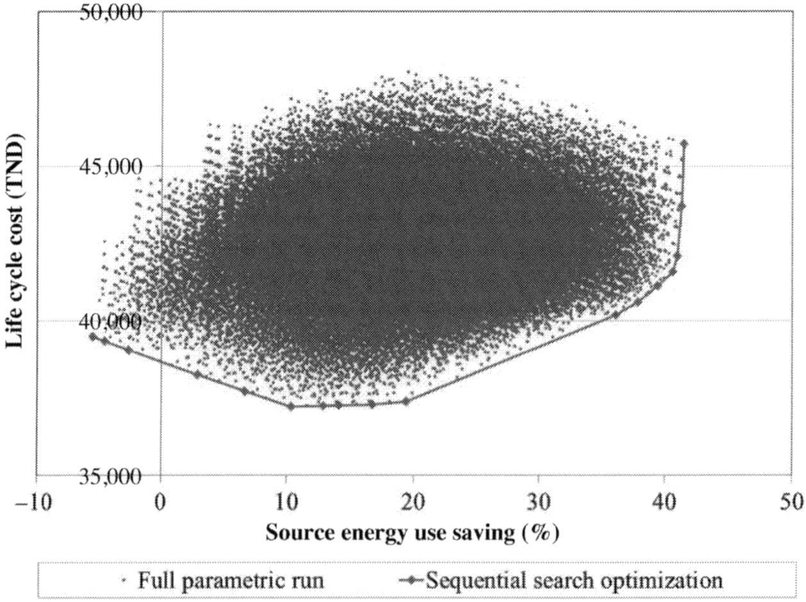

FIGURE 6.34 Comparison of optimal results obtained by the brute-force analysis approach and the sequential search optimization for the 8-EEM package for a villa.

TABLE 6.5 Summary of Comparative Results Between Brute-Force Analysis and Sequential Search Optimization for Three Analysis Cases for a Villa

Number of EEMs	Number of possible building design options	Computing time for brute-force analysis (min)	Computing time for sequential search (min)	Reduced CPU time (%)
4	480	7.0	2.1	69.9
6	7680	123.5 (2.1 h)	3.1	97.5
8	92,160	1732.8 (28.9 h)	4.6	99.7

EEMs, Energy efficiency measures.

those found by the sequential search optimization approach. The results in Figs. 6.32–6.34 are presented in terms of life-cycle cost as a function of percent savings of total source building energy use. Table 6.5 summarizes the comparative results for the brute-force analysis and the sequential search optimization approach. As indicated in Figs. 6.32–6.34 and Table 6.15, the sequential search optimization technique finds the same optimum solutions found through the brute-force technique for the three analysis cases. The computational time of the sequential search technique (4.6 min) is significantly lower by up to 99.7% when compared to the brute-force analysis approach (28.9 h) for the 8-EEM package analysis case using a 2.8-HGZ processor.

6.6 CASE STUDY: OPTIMAL DESIGNS OF OFFICE BUILDINGS

In this section, a comparative evaluation using US and French building energy efficiency standards is presented using the detailed analysis completed by Deneuville and Krarti (2015). The main objective of the comparative analysis is to identify design features that can achieve higher performance buildings including net zero and positive energy buildings. The comparative evaluation is based on detailed energy and economical optimal analysis for select representative climatic conditions for the United States and France. The analysis presented in this case study focuses on office buildings. Indeed, office buildings represent nearly 19% of building primary energy consumption and 17% of total floor area in the commercial building sector in the US (EIA, 2005). Similarly, office buildings account for 45% of total building energy use and 22% of commercial heated building floor area in France (Ademe, 2005). First, the general analysis approach is briefly presented including the characteristics of the baseline office building energy model as well as the optimization technique used. Then, selected optimal analysis results are discussed for representative climatic conditions associated with the US and France. Finally, some recommendations are provided to improve the design of office buildings in the US and France and the HQE label system in France.

In the United States, buildings consume over 41% of total primary energy and over 74% of total electrical energy (EIA, 2011). To reduce significantly the energy needs for buildings, the Department of Energy (DOE) has set a strategic goal to encourage the design and construction of ZNE buildings through initially a wide range of research initiatives (DOE, 2012). A ZNE building is characterized by reduced overall energy needs with the balance supplied by renewable technologies. To reach ZNE designs for buildings by 2025, the DOE has developed a series of technical guidelines for the building industry that can be used to progressively achieve building site energy savings of 30%, 50%, and 70% relative to ANSI/ASHRAE/IESNA Standard 90.1 (ASHRAE, 2004). In particular, technical guidelines have been made available for various prototypical building types and include comprehensive sets of modeling assumptions, analysis methods, and recommendations for building design packages that can achieve the aimed energy savings levels in a wide range of building categories and climates. Specifically, three guidelines have been developed for site energy savings up to 50% in small and medium offices (Thornton et al., 2009), as well as large offices (Leach et al., 2010). Torcellini and Pless (2010) have reported that the definition of zero energy buildings can greatly impact the design choices and whether these buildings can actually achieve ZNE levels.

Similarly in France, buildings account for nearly 45% of primary energy consumption and 23% of carbon dioxide emissions (Orselli, 2007; De Boissieu, 2006). The strategic goal of Bâtiment à Energie Positive (BEPOS) was developed to foster the design of high-performance buildings and is expected to be the minimum energy efficiency standard by 2020 in France (Legifrance, 2010). A BEPOS is a building that produces more energy through renewable energy systems than its primary energy needs. The current building energy efficiency code in France, RT 2012, defines a specific metric value to evaluate the primary energy use intensity of a building and is referred to as Coefficient en énergie primaire (Cep). The energy efficiency code requires that the energy performance of a building cannot exceed a maximum energy use intensity level (Cep max).

The requirements set by the Réglementation Thermique (RT) are aimed to be the path to achieve ZNE buildings by 2020 with an intermediate goal of 50% savings set for new buildings in France. However, these objectives are being questioned due to concerns about the scope of analysis considered by developing RT requirements including its assumptions. The main approach of the analysis methodology used for RT development, outlined by CSTB (CSTB, 2011), defines the overall building energy performance through three metrics (CSTB, 2011):

- primary energy intensity or "Coefficient en Energie Primaire" (Cep),
- energy load intensity or "Besoin bioclimatique" (Bbio), and
- indoor temperature setting or "Température intérieure conventionnelle" (Tic).

Maximal threshold levels are established for these three metrics depending on the building location and type. Using the RT approach, the building performance has to be evaluated using a predefined building operation that may be different from the actual building usage. For instance, the heating set-point has to be maintained at 19°C (66°F), which may not satisfy occupants' thermal comfort in some conditions. Indeed, heating set-points are kept actually at the 20–21°C (68–70°F) range in most office buildings in France. Moreover, plug loads are simply excluded from the RT energy balance analysis used to estimate the building performance, even though these loads can significantly affect heating and cooling loads even for highly energy-efficient buildings including those with the label of Haute Qualité Environnementale (HQE). In France, the HQE label is similar to the US Leadership in Energy and Environmental Design label for sustainable buildings. In particular, HQE typically prescribes energy intensity reduced by an additional 30%, and to 60% compared to minimal requirements set by RT. Thus, the RT methodology can lead to a significant discrepancy between targeted and actual energy performance of buildings as documented by some reported studies in France. Indeed, Catarina and Illouz (2009) documented actual energy consumption of three commercial buildings and found a large discrepancy between modeled and measured energy consumption. In addition, Alessandrini (2008a, 2008b) has studied 28 exemplary buildings in Programme de Recherche sur l'Energie dans le Bâtiment, including five office buildings and has concluded that the energy performance of the buildings did not meet the targeted levels.

6.6.1 Analysis Approach

For the analysis carried out for this study, a medium-sized office building is considered for the US and France. The basic design features of the office building energy model are detailed in Tables 6.6 and 6.7. The 3-D rendering of the building model is shown in Fig. 6.35. The main physical features of the prototype building are based on DOE medium-sized office benchmark (DOE, 2012). The temperature settings, ventilation rates, as well as internal gains attributed to people, lighting, and plug loads are set according to DOE building benchmark for US locations and to RT 2012 specifications for French locations (CSTB, 2011). Two energy sources supplied to the building are considered for the US and France including natural gas and electricity. To estimate the primary energy consumption in the US and France, site-to-source conversion factors are, respectively, 1 and 1.1 for natural gas, 3 and 2.58 for electricity.

6.6.2 Representative Climatic Zones

In the US, ASHRAE climate zones are commonly used for building energy analysis including building energy efficiency standards. These climate zones, depicted in Fig. 6.36, range from 1 to 8 with 15 subzones, and are characterized by their heating and cooling degree-days while subzones indicate varying moisture conditions (ASHRAE, 2004). Humid subzones are designated by the

TABLE 6.6 Basic Features of the Office Building Baseline Model

Floor area	4982 m^2 (53,630 ft^2)	
Floors number	3	
Footprint	50 m × 33.2 m (164 ft × 109 ft)	
WWR	33%	
Envelope features		**Units**
Exterior wall *U*-value	0.704 W/m^2°C	(0.124 Btu/ft^2 F h)
Roof *U*-value	0.358 W/m^2°C	(0.063 Btu/ft^2 F h)
Windows *U*-value	3.066	(0.540 Btu/ft^2 F h)
Windows SHGC	0.40	
Air tightness (ACH)	0.20	
Equipment type	Packaged rooftop VAV gas furnace + electric reheat DX cooling	
Heating efficiency (AFUE)	78%/80%	
Cooling efficiency (EER)	9.5/10.1	
Fan efficiency	62%	
Fan pressure drop	1.574/1.389 kPa	(6.32/5.58 in-water)

ACH, Air-changes per hour.

TABLE 6.7 Temperature Settings, Ventilation Rates, and Internal Loads for the Office Building Baseline Model

Climate zone	Units	US	France
Peak occupancy	Occ/m^2 Occ/ft^2	0.054 0.005	0.060 0.0056
Lighting power density	W/m^2 W/ft^2	10.76 1.0	8.29 0.77
Plug load density	W/m^2 W/ft^2	8.07 0.75	8.07 0.75
Heating set-point (setback)	°C °F	21.1 (15.6) 70 (60)	18.9 (16.1) 66 (61)
Cooling set-point (setup)	°C °F	23.9 (26.7) 75 (80)	26.1 (30.0) 79 (86)
Outdoor air ventilation	l/m^2 s cfm/ft^2	0.560 0.112	0.370 0.074

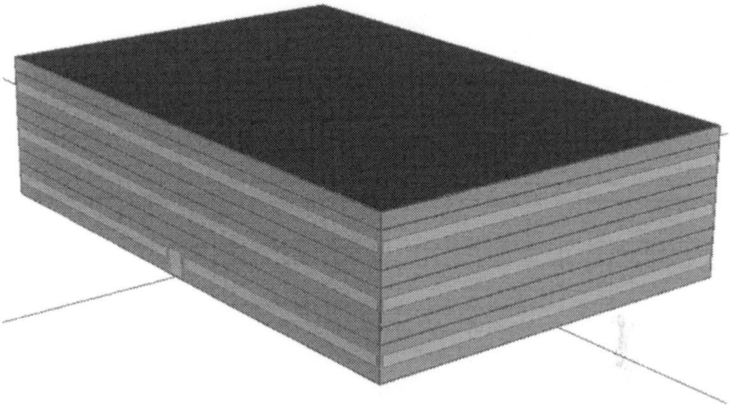

FIGURE 6.35 Isometric view of medium office model.

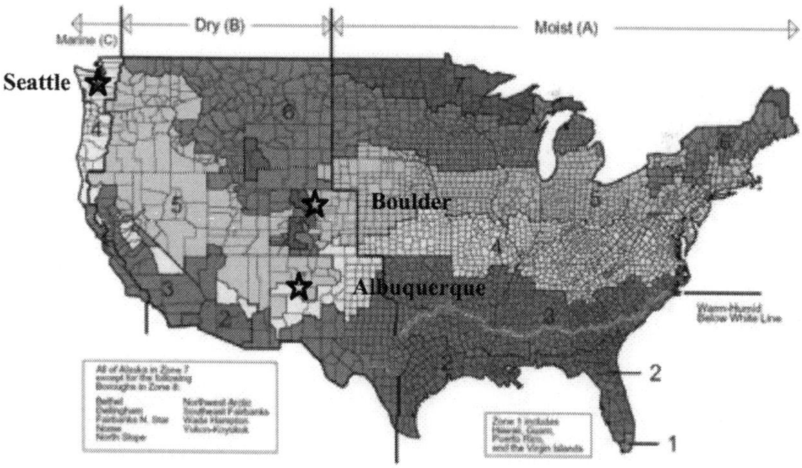

FIGURE 6.36 ASHRAE climate zones in the United States. *(Source: ASHRAE, 2004).*

letter A, dry subzones by the letter B, and marine subzones by the letter C as outlined in Fig. 6.36.

In France, three climate zones are considered in the RT including H1, H2, and H3 with eight subzones (H1 a/b/c, H2 a/b/c/d, H3) as depicted in Fig. 6.37. Compared to the US climatic conditions, French climates are typically mild and fall within ASHRAE climate zones 4 and 5.

To develop a meaningful comparison between energy efficiency measures, the analysis performed in the study presented in this paper consider US and French climates with similar characteristics. In particular, six sites with different climates (three from the US and three from France) have been selected to

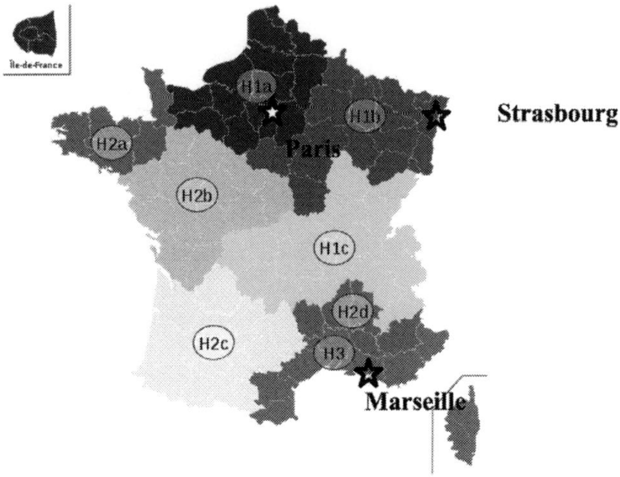

FIGURE 6.37 Réglementation Thermique climatic zones in France.

perform the comparative analysis. Table 6.8 summarizes the climate zones as well as the heating degree-days (HDD) and cooling degree-days (CDD) for the selected sites.

6.6.3 Energy Efficiency Measures

The impact of selected EEMs on the overall primary energy consumption of the prototypical office building of Fig. 6.35 has been evaluated using an

TABLE 6.8 Climate Zones and Heating and Cooling Degree-Days for US and French Selected Sites

Zones	4B	H3	H1a	4C	5B	H1b
	Albuquerque	Marseille	Paris	Seattle	Boulder	Strasbourg
HDD (12°C) HDD (50°F)	236 425	1774 3194	2803 5046	2727 4908	3086 5554	3074 5533
CDD (18°C) CDD (65°F)	2171 3908	2185 3933	1228 2211	1178 2120	1567 2820	1218 2193

CDD, Cooling degree-days; *HDD*, heating degree-days.

optimization approach based on a life-cycle cost analysis. Specifically, 14 EEMs have been considered among eight building energy system categories as detailed in Table 6.9.

To generate electrical energy, PV panels are considered for installation on the building roof and/or off-site (in ground-mounted systems located close to the building) depending on the requirements to achieve ZNE conditions.

TABLE 6.9 Characteristics of EEMs for Various Building Energy Systems

EEM	SI units	IP units
Additional wall insulation	W-EEM-1: R-1.32 c.i. m^2°C/W	R-7.5 c.i. ft^2°F h/Btu
	W-EEM-2: R-3.96 c.i. m^2°C/W	R-22.5 c.i. ft^2°F h/Btu
	W-EEM-3: R-5.02 c.i. m^2°C/W	R-28.5 c.i. ft^2°F h/Btu
Additional roof insulation	R-EEM-1: R-5.28 c.i. m^2°C/W	R-30 c.i. ft^2°F h/Btu
	R-EEM-2: R-7.04 c.i. m^2°C/W	R-40 c.i. ft^2°F h/Btu
	R-EEM-3: R-10.57 c.i. m^2°C/W	R-60 c.i. ft^2°F h/Btu
Windows *U*-value/SHGC/VLT	G-EEM-1: 2.50/0.38/0.44	0.44/0.38/0.44
	G-EEM-2: 1.48/0.56/0.75	0.26/0.56/0.75
	G-EEM-3: 1.19/0.36/0.50	0.21/0.36/0.54
Overhangs Sun shading	Projection factor	0.5
Lighting efficiency and controls	8.07 W/m^2	0.75 W/ft^2
Plug load efficiency and controls	5.92 W/m^2	0.55 W/ft^2
Daylighting controls 250 lux (25 fc) setpoint, continuous dimming)		
HVAC efficiency and controls		
Cooling efficiency (EER)		11.5/11.3/10.5
Exhaust air energy recovery		67%/62% (sensible/latent)
Supply air temperature reset		12.8°C to 16.1°C (55°F to 61°F)
Demand control ventilation		CO_2 sensor in zone return airstream

EEMs, Energy efficiency measures; *HVAC*, heating, ventilation, and air conditioning.

Typically, the PV panels are integrated on the top of the roof or mounted on surfaces tilted at the latitude angle of the building site location.

The capital costs of the EEMs and PV system considered in the analysis are accounted for as incremental costs over the baseline construction global costs of the prototypical office building of Fig. 6.35. Table 6.10 summarizes the incremental costs for all EEM categories and PV systems considered in the optimization analysis.

TABLE 6.10 Incremental Costs for EEM Categories and PV Systems

EEM	Unit	US[a] ($/unit)	France[b] ($/unit)
Wall insulation	Wall area, m^2 (ft^2)	W-EEM-1 $\$15.07/\text{m}^2$ ($\$1.40/\text{ft}^2$)	$\$10.3/\text{m}^2$ ($\$0.96/\text{f}^2$)
		W-EEM-2 $\$44.14/\text{m}^2$ ($\$4.10/\text{ft}^2$)	$\$44.35/\text{m}^2$ ($\$4.12/\text{ft}^2$)
		W-EEM-3 $\$54.90/\text{m}^2$ ($\$5.10/\text{ft}^2$)	$\$52.86/\text{m}^2$ ($\$4.91/\text{ft}^2$)
Roof insulation	Roof area, m^2 (ft^2)	R-EEM-1 $\$14.21/\text{m}^2$ ($\$1.32/\text{ft}^2$)	$\$23.90/\text{m}^2$ ($\$2.22/\text{ft}^2$)
		R-EEM-2 $\$23.79/\text{m}^2$ ($\$2.21/\text{ft}^2$)	$\$45.97/\text{m}^2$ ($\$4.27/\text{ft}^2$)
		R-EEM-3 $\$42.74/\text{m}^2$ ($\$3.97/\text{ft}^2$)	$\$78.26/\text{m}^2$ ($\$7.27/\text{ft}^2$)
Windows *U*-value/SHGC/ VLT	Window area, m^2 (ft^2)	G-EEM-1 $\$7.32/\text{m}^2$ ($\$0.68/\text{ft}^2$)	$\$12.06/\text{m}^2$ ($\$1.12/\text{ft}^2$)
		G-EEM-2 $\$37.25/\text{m}^2$ ($\$3.46/\text{ft}^2$)	$\$140.38/\text{m}^2$ ($\$13.04/\text{ft}^2$)
		G-EEM-3 $\$275.15/\text{m}^2$ ($\$25.56/\text{ft}^2$)	$\$\$348.68/\text{m}^2$ ($\$32.39/\text{ft}^2$)
Overhangs	Overhang area, m^2 (ft^2)	$\$78.91/\text{m}^2$ ($\$7.33/\text{ft}^2$)	$\$242.75/\text{m}^2$ ($\$22.55/\text{ft}^2$)
Lighting efficiency and controls	Floor area, m^2 (ft^2)	$\$11.09/\text{m}^2$ ($\$1.03/\text{ft}^2$)	$\$11.63/\text{m}^2$ ($\$1.08/\text{ft}^2$)
Daylighting controls	Perimeter floor area, m^2 (ft^2)	$\$2.15/\text{m}^2$ ($\$0.20/\text{ft}^2$)	$\$2.37/\text{m}^2$ ($\$0.22/\text{ft}^2$)
HVAC efficiency and controls	Package	$29,480	$30,525
Plug load efficiency and controls	Floor area, m^2 (ft^2)	$\$17.44/\text{m}^2$ ($\$1.62/\text{ft}^2$)	$\$17.44/\text{m}^2$ ($\$1.62/\text{ft}^2$)
PV	W_{DC}, Watt	6.18	7.36

EEM, Energy efficiency measure; *HVAC*, heating, ventilation, and air conditioning; *PV*, photovoltaic.
[a]*The US costs are obtained from the analysis by Thornton et al. (2009).*
[b]*The costs in France are obtained from the study conducted by Riser and Sidler (2010).*

6.6.4 Life-Cycle Cost Analysis

To assess the cost-effectiveness of various EEMs and PV systems, an *LCC* analysis approach is considered as outlined in Chapter 5. The *LCC* analysis accounts for the total energy related costs of the prototypical office building including capital, maintenance, and energy consumption costs (Krarti, 2010; Leach et al., 2010; Rieser and Sidler, 2010; Thornton et al., 2009). In particular, *LCC* includes annual cash flows (*CF*s) that have been evaluated for a life cycle of *N* years and actualized using a real discount rate, r_d:

$$LCC = \sum_{n=1}^{N} \frac{CF_n}{(1+r_d)^n} \tag{6.23}$$

with the annual cash flow estimated as follows:

$$CF_n = \left[C_n^B + O\&M_N^B\right] + \left[(C_n + O\&M_n)\right] + \left[C_e + C_g\right] \tag{6.24}$$

where CF_n is cash flow in year n, C_e is annual cost of electric power consumption, C_g is annual cost of gas consumption, C_n^B is baseline capital cost in year n, $O\&M_n^B$ is baseline operation and maintenance cost in year n, C_n is incremental capital cost in year n, $O\&C_n$ is incremental O&M cost in year n.

The baseline building capital costs, operation and maintenance costs, as well as the utility rates for electricity and natural gas considered in the optimization analysis are summarized for the US and France in Table 6.11. Typically, a life cycle of $N = 25$ years and a discount rate of $r_d = 2.2\%$ is considered as the baseline values for the comparative analysis. However, a sensitivity analysis has been carried out to determine the impact of select parameters (including N and r_d) on the optimization results. Some of the results of the sensitivity analysis are discussed in the following sections.

TABLE 6.11 Capital Costs, O&M Costs, and Utility Rates Used in the Optimization Analysis

	US	France
Baseline capital cost	\$1249.7/m² (\$116.1/ft²)	\$1892.3/m² (\$175.8/ft²)
Baseline O&M cost	\$1.18/m² yr (\$0.11/ft² yr)	\$1.08/m² yr (\$0.10/ft² yr)
Gas rate (\$/1000 ft³)	\$0.3705/m³ (\$0.01049/ft³)	\$0.5541/m³ (\$0.01569/ft³)
Electricity rate (¢/kWh)	7.89	12.03
PV sales rate (¢/kWh)	7.89	25.14[a] 14.03[b]

[a]*If integrated to flat roof.*
[b]*For other PV locations.*

6.6.5 Optimization Analysis Approach

For the analysis carried out for this case study, the sequential search methodology outlined in Section 6.5 is utilized to find the best combinations of EEMs and PV size to minimize the life-cycle cost function as expressed by Eqs. (6.23) and (6.24). It should be noted that *LCC* accounts for the implementation costs of the EEMs and PV systems as well as the operating energy costs over the life of the building. As noted earlier, the sequential search optimization technique has the advantage of identifying the path to reach the optimal solution as outlined in Fig. 6.38 illustrating the typical optimal path obtained in the analysis performed for this case study (Deneuville and Krarti, 2015). In particular, six design options for building designs can be identified as part of the optimal path toward achieving a ZNE building. These design options are as follows:

- The first design option, referred to as the optimal design, is depicted in Fig. 6.38 at the bottom of the optimal path. By definition, it represents the design option that has the lowest life-cycle costs while providing some primary energy savings due to the application of the optimal package of energy efficiency measures. This option is typically chosen by building owners and designers to reduce the project total operating and energy costs.
- The switchover design option is associated with the package of energy efficiency measures that achieves the maximum energy savings obtained without using any PV system. With this option, using PV becomes more cost effective than implementing any additional EEMs. As noted in

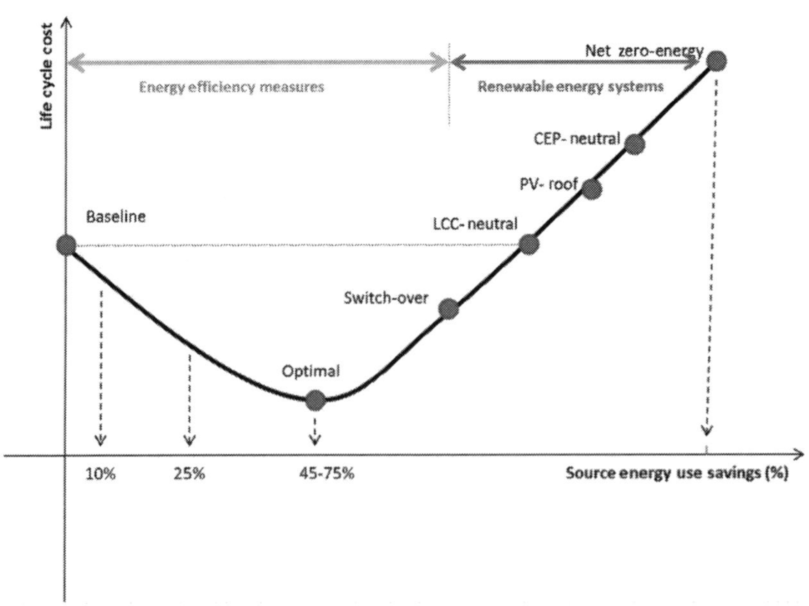

FIGURE 6.38 Typical optimal path and high-performance building design options.

Fig. 6.38, *LCC* increases linearly with the PV system size after the switchover design option.

- The *LCC*-neutral design option or neutral point representing the option where the *LCC* is the same as the baseline but with lower primary energy use. It represents the design option that achieves significant primary energy savings while maintaining *LCC* the same as that of the baseline design option. Depending on the building location and on the cost-effectiveness of EEMs, this option may be reached before switching to the PV system.
- The PV-roof design option corresponds to the option with the lowest *LCC* value and highest primary energy intensity achievable with PV installed on the entire building roof footprint area.
- The Cep-neutral design option refers to the option where the PV system is sized to generate electrical energy to offset building energy consumption equivalent to the Cep value. In Paris, for instance, this target requires the installation of PV on the entire available roof as well as in an off-site.
- The ZNE design option presents the option combining EEMs and PV panels with the annual building primary energy consumption entirely compensated by the PV electricity production.

6.6.6 Discussion of Selected Results

In this section, selected results of the optimization and sensitivity analyses results are presented. First, optimization results obtained for one US climate (Seattle) and French climate (Paris) are presented in some detail. Then, a brief summary is provided for the optimization results associated with all US and French sites. Finally, the sensitivity results are discussed.

The optimal path obtained for the prototypical office building using the sequential search technique is shown in Figs. 6.39 and 6.40, for respectively, Seattle and Paris. Tables 6.12 and 6.13 summarize the list of the EEMs and PV system sizes as well as *LLC* values, PV capital costs, and primary energy consumption, and Cep values for high-performance building design options including optimal and *LCC*-neutral obtained for Seattle and Paris, respectively.

For the optimal design option, primary energy savings of 29% and 41% can be achieved with 2.9% and 3.8% *LCC* savings when the building is located in Seattle and Paris, respectively. The higher utility energy rates in France as indicated in Table 6.11 provide a strong financial incentive to invest in additional energy efficiency than in the US. Indeed, the optimal design option in Paris includes more EEMs than that in Seattle. The additional EEMs consist of reducing the electric lighting loads and highly insulated building walls.

RT 2012 requires that Cep value should be less than 110 kWh/m^2 for Paris. As indicated in Table 6.13, this threshold is achieved at the switchover design option using all the cost-effective EEMs when the building is located in Paris. The same Cep value is reached in Seattle by adding higher insulation in the walls and roof to the optimal design option as outlined in Table 6.13. In Paris

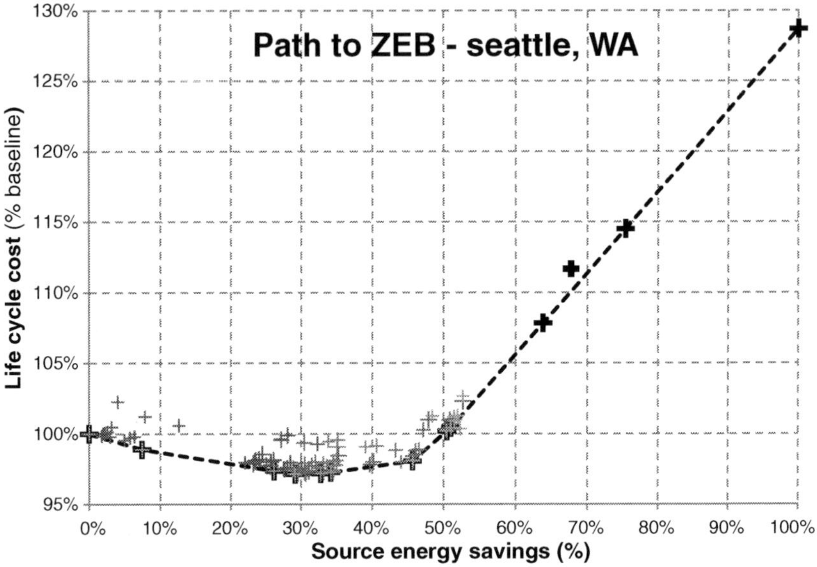

FIGURE 6.39 Optimal path toward zero energy buildings (ZEB) for the prototypical office building in Seattle.

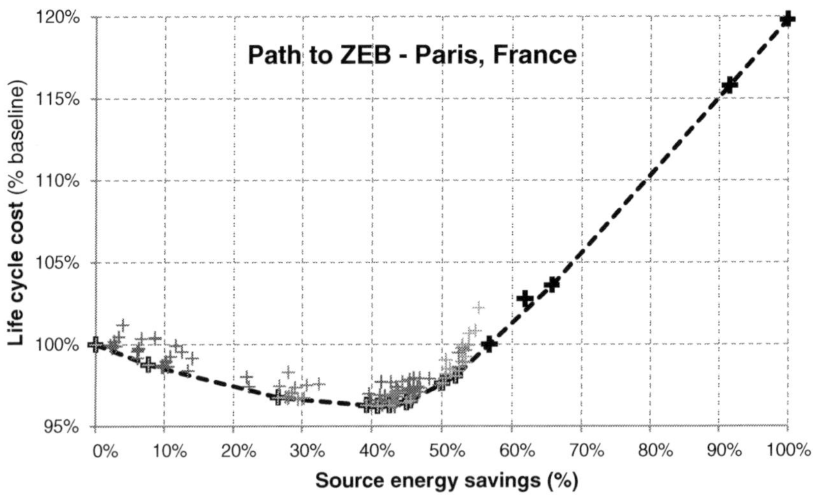

FIGURE 6.40 Optimal path toward ZEB for the prototypical office building in Paris.

TABLE 6.12 Summary of Characteristics of Office Building Design Options for Seattle

Design	Option	Life-Cycle Cost ($ LCC)	Life-Cycle Cost (% baseline)	Primary Energy (kBtu/ft² yr)	Primary Energy Savings (%)	Cep (kWh$_{EP}$/ m$^2_{SHON}$ an)
Baseline	EEM	$7,564,885	100	146.8	0	200.2
	Win-dows-1	$7,479,798	98.9	135.9	7.4	190.9
	HVAC	$7,371,616	97.4	108.5	26.1	129.4
Optimal	Daylight-ing	$7,348,834	97.1	104.0	29.1	117.6
	Wall.Ins-2	$7,351,910	97.2	98.7	32.8	112.1
	Roof.Ins-1	$7,358,811	97.3	96.6	34.2	109.9
	LPD	$7,418,786	98.1	79.7	45.7	96.8
LCC-neutral	Plug load	$7,581,867	100.2	72.7	50.5	96.8
Switcho-ver	Win-dows-2	$7,600,291	100.5	71.9	51.0	98.4
PV-roof	PV-roof (tilt)	$8,158,501	107.8	53.1	63.8	54.3
Cep-neutral	PV-roof (tilt)	$8,664,060	114.5	36.1	75,4	0
ZEB	PV-offsite (tilt)	$9,737,277	128.7	0	100	−91.5

Cep, Coefficient en énergie primaire; *EEM*, energy efficiency measure; *HVAC*, heating, ventilation, and air conditioning; *LCC*, life-cycle cost; *PV*, photovoltaic; *ZEB*, zero energy buildings.

and Seattle, Cep requirements can be reached at *LCC* lower than baseline by, respectively, 1.9% and 2.7%. A reduction of primary energy use of 34.2% is needed in Seattle and of 51.9% is required for Paris to meet the RT 2012 Cep threshold of 110 kWh/m^2.

The *LCC*-neutral design option allows the reduction in primary energy use by over 50% in Seattle (50.5%) and Paris (56.7%). The higher reduction in Paris is attributed to the contribution of the PV system (52.4 kWDC) installed in the building roof. When PV panels are installed on the entire roof surface area, a reduction of primary energy of 63.8% and 65.7% are achieved in Seattle and Paris, respectively. The required PV system size is 101 kWDC in Seattle and 151 kWDC in Paris leading to higher *LCC* relative to the baseline design options of, respectively, 7.8% and 3.6%.

TABLE 6.13 Summary of Characteristics of Office Building Design Options for Paris

Design	Option	Life-cycle cost ($ *LCC*)	(% base-line)	Primary energy (kBtu/ft^2 yr)	Savings (%)	Cep (kWh$_{EP}$/m^2$_{SHON}$ an)
Baseline	EEM	$11,495,605	100	133.1	0	217.7
	Windows-1	$11,353,680	98.8	123.0	7.6	205.0
	HVAC	$11,121,445	96.7	97.8	26.6	136.4
	LPD	$11,067,095	96.3	80.8	39.3	121.0
Optimal	Wall.ins-1	$11,060,538	96.2	78.7	40.9	118.4
	Roof.ins-1	$11,063,956	96.2	76.5	42.5	115.6
	Wall.ins-2	$11,085,982	96.4	73.2	45.0	111.6
	Sun shading	$11,098,971	96.5	72.4	45.6	110.7
	Plug load	$11,223,554	97.6	66.4	50.1	110.7
Switchover	Windows-2	$11,274,487	98.1	64.0	51.9	109.5
LCC-neutral	PV roof (flat)	$11,495,605	100	57.6	56.7	89.1
PV roof	PV roof (flat)	$11,908,481	103.6	45.6	65.7	50.9
Cep-neutral	PV offsite (tilt)	$12,559,244	109.3	29.7	77.7	0
ZEB	PV-Offsite (tilt)	$13,769,475	119.8	0	100	−35.9

Cep, Coefficient en énergie primaire; *EEM*, energy efficiency measure; *HVAC*, heating, ventilation, and air conditioning; *LCC*, life-cycle cost; *PV*, photovoltaic; *ZEB*, zero energy buildings.

The Cep-neutral design option cannot be achieved in Paris without an off-site PV system in addition to the PV-roof system. An additional 119 kWDC PV system is needed to be installed off-site with the 151 kWDC roof PV panels. The annual electrical energy that the additional off-site PV system generates amount to 112 MWh/yr resulting to a total PV electrical energy generated for roof and off-site system of 321 MWh/yr. The Cep-neural design option for Paris helps reduce the total primary energy source by 77.7% with an increase of 9.3% relative to the baseline design option.

The ZNE design option for Seattle and Paris requires off-site PV systems to be installed. In addition to all the EEMs part of the switchover design option, the total PV system required to achieve ZNE is 389 kWDC and 491 kWDC in Seattle and Paris, respectively, resulting in an increase of *LCC* by 28.7% and 19.8%. The electrical energy generated by the total PV system is estimated at 377 MWh/yr and 380 kWh/yr in Seattle and Paris, respectively.

The optimization results for all the US and French sites obtained for the optimal design option and the *LCC*-neutral design option are summarized in, respectively, Tables 6.14 and 6.15. Some general comments can be gleaned from the optimization results as noted next.

For all locations, using high-performance window glazing and VAV system is common to achieve optimal building designs in addition to daylighting controls specific to US locations and lower LPD specific to French sites.

- *LCC*-neutral design options can achieve an additional 15%–23% primary energy savings relative to the optimal design options resulting in overall energy savings relative to the baseline design options ranging from 43.6% to 51.6% in the US and 56.7% to 62.2% in France.

TABLE 6.14 Summary of EEMs and Performance Metrics for Optimal Design Options

Energy efficiency measure	Boulder	Albuquerque	Seattle	Strasbourg	Marseille	Paris
Roof insulation	RSI-2.64	RSI-2.64	RSI-2.64	RSI-5.28	RSI-2.64	RSI-2.64
Wall insulation	RSI-2.29	RSI-2.29	RSI-2.29	RSI-2.29	RSI-2.29	RSI-2.29
	+RSI-1.32 c.i.	+RSI-1.32 c.i.		+RSI-1.32 c.i.		+RSI-1.32 c.i.
Window-Uw (W/m^2°C)	2.5	2.5	2.5	2.5	2.5	2.5
Window-SHGC	0.39	0.39	0.39	0.39	0.39	0.39
South overhang		✓		✓		
Dimming	✓	✓	✓			
LPD (W/m^2)	10.76	10.76	10.76	8.07	8.07	8.07
Premium VAV		✓	✓	✓	✓	✓
LCC (% of baseline)	95.30	97.20	97.10	96.10	96.90	96.20
Source energy savings (%)	31.70	28.60	29.10	44.80	39.40	40.90
Source energy savings (kBtu/ft^2 yr)	53.6	45.1	42.8	60.7	45.9	54.4
RT Cep (kWh$_{EP}$ m^{-2} yr^{-1})	–	–	–	115.9	112.9	118.4

Cep, Coefficient en énergie primaire; *EEM*, energy efficiency measure; *LCC*, life-cycle cost; *RT*, Réglementation Thermique.

TABLE 6.15 Summary of EEMs and Performance Metrics for LCC-Neutral Design Options

Energy efficiency measure	Boulder	Albuquerque	Seattle	Strasbourg	Marseille	Paris
Roof insulation	RSI-5.28	RSI-5.28	RSI-5.28	RSI-5.28	RSI-5.28	RSI-5.28
Wall insulation	RSI-2.29	RSI-2.29	RSI-2.29	RSI-2.29	RSI-2.29	RSI-2.29
	+RSI-3.96 c.i.	+RSI-1.32 c.i.	+RSI-1.32 c.i.	+RSI-1.32 c.i.	+RSI-3.96 c.i.	+RSI-3.96 c.i.
Window-Uw ($W/m^2{}^\circ C$)	2.5	1.5	2.5	1.5	2.5	1.5
Window-SHGC	0.39	0.56	0.56	0.56	0.39	0.56
South overhang	✓	✓		✓		
Dimming	✓	✓	✓			
LPD (W/m^2)	10.76	10.76	8.07	8.07	8.07	8.07
Plug load (W/m^2)	5.92	5.92	5.92	5.92	5.92	5.92
Premium VAV		✓	✓	✓	✓	✓
PV size (kW)	57.8 kWDc	16.6 kWDc	–	63.8 kWDc	71.3 kWDc	52.4 kWDc
Roof area occupied by PV (%)	26	7.50	–	28	32	2
Source energy savings (%)	51.60	43.60	50.50	57.90	62.20	56.70
Source energy savings ($kBtu/ft^2\ yr$)	87.2	68.7	74.1	75.5	72.5	77.5
RT Cep ($kWh_{EP}\ m^{-2}\ yr^{-1}$)	–	–	–	88.5	109.5	67

Cep, Coefficient en énergie primaire; *LCC*, life-cycle cost; *PV*, photovoltaic; *RT*, Réglementation Thermique.

- A reduction of primary energy source by 50% can be achieved in five locations without increasing the *LCC* over the baseline design option. The exception is in Albuquerque where a reduction of 50% of total office building primary energy use requires an installation of PV panels over 22% of the building roof area resulting in an increase in *LCC* by 2.1% over baseline design option.
- Both optimal and *LCC*-neutral design options do not achieve RT 2012 requirements for Cep threshold of 110 kWh m^{-2} yr^{-1} for Paris and Strasbourg. However, the optimal design option in Marseille has easily met the Cep threshold of 132 kWh m^{-2} yr^{-1}.

- The EEM lists, *LCC* values, primary energy savings, and Cep values for various design options obtained for Marseille (French Zone H3), Strasbourg (French zone H1b), Albuquerque (ASHRAE zone 4B), and Boulder (ASHRAE zone 5B) are summarized in, respectively, Tables 6.16–6.19. A few findings can be highlighted from the results:
  - Optimal office building designs achieve 39.4% and 44.8% in primary energy savings and *LCC* reductions of 3.1% and 3.9% for, respectively, Marseille and Strasbourg. As noted in the results obtained for Paris, for both French locations, the optimal designs are barely compliant with the French standard RT 2012.
  - When all the cost-effective EEMs are considered (i.e., the switch design case) up to 51.2%, 52.0%, 40.1%, and 42.1% of primary energy savings can be achieved with an *LCC* reduction of 1.7%, 2.3%, 1.1%, and 3.2% for, respectively, Marseille, Strasbourg, Albuquerque, and Boulder.

TABLE 6.16 Summary of Characteristics of Office Building Design Options for Marseille

Design	Option	Life-cycle cost ($ *LCC*)	(% baseline)	Source energy (kBtu/ft^2 yr)	Savings (%)	Cep (kWh$_{EP}$/m^2$_{SHON}$ an)
Baseline	EEM	$11,253,183	100	116.6	0	195.3
	Windows-1	$11,156,403	99.1	109.6	6.1	185.4
	HVAC	$11,022,040	97.9	91.8	21.3	131.7
Optimal	LPD	$10,905,614	96.9	70.7	39.4	✓ 112.9
	Wall.Ins-1	$10,907,509	96.9	69.3	40.6	✓ 111.5
	Roof.Ins-1	$10,921,938	97.1	67.8	41.8	✓ 110.0
	Wall.Ins-2	$10,949,025	97.3	65.7	43.7	✓ 107.6
Switchover	Plug load	$11,060,536	98.3	56.9	51.2	✓ 107.6
LCC-neutral	PV-roof (flat)	$11,253,183	100	44.1	62.2	✓ 67.0
PV-roof	PV-roof (flat)	$11,467,681	101.9	29.8	74.4	✓ 21.5
Cep-neutral	PV-offsite (tilt)	$11,604,741	103.1	23.1	80.2	✓ 0
ZEB	PV-offsite (tilt)	$12,074,867	107.3	0	100	✓ −73.6

Cep, Coefficient en énergie primaire; *EEM*, energy efficiency measure; *HVAC*, heating, ventilation, and air conditioning; *LCC*, life-cycle cost; *PV*, photovoltaic; *ZEB*, zero energy buildings.

TABLE 6.17 Summary of Characteristics of Office Building Design Options for Strasbourg

		Life-cycle cost		Source energy		Cep
Design	Option	($ *LCC*)	(% baseline)	(kBtu/ft^2 yr)	Savings (%)	kWh_{EP}/ (m^2_{SHON} an)
Baseline	EEM	$11,528,528	100	135.4	0	231.5
	Windows-1	$11,387,087	98.8	125.2	7.5	219.4
	Wall.Ins-1	$11,352,165	98.5	122.0	9.9	214.9
	HVAC	$11,136,056	96.6	97.6	27.9	139.3
	LPD	$11,084,561	96.1	80.9	40.3	123.9
	Roof.Ins-1	$11,083,823	96.1	78.4	42.1	120.5
Optimal	Wall.Ins-2	$11,083,410	96.1	74.7	44.8	115.9
	Plug Load	$11,209,099	97.2	68.8	49.2	115.9
	Windows-2	$11,259,666	97.7	66.4	51.0	115.0
Switchover	Sun shading	$11,263,185	97.7	64.9	52.0	113.9
LCC-neutral	PV-roof (flat)	$11,528,529	100	57.0	57.9	✓ 88.5
PV-roof	PV-roof (flat)	$11,887,621	103.1	46.2	65.9	✓ 54.0
Cep-neutral	PV-offsite	$12,567,914	109.0	29.2	78.4	✓ 0
ZEB	PV-offsite	$13,741,762	119.2	0	100	✓ −93.2

Cep, Coefficient en énergie primaire; *EEM*, energy efficiency measure; *HVAC*, heating, ventilation, and air conditioning; *LCC*, life-cycle cost; *PV*, photovoltaic; *ZEB*, zero energy buildings.

- In all locations, the ZNE conditions cannot be achieved using PV panels located on the available roof area. The size of the PV system needed to achieve ZNE is 290, 491, 294, and 357 kWDC for, respectively, Marseille, Strasbourg, Albuquerque, and Boulder.

6.6.7 Sensitivity Analysis

A sensitivity analysis is conducted to assess the impact of the economical parameters on the optimization results (Deneuville and Krarti, 2015). The parameters that have been considered in the sensitivity analysis include discount rate, life-cycle period, and capital costs. For the sensitivity analysis, each parameter has been varied between a lower and an upper bound value:

TABLE 6.18 Summary of Characteristics of Office Building Design Options for Albuquerque

Design	Option	Life-cycle cost ($ *LCC*)	(% baseline)	Source energy (kBtu/ft^2 yr)	Savings (%)
Baseline	EEM	$6,807,857	100	157.8	0
	Windows-1	$6,714,754	98.6	146.2	7.4
	Sun shading	$6,679,726	98.1	140.5	10.9
	Daylighting	$6,670,899	98.0	138.9	12.0
	Wall.Ins-1	$6,627,035	97.3	130.3	17.5
Optimal	HVAC	$6,615,327	97.2	112.7	28.6
	Plug load	$6,689,122	98.3	101.0	36.0
	Roof.Ins-1	$6,705,374	98.5	98.5	37.6
Switchover	Windows-2	$6,733,736	98.9	94.5	40.1
LCC-neutral	PV-roof (tilt)	$6,807,857	100	89.1	43.6
PV-roof	PV-roof (tilt)	$7,276,489	106.9	55.0	65.2
ZEB	PV-roof (tilt)	$8,031,901	118.0	0	100

EEM, Energy efficiency measure; *HVAC*, heating, ventilation, and air conditioning; *LCC*, life-cycle cost; *PV*, photovoltaic; *ZEB*, zero energy buildings.

- Discount rate is varied from 0% to 4% based on the historic values observed for interest and inflation rates in both the US and France (Krarti, 2010).
- Life-cycle period is set between 10 years, which would account for short-term budget constraints, and 50 years, which would tend to match the entire building life.
- EEM capital costs are set to vary in a range from −50% to +50% to account for any changes in capital investments.

Little variations in the optimization results are found when discount rate and capital costs are changed. In particular, these variables have no impact on the nature and the ranking of the selected EEMs but have slightly affected the estimated life-cycle costs for various design options for the office building. However, the life-cycle period has a significant impact on the optimization results as outlined in Table 6.20 for the case where the office building is located in Paris. Fig. 6.41 illustrates the optimization path for three life-cycle period when the office building is located in Paris. In a short-term scenario of 10-year *LCC* analysis, the optimal design option includes a restricted number of EEMs compared to the 25-year *LCC* analysis, with only insulated and tinted windows, improved VAV, and lighting power density reduction. However, the *LCC*-neutral

TABLE 6.19 Summary of Characteristics of Office Building Design Options for Boulder

		Life-cycle cost		Source energy	
Design	**Option**	**($ *LCC*)**	**(% baseline)**	**(kBtu/ft^2 yr)**	**Savings (%)**
Baseline	EEM	$7,442,664	100	169.1	0
	Windows-1	$7,322,655	98.4	156.4	7.5
	Wall.Ins-1	$7,297,798	98.1	152.5	9.8
	Daylighting	$7,268,931	97.7	147.8	12.6
	HVAC	$7,100,162	95.4	119.9	29.1
Optimal	Wall.Ins-2	$7,093,276	95.3	115.5	31.7
	Sun shading	$7,094,122	95.3	115.0	32.0
	Roof.Ins-1	$7,099,622	95.4	112.0	33.8
	Windows-2	$7,107,799	95.5	108.9	35.6
Switchover	Plug load	$7,202,210	96.8	98.0	42.1
LCC-neutral	PV-roof (tilt)	$7,442,664	100	81.9	51.6
PV-roof	PV-roof (tilt)	$7,680,716	103.2	65.9	61.0
ZEB	PV-offsite (tilt)	$8,665,943	116.4	0	100

EEM, Energy efficiency measure; *HVAC*, heating, ventilation, and air conditioning; *LCC*, life-cycle cost; *PV*, photovoltaic; *ZEB*, zero energy buildings.

TABLE 6.20 Impact of Life Cycle Period on Primary Energy Use Savings for High Performance Design Options

Package	**10 years (%)**	**25 years (%)**	**50 years (%)**
Optimal	39.3	40.9	45.0
Switchover	52.4	51.9	49.6
LCC neutral	51.9	56.7	69.1
PV-roof	66.2	65.7	63.5
Zero Cep	78.2	77.7	75.9

Cep, Coefficient en énergie primaire; *LCC*, life-cycle cost; *PV*, photovoltaic.

design option features more EEMs (no PV). As expected, the high initial costs of PV reduced its attractiveness for a 10-year *LCC* analysis.

On the other hand, in a long-term scenario of 50-year *LCC* analysis, more EEMs become economically feasible and are included in the optimal design option. PV is more competitive than some EEMs under the 50-year *LCC* analysis

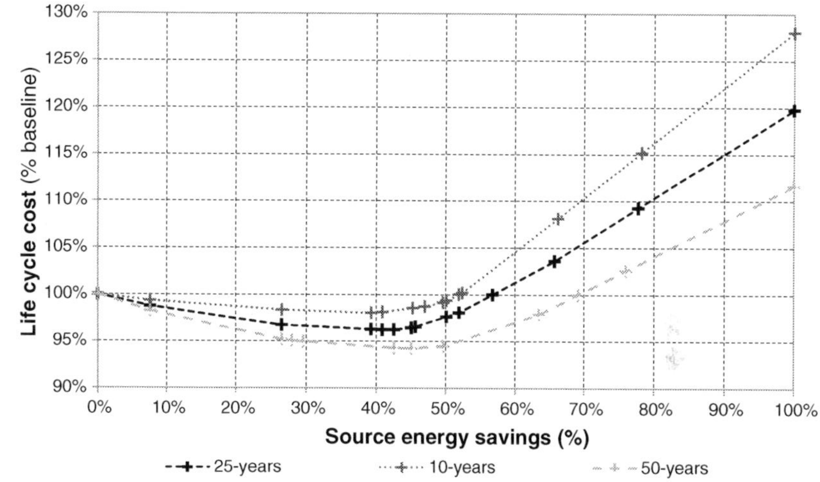

FIGURE 6.41 Optimization path for life-cycle periods when the office building is in Paris.

since it is assumed that no replacement would be required for PV modules. However, a generation reduction of 0.5% per year is assumed for the PV system during the 50-year period with maintenance costs of \$0.36/m^2 year (3.5 ¢/ft^2 year).

6.6.8 General Discussion

Based on the RT 2012 requirements, the targeted level of source of energy intensity, Cep, differs between Paris and Strasbourg (110 kWh/m^2 yr or 35 kBtu/ft^2 yr) and Marseille (132 kWh/m^2 yr or 42 kBtu/ft^2 yr). The results of the optimization analysis, however, indicate that the medium office building performs generally better in Marseille than in the other two French locations, Paris and Strasbourg. Indeed, the optimal building design in Marseille requires only three energy efficiency measures to meet the RT 2012 Cep threshold as summarized in Table 6.21. The three measures include low-SHGC windows, improved VAV system, and reduced lighting power. When the office building is located in Paris, six different upgrades are needed to reach the Cep target of 132 kWh/m^2 yr (42 kBtu/ft^2 yr). In addition to the three measures found for Marseille, higher insulation in the walls and roof as well as reduced plug load are needed to meet the Cep target of 110 kWh/m^2 yr (35 kBtu/ft^2 yr). Overhangs and PV panels have to be added to the six energy efficiency measures identified for Paris to meet RT 2012 Cep requirements for Strasbourg as indicated in Table 6.21.

The Cep target set by RT 2012 seems to be stringent for heating dominated climates such as that of Strasbourg and Paris and lenient for mild climates such as that of Marseille. Based on the results of the energy analysis performed in this study, it is proposed that the Cep targets be reversed and be slightly more

TABLE 6.21 Summary of Office Building Designs Compliant With RT 2012

	Marseille	Paris	Strasbourg
Selected EEMs	Windows $Uw = 0.44$ Btu/h ft^2°F/ SHGC = 0.39 Improved VAV LPD = 0.75 W/ft^2	Windows $Uw = 0.26$ Btu/h ft^2°F/ SHGC = 0.56 Improved VAV LPD = 0.75 W/ft^2 R-roof = R-30 R-wall: R-13 + R-22.5 c.i. Plug load = 0.55 W/ft^2	Windows $Uw = 0.26$ Btu/h ft^2°F/ SHGC = 0.56 Improved VAV R-Wall: R-13 + R-22.5 c.i. LPD = 0.75 W/ft^2 R-roof = R-30 Plug load = 0.55 W/ft^2 South overhang Rooftop PV
LLC (% baseline)	97.9%	98.1%	100%
Source energy [kBtu/ft^2 yr] (% savings)	70.7 (39.4)	64.0 (51.9%)	57.0 (57.6%)

EEMs, Energy efficiency measures.

stringent for Marseille than for Paris and Strasbourg whether or not the plug loads are considered. Specifically, Cep targets of 100 kWh/m^2 yr (31.8 kBtu/ft^2 yr) and 110 kWh/m2 yr (35 kBtu/ft^2 yr) are recommended for respectively, Marseille and Paris/Strasbourg when plug loads are not considered in the overall building energy performance following the RT 2012 approach. Moreover, Cep targets of 160 kWh/m^2 yr (51 kBtu/ft^2 yr) and 170 kWh/m^2 yr (54 kBtu/ft^2 yr) are proposed for, respectively, Marseille and Paris/Strasbourg when the plug loads are considered to account for actual building operation conditions. Indeed, it is estimated that without considering their impact on heating, cooling or ventilation, the plug loads are roughly equivalent to 60 kWh/m^2 yr (19 kBtu/ft^2 yr).

While in the US, the concept of ZNE building is clearly defined and is consistent with the performance compliance approach using ASHRAE energy efficiency standards, the concept of positive energy buildings or BEPOS is not well aligned with RT 2012 requirements. Indeed, BEPOS is defined as a building that produces more renewable energy than its primary energy needs and is therefore equivalent to the ZNE building defined in the US. In France, BEPOS is expected to be the minimum standard for buildings by 2020. However, the compliance approach to evaluate RT 2012 requirements is not based on actual building performance. Consequently, alternative approaches other than RT 2012 are needed to develop effective design guidelines for ZNE buildings.

The results of the energy analysis conducted in this study have indicated that office buildings with zero-Cep level do not achieve ZNE and subsequently BEPOS status in all three French locations. Indeed, buildings with zero-Cep level fall short of reaching ZNE design by 72–91 kWh/m^2 yr (23–29 kBtu/ft^2 yr) but achieving 77%–80% primary energy savings relative to the baseline design option considered for the prototypical office building considered in the analysis.

6.7 SUMMARY

In this chapter, a sequential selection methodology is introduced to select an optimal package of energy efficiency measures that minimize life-cycle cost for designing and retrofitting individual commercial and residential buildings. The methodology is based on a sequential search optimization technique and can be utilized to find the path toward ZNE buildings. In addition to the minimum-cost and ZNE designs and retrofits, the methodology can provides insights on how to improve progressively the energy performance of buildings while reducing life-cycle costs. Based on a detailed energy simulation analysis, it is found that up to 50% of energy use savings can be obtained when minimum life-cycle costing is considered for building design and retrofit applications. Even higher energy savings can be achieved when neural life-cycle costs or ZNE designs and retrofits are considered. In these cases, renewable energy systems are typically required including solar domestic hot water and PV systems.

PROBLEMS

5.1 A homeowner is considering adding a sunspace to his home. The project is carried out during 4 years and has to spend \$20,000 the first year, \$15,000 the second year, \$10,000 the third year, and \$5000 the fourth year. Determine how much money the homeowner needs to set aside to pay for the sunspace addition project:

(a) If the interest rate is 4% per year, compounded annually.

(b) If the interest rate is 8% per year, compounded annually.

5.2 You want to buy a new home at a cost of \$250,000. Determine how much money you should pay every year for 20 years to pay for the house based on equal annual payments:

(a) If the interest rate is 5% per year, compounded annually.

(b) If the interest rate is 10% per year, compounded annually.

5.3 An energy audit of a residential heating system reveals that the boiler-burner efficiency is only 60%. In addition, the energy audit showed that an electric water heater is used. The house uses annually 1500 gallons of oil at a cost of \$1.40/gallon. The total electric bill averaged \$84.50 per month for an average monthly consumption of 818 kWh. Of this total, about 35% was for domestic water heating. It is suggested to the owner of the house to equip the boiler with a tank-less domestic water heater replacing the existing electric water heater. The existing water heater is 12 years old and has no resale value and little expected life. The owner of the house is expected to spend \$300 within the coming year to replace the electric water heater. The cost of the tank-less heater is \$400.

To improve heating efficiency, it is suggested to

(i) Replace the existing burner with a new one (this burner replacement will improve efficiency to 65%) costing $520.

(ii) Replace the entire heating plant with higher efficiency (85%) with a cost of $2000.

For two discount rates (5% and 10%), provide the *LCC* analysis of the following options and make the appropriate recommendations:

- Keep the boiler-burner and replace the electric water heater with like system.
- Replace burner and electric water heater with like systems.
- Replace boiler-burner with efficient boiler and replace electric heater with like system.
- Replace existing boiler-burner and electric water heater with new boiler/tank-less domestic water heater.

Note: State all the assumptions made in your calculations.

5.4 An energy audit indicates that a home consumption is 20,000 kWh/yr. By using Energy Star appliances and CFLs, a savings of 20% can be achieved for the electrical energy use. The cost for the home retrofit is about $15,000. Assuming that the average energy charge is $0.09/kWh, is the expenditure justified based on a minimum rate of return of 18% before taxes. Assume a 20-year life cycle and use the present worth, annual cost, and rate of return methods.

5.5 An electrical energy audit indicates that lighting consumes 5000 kWh per year. By using more efficient lighting fixtures, electrical energy savings of 15% can be achieved. The additional cost for these energy efficient lighting fixtures is about $900. Assuming that the average energy charge is $0.08/kWh, is the expenditure justified based on a minimum rate of return (i.e., a discount rate) of 7.0%. Assume a 10-year life cycle and use the present worth, annual cost, and rate of return methods.

5.6 A residential building is heated with a 78% efficient gas-fired furnace (costing $1600). The annual heating load is estimated at 160 MMBtu. The cost of gas is $6.00/MMBtu. For the discount rate of 5% and 10-year cycle, determine the following:

(a) The life cycle of the existing heating system.

(b) If it is worth considering a 90% efficient gas-fired furnace that costs $2800.

(c) The variation of the life-cycle cost savings between the two systems [the existing system and the proposed in (b)] versus the life of the system for life cycles ranging from 3 to 20 years. Conclude.

5.7 An air conditioner (AC) consumes 7500 kWh annually with an overall efficiency of COP = 3.0. If this AC is replaced by a more energy efficient AC (COP = 3.7) at a cost of $3500.

(a) Determine the simple payback period of the chiller replacement.
(b) If the expected life of the old chiller is 15 years, is it cost-effective to replace the chiller in an economy with a discount rate of 7%?

5.8 Two air conditioner systems are proposed to cool a house. Each AC has a rated capacity of 5 tons and is expected to operate 500 full-load equivalent hours per year. Air conditioner A has a standard efficiency with a COP of 2.8 and costs $1600 while AC B is more efficient with a COP of 3.5 at a cost of $2750. The electricity cost is estimated to be $0.09/kWh. For a lifetime of 20 years and a discount rate of 5%:
(a) Estimate the payback time for using air conditioner B instead of A.
(b) Calculate the life cycle costs for both air conditioners. Conclude.
(c) Estimate the rate of return for using air conditioner B instead of A.
(d) Determine the highest cost for air conditioner B at which air conditioner A is more competitive.
(e) Determine the cut-off electricity price at which both air conditioners A and B have the same life-cycle cost.

Assume that the average cost of electricity is $0.08/kWh.

REFERENCES

ADEME, 2005. Stratégie Utilisation rationnelle de l'énergie. Agence De l'Environnement et de la Maîtrise de l'Energie (ADEME), Paris, France.

Alessandrini, J.-M., 2008a. Note de synthèse du descriptif de bâtiments basse consommation. Centre Scientifique et Thermique des Bâtiments, ESE/DE/PEB-08. 156N, Marne-La-Vallée, France.

Alessandrini, J.-M., 2008b. Description de bâtiments exemplaires à basse consommation réalisés en France FICHE-BBC – Rapport Final. Centre Scientifique et Thermique des Bâtiments, ESE/DE/PEB-08.185R, Marne-La-Vallée, France.

Allocca, C., Chen, Q., Glicksman, L., 2003. Design analysis of single-sided natural ventilation. Energy Build. 35 (8), 785–795.

Axley, J.W., 2001. Application of Natural Ventilation for U.S. Commercial Buildings Climate Suitability Design Strategies and Methods Modeling Studies. GCR-01-820, National Institute of Standards and Technology, Gaithersburg, MD.

Anderlind, G., 1985. Energy consumption due to air infiltration. In: Proceedings of the 3rd ASHRAE/DOE/BTECC Conference on Thermal Performance of the Exterior Envelope of Buildings. Clearwater Beach, FL. .

ASHRAE, 2004. ANSI/ASHRAE/IESNA Standard 90.1-2004. Energy Standard for Buildings Except Low-Rise Residential Buildings. American Society of Heating, Refrigerating and Air-Conditioning Engineers, Atlanta, GA.

ASHRAE, 2008. Handbook of HVAC Systems and Equipment. American Society of Heating, refrigerating and Air-Conditioning Engineers, Inc, Atlanta, GA.

Balcomb, J.D., Barley, D., McFarland, R., Perry, J., Wray, W., Noll, S., 1980. Passive Solar Design Handbook, Volume Two: Passive Solar Design Analysis DOE/CS-0127/2. U.S. Department of Energy, Washington, DC.

Berggren, B., Hall, M., Wall, M., 2013. LCE analysis of buildings – taking the step towards net zero energy buildings. Energy Build. 62, 381–390.

Bichiou, Y., Krarti, M., 2011. Optimization of envelope and HVAC systems selection for residential buildings. Energy Build. 43 (12), 3373–3382.

Butti, K., Perlin, J., 1980. A Golden Thread: 2500 Years of Solar Architecture and Technology. Van Nostrand Reinhold Company, Palo Alto.

CADDET, 1999. Radiant Heating Panels Save Energy in Homes. Centre for the Analysis and Dissemination of Demonstrated Energy Technologies, US-1999-502, 337.

Caldas, L.G., Norford, L.K., 2003. Genetic algorithms for optimization of building envelopes and the design and control of HVAC systems. ASME J. Solar Energy Eng. 125, 343–351.

Catarina, O., Illouz, S., 2009. Retour d'expérience de bâtiments certifiés HQE: Dynamiser l'efficacité énergétique des gestionnaires de patrimoine du secteur privé. Agence De l'Environnement et de la Maîtrise de l'Energie, Paris, France.

Christensen, C., Barker, G., Horowitz, S., 2004. A sequential search technique for identifying optimal building designs on the path to zero net energy. In: Proceedings of the Solar 2004. Portland, OR. American Solar Energy Society.

Cockroft, J.P., Robertson, P., 1976. Ventilation of an enclosure through a single opening. Build. Environ. 11 (1), 29–35.

CSTB, 2011. Méthode de calcul Th-BCE, Réglementation thermique du bâtiment 2012 (RT 2012). Centre Scientifique et Thermique des Bâtiments, Marne-La-Vallée, France.

De Boissieu, C., 2006. Rapport du Groupe de travail "Division par quatre des émissions de gaz à effet de serre de la France à l'horizon 2050". Ministère de l'Economie, des Finances, de l'Industrie, Ministère de l'Ecologie et du Développement Durable, Paris, France, ISBN 2-11-006280-0.

DEG, 1994. Coachella Valley Project: La Paloma Site Final Design Report. Davis Energy Group, Davis, CA.

De Dear, R.J., Brager, G.S., 2002. Thermal comfort in naturally ventilated buildings: revisions to ASHRAE standard 55. Energy Build. 34 (6), 549–561.

Deneuville, A., Krarti, M., 2015. Comparative analysis of optimal designs for French and US office buildings. Energy Build. 93, 332–344.

DOE, 2005. A Look at Residential Energy Consumption in 1997. DOE/EIA-0632 (97). U.S. Department of Energy, Washington, DC.

DOE, 2009a. Energy Star Qualified Water Heaters-Partner Guide. US Department of Energy, Washington, DC.

DOE, 2009b. Water Heater Market Profile. US Department of Energy, Washington, DC.

DOE, 2012. Commercial Buildings Initiative. U.S. Department of Energy, http://www1.eere.energy.gov/buildings/commercial_initiative/new_construction.html.

EIA, 2005. Commercial Buildings Energy Consumption Survey 2003, Energy Information Administration of U.S. Department of Energy, Washington, DC, http://www.eia.doe.gov/emeu/cbecs/contents.html.

EIA, 2011. State Energy Consumption Database. Energy Information Administration of U.S. Department of Energy, Washington, DC, http://www.eia.gov/state/seds/.

Emmerich, S.J., 2006. Simulated performance of natural and hybrid ventilation systems in an office building. HVAC&R Res. 12 (4), 975–1004.

Ezzeldin, S., Rees, S., Cook, M., 2009. Performance of mixed-mode cooling strategies for office buildings in arid climates. In: Proceedings of the 11th International IBPSA Conference. Glasgow, Scotland.

Fong, K., Hanby, V., Chowa, T., 2006. HVAC system optimization for energy management by evolutionary programming. Energy Build. 38, 220–231.

FSEC, 2007. EnergyGauge Pro Software. Florida Solar Energy Center, Cocoa, FA, http://energygauge.com/FlaRes/features/pro.htm.

Ghiaus, C., Allard, F., Santamouris, M., Georgakis, C., Nicol, F., 2006. Urban environment influence on natural ventilation potential. Build. Environ. 41 (4), 395–406.

Goswami, Y., Kreith, F., Kreider, J., 2000. Principles of Solar Engineering. Taylor & Francis Group, New York.

Haase, M., Amato, A., 2009. An investigation of the potential for natural ventilation and building orientation to achieve thermal comfort in warm and humid climates. Solar Energy 83 (3), 389–399.

Hernandez, P., Kenny, P., 2010. From net energy to zero energy buildings: defining life cycle zero energy buildings (LC-ZEB). Energy Build. 42 (6), 815–821.

Huang, W., Lam, H.N., 1997. Using genetic algorithms to optimize controller parameters for HVAC systems. Energy Build. 26 (3), 277–282.

Ihm, P., Krarti, M., 2012. Design optimization of energy efficient residential buildings in Tunisia. Build. Environ. 58, 81–90.

Ihm, P., Krarti, M., 2005. Controlling hydronic radiant heating and cooling systems: optimal control strategies for heated radiant floor systems. ASHRAE Trans 111 (1), 535–546.

Krarti, M., 1994. Effect of air flow on heat transfer in walls. ASME J. Solar Energy Eng. 116 (1), 35–42.

Krarti, M., Lopez-Alonzo, C., Claridge, D.E., Kreider, J.F., 1995. Analytical model to predict annual soil surface temperature variation. ASME Trans. J. Solar Energy Eng. 117 (2), 91–99.

Krarti, M., Kreider, J., 1996. Analytical model for heat transfer in an underground air tunnel. Energy Convers. Manage. J. 37 (10), 1561–1574.

Krarti, M., 2010. Energy Audit of Building Systems: An Engineering Approach, second ed. Taylor & Francis Publishing/CRC Press, Boca Raton, FL.

Kusuda, T., Achenbach, P.R., 1965. Earth temperatures and thermal diffusivity at selected stations in the United States. ASHRAE Trans. 71 (1), 61–70.

Leach, M., Lobato, C., Hirsch, A., Pless, S., Torcellini, P., 2010. Technical Support Document: Strategies for 50% Energy Savings in Large Office Buildings. National Renewable Energy Laboratory, NREL/TP-550-49213, Golden, CO.

Legifrance, 2010. Arrêté du 26 Octobre 2010 relatif aux caractéristiques thermiques et aux exigences de performance énergétique des bâtiments nouveaux et des parties nouvelles de bâtiments. Ministère de l'écologie, de l'énergie, du développement durable et de la mer, en charge des technologies vertes et des négociations sur le climat. NOR: DEVU1026270A, Paris, France.

Meier, A.K., 1982. Supply Curves of Conserved Energy. Lawrence Berkeley National Laboratory, LBL-Report No.14686, Berkeley, CA.

Noris, F., Musall, E., Salom, J., Berggren, B., Østergaard Jensen, S., Lindberg, K., Sartori, I., 2014. Implications of weighting factors on technology preference in net zero energy buildings. Energy Build. 82, 250–260.

Op't Veld, P., 2008. Introduction to EC RESHYVENT–EU cluster project on demand controlled hybrid ventilation for residential buildings. Build. Environ. 43 (8), 1342–1349.

Orselli, J., 2007. Economies et substitutions d'énergie dans les bâtiments. Commission Urbanisme et habitat. Commission Energie et changement climatique. Académie des Technologies, Paris, France.

Phaff, J.C., de Gids, W.F., Ton, D.V., van der Ree, Schijndel, L.L.M., 1908. The Ventilation of Buildings: Investigation of the Consequences of Opening One Window on the Internal Climate of a Room. TNO Institute of Environmental Hygiene and Health, Delft, The Netherlands, Report C 448.

Ramlow, B., Nusz, B., 2006. Solar Water Heating: A Comprehensive Guide to Solar Water Heating and Space Heating Systems. New Society Publishers, Gabriola Island.

Rieser, T., Sidler, O., 2010. Coûts des premières rénovations « basse consommation » en France Perspectives. Technical Report by Enertech, www.enertech.fr.

Roth, K., Dieckmann, J., Brodrick, J., 2006. Natural and hybrid ventilation. ASHRAE J. 48 (6), H37–H39.

Sartori, A., Napolitano, K., Voss, K., 2012. Net zero energy buildings: a consistent definition framework. Energy Build. 48, 220–232.

Seppanen, O., Fisk, W.J., 2002. Association of ventilation system type with SBS symptoms in office workers. Indoor Air 12 (2), 98–106.

Seo, D., Krarti, M., 2010. Energy Analysis of Cecil Residential Complex (Final Report). Building Systems Program, University of Colorado, Boulder, CO.

Solupe, M., Krarti, M., 2014. Assessment of air infiltration heat recovery and its impact on energy consumption for residential buildings. Energy Convers. Manage. 78, 316–323.

Siegenthaler, J., 2008. Active Solar Heating: New Opportunities for Hydronics Pros (Webinar).

Simmonds, P., 1993. Thermal comfort and optimal energy use. ASHRAE Trans 99 (1), 1037–1048.

SRCC, 1994. Solar Rating and Certification Corporation, Methodology for Determining the Thermal Performance Rating for Solar Collectors. Cocoa.

Torcellini, P., Pless, S., 2010. Net Zero Energy Buildings: A Classification System Based on Renewable Energy Supply Options. National Renewable Energy Laboratory, Golden, CO, NREL/TP-550-44586.

Thornton, B.A., Wang, W., Lane, M.D., Rosenberg, M.I., Liu, B., 2009. Technical Support Document: 50% Energy Savings Design Technology Packages for Medium Office Buildings. Pacific Northwest National Laboratory, Richland, WA, PNNL-16031.

Tuhus-Dubrow, D., Krarti, M., 2009. Comparative analysis of optimization approaches to design building envelope for residential buildings. ASHRAE Trans. 115 (2), 205–219.

Tuhus-Dubrow, D., Krarti, M., 2010. Genetic algorithm based approach to optimize building envelope design for residential buildings. Build. Environ. 45 (7), 1574–1584.

Walton, G.N., Dols, S., 2006. CONTAM 2.4 User Guide and Program Documentation. NIST, Gaithersburg, MD.

Wanga, W., Zmeureanu, R., Rivard, H., 2005. Applying multi-objective genetic algorithms in green building design optimization. Build. Environ. 40, 1512–1525.

Warren, P.R., 1978. Ventilation through openings on the one wall only. Energy Conserv. Heat. Cool. Ventil. Build. 1 (1), 186–206.

Watson, R.D., Chapman, K.S., 2002. Radiant Heating and Cooling Handbook. McGraw-Hill, New York, NY.

WBCSD, 2009. Energy Efficiency in Buildings: Transforming the Market. Report by the World Business Council for Sustainable Development, Washington, DC.

Weiss, W., 2003. Solar Heating Systems for Houses: A Design Handbook for Solar Combisystem. James & James (Science Publishers) Ltd, London, UK.

Weiss, W., Bergmann, I., Faninger, G., 2006. Solar Heat Worldwide: Markets and Contribution to the Energy Supply 2006. IEA, Gleisdorf.

Wetter, M., 2004. GenOpt®, Generic Optimization Program. Lawrence Berkeley National Laboratory, Berkeley, CA, http://gundog.lbl.gov/GO/download/documentation.pdf.

Wright, J.A., Loosemore, H.A., Farmani, R., 2002. Optimization of building thermal design and control by multi-criterion genetic algorithm. Energy Build. 43, 959–972.

Yamanaka, T., Kotani, H., Iwamoto, K., Kato, M., 2006. Natural, wind-forced ventilation caused by turbulence in a room with a single opening. Int. J. Ventil. 5 (1), 179–187.

Chapter 7

Integrated Design of Communities

Chapter outline

Optimal Design and Retrofit of Energy Efficient Buildings, Communities, and Urban Centers.
http://dx.doi.org/10.1016/B978-0-12-849869-9.00007-7

7.1 INTRODUCTION

Supplying cost-effective and sustainable energy to buildings and communities is one of the most important goals of almost all nations, governments, and policymakers. In particular, the United Nations (UN) initiative for sustainable energy for all (SE4All) was initiated in 2015 to secure access to affordable, reliable, and sustainable energy for all communities worldwide by 2030 (SE4All, 2017). It is estimated that currently over 1.3 billion people are still without access to utility grid and energy services (Akinyele and Rayudu, 2016). The three objectives of the SE4All initiative are the following: (1) providing universal access to modern energy services, (2) doubling energy efficiency, and (3) double the share of renewable energy in the global energy mix (SE4All, 2017). Thus, the integration of energy efficiency and renewable energy technologies are key to develop sustainable communities. Several aspects should be considered when designing sustainable communities including technical, social, environmental, and economic. In this chapter, the design guidelines of cost-effective sustainable power systems for off-grid and grid-connected communities are presented with a special emphasis on distributed generation technologies combining renewable and nonrenewable systems.

Several technologies are available to generate electricity to serve the energy needs for buildings and communities. These technologies include the conventional fuel-based generators, nuclear reactors, and renewable energy technologies. Table 7.1 compares the capital and operating costs of various power plant capacities and technologies based on EIA estimates (EIA, 2013). As indicated in Table 7.1, while the capital and operation and maintenance (O&M) costs for hydroelectric and wind are competitive compared to those of fuel gas–based generating technologies, other generating technologies remain expensive to install and operate. Table 7.1 also indicates that most fuel-based generating technologies can be utilized for combined heat and power (CHP) production including fuel cells whose capital costs are still high. CHP systems offer energy efficient alternative options to serve both the electricity and thermal needs of buildings and should be considered especially for large facilities such as hospitals and university campuses.

As noted in Chapter 5, the cost-effectiveness of various generating technologies can be assessed through a wide range of economic analysis methods. A recent EIA evaluation has compared the levelized cost of energy or LCOE (refer to Chapter 5) for a wide range of generating technologies in the United States. Table 7.2 presents a summary of the total and specific components of LCOE values as well as capacity factors for utility scale US power plants listed in Table 7.1. Note that the total LCOE values listed in Table 7.2 include the

TABLE 7.1 Capital and O&M Cost Estimates for US Generating Power Plants (EIA, 2013)

Fuel/power plant type	Electrical capacity (MW)	Heat output rate (Btu/kWh)	Capital costs ($/kW)	Fixed O&M cost ($/kW·yr)	Variable O&M ($/kW·yr)
Coal					
Pulverized coal (single unit)	650	8,800	$3,246	$37.80	$4.47
Integrated gasification Combine cycle with carbon Capture	520	10,700	$6,599	$72.83	$8.45
Natural gas					
Conventional Combined Cycle	620	7,050	$917	$13.17	$3.60
Advanced Combustion Turbine	210	9,750	$676	$7.04	$10.37
Fuel cells	10	9,500	$7,108	$0.00	$43.00
Uranium					
Dual unit nuclear reactor	2,234	N/A	$5,530	$93.28	$2.14
Biomass					
Combined cycle	20	12,350	$8,180	$356.07	$17.49
Wind					
Onshore turbine	100	N/A	$2,213	$39.55	$0.00
Offshore turbine	400	N/A	$6,230	$74.00	$0.00
Solar					
Solar thermal (CSP)	100	N/A	$5,067	$67.26	$0.00
Photovoltaic (PV)	150	N/A	$3873	$24.69	$0.00
Geothermal					
Dual flash	50	N/A	$8312	$132.00	$0.00
Hydroelectric					
Conventional	500	N/A	$2939	$14.13	$0.00
Pumped storage	200	N/A	$5288	$18.00	$0.00

TABLE 7.2 Estimates for Average LCOE Values and Capacity Factors for US Generating Power Plants (EIA, 2015)

Fuel/power plant type	LCOE capital costs ($/MWh)	LCOE fixed O&M costs ($/MWh)	LCOE variable O&M costs ($/MWh)	Total system LCOE ($/MWh)	Capacity factor (%)
Coal					
Pulverized coal (single unit)	60.4	4.2	29.4	95.1	85
Integrated gasification Combine cycle with carbon capture	97.3	9.8	36.1	144.4	85
Natural gas					
Conventional combined cycle	14.4	1.7	57.8	75.2	87
Advanced combustion turbine	27.8	2.7	79.6	113.5	30
Uranium					
Dual unit nuclear reactor	70.1	11.8	12.2	95.2	90
Biomass					
Combined cycle	47.1	14.5	37.6	100.2	83
Wind					
Onshore turbine	57.7	12.8	0.0	73.6	36
Offshore turbine	168.6	22.5	0.0	196.9	38
Solar					
Solar thermal (CSP)	191.6	42.1	0.0	239.7	20
Photovoltaic (PV)	109.8	11.4	0.0	125.3	25
Geothermal					
Dual flash	34.1	12.3	0.0	47.8	92
Hydroelectric					
With seasonal storage	70.7	3.9	7.0	83.5	54

transmission costs ranging from \$5.8/MWh for offshore wind farms to \$1.2/MWh for conventional fuel-based power plants. As noted in Table 7.2, renewable energy power plants have typically low capacity factors and are considered as nondispatchable systems and thus cannot be relied on as emergency systems for buildings unless combined with storage systems (i.e., electrical batteries).

In this chapter, renewable and fuel-based generation technologies are presented to provide energy needs to buildings. First, the basic design approach of sustainable power systems for grid-connected and off-grid communities is presented. Then, the design structure of microgrids is outlined as an efficient and cost-effective alternative to the main grid to power communities. Then, overview and analysis approaches of solar and wind generation technologies are presented in detail. In addition, a detailed description is provided for fuel-based generation and cogeneration technologies that can be used to generate power for communities. The integrative analysis approach of renewable energy to design net zero-energy buildings and carbon neutral communities is highlighted using two case studies outlined at the end of this chapter.

7.2 GENERAL DESIGN APPROACH

Several studies have been reported to estimate the potential benefits of using renewable energy technologies and hybrid systems to design power systems to meet the need of communities. However, most of the studies consider standalone hybrid systems. A set of best practices and approaches for designing power systems for remote communities should be used to improve their overall reliability and sustainability (Akinyele and Rayudu, 2016). Several studies for hybrid power systems have been reported for individual households or remote residential communities and rural villages worldwide (Adaramola et al., 2014; Bhattacharyya, 2015; Domenech et al., 2014; Grande et al., 2015; Kanase-Patil et al., 2010; Li et al., 2013; Posadillo and Luque, 2008; Ravindra and Iyer, 2014; Rezzouk and Mellit, 2015).

Fig. 7.1 outlines the main phases of the analysis approach as well as some of the simulation tools that can be utilized to perform the tasks required to determine the design specifications for both building energy systems and electrical power generation technologies suitable for large communities. In particular, three phases are typically considered when designing hybrid power systems for communities including data collection (phase 1), energy analysis of both energy efficiency technologies and power systems (phase 2), and economic analysis for making decisions on the best design options (phase 3).

As part of the first phase, specifications and data are collected to define the size of the community and the energy efficiency standards that buildings within the community have to meet. Moreover, long-term weather data are obtained to perform building energy analysis and assessment of available renewable energy resources. As a second phase of the analysis approach, a series of detailed energy analysis is carried out using data collected in phase 1 to determine the set of

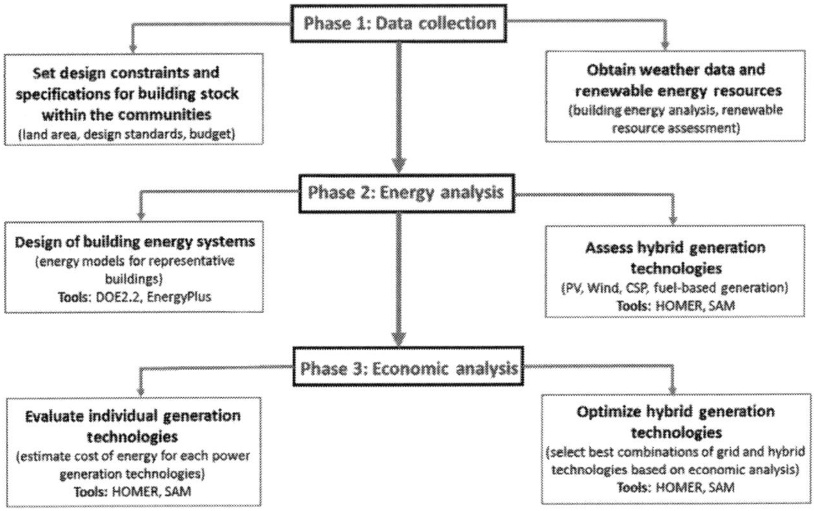

FIGURE 7.1 Flowchart of the design approach for hybrid power systems to serve communities.

energy efficiency measures that can be integrated with the design specifications of the residential buildings to meet performance targets set by energy conservation codes or cost optimizations. In addition, assessment of the renewable energy technologies is carried out as part of phase 2 to compare their suitability to meet the electricity demands for the residential communities. Finally, economic and environmental benefits of hybrid generation systems including renewable energy and fuel-powered generators are evaluated and compared relative to a grid-only option considered as the benchmark. Several simulation analysis tools are considered in the analysis. In particular, DOE2.2 (eQUEST, 2015) and EnergyPlus (2017), whole-building energy simulation tools, can be utilized to determine the best energy efficiency measures to reduce the energy demand of all buildings located within the communities to achieve desired targets such as cost-optimal designs or zero-net energy designs. Moreover, hourly simulation analysis tools such as HOMER (2016) and SAM (2016) can be used to evaluate the performance of the grid-connected hybrid generation systems.

In the following sections, several distribution generation technologies are presented including the required resources, analysis approaches, and performance metrics. Two case studies are outlined in the last two sections of this chapter to illustrate the design approach outlined in Fig. 7.1, which can be used to design hybrid power systems for grid-connected and remote communities.

7.3 RENEWABLE POWER GENERATION SYSTEMS

In this section, renewable generation systems using wind and solar recourses are discussed since they are commonly considered for supplying power to microgrids and remote communities.

7.3.1 Wind Turbines

Wind associated with the movement of air due to pressure differential harness a significant amount of energy that can be partially captured using turbines. Specifically, wind turbines can use the wind to induce rotation of large blades. The mechanical power associated with the blades' rotation is then converted to electricity using a generator. Fig. 7.2 illustrates the main components of a turbine including the foundation, tower, rotor blades, the bull, and the nacelle.

Wind turbines are generally categorized into two configurations depending on the axis of rotation: (1) horizontal-axis or HAWT and (2) vertical-axis or VAWT. The HAWTs are common and rely on the prevailing wind direction to generate power. As illustrated in Fig. 7.2, HAWTs have typically three blades facing the wind direction. Most of the wind turbines installed in the United States are HAWTs with electrical power generating capacities ranging from 100 kW to 5 MW. VAWTs are less efficient than HAWTs even though they can generate power from multidirectional wind speeds. Structural specifications required to ensure rigid vertical shaft limit the height and capacity ranges for VAWTs

The power generated by an air mass flowrate, $\dot{m}$, at a wind speed, V, can be estimated from the kinetic energy:

$$P_{wind} = \frac{1}{2}\dot{m}V^2 \tag{7.1}$$

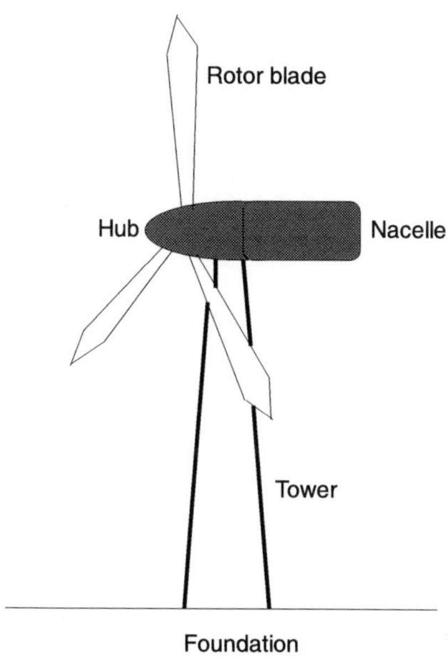

FIGURE 7.2 Typical components of a wind turbine.

The mass flowrate can be estimated based on the air density, ρ, the wind speed, V, and the area swept by the rotor blades, A:

$$\dot{m} = \rho VA \tag{7.2}$$

Thus, the wind power generated can be estimated as follows:

$$P_{wind} = \frac{1}{2} \rho A V^3 \tag{7.3}$$

Table 7.3 lists the magnitude of the wind power per unit of rotor blade area at various velocities at sea level and 1600 m altitude for ambient temperatures of 15°C and 30°C. It is clear that that wind velocity affects significantly the wind power. Typically, wind power of at least 10 m/s is desirable to harness high power densities. The altitude can affect slightly the amount of wind power that can be collected while the ambient temperature has limited impact.

Fig. 7.3 illustrates a map that provides the annual average wind velocity at 100 m for all continental US locations (NREL, 2016a). The wind speeds are estimated based on data obtained for the 1998–2015 period. The map in Fig. 7.3 indicates that the wind speed can exceed 10 m/s in the middle-section of the United States. Therefore, regions in North and South Dakota, Nebraska, Kansas, Oklahoma, and Texas, have a significant potential for generating cost-effective electricity using wind turbines.

TABLE 7.3 Wind Power Density for Various Wind Speed and Ambient Temperatures at Sea Level and Altitude

	Wind power (W/m²)			
	Sea level		1600-m altitude	
Wind speed (m/s)	T = 15°C	T = 30°C	T = 15°C	T = 30°C
2	5	5	4	4
4	39	37	32	31
6	132	126	110	104
8	314	298	260	247
10	613	582	507	482
12	1058	1006	876	833
14	1681	1598	1391	1322
16	2509	2385	2077	1974
18	3572	3395	2957	2810
20	4900	4658	4056	3855

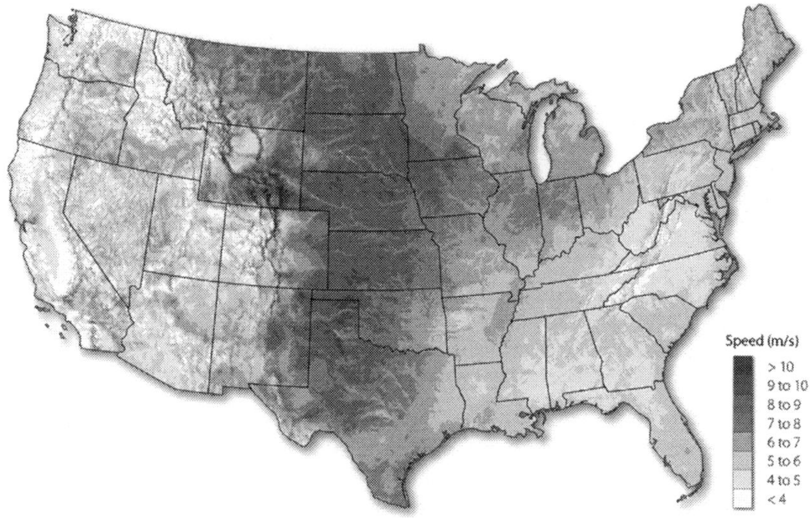

FIGURE 7.3 US map for wind resources at 100-m altitude (NREL, 2016a).

The power, $P_{Turbine}$, that any turbine can harvest is a fraction of the available wind power, P_{wind}:

$$P_{Turbine} = C_P P_{wind} \tag{7.4}$$

The maximum value of any turbine power coefficient C_P can be determine using the Betz's law as outlined in the following section.

Fig. 7.4 illustrates the variation of the wind speed as it passes through wind rotor blades. The undistributed wind speed, V, is reduced by a fraction, a, as it goes through the rotor blades, and is further decreased by the fraction, $2a$, after the passage through the turbine. Indeed, the wind speed within the rotor blades is equal to the average between upstream and downstream wind speed values by conservation of momentum.

The power collected by the turbine is the difference between the kinetic power obtained upstream and downstream:

$$P_{Turbine} = \frac{1}{2}\dot{m}V^2 - \frac{1}{2}\dot{m}(V - 2aV)^2 \tag{7.5}$$

The mass flowrate is estimated at the turbine disk where the wind velocity is $V(1 - a)$:

$$\dot{m} = \rho V(1 - a)A \tag{7.6}$$

After combining Eqs. (7.5) and (7.6), it can be shown that

$$P_{Turbine} = \frac{1}{2}\rho A V^3 (4a - 4a^2)(1 - a) = 4a(1 - a)^2 P_{wind} \tag{7.7}$$

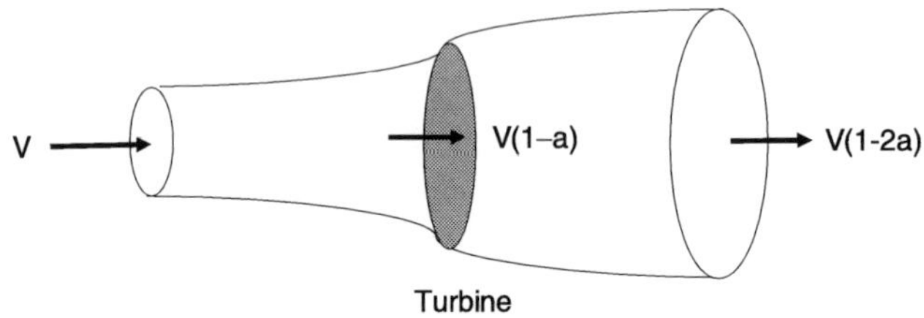

FIGURE 7.4 The wind speed variation upstream and downstream an ideal turbine.

Thus, the power coefficient, C_P, introduced in Eq. (7.4), can be estimated as

$$C_P = 4a(1-a)^2 \tag{7.8}$$

The variation of C_P as function of the fraction, a, is shown in Fig. 7.5. It is clear that the maximum value of C_P is 0.59, which occurs when $a = 1/3$.

The wind speed, V, generally increases with the height, z, depending on the terrain topography according to either the power law or the logarithm law as outlined by, respectively, Eqs. (7.9) and (7.10):

$$V(z) = V_0\left(\frac{z}{z_0}\right)^a \tag{7.9}$$

where V_o is the wind speed measured at the reference height z_0 typically 10 m. The coefficient, a, depends on the roughness of the terrain. For instance, $a = 0.10$ for water surfaces, $a = 0.20$ for grassy terrains, $a = 0.25$ for trees, and $a = 0.40$ for tall buildings.

$$V(z) = C_K V_0 Ln\left(\frac{z}{z_0}\right) \tag{7.10}$$

where the Von Karman constant, C_K, depends on the type of the terrain. Typically, $C_K = 0.42$.

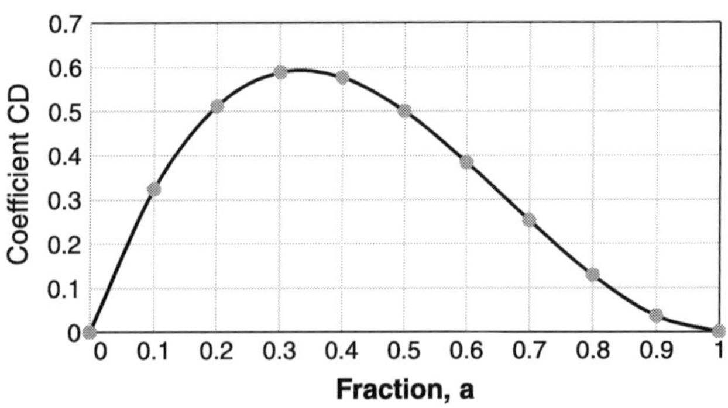

FIGURE 7.5 Variation of the power coefficient, C_P, as a function of the fraction a.

In several sites, measured wind data are not available. In these sites, Weibull frequency distribution is typically used to determine the probability that the wind speed is equal to a specific value V. The Weibull probability distribution function is expressed as follows:

$$p(V) = \left(\frac{K}{C}\right)\left(\frac{V}{C}\right)^{K-1} exp\left[-\left(\frac{V}{C}\right)^{K}\right]\left\{-\left(\frac{V}{C}\right)^{K}\right\} \qquad (7.11)$$

where K is the Weibull shape factor and is typically set to 2 m/s and C is the scale parameter that depends on the annual average wind speed, $\overline{V}$:

$$C = \frac{\overline{V}}{\Gamma\left(1+\frac{1}{K}\right)} \qquad (7.12)$$

with the function, Γ, defined as

$$\Gamma(x) = 3.0867 - 4.4289x + 3.3334\,x^2 - 1.1629x^3 + 0.1712x^4 \qquad (7.13)$$

Fig. 7.6 illustrates the Weibull probability distribution functions for shape factor, K, of 2 m/s and three annual average wind speed values, $\overline{V}$, of 2, 6, and 12 m/s. Fig. 7.7 shows the Weibull cumulative distribution functions for the same shape factor and annual average wind speed values.

Fig. 7.8 shows a typical power curve for a wind turbine and its operation regions. The theoretical wind power obtained from Eq. (7.3) is also provided. As shown in Fig. 7.8, a wind turbine starts operating when the wind speed is

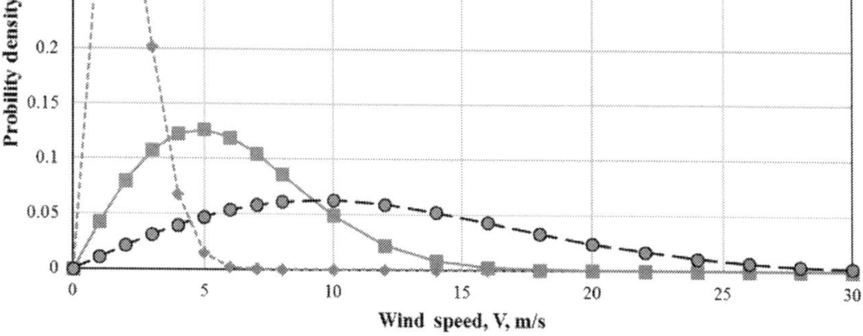

FIGURE 7.6 Weibull probability distributions for three values of annual average wind speed.

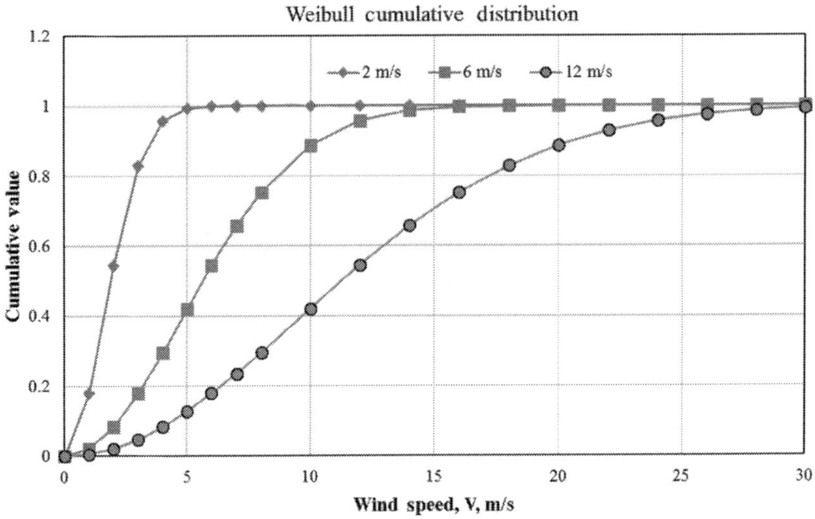

FIGURE 7.7 Weibull cumulative distributions for three annual average wind speed values.

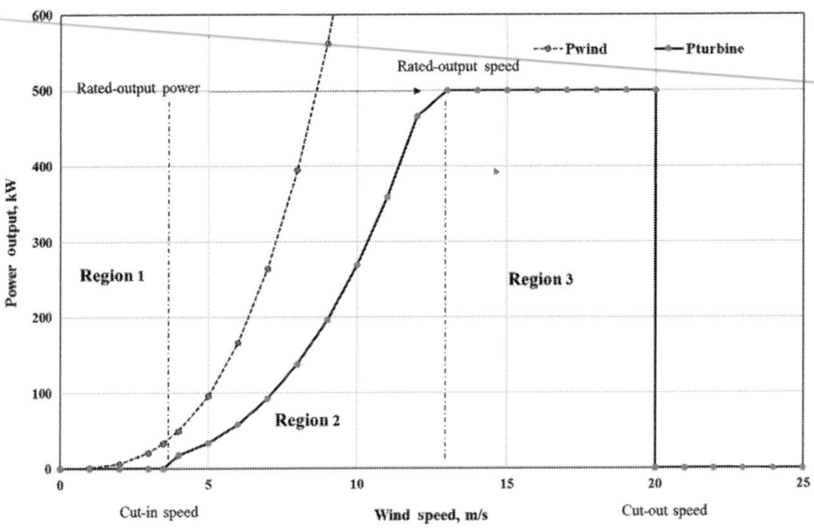

FIGURE 7.8 Power curve and operation regions for a wind turbine.

higher than the cut-in speed of about 3–4 m/s. Indeed, at low wind speeds, commonly known as operation region 1, there is not enough torque to make the blades rotate. Moreover, and even when the blades are able to rotate at these low speeds, the turbine generates only small electrical power amounts that are not sufficient to compensate for the inherent losses. As the wind speed becomes higher, the turbine output is increased significantly until it reaches its rating capacity. A control mechanism is used to limit its output to this level when the wind speed exceeds the rated-output speed (about 13.5 m/s in Fig. 7.8). The operation range of the turbine between the cut-in and rated-output speed values is referred to as region 2. When wind velocity exceeds the cut-out speed (typically 20–25 m/s), the turbine is stopped to prevent any structural damages to mostly its rotor blades. Region 2 refers to the operation of the turbine between the rated-output and cut-out speed values when the power generated by the turbine is typically constant.

7.3.2 Flat PV Systems

The electrical performance of a photovoltaic (PV) solar cell is often represented and modeled using a diode as illustrated in Fig. 7.9. The diode current, I_o, present when the cell is not exposed to solar radiation (i.e., in the dark) is small compared to light-induced current, I_L. The equivalent current, I, that can be supplied by the cell can be estimated using the Shockley diode equation:

$$I = I_L - I_o \left[\exp\left(\frac{qV}{kT} \right) - 1 \right] \qquad (7.14)$$

where q is the charge of one electron ($q = 1.6 \times 10^{-19}$ C), V is the voltage of the solar cell, k is the Boltzmann constant ($k = 1.38 \times 10^{-23}$ J/K), and T is the absolute temperature of the solar cell.

Eq. (7.14) provides the main correlation between the current and voltage variations for a solar PV cell. In particular, Fig. 7.10 shows five key values that are required to be specified on PV cells by the Underwriters Laboratories (UL) as part of the listing process. These include open circuit voltage (V_{OC}), short-circuit current (I_{SC}), maximum power voltage (V_{MP}), maximum power current (I_{MP}), and maximum power (P_{MAX}). The open circuit voltage and short-circuit current represent the maximum values achieved under the STC. Under normal operating conditions, the array can experience higher values of both voltage and current.

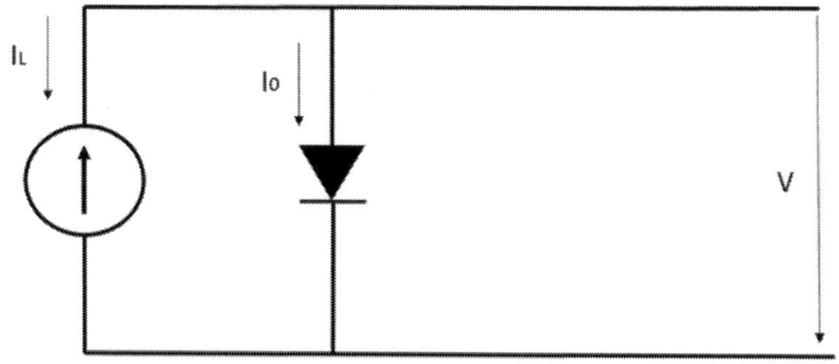

FIGURE 7.9 **A diode representation of a solar cell.**

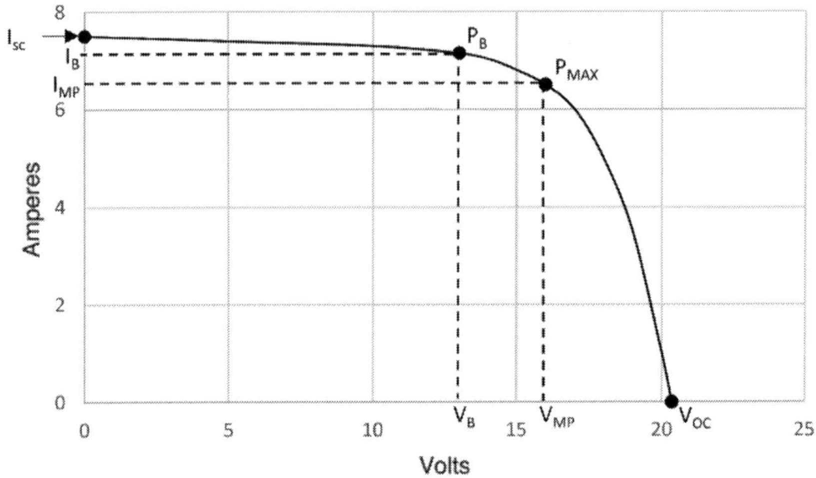

FIGURE 7.10 Typical current–voltage curve for a photovoltaic module.

Table 7.4 presents typical PV module characteristics for common cell types including rated power, short-circuit current, open circuit voltage, and efficiency.

As noted in Table 7.4, the conversion efficiency of commercially available PV modules is in the 10%–13% range under rated conditions. It should be noted that thin-film PV modules are typically used for building integrated PV applications (such as shingle-PV systems) and their efficiency is improving.

The open circuit voltage, V_{OC}, can be estimated from Eq. (7.14) by setting the current, I, to zero (i.e., $I = 0$):

$$V_{OC} = \frac{kT}{q} Ln\left(1 + \frac{I_L}{I_0}\right) \tag{7.15}$$

The short-circuit current, I_{sc}, is obtained when the voltage is set to zero (i.e., $V = 0$). In the ideal case, the short-circuit current, I_{sc}, is equal to the light-induced current, I_L:

$$I_{SC} = I_L \tag{7.16}$$

TABLE 7.4 Typical PV Module Characteristics

PV cell type	Rated power (W)	Short-circuit current, I_{sc} (A)	Open circuit voltage, V_{oc} (V)	Efficiency (%)
Single crystal	175	5.55	44.4	13.5
Multi crystal	75	4.75	21.8	11.6
Thin film	40	2.68	23.3	9.4

The electric power, P_B, obtained from the solar cell as any operation point, B, can be obtained as noted in Fig. 7.9 as the product of the current, I_B, and the voltage, V_B:

$$P_B = V_B I_B \tag{7.17}$$

The maximum power, P_{MAX}, occurs for specific set of voltage and current values, V_{MP} and I_{MP}:

$$P_{MAX} = V_{MP} I_{MP} \tag{7.18}$$

The ratio of the maximum power voltage and current is typically referred to as the characteristic resistance of the PV cell, R_{ch}:

$$R_{ch} = \frac{V_{MP}}{I_{MP}} \tag{7.19}$$

The ratio of the maximum power, P_{MAX}, to the product of open circuit voltage, V_{OC}, and short-circuit current, I_{sc}, is often called the fill factor or FF:

$$FF = \frac{P_{MAX}}{V_{OC} \cdot I_{SC}} \tag{7.20}$$

Thus, the maximum power generated by a solar cell can be estimated from its characteristics, V_{OC} and I_{sc}:

$$P_{MAX} = FF \cdot . V_{OC} \cdot . I_{SC} \tag{7.21}$$

To maximize the power output of PV panels, tracking and/or concentrating systems are utilized. Tracking systems allow solar panels or reflectors to be oriented toward the sun to ensure optimal collection of solar radiation. The main goal of a tracking system is to minimize the incidence angle of solar rays hitting a PV panel. Two types of tracking systems are commonly used for PV panels: single-axis and dual-axis trackers.

- Single-axis trackers: These systems allow the PV panels to rotate around one axis: either vertical or horizontal. The vertical axis trackers are recommended for high latitude locations where the sun position is low and the summer days are long. The horizontal axis trackers are more suitable for tropical regions.
- Dual-axis trackers: These systems can allow the PV panels to rotate around both vertical and horizontal axes. Therefore, the dual-axis trackers can follow the sun position independently of the PV system location.

Fig. 7.11 compares the monthly electrical energy generated from a 500-kW PV array with a fixed tilt, one-axis tracking system, and two-axis tracking system located in three sites: Boulder, Berlin, and Riyadh. Fig. 7.12 compares for each of three locations (i.e., Boulder, Berlin, and Riyadh), the monthly

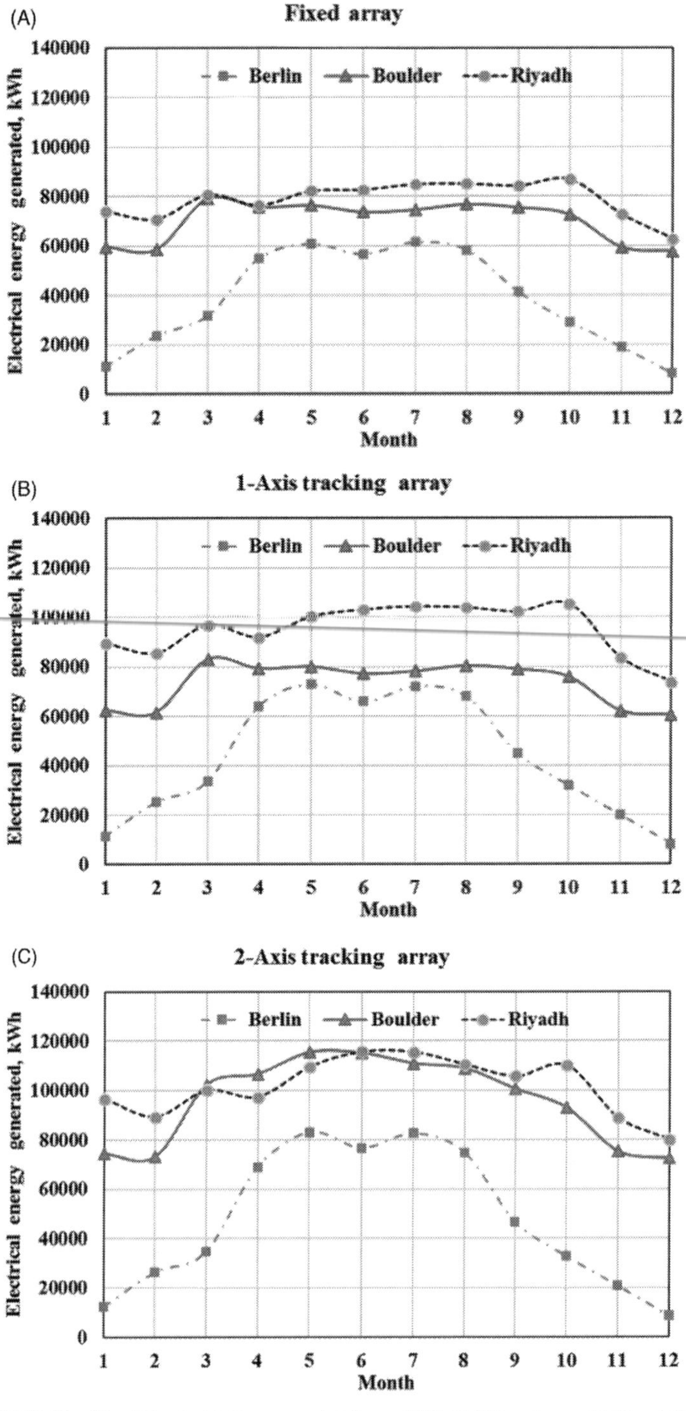

FIGURE 7.11 Monthly electricity generated from 500-kW PV systems. (A) Fixed tilt, (B) one-axis tracking system, and (C) two-axis tracking system for three locations: Boulder, Berlin, and Riyadh.

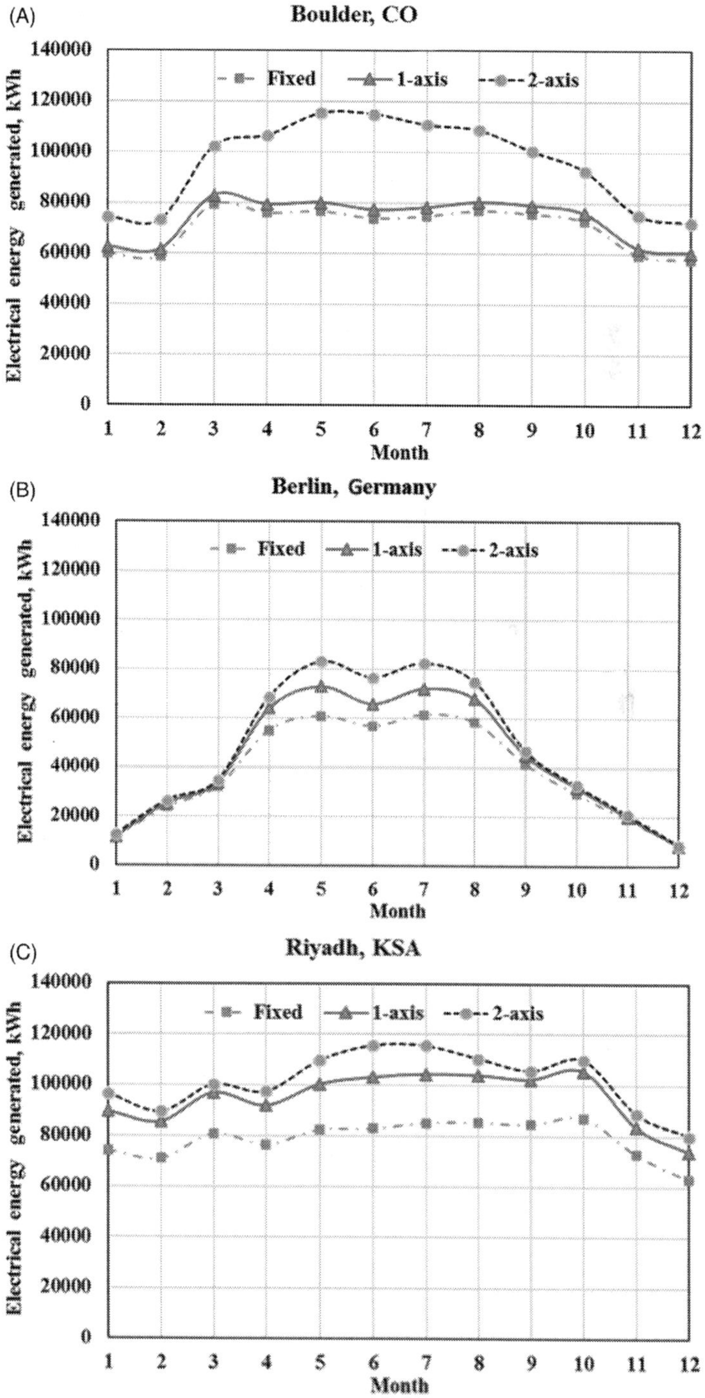

FIGURE 7.12 Monthly electricity generated from flat, one-axis tracking, and two-axis tracking 500-kW PV systems located in (A) Boulder, (B) Berlin, and (C) Riyadh.

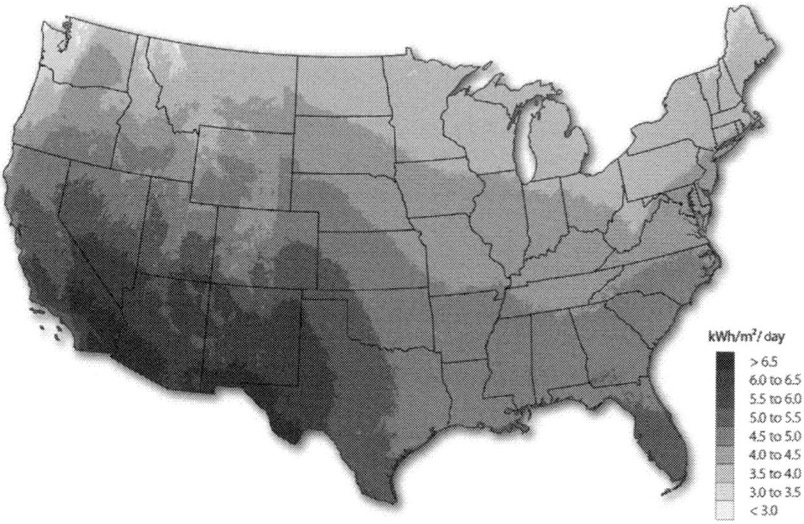

FIGURE 7.13 US map for global solar radiation resources suitable for PV systems (NREL, 2016b).

electricity generated by a 500 kW PV array with a fixed tilt, one-axis tracking system, and two-axis tracking system.

The PV solar resources for the United States are highlighted in Fig. 7.13. Indeed, Fig. 7.13 shows a map that provides the daily average global horizontal solar irradiation expressed in $kW/m^2/day$[1] for all continental US locations (NREL, 2016b). The global solar irradiation values of Fig. 7.13, including direct and diffuse solar radiation components, are determined based on data obtained for the 1998–2015 period. As expected, the map of Fig. 7.13 indicates that the southwest of the United States has the highest solar resources and has the potential for generating electricity cost-effectively from PV systems.

7.3.3 Concentrated PV Systems

In addition to tracking systems, concentrators can enhance the power output of PV panels. The concentrated PV (CPV) systems collect beam solar radiation using lenses or mirrors into a small area where PV cells are located. Fresnel lenses combined with two-axis trackers are often used to increase the performance of CPV systems as outlined in Fig. 7.14.

It should be noted that CPVs are mostly suitable for locations with high direct normal irradiance since they cannot collect diffuse radiation. Depending on the concentration level of CPVs, some passive or active cooling systems may be needed to dissipate heat from the solar cells to enhance their energy efficiency. Moreover, multijunction cells instead of conventional silicon cells are considered especially for CPV with high concentration levels.

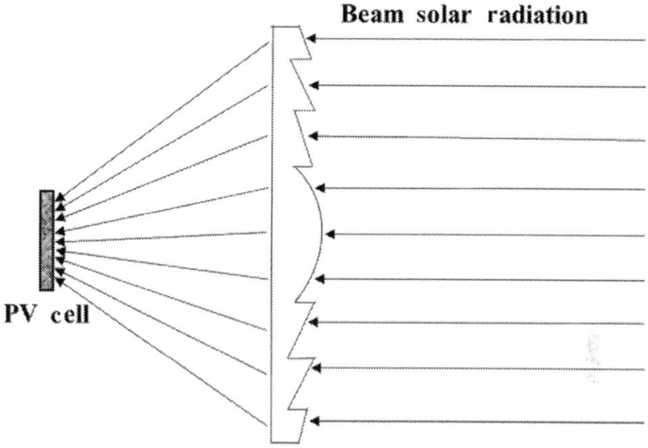

FIGURE 7.14 A concentrated PV system using Fresnel lens.

The impact of the concentration ratio, C, on the performance of a CPV system can be evaluated using the diode representation of a PV solar cell as shown in Fig. 7.9. Indeed, the flow of electrons increases due to higher concentrated solar radiation. Thus, the values of currents including short-circuit current, $I_{sc,C}$, increase proportionally with the concentration ratio, C, relative to the values obtained for nonconcentrated solar cells, I_{sc}:

$$I_{sc,C} = C\ I_{sc} \tag{7.22}$$

The open circuit voltage for the concentrated solar cell, $V_{OC,C}$ can be estimated using the Shockley equation. Specifically, Eqs. (7.15), (7.16), and (7.21) provide

$$V_{OC,C} = \frac{kT}{q} Ln\left(1 + \frac{C.I_{SC}}{I_0}\right) \tag{7.23}$$

Since the ratio is $C.I_{sc}/I_o$ is significantly higher than unity (i.e., $C.I_{sc}/I_o >> 1$), Eq. (7.22) can be approximated by

$$V_{OC,C} \approx \frac{kT}{q} Ln\left(\frac{C.I_{SC}}{I_0}\right) \tag{7.24}$$

Thus, the open circuit voltage for CPV solar cell can be estimated as follows:

$$V_{OC,C} \approx \frac{kT}{q}\left[Ln(C) + Ln\left(\frac{I_{SC}}{I_o}\right)\right] \approx V_{OC} + \frac{kT}{q} Ln(C) \tag{7.25}$$

The term $k.T.Ln(C)/q$ depends on the concentration ratio and range from 0.06 to 0.18 V when C changes from 10 to 1000 and the cell temperature is 25°C.

The maximum power output of the concentrated solar cell, $P_{MAX,C,}$ can be estimated from its characteristic values including the fill factor FF, the open circuit voltage, $V_{OC,C}$, and short-circuit current, $I_{SC,C}$ as noted in Eq. (7.21):

$$P_{MAX,C} = FF.V_{OC,C}.I_{SC,C} \tag{7.26}$$

Therefore, and by combining Eqs. (7.22), (7.25), and (7.26)

$$P_{MAX,C} = FF.C.\left[V_{OC} + \frac{kT}{q}Ln(C)\right].I_{SC} \tag{7.27}$$

Using the maximum power, P_{MAX}, generated by the nonconcentrated solar cell as expressed in Eq. (7.21), it can be shown that

$$P_{MAX,C} = C.P_{MAX}\left[1 + \frac{kT}{qV_{OC}}Ln(C)\right] \tag{7.28}$$

Example 7.1 shows the impact of using a concentrated PV system on the maximum power output of the solar cell.

Example 7.1

Problem: A solar cell generates maximum power of 2.4 W at regular solar conditions at 25°C. The open circuit voltage is measured at 0.55 V. Determine the maximum power of the solar cell if a concentrated PV system is considered with C = 10.

Solution:

Using Eq. (28) with P_{MAX} = 2.4 W, T = 273 + 25 = 298 K, V_{OC} = 0.55 V, and C = 10, the maximum power of the concentrated solar cell can be estimated:

$$P_{MAX,C} = (10)(2.4)\left[1 + \frac{(1.38\ 10^{-23} * 298)}{(1.6\ 10^{-19} * 0.55)}Ln(10)\right]$$

or

$$P_{MAX,C} = 2.26\text{W}$$

The power output of the CPV cell is slightly more than 11 times that of conventional PV cell.

Since the concentration system increases the solar irradiance proportionally with the concentration ratio, C, the efficiency of the concentrated solar cell, η_C, can be estimated from the efficiency from the nonconcentrated cell, η:

$$\eta_C = \eta\left[1 + \frac{kT}{qV_{OC}} Ln(C)\right] \tag{7.29}$$

Example 7.2 shows the variation of the solar cell efficiency when a concentrated PV system is used.

Example 7.2

Problem: Consider the solar cell considered in Example 7.1, determine the percent change in the efficiency of the solar cell if a concentrated PV system is considered with $C = 10$.

Solution:

Using Eq. (7.29) with $P_{MAX} = 2.4$ W, $T = 273 + 25 = 298$ K, $V_{OC} = 0.55$ V, and $C = 10$, the maximum power of the concentrated solar cell can be estimated:

$$\frac{\Delta\eta}{\eta} = \frac{kT}{qV_{OC}} Ln(C) = \frac{\left(1.38\ 10^{-23} * 298\right)}{\left(1.6\ 10^{-19} * 0.55\right)} Ln(10)$$

Thus,

$$\frac{\Delta\eta}{\eta} = 10.8\%$$

Therefore, if the nonconcentrated solar cell has an efficiency of 15%, the CPV cell would have slightly improved efficiency of 16.6%.

Based on Eqs. (7.28) and (7.29), it appears that the power output and the efficiency increases indefinitely with the concentration ratio C. However, higher concentration ratios induce higher wire losses (i.e., RI^2) through the series resistance of the solar cell, R_S. It has been shown that an optimal concentration ratio, C_{opt}, can be estimated depending on the characteristics of the solar cell (Luque, 1989):

$$C_{opt} \approx \frac{kT}{qI_{SC}R_s} \tag{7.30}$$

Example 7.3 determines the optimal concentration ratio for a specific set of a solar cell.

Example 7.3

Problem: Determine the optimal concentration ratio for a solar cell that has a series resistance of 0.012 Ω and a short-circuit current of 0.16 A at 25°C.

Solution:

Using Eq. (7.30) with $R_S = 0.012\ \Omega$ and $I_{SC} = 0.16$ A, the optimal concentration ratio, C_{opt}, can be estimated:

$$C_{opt} = \frac{(1.38\ 10^{-23} * 298)}{(1.6\ 10^{-19} * 0.012 * 0.16)} = 13.4$$

The optimal concentration ratio for the CPV cell is about 13.5.

Currently, few utility-scale CPV systems are operating throughout the world. Table 7.5 lists the capacity and the location of some of the major CPV systems completed and operational to date worldwide (CPV Consortium, 2017).

7.3.4 Concentrated Solar Power Systems

Concentrated solar power (CSP) plants convert solar energy into electricity through heat engines using working fluids. Depending on the temperature of the working fluid, solar thermal power systems are classified into three types (Pavlović et al., 2012):

- Low-temperature plants when the working fluid temperatures are below 100°C: These plants typically use flat collectors to capture solar radiation and water or air to transfer heat.
- Medium-temperature plants when the working fluid temperatures are between 100 and 400°C: These plants utilize generally vacuum collectors and collectors with concentrators.
- High-temperature plants when the working fluid temperatures are above 400°C: These plants typically use solar furnaces and concentrated solar collectors including CSP systems.

CSP plants have three main components: (1) collectors, (2) steam turbines, and (3) generators. Using mirrors or lenses, sunlight is reflected to receivers where heat is collected by a thermal energy carrier (primary circuit), and

TABLE 7.5 Basic Characteristics of the Major CPV Projects Worldwide

Project name	Location	Capacity (MW)	Number of CPV systems	Year of operation
Golmut 2	Qinghai, China	79.83	3168	2013
Golmut 1	Qinghai, China	57.96	2300	2012
Touwsrivier	Breede Valley, South Africa	44.19	1500	2014
Alamosa Solar Project	Alamosa, CO, US	35.3	504	2012
Hami II	Xinjiang, China	5.88	300	2014

subsequently used directly or through a secondary circuit to power turbines and generate electricity (Zhang et al., 2013). The generation of electricity through CSP plants is highly dependent on the available direct solar irradiation. Thus, electricity production may vary significantly during cloudy days unless thermal storage or auxiliary power back-up systems are considered. There are four main types of CSP technologies depending on the collector system used to capture solar radiation:

1. *Parabolic Trough Collector:* Parabolic trough collectors or PTCs are made of curved mirrors with a thermally efficient absorber tube, placed along the focal line of the parabolas of the collectors. A synthetic oil flows through the absorber tube and can be heated approximately up to 400°C (Pavlović et al., 2012). The curved mirrors, typically called reflectors, are generally aligned in a north–south direction as they move in tandem with the sun as it rises from the east and sets in the west. An assembly of such reflectors and absorbers is called the collector field or the solar field. The oil transfers from the collector field to a heat exchanger, where water is first preheated, evaporated, and then superheated. The superheated steam is fed to a turbine that drives a generator to produce electricity following a Rankine power cycle. Indeed, the superheated steam is then cooled and condensed and sent back to the heat exchangers. For typical PTCs, collectors can have a concentration ratio of about 70 to 100 and operating temperatures in the range of 350–550°C resulting in an average solar to electric efficiency of 15% (Ummadisingu and Soni, 2011). The absorber tube is also called the heat collecting element (HCE) of the CSP plant. Specifically, the HCE is generally made-up of a metal tube with a glass envelope cover separated by either air or vacuum to reduce convective heat losses and allow for thermal expansion. The metal tube is typically coated with a selective colored membrane with high solar irradiation absorbance and low thermal remittance (Zhang et al., 2013). The high solar absorption coefficient of the HCE combined with the parabolic concentrator provides an efficient heating mechanism for the working fluid and increases the solar to thermal efficiency (Pavlović et al., 2012). Fig. 7.15 shows a basic solar collector layout for a PTC system.

   PTC plants can have several designs and configurations. Fig. 7.16 presents a typical design configuration for a PTC plant with three blocks: (1) solar field with the parabolic trough collectors, (2) thermal energy storage (TES) system, and (3) a power generation plant. Some PTC plants use molten salts as the heat transfer fluid. Other configurations use water in the absorber tube such as the direct steam generation systems. Some PTC plants use backup systems or thermal storage systems to ensure continuous supply of power. Fig. 7.16 shows a basic layout of a PTC plant with a TES system.
2. *Solar Power Towers*: Solar power tower systems consists of a solar field, which can be flat or concave-shaped mirrors called as heliostats. A solar power tower is also called a central receiver system because the receiver is

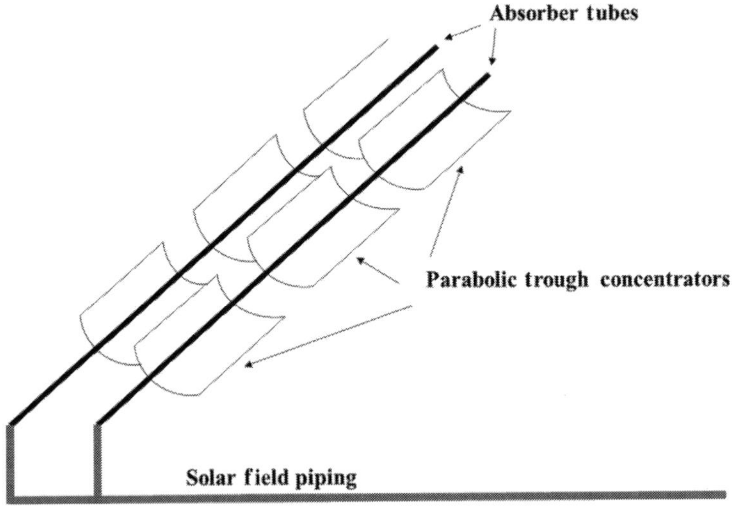

FIGURE 7.15 A parabolic trough collector.

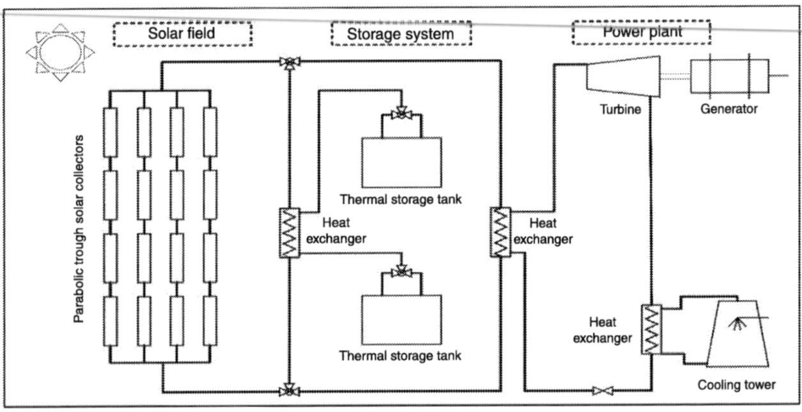

FIGURE 7.16 Schematic diagram for the PTC plant with a thermal energy storage system.

located on the top of a tower facing the heliostats. All the heliostats use a two-axis tracking system to concentrate direct solar radiation onto the receiver as illustrated in Fig. 7.17. The receiver is generally made up of ceramic materials or metals that are stable at high temperatures. The working fluid can be air, molten nitrate salt, liquid sodium, or water. The heliostats concentrate all the solar energy onto the receiver where the working fluid can reach temperatures of up to 1500°C. The solar to electric conversion efficiencies of solar power towers are in the range of 20%–35% (Ummadisingu and Soni, 2011). Several existing solar power tower systems use molten salts as the working fluid in the central receiver. Indeed, molten salts are nonflammable and

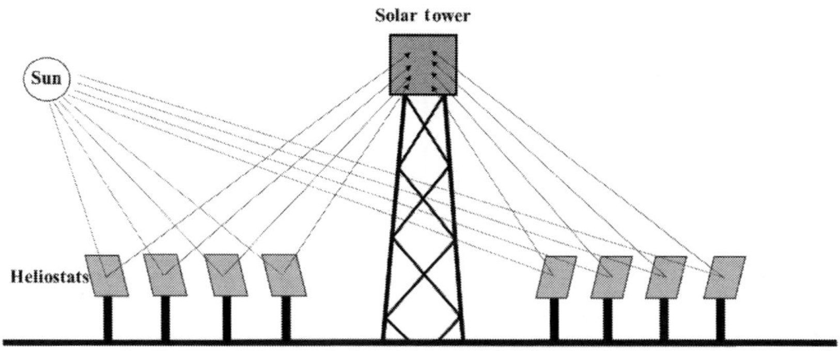

FIGURE 7.17 A solar power tower collector with heliostats.

nontoxic and have high capacities for thermal storage. It should be noted that solar power towers have the highest land requirements among all the CSP technologies. In addition, their overall efficiency can be affected by several factors including the optical characteristics of heliostats, cleanliness of mirrors, and precision of tracking systems. These plants are suitable to be combined with fuel backup systems in addition to TES systems (Pavlović et al., 2012).

3. *Parabolic Dish*: A parabolic dish reflector is a point-focus collector that concentrates direct solar irradiation at the focal point of the dish as illustrated in Fig. 7.18. The fluid or air in the receiver, located at the focal point, can be heated to about 750°C. At the focal point of the receiver, a heat engine/generator unit is attached to produce electricity using typically a Stirling engine cycle (Ummadisingu and Soni, 2011). The parabolic dish can have a concentration ratio up to 2000 and can achieve a pressure in the receiver

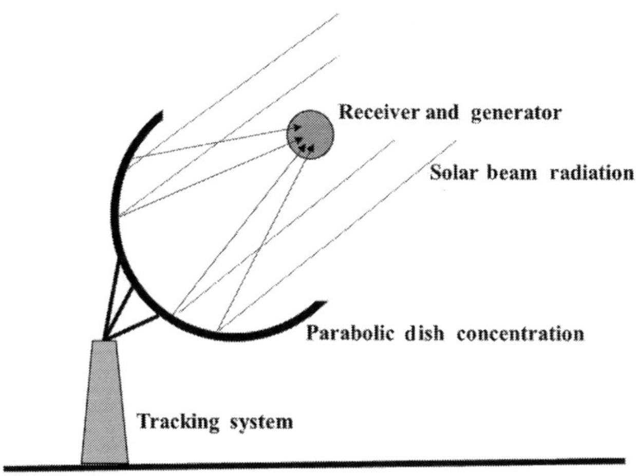

FIGURE 7.18 A parabolic dish CSP system.

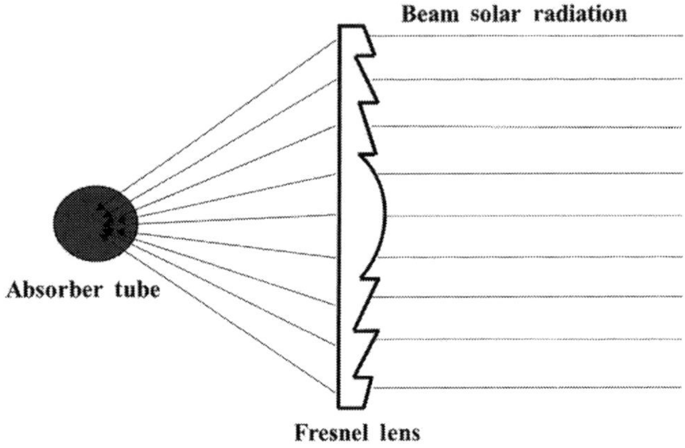

FIGURE 7.19 **A linear Fresnel mirror-based CSP system.**

up to 200 bars. Typically, parabolic dishes have a diameter of 5–10 m and a surface area of 40–120 m^2. Their reflecting surface can be made of aluminum, silver, glass, or plastic. The power produced through a parabolic dish power plant ranges from 100 to 400 kW (Pavlović et al., 2012). It should be noted that the parabolic dish system tracks the sun while the engine generates electricity simultaneously with no need to have a cooling water and a heat rejection system. However, parabolic dish systems are expensive and have low compatibility with thermal energy storage, hybridization, and large-scale deployment. Their main applications are for isolated and off-grid power systems. Promoters do claim that large-scale deployment and production of parabolic dishes can reduce their costs to be competitive with other CSP systems (Zhang et al., 2013).

4. *Linear Fresnel Reflectors*: Linear Fresnel reflectors with flat or slightly curved mirrors can focus solar beam radiation to be directly captured by the receiver absorber tubes as illustrated in Fig. 7.19. The receiver is a fixed structure mounted over a tower at a very low height and along the rows of collectors. The mirrors are mounted on single or double axis tracking systems. The main advantage of the linear Fresnel reflectors is their low installation cost. However, these systems are not efficient as other CSP plants in converting solar energy to electricity, but are efficient in direct steam generation with no need for heat transfer fluids and heat exchangers. One type of linear Fresnel reflectors, called compact linear Fresnel reflector, has two parallel receivers for a given row of mirrors and can produce more steam by using less land (Zhang et al., 2013). The reported capacity of linear Fresnel plants is in the range of 10–200 MW with a solar-to-electric efficiency of 8%–10% (Ummadisingu and Soni, 2011).

TABLE 7.6 Comparative Characteristics of Common CSP Technologies

CSP system	Parabolic trough	Solar tower	Parabolic dish	Linear Fresnel
Capacity range (MW)	10–200	10–200	0.01–0.04	10–200
Concentration ratio	70–80	300–1000	1000–3000	25–100
Thermal efficiency (%)	30–40	30–40	Higher	30–40
Annual solar to electric efficiency (%)	15	8–10	16–18	8–10
Capital cost ($/kW)	3972	4000+	12578	Low
Capital cost ($/m^2)	424	476	Low	235
O&M cost ($/kWh)	0.012–0.02	0.034	0.21	Low
Land use (m^2/MWh)	6–8	8–12	8–12	4–6
Specific power (W/m^2)	300	300	200	Low
Operating temperature range (°C)	20–400	300–565	120–1500	50–300
Cooling water requirement(L/MWh)	3000 or dry	1500 or dry	None	3000 or dry
Outlook for improvement	Limited	Very significant	High through mass production	Significant

Table 7.6 gives a comparative summary of the major features, performance indicators, and costs for all the types of CSP plants outlined.

Some of the important design and operating features of the CSP plants, outlined in Table 7.6, are discussed in more detail in the following sections.

7.3.4.1 Installation Costs

The large-scale deployment of CSP plants is largely dependent on the cost of various components. The competitiveness of the CSP plants compared to other renewable technologies would depend on their installation costs. However, these costs would also depend on the number of CSP plants built and operated (IEA, 2010). Indeed, analyses by the National Renewable Energy Laboratory (Turchi et al., 2010) and the International Renewable Energy Agency (IRENA, 2012) concur that significant cost reduction of CSP plants will only come from economies of scale, that is, a large-scale implantation. The impacts of the installation costs are mainly analyzed through the calculation of the LCOE as presented in Chapter 5.

The direct installation costs for PTC plants and their components are itemized in Table 7.7 based on data reported by NREL (Kutscher et al., 2010) and IRENA (2012).

TABLE 7.7 Installation Cost of PTC Plants and Their Components

Cost component	IRENA data	NREL data
Total cost (million USD 2010)	364	658
Cost per kW (USD/kW)	7,280	8,250
Labor cost: solar field	11.40%	6.60%
Labor cost: piping	1.80%	1.80%
Labor cost: electric installations and others	4.00%	0.40%
Mirrors	6.40%	5.00%
Receivers	7.10%	7.20%
Steel construction	10.70%	8.20%
Pylons	1.10%	0.20%
Foundations	2.10%	1.90%
Trackers (hydraulics and electrical motors)	0.40%	1.40%
Swivel joints	0.70%	0.20%
HTF System (piping, insulation, heat exchanges, pumps)	5.40%	2.60%
Heat transfer fluid	2.10%	2.30%
Electronics, controls, electrical and solar equipment	2.50%	3.50%
Storage Fluid	5.10%	7.50%
Storage tanks	1.80%	5.60%
Heat exchangers and pumps	1.80%	3.40%
Balance of storage system	1.80%	0.40%
Power block, grid connection and other direct misc. costs	14.30%	20.10%
Project development, EPC, financing and allowances	19.50%	22.80%

It should be noted that the baseline CSP plants for the cost data provided in Table 7.7 have different capacities and design configurations. Specifically, the CSP plant used to generate for IRENA data is a 50-MW Andasol-type plant with 7.5 h of storage, while the plant considered to estimate NREL data is a 100-MW plant with 6 h of storage. In general, the cost components given in Table 7.7 include all the direct and indirect costs of a typical CSP plant. One of the large differences between IRENA and NREL data is the allocation of the labor costs. The IRENA labor costs are based on several international case studies and field data, whereas those for the NREL are based on a study specific to the SEGS plants located in the United States.

7.3.4.2 Operation and Maintenance Costs

The O&M maintenance costs for CSP plants are related to various costs such as replacement of mirrors and receivers associated with glass breakage, mirror washing including water costs, and plant insurance fees (IRENA, 2012). For recently built and operated CSP plants, O&M costs are significantly lower than those of the SEGS plants built in California during the 1980s. Indeed, advances in the construction and operation technologies have reduced the replacement requirements of mirrors and more automated O&M procedures. The IRENA study has estimated that O&M costs for PTC, the most common CSP plants, range between \$0.02/kWh and \$0.03/kWh (IRENA, 2012). NREL estimates for typical CSP plants indicate that O&M costs can be lowered from \$70/kW·yr in 2010 to \$45/kW·yr in 2020. For PTC plants, O&M costs of \$45/kW·yr are expected to be lower than \$0.02/kWh (Turchi et al., 2010). O&M costs can be further reduced by improving CSP plant reliability, operation techniques, and automated cleaning procedures (Kutscher et al., 2010). Installation and O&M costs for specific CSP plant designs obtained from various references are summarized in Table 7.8.

TABLE 7.8 Installation and O&M Costs for CSP Plants

Plant size (MW)	Installation cost (million USD)	O&M cost	Country	Notes	Reference
50	240	5.44 (million USD)	Jordan	Pure CSP	Al-Soud and Hrayshat (2009)
50	380	3% of equipment costs	India	Base case of a hypothetical plant	Jain et al. (2013)
30	118.5	\$0.00398/kWh	Egypt	Integrated solar combined cycle plant with only 127 MW of total capacity. The cost is only for the solar field.	Horn et al. (2004)
25	432.6	Not Available	Algeria	Integrated solar combined cycle plant with 70 MW total capacity.	Boukelia and Mecibah (2013)
200	Not available	\$70/kW·yr	Chile	It is a simulation with actual total cost not available.	Corral et al. (2012)

7.3.4.3 Solar Radiation Requirements

Actual CSP plant power generation and economic feasibility are highly dependent on the availability of direct solar irradiation. Therefore, only locations with high solar direct normal irradiation (DNI) levels are suitable for CSP plants. Indeed, only direct solar radiation can be used to maintain operating temperatures sufficiently high for CSP plants to generate electricity (IRENA, 2012). Various references suggest threshold values for the minimum DNI requirements for CSP plants as summarized for select locations in Table 7.9. Solar resource data are not readily available for all locations throughout the world. Generally, datasets collected by organizations such as NASA and NREL are used to perform preliminary analysis and assess potential sites for CSP plants in a specific country (Fluri, 2009). Typically, CSP analysis tools require hourly data for solar

TABLE 7.9 Desired Threshold Values for Direct Normal Irradiation at Various Countries

Country/reference/report	Annual DNI ($kWh/m^2/yr$)	Notes
Any location (IRENA, 2012)	2,000	A general value for achieving attractive LCOE values.
Chile (Corral et al., 2012)	2,190	This DNI value is for northern Chile where a CSP is proposed.
Turkey (Kaygusuz, 2011)	1,800	This DNI value is considered for economic feasibility in Turkey. Some locations have as high as 9 $kWh/m^2/day$ average DNI.
Qatar (Weber, 2013)	2,200	Assume for CSP economic feasibility.
India (Jain et al., 2013)	2,100	This DNI value is specific to Jodhpur City.
South Africa (Fluri, 2009)	2,100	Only sites having DNI above this value are considered suitable for CSP plants.
Australia (Hinkley et al., 2013)	2,564	At hypothetical plant location in Longreach, Queensland.
Cyprus (Poullikkas, 2009)	2,000	This DNI is assumed for analysis.
Jordan (Al-Soud and Hrayshat, 2009)	2,400	This DNI value is specific for Kharana. Economic viability is assumed to be above 2000 $kWh/m^2/yr$
Egypt (Blanco et al., 2013)	2,496	This DNI value is specific to Port Safaga, on the west side of Red Sea.
Algeria (Boukelia and Mecibah, 2013)	1,800	Sites with as high as 2650 $kWh/m^2/yr$ DNI values are available.
Serbia (Pavlović et al., 2012)	1,400	This DNI value represents the mean of 3.3–4.3 $kWh/m^2/day$.

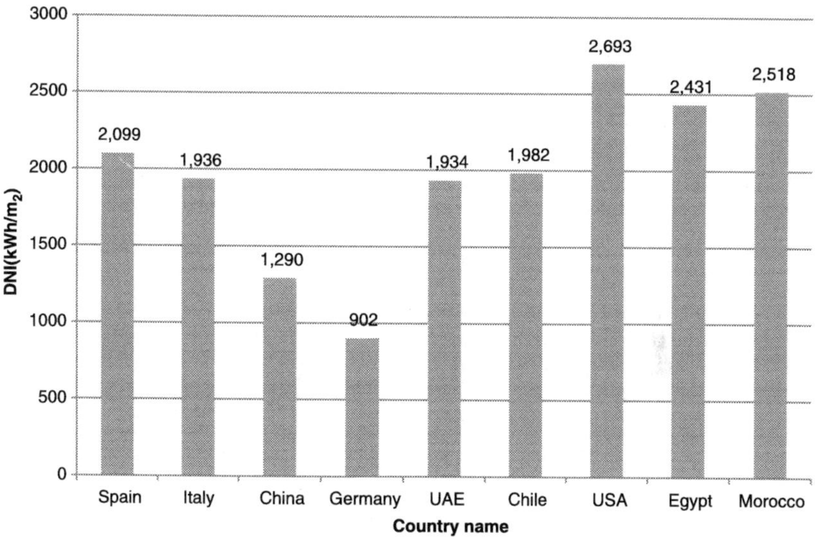

FIGURE 7.20 **Average reported DNI for select countries.**

irradiation. However, only monthly extraterrestrial solar radiation is generally available for remote locations using mostly satellite data. The extraterrestrial radiation data obtained using satellite imaging are usually specific to large areas rather than specific locations. For example, data available from NREL for most African countries are spread over a 40 × 40 km area and that from NASA covers 80 × 110 km area. Models to convert monthly satellite-based extraterrestrial solar radiation to hourly DNI data do exist and typically require hourly cloud cover or clearness index values (Corral et al., 2012). Fig. 7.20 provides the annual average DNI values for countries where CSP plants have been constructed and operated.

Fig. 7.21 presents the US solar resources suitable for CSP pants. Specifically, Fig. 7.21 shows a map that provides the daily average normal direct solar irradiation expressed in $kW/m^2/day$[1] for all continental US locations (NREL, 2016c). The direct solar irradiation values are estimated based on data obtained for the 1998–2015 period. Similar to the solar resources shown for PV applications, the map of Fig. 7.21 indicates that the southwest of the United States has the highest solar resources suitable for generating electricity cost-effectively from CSP plants.

7.3.4.4 Site Selection for CSP Plants

Other factors have to be considered when selecting a location for a CSP plant in addition to high solar resources. Some of these factors include the following:

- *Land Use*: Compared to conventional power plants, large land areas are required to construct and operate CSP plants. The typical land area required for generating 1 MW of power is generally 5 acres or 20,234 m^2 (Boukelia and Mecibah, 2013). For larger CSP plants with storage or hybridization, the land area required is even larger. Fig. 7.22 correlates the required aperture area for the solar field as a function of the power-generating capacity of the CSP plant based on the design specifications of actual projects. Before the site is selected, it is important to determine the cost to acquire a large land area suitable for the CSP plant. Indeed, the land acquisition cost is a crucial factor in estimating the LCOE for the CSP and thus its economic feasibility.

- *Land Cover*: The best sites for CSP plants are usually hot and arid desert regions where CSP plants (and any other power plants) would have minimal impacts. Fluri (2009) has proposed an approach to categorize lands as critically endangered, endangered, and least endangered. He suggested that only those sites with a land categorized as least endangered can be selected for CSP plants. Other land suitable for CSP plants can include wasteland with low residential value, low agricultural value, and low biological habitat. Moreover, the selected land should not have any value in terms of fossil recovery but should have good soil quality to enable the construction of the CSP plant infrastructure (Boukelia and Mecibah, 2013).
- *Topography*: Topography usually refers to the slope of the land. For CSP applications, a slope of 1% or a flat land is the most economical. Sites with up to 3% slope are reasonable as long as the site preparation costs including labor costs are low. It is estimated that every 3% slope increase results in a 10% increase in the plant installation costs (Boukelia and Mecibah, 2013). Apart from the slope, topographical factors to be considered for CSP installations are flood potential, seismic history, soil stability, potential sun obstruction, and existence of dust and other debris that can degrade the quality of mirrors and lenses for the CSP plants.

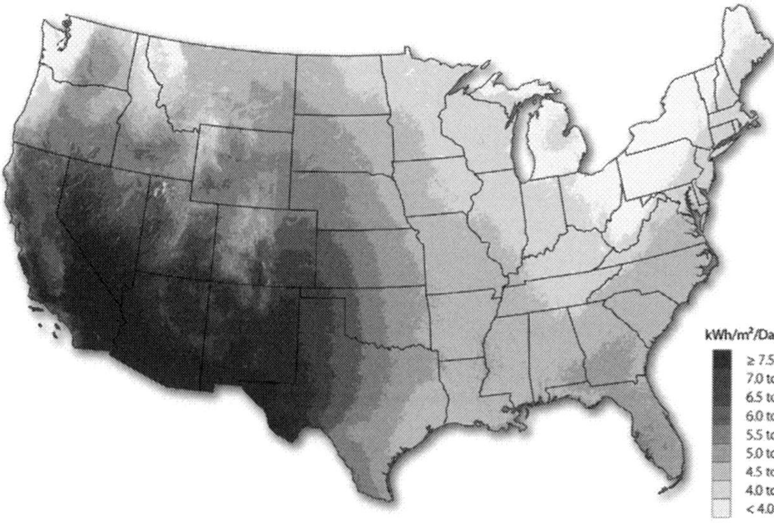

FIGURE 7.21 US map for global solar radiation resources suitable for PV systems (NREL, 2016c).

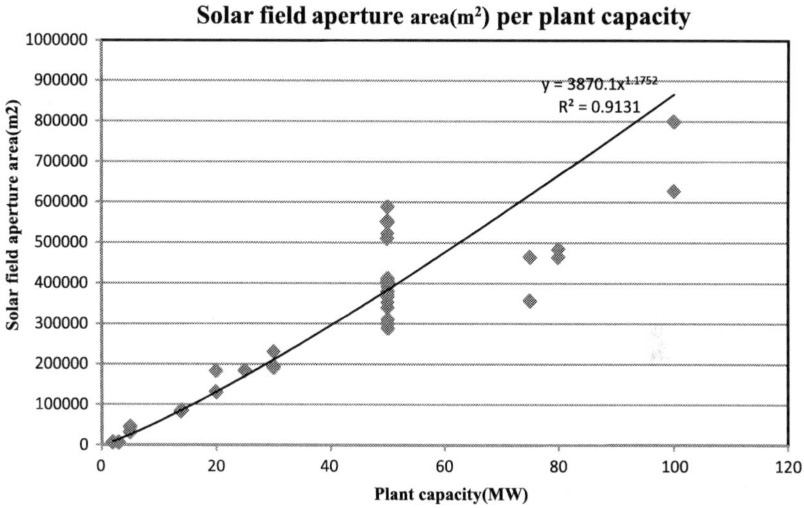

FIGURE 7.22 **Solar field aperture area for given plant capacity.**

- *Infrastructure*: The selected site should be close to the power transmission lines. Proximity to the grid transmission lines is highly desirable since the cost of adding new power transmission lines can be significantly high and therefore can reduce the economic feasibility of the CSP plant. For countries in the MENA region, it can be also highly beneficial to have the CSP plant site near natural gas or oil pipelines. Indeed, such sites would reduce natural gas transmission costs when integrated combined cycle or solar-gas hybrid type CSP plant are considered.
- *Water Availability*: Water is mainly required for a CSP plant to cool the steam cycle and to replenish the working HTF fluid. In addition, water may be required for cleaning CSP mirrors and for maintenance purposes. Typically, water requirement is estimated between 3 and 3.5 m^3/kWh for CSP plants with 95% of the water for plant operation and 5% for plant maintenance. As most suitable sites for CSP plants are located in arid regions, an analysis has to be performed to assess whether water can be cost-effectively transported or dry cooling is to be considered. Some desirable sites can be closer to the sea lines. In such cases, desalination plants can be built along with the CSP plants to provide water needs (Blanco et al., 2013). A feasibility analysis for a CSP plant in Chile indicated that transporting water from the sea to the power plant can be a cost-effective option (Corral et al., 2012).
- *Wind Assessment*: Wind speed data are required for selecting a site for a CSP plant to perform a proper design of the collector structure. Indeed, the structure of the CSP collector can represent 40% of the solar field costs. To reduce these costs, it is preferable to have a site with low wind speeds. The SEGS plants in California were built to resist wind speeds of up to 35 mph. Several sites located in the Sahara Desert region have wind speeds that can reach up to 40 mph, and the structural support for collectors has to be capable of withstanding these speeds (Boukelia and Mecibah, 2013).

7.3.4.5 Operation Modes

The main operation modes for CSP plants include generally three strategies (Boukelia and Mecibah, 2013):

1. *Solar Only Mode*: In this mode, only solar radiation is used for electricity generation. There is no additional resource or backup system is used to produce electrical power. In this mode, a TES system can be utilized for longer operation of the CSP plant. During hours when direct solar radiation can be captured, the plant uses only solar energy to generate electricity and to charge the TES. During afternoon hours with little solar energy resources, electricity is generated from both the solar field and the TES system. After sunset, the TES system is used as the source of heat to generate electricity.
2. *Direct Steam Generation Mode*: There are several applications for CSP plants other than generating electricity. Indeed, the solar field can be used simply to provide heat for steam generation that can be used for various thermal processes. Moreover, using steam instead of synthetic oil in the HTF system can provide several advantages. In a CSP plant with direct steam generation, water is first preheated, evaporated, and then superheated. This operation can increase the thermal efficiency and reduce the CSP system costs by eliminating the need for heat exchangers.
3. *Hybrid System Mode*: Similar to a TES, a hybrid system can be considered to fully or partially supply the CSP plant with energy from a fossil fuel to generate electricity. This energy is typically used during night hours or when direct normal solar radiation is not available to ensure a full day operation. This operation improves the reliability of the CSP system, and can help in reducing the LCOE by using fossil fuel if it is readily available. The main disadvantage of a hybrid system is it does benefit from most rebates and incentives offered for renewable energy systems. Moreover, it may not result in as much reduction in the carbon footprint compared to a totally solar-operated CSP plant. There are several CSP plants in the MENA region that are built as integrated combined cycle solar plants with hybrid systems (Al-Soud and Hrayshat, 2009; Boukelia and Mecibah, 2013; Brand et al., 2012). Indeed, these systems improve the cost-effectiveness of CSP for MENA countries where natural gas is generally readily available.

7.3.4.6 Performance Analysis of CSP Systems

A simplified analysis of the thermal performance of CSP systems is carried out using the Muezer engine model as noted in Fig. 7.23. The efficiency of the CSP system is defined as the ratio of the work output, W_{net}, over the power input to the system, Q_{in}:

$$\eta_{CSP} = \frac{W_{net}}{Q_{in}} \tag{7.31}$$

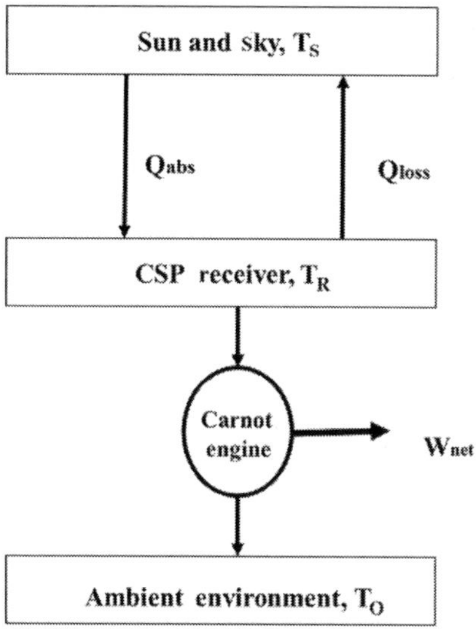

FIGURE 7.23 **Simplified thermal model for CSP system.**

The CSP consists of two main systems: (1) a receiver that converts the incoming solar irradiance into heat with an efficiency, η_R, and (2) a heat engine that transforms the heat collected by the receiver into power with an efficiency, η_E. Therefore, the CSP efficiency can be estimated from the product of both receiver and heat engine efficiencies:

$$\eta_{CSP} = \eta_R \eta_E \tag{7.32}$$

The receiver efficiency depends on the heat absorbed, Q_{abs}, from the solar radiation, Q_{solar}, as well as the losses mainly due to radiation, Q_{loss}:

$$\eta_R = \frac{Q_{abs} - Q_{loss}}{Q_{solar}} \tag{7.33}$$

The heat engine efficiency can be approximated using the Carnot efficiency between the ambient temperature, T_o, and the CSP receiver at the temperature T_R:

$$\eta_R = \frac{T_R - T_o}{T_R} \tag{7.34}$$

The solar irradiation collected by the CSP receiver can be approximated using the following equation:

$$Q_{solar} = \eta_{optic} \cdot C \cdot I_{solar} \cdot A_C \tag{7.35}$$

where C is the concentration ratio of the collector, η_{optic} is the optical efficiency of the receiver, A_C is the collector surface area, and I_{solar} is the solar irradiance hitting the collector surface.

The heat absorbed by the receiver depends on its absorptivity, α, and the collected solar irradiance, Q_{solar}:

$$Q_{abs} = \alpha \cdot Q_{solar} \tag{7.36}$$

Due to the high temperature at the receiver, it can be assumed that the energy lost by the receiver is limited to radiation heat transfer:

$$Q_{loss} = \sigma \cdot \varepsilon A_C T_R^4 \tag{7.37}$$

where T_R is the receiver temperature, σ is the Boltzmann constant (i.e., $\sigma = 5.67 \cdot 10^{-8}$ W/m^2·K^4), and ε is the emissivity of the receiver.

By combining Eqs. (7.32) through (7.37), it can be shown that the CSP efficiency can be determined as follows:

$$\eta_{CSP} = \left(1 - \frac{T_o}{T_R}\right)\left(\alpha - \frac{\sigma \cdot \varepsilon \cdot T_R^4}{\eta_{optic} \cdot C \cdot I_{solar}}\right) \tag{7.38}$$

Fig. 7.24 illustrates the variation of the CSP efficiency for various receiver temperatures, T_R, and concentration ratios, C, when I_{solar} = 1000 W/m^2, T_o = 25°C = 298K, and $\eta_{optic} = \alpha = \varepsilon = 1.0$.

As illustrated in Fig. 7.24, the CSP efficiency initially increases with the receiver temperature, T_R, then starts to decrease after passing through an optimal value before it reaches an efficiency of zero. In fact, while the efficiency of the

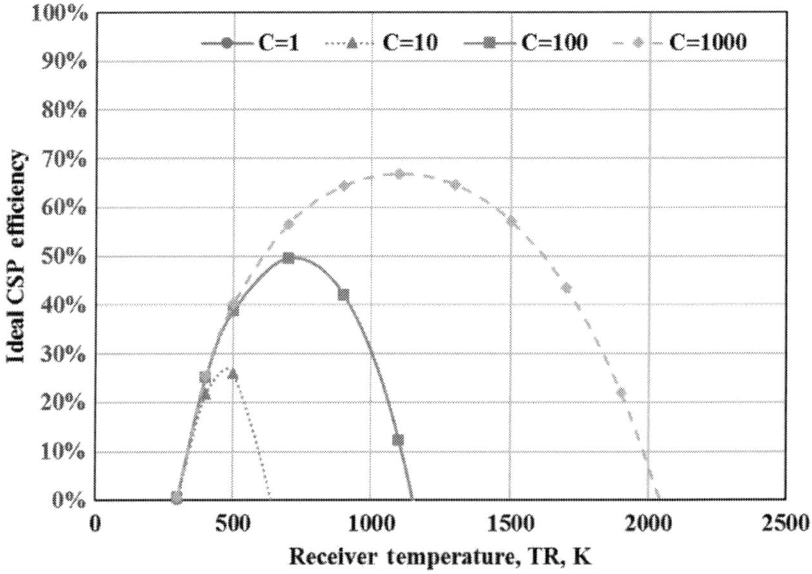

FIGURE 7.24 Variation of ideal CSP efficiency as a function of concentration ratio, *C*, and receiver temperature, T_R.

heat engine always increases with T_R, the receiver efficiency varies significantly with T_R. Indeed, the losses to the environment outweighs the amount of heat that can be absorbed by the receiver at high temperatures. The maximal receiver temperature, $T_{R,max}$, can be estimated from Eq. (7.38) when the CSP efficiency is set to zero (i.e., $\eta_{CSP} = 0$):

$$T_{R,max} = \left[\frac{\alpha \cdot \eta_{optic} \cdot C \cdot I_{solar}}{\sigma \cdot \varepsilon} \right]^{1/4} \tag{7.39}$$

The receiver temperature, $T_{R,opt}$, when the CSP efficiency becomes optimal is determined by setting the derivative of η_{CSP} defined by Eq. (7.38) to zero (i.e., $d\eta_{CSP}/dT_R = 0$):

$$4T_{R,opt}^5 - 3T_o T_{R,opt}^4 - \frac{\alpha \cdot \eta_{optic} \cdot C \cdot I_{solar} \cdot T_o}{\sigma \cdot \varepsilon} = 0 \tag{7.40}$$

The values for both temperatures $T_{R,opt}$ and $T_{R,max}$ depend on the concentration ratio, C. Fig. 7.25 shows the variation of $T_{R,opt}$ and $T_{R,max}$ as a function of the concentration ratio, C, when I_{solar} = 1000 W/m^2, T_o = 25°C = 298K, and $\eta_{optic} = \alpha = \varepsilon = 1.0$.

Example 7.4 indicates the maximal efficiency level for an ideal CSP system with a concentration of $C = 100$ as well as the required receiver temperature.

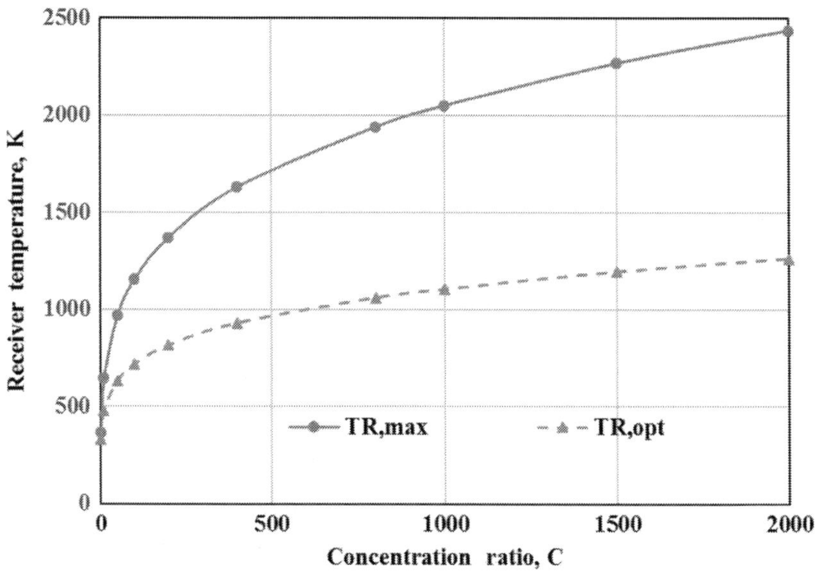

FIGURE 7.25 Variation of maximal and optimal receiver temperatures with the concentration ratio, *C*.

Example 7.4

An ideal CSP system is designed to have a concentration ratio of $C = 100$. Determine the maximum efficiency of the CSP system and the associated receiver temperature when it is operated when the ambient temperature is 25°C and the solar irradiance is $I_{solar} = 1000\ W/m^2$. For the calculations, assume $\eta_{optic} = \alpha = \varepsilon = 0.9$.

Solution:

By solving numerically Eq. (7.40) with $C = 100$, $I_{solar} = 1000\ W/m^2$, $T_o = 273 + 25 = 298$ K, and $\eta_{optic} = \alpha = \varepsilon = 0.9$, the optimal receiver temperature, $T_{R,opt}$, can be determined:

$$T_{R,opt} = 704\text{K}$$

Then, the maximal efficiency for the CSP system with $C = 100$ can be estimated from Eq. (8):

$$\eta_{CSP,opt} = \left(1 - \frac{298}{704}\right)\left(0.9 - \frac{(5.67 \cdot 10^{-8}) \cdot (0.9) \cdot (704)^4}{(0.9) \cdot (100) \cdot (1000)}\right) = 43.9\%$$

The CSP system efficiency does not exceed 43.9%. When the receiver temperature is allowed to reach $T_{R,max} = 1122\text{K}$, the CSP system efficiency is reduced to zero.

7.4 FUEL-BASED GENERATION SYSTEMS

There are several types of power generation systems that are commercially available and can be utilized for CHP for building applications. In general, three categories of cogeneration systems can be considered:

1. *Conventional Cogeneration Systems*: These systems consist of large cogeneration units (more than 1000 kW) and require a thorough design process to select the size of all equipment and components (i.e., prime movers, electrical generators, and heat recovery systems).
2. *Packaged Cogeneration Systems*: These systems are small (below 1000 kW) and are easy to design and install since they are preengineered and preassembled units.
3. *Distributed Generation Technologies*: Some cogeneration systems can use a number of technologies that have been recently developed to produce both electricity and heat including fuel cells.

7.4.1 Conventional Cogeneration Systems

A typical cogeneration plant consists of several pieces of equipment to produce electricity and heat (in the form of either steam or hot water). The number and the type of equipment in a cogeneration plant depend on the size of the system and the procedure used to generate electricity and heat. Generally, a conventional cogeneration system includes the following components:

1. *A Prime Mover*: This is the most important equipment in a cogeneration system. It is typically a turbine that generates mechanical power using a primary source of fuel. There are three turbine types that are commonly used in cogeneration plants: turbines operated by steam generated from boilers, gas turbines fueled by natural gas or light petroleum products, and internal combustion engines fueled by natural gas or distillate fuel oils.
2. *A Generator*: A device that converts the mechanical power to electrical energy.
3. *A Heat Recovery System*: A set of heat exchangers that can recover heat from exhaust or engine cooling and convert it into a useful form—typically hot water.

To operate a cogeneration plant, a robust control system is needed to ensure that all the individual pieces of equipment provide the expected performance. Two basic operation cycles are used to generate electricity and heat: (1) a bottoming cycle or (2) a topping cycle:

1. *Bottoming Cycle*: In this cycle, the generation of heat is given the priority to supply process heat to the facility. Thermal energy is produced directly from fuel combustion (in the prime mover). Heat is then recovered and fed to the generator to produce electricity as illustrated in Fig. 7.26A. Industrial plants characterized by high-temperature heat requirements (such as steel, aluminum, glass, and paper industries) use typically bottoming cycle cogeneration systems.
2. *Topping Cycle*: Unlike the bottoming cycle, the generation of electricity takes precedence over the production of heat as indicated in Fig. 7.26B. The waste heat is then recovered and converted to either steam or hot water. Most existing cogeneration systems are based on topping cycles. A hybrid of a topping cycle commonly used by several industrial facilities and even by electrical utilities is the combined cycle as depicted in Fig. 7.26C. In this cycle, a gas turbine is typically used to produce electricity. The exhaust gas is then fed to a heat recovery steam generator to generate more electricity—using a steam turbine. For a cogeneration plant, a small portion of steam can be converted into a useful form of thermal energy.

Three types of prime movers are generally considered for large and medium conventional cogenerations systems: (1) steam turbines, (2) reciprocating engines, and (3) gas turbines. A brief overview of each prime mover type is provided in the following sections.

7.4.1.1 Steam Turbines

Steam turbines are the oldest and most versatile prime movers used in power generation. In the United States, most of the electricity generated from power plants use steam turbines. However, steam turbines are also utilized in CHP systems, combined cycle power plants, and district heating systems. The capacity of steam turbines ranges from 50 kW to several hundred megawatts. There are several types of steam turbines used today in power generation applications including the following:

- Condensing turbines: These are power-only utility turbines. They exhaust directly to condensers that maintain vacuum conditions at the discharge.

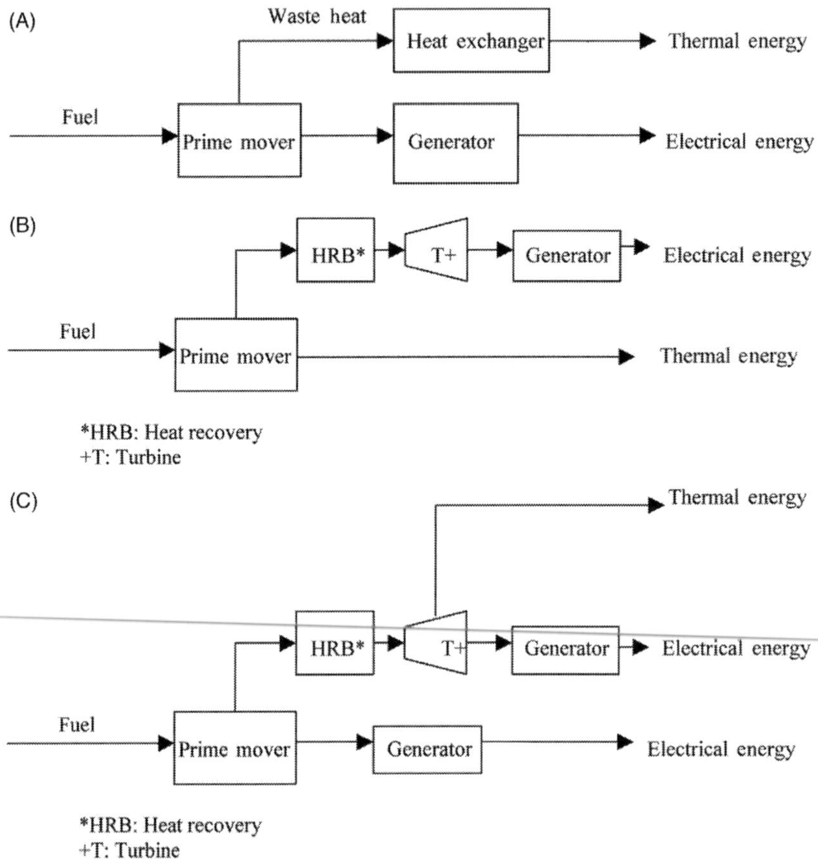

FIGURE 7.26 Different types of cogeneration systems. (A) Topping cycle cogeneration system. (B) Bottoming cycle cogeneration system. (C) Combined cycle cogeneration system.

- Noncondensing turbines: Also referred to as back-pressure turbines. They exhaust steam to the facility mains at conditions close to the process heat requirements.
- Extraction turbines: These have openings in their casing for extraction of a portion of the steam at some intermediate pressure before condensing the remaining of the steam.

The electrical generating efficiency of steam turbines varies from 37% for large electric utility plants to 10% for small plants that produce electricity as a byproduct of steam generation. The common applications of steam turbines for CHP systems involve industrial processes where solid or waste fuels are readily available. In the United States, it is estimated that over 580 industrial and institutional facilities use steam engines to produce about 19,000 MW of electric capacity. Table 7.10 lists typical cost and performance parameters for cogeneration systems using steam turbines.

TABLE 7.10 Cost and Performance Characteristics for Selected Cogeneration Systems Using Steam Turbines

Cost and Performance Parameters	System capacity in kW		
	500	3000	15,000
Total installed cost ($/kW)	$918	$385	$349
Fuel input (MMBtu/h)	26.7	147.4	549.0
Electric efficiency (%)	6.4	6.9%	9.3%
Steam to process (MMBtu/h)	19.6	107.0	386.6
Overall efficiency (%)	79.6	79.5	79.7

Source: Energy Nexus Group, 2002a. Technology Characterization: Steam Turbines, A report prepared for EPA, Climate Protection Partnership Division, Washington, D.C.

7.4.1.2 Reciprocating Engines

Reciprocating internal combustion engines are a widely used technology to generate power for several applications including institutional and industrial applications, and CHP systems. The capacity of reciprocating engines varies from a few kilowatts to over 5 MW. Two basic types of reciprocating engines are utilized for power generation applications:

1. Spark ignition engines use generally natural gas. Spark ignition engines are now commonly used for duty-cycle stationary power applications.
2. Compression ignition or diesel engines have been historically popular for small and large power generation applications. However, diesel engines are now restricted to emergency and standby applications due to emission concerns.

Electric efficiency of reciprocating engines ranges from 28% for small engines (less than 100 kW) to 40% for large engines (above 3 MW). Waste heat can be recovered from four sources in reciprocating engines: (1) exhaust gas, (2) engine jacket cooling water, (3) lube oil cooling water, and (4) turbocharger cooling. High- and low-pressure steam can be produced from reciprocating engines for CHP applications. Overall efficiency for a CHP system using a reciprocating engine fueled by natural gas can exceed 70%.

Reciprocating engine technology has improved significantly over the last decades with increased fuel efficiency, reduced emissions, improved reliability, and low first cost. The use of reciprocating engines for CHP generation applications is expected to continue to grow in the next decade. Currently, it is estimated that 1055 CHP systems operating in the United States use reciprocating engines with an overall power capacity of 800 MW. Table 7.11 provides typical cost and performance parameters for commercially available reciprocating engines suitable for cogeneration applications.

TABLE 7.11 Cost and Performance Characteristics for Selected Cogeneration Systems Using Reciprocating Engines

Cost and performance parameters	System capacity in kW				
	100	300	800	3000	5000
Total installed cost (2001 $/kW)	$1515	$1200	$1000	$920	$920
Fuel input (MMBtu/h)	1.11	3.29	8.20	28.48	43.79
Electrical efficiency (%)	30.6	31.1	33.3	36.0	39.0
Total heat recovered (MMBtu/h)	0.57	1.51	3.50	11.12	15.28
Overall efficiency (%)	81.0	77.0	76.0	75.0	74.0

Source: Energy Nexus Group, 2002b. Technology Characterization: Reciprocating Engines, A report prepared for EPA, Climate Protection Partnership Division, Washington, D.C.

7.4.1.3 Gas Turbines

Gas turbines provide one of the cleanest means for electric power generation with very low emissions of carbon dioxide (CO_2) and oxides of nitrogen (NO_x). The available capacity of gas turbines ranges from 500 kW to 250 MW. Gas turbines are well suited for CHP applications because high-pressure steam (as high as 1200 psig) can be generated from their high-temperature exhaust using heat recovery steam generators. It is estimated that over 575 industrial and institutional facilities in the United States use gas turbines to generate power and heat with a total capacity of 40,000 MW (Energy Nexus Group, 2002c). Table 7.12 lists the performance characteristics of selected commercially available gas turbine CHP systems. As indicated in Table 7.12, electrical efficiency and overall CHP efficiency increase with the size of the gas turbine.

7.4.2 Packaged Cogeneration Systems

For cogeneration facilities requiring small systems ranging from less than 50 kW to about 1 MW, preengineered and factory-assembled cogeneration units are currently available with reduced construction, installation, and operation costs. In addition, small packaged systems with capacities ranging from 4 to 25 kW have been developed and can be installed in a short period of time with little interruption of service. Almost all packaged cogeneration systems are equipped with advanced controls to improve the reliability and the energy efficiency of the units.

Packaged cogeneration units are often sold as turnkey installations. In particular, manufacturers are generally responsible for the testing and installation of the complete cogeneration systems. The development of packaged systems has enlarged the appeal of cogeneration to a wide range of facility types including office buildings, restaurants, homes, and multifamily complexes. In

TABLE 7.12 Cost and Performance Characteristics for Selected Cogeneration Systems Using Gas Turbines

Cost and performance parameters	System capacity in kW				
	1000	5000	10,000	25,000	40,000
Total installed cost (2000 $/kW)	$1780	$1010	$970	$860	$785
Fuel input (MMBtu/h)	15.6	62.9	117.7	248.6	368.8
Electrical efficiency (%)	21.9	27.1	29.0	34.3	37.0
Steam output (MMBtu/h)	7.1	26.6	49.6	89.8	128.5
Overall efficiency (%)	68.0	69.0	71.0	73.0	74.0

Source: Energy Nexus Group, 2002c. Technology Characterization: Gas Turbines, A report prepared for EPA, Climate Protection Partnership Division, Washington, D.C.

addition, packaged cogeneration systems are now typically more cost-effective than conventional systems for small and medium hospitals, schools, and hotels. However, each facility has to be thoroughly evaluated to determine the economic feasibility of any packaged cogeneration system.

Packaged cogeneration systems use typically reciprocating engines. Recently, microturbines have been developed and have been commercially available since 2000. Microturbines are small electricity generators that operate at very high speeds (over 60,000 rpm). They are available in sizes ranging from 30 to 350 kW. They have been promoted as ideal generators for distributed generation applications including cogeneration systems due to their connection flexibility (they can be stacked in parallel), their reliability, and their low emissions. Table 7.13 summarizes performance characteristics of selected commercially available microturbines suitable for CHP applications.

7.4.3 Distributed Generation Technologies

Distributed generation systems are considered to produce electricity using small and modular generators. The small generators with capacities in the range of 1 kW to 10 MW can be assembled and relocated in strategic locations (typically near customer sites) to improve power quality, reliability, and flexibility to meet a wide range of customer and distribution system needs. Some technologies have emerged in the last decade that allow to generate electricity with reduced waste, cost, and environmental impact, which may make the future of distribution generation promising especially in a competitive deregulated market. Among these technologies are renewable energy sources (wind and solar), fuel cells, as well as microturbines, combustion turbines, gas engines, and diesel engines. Table 7.14 summarizes some distribution generation technologies with their average efficiency, capacity, and applications

TABLE 7.13 Cost and Performance Characteristics for Selected Cogeneration Systems Using Microturbines

Cost and performance parameters	System capacity in kW			
	30	70	100	350
Total installed cost (2000 $/kW)	$2516	$2031	$1561	$1339
Fuel input (MMBtu/h)	0.437	0.948	1.264	4.118
Electrical efficiency (%)	23.4	25.2	27.0	29.0
Heat output (MMBtu/h)	0.218	0.369	0.555	1.987
Overall efficiency (%)	73.0	64.0	71.0	77.0

Source: Energy Nexus Group, 2002d. Technology Characterization: Micro-Turbines, A report prepared for EPA, Climate Protection Partnership Division, Washington, D.C.

(American Gas, 1999). Perhaps the recent developments in the fuel cells represent the best opportunity for distributed generation and CHP applications. It is expected that fuel cells will play a significant part in the 21st century electricity market.

The principle of the fuel cell was first demonstrated over 150 years ago. In its simplest form, the fuel cell is constructed similar to a battery with two electrodes in an electrolyte medium, which serves to carry electrons released at one electrode (anode) to the other electrode (cathode). Typical fuel cells use hydrogen (derived from hydrocarbons) and oxygen (from air) to produce electrical power with other by-products (such as water, carbon dioxide, and heat). High efficiencies (up to 73%) can be achieved using fuel cells.

Table 7.15 summarizes various types of fuel cells that are commercially available or under development. Each fuel cell type is characterized by its electrolyte, fuel (source of hydrogen), oxidant (source of oxygen), and operating temperature range. CHP systems using fuel cells are available in sizes ranging from a few kilowatts to thousands of kilowatts. Table 7.16 lists some performance characteristics of selected commercially available CHP systems using the fuel cell technology.

To evaluate the technical and the economic feasibility of a cogeneration system, it is important to collect accurate data about the facility and its energy consumption. In particular, current and projected future energy consumption and costs need to be available. For a detailed evaluation analysis, hourly electrical and thermal energy data are required. However, monthly and even yearly energy data can be sufficient for a preliminary feasibility analysis of cogeneration systems. In this section, basic considerations are discussed to evaluate the feasibility of cogeneration systems.

To account for the fact that a typical cogeneration system produces both electrical power, E_e, and thermal energy, E_t, from fuel energy, FU, as illustrated

TABLE 7.14 Characteristics of Current Distributed Generation Technologies

Type	Efficiency (%)	Size	Applications
Combustion turbine	24–40	500 kW to 30 MW	Cogeneration (commercial/industrial), transmission and distribution support
Diesel engine	36–42	50 kW to 6 MW	Standby, remote and peak shaving power
Gas engine	28–38	5 kW to 2 MW	Cogeneration (commercial/industrial), peak shaving and primary power
Microturbine	21–40	25 kW to 300 kW	Cogeneration (commercial/light industrial), primary power
Fuel cell	40–65	1 kW to 3 MW	Cogeneration (residential/commercial), primary power

TABLE 7.15 Basic Characteristics of Selected Types of Fuel Cells

Fuel cell name	Electrolyte	Fuel	Oxidant	Operating temperatures (°C)
PAFC	Phosphoric acid	Pure hydrogen	Clear air (without CO_2)	200
AFC	Alkaline	Pure hydrogen	Pure oxygen and water	60–120
SPFC	Solid polymer	Pure hydrogen	Pure oxygen	60–100
MCFC	Molten Carbonate Hydrocarbons	Air and oxygen	650	
SOFC	Solid oxide	Any fuel	Air	900–1000

TABLE 7.16 Cost and Performance Characteristics of Selected CHP Using FUEL Cells

	System capacity in kW			
Cost and performance parameters	10	100	200	2,000
Fuel cell type	PEM	SOFC	PAFC	MCFC
Total installed cost (2002 $/kW)	$5500	$3500	$4500	$2800
Fuel input (MMBtu/h)	0.10	0.80	1.90	14.80
Electrical efficiency (%)	30.0	45.0	36.0	46.0
Heat output (MMBtu/h)	0.04	0.19	0.74	3.56
Overall efficiency (%)	68.0	70.0	75.0	70.0

Source: Energy Nexus Group (2002e).

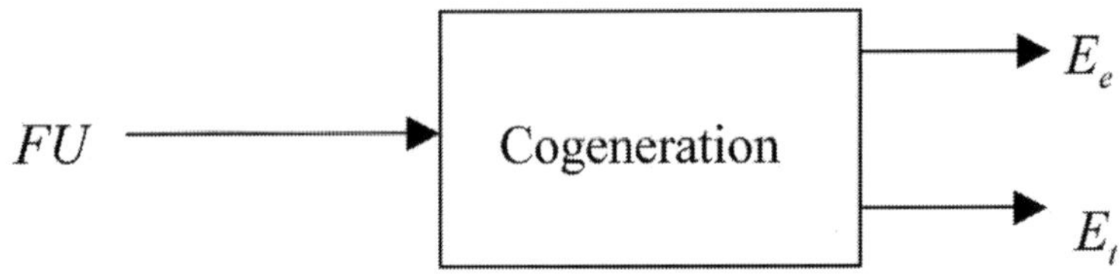

FIGURE 7.27 Energy input/output for a typical cogeneration system.

in Fig. 7.27, the overall thermodynamic efficiency, $\eta_{overall}$, of the cogeneration system is defined as follows:

$$\eta_{overall} = \frac{E_e + E_t}{FU} \tag{7.41}$$

It should be noted that for a cogeneration facility to meet the criteria specified by the PURPA, it has to comply with certain efficiency standards. These standards use a "PURPA efficiency," η_{PURPA}, which is defined by Eq. (7.42):

$$\eta_{PURPA} = \frac{E_e + E_t / 2}{FU} \tag{7.42}$$

PURPA states that to be a Qualified Facility (QF) for cogeneration, the efficiency, η_{PURPA}, has to be at least

1. 45% for the cogeneration facilities with a useful thermal energy fraction that is larger than 5%.
2. 42.5% for the cogeneration facilities with a useful thermal fraction that is larger than 15%.

The main purpose of the PURPA efficiency is to ensure that a sufficient amount of thermal energy is produced so that the cogeneration facility is more efficient than the electric utility. A number of regulations within PURPA provide incentives to develop cogeneration facilities. Among these incentives are the legal obligations of the electric utilities toward the cogeneration facilities set by section 210 of PURPA (FERC, 1978). For instance, the electric utility has to

- purchase cogenerated electrical energy from the QFs,
- sell electrical energy to the QFs, and
- provide access for the QFs to transmission grid.

Example 7.5 illustrates an estimation of the overall thermodynamic and PURPA efficiencies for a conventional cogeneration system.

Example 7.5

Problem: Consider a 20-MW cogeneration power plant in a campus complex. An energy balance analysis indicates the following energy fluxes for the power plant:

- electricity generation 33%
- condenser losses 30%
- stack losses 30%
- radiation losses 7%

It is estimated that all the condenser losses but only 12% of the stack losses can be recovered. Determine the overall thermodynamic efficiency as well as the PURPA efficiency of the power plant.

Solution: First, the recovered thermal energy is determined:

$$E_t = \text{Condenser Losses} + \text{Part of the Stack Losses}$$

This thermal energy output can be expressed in terms of the fuel use (FU) of the power plant:

$$E_t = 30\% * \text{FU} + 12\% * [30\%\text{FU}] = 0.34 * \text{FU}$$

The energy flow for the campus power plant is summarized in the diagram:

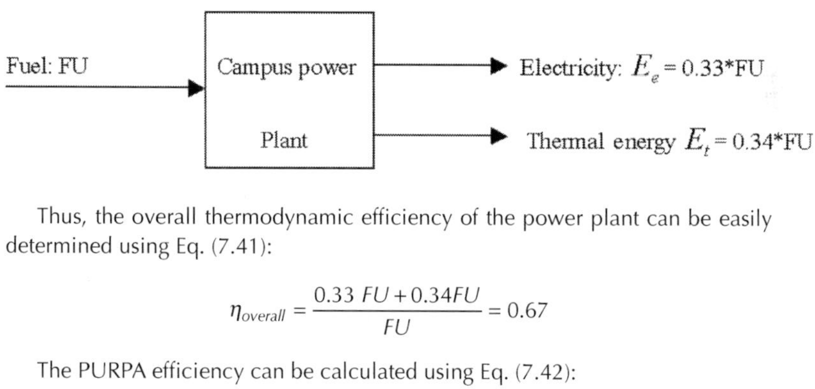

Thus, the overall thermodynamic efficiency of the power plant can be easily determined using Eq. (7.41):

$$\eta_{overall} = \frac{0.33\ FU + 0.34FU}{FU} = 0.67$$

The PURPA efficiency can be calculated using Eq. (7.42):

$$\eta_{PURPA} = \frac{0.33\ FU + (0.34FU)/2}{FU} = 0.50$$

Therefore, the campus power plant meets the PURPA criteria ($\eta_{PURPA} > 45\%$) and is thus a qualified cogeneration facility.

7.5 CASE STUDY 1: CARBON-NEUTRAL RESIDENTIAL COMMUNITIES IN KUWAIT

Kuwait relies on an increasingly significant portion of its fossil fuel production to generate and meet its electricity needs. Moreover, electricity and other energy resources are supplied to customers using heavily subsidized prices resulting in significant financial burdens to the Kuwaiti government. According to the Kuwait budget for 2012/2013, KD3.1 billion ($10.9 billion) of the total 6.3 billion ($22.1 billion) subsidization for the consumer services is attributed to electricity subsidies (Ansari, 2013). In 2010/2011, electricity production cost the Kuwaiti government about 29.5 fils/kWh ($0.103/kWh), while the total generation and distribution cost is 37.89 fils/kWh ($0.133/kWh) (MEW, 2012). The fuel cost is about 24.01 fils/kWh ($0.084/kWh), which is about 81% of the production cost and 63% of the total generation and distribution cost (MEW, 2012). On the other hand, electricity price to the end-users is only 2 fils/kWh ($0.007/kWh). Thus, the Kuwaiti government subsidizes about 93% of the production cost and 94.7% of the total costs of electricity generation and distribution. These low electricity prices sold to the end-users make most investments by the private sector in energy efficiency and renewable technologies not cost effective. The study presented in this chapter investigates from the perspective of the Kuwaiti government the economic and environmental benefits of utilizing renewable energy resources to generate at least part of the electricity needs for residential communities with different energy efficiency designs and sizes. There are several potential benefits and significant challenges of integrating renewable energy technologies in the Kuwaiti energy mix (Munawwar and Ghedira, 2014). First, the use of the abundant wind and solar energy resources can increase the revenues associated with exporting more fossil fuel and can assist in reducing carbon emissions to mitigate for climate change. Without any policy changes, it is expected that 20% of the country's oil production would have been utilized for generating electricity consumed in Kuwait as early as 2017 (Ansari, 2013). However, subsidized fossil fuel power generation makes the cost-effectiveness associated with any private sector investments in renewable energy technologies difficult to achieve. In addition, the lack of local renewable energy experts and awareness of renewable energy benefits are cited as among the challenges noted for Kuwait (Munawwar and Ghedira, 2014).

In Kuwait and in the Gulf Cooperation Council (GCC) region, limited analyses for designing power systems for communities have been reported. However, several studies have been carried out to estimate the potential benefits of using renewable energy technologies including solar and wind to generate electricity in GCC countries. Most of these studies have considered generation systems that are not directly connected to the grid suitable to serve individual buildings. In particular, Al-Enezi et al. (2011) have estimated solar resources in Kuwait suitable for PV systems. They found that a 2.4-m^2 silicon mono-semicrystalline PV panel can generate more than 300 W of electrical power in Kuwait. Al-Nassar et al. (2005) carried out a detailed analysis of the potential wind resources in six Kuwaiti sites. Fig. 7.28 summarizes the annual average wind speed at 10 m height for the six sites. Radhi (2011) evaluated the impact of roof-mounted PV systems on net energy consumption of prototypical residential buildings in GCC countries. Their analysis results showed that due to the high temperatures prevalent in the GCC region, the efficiency of the PV module drops from a rated

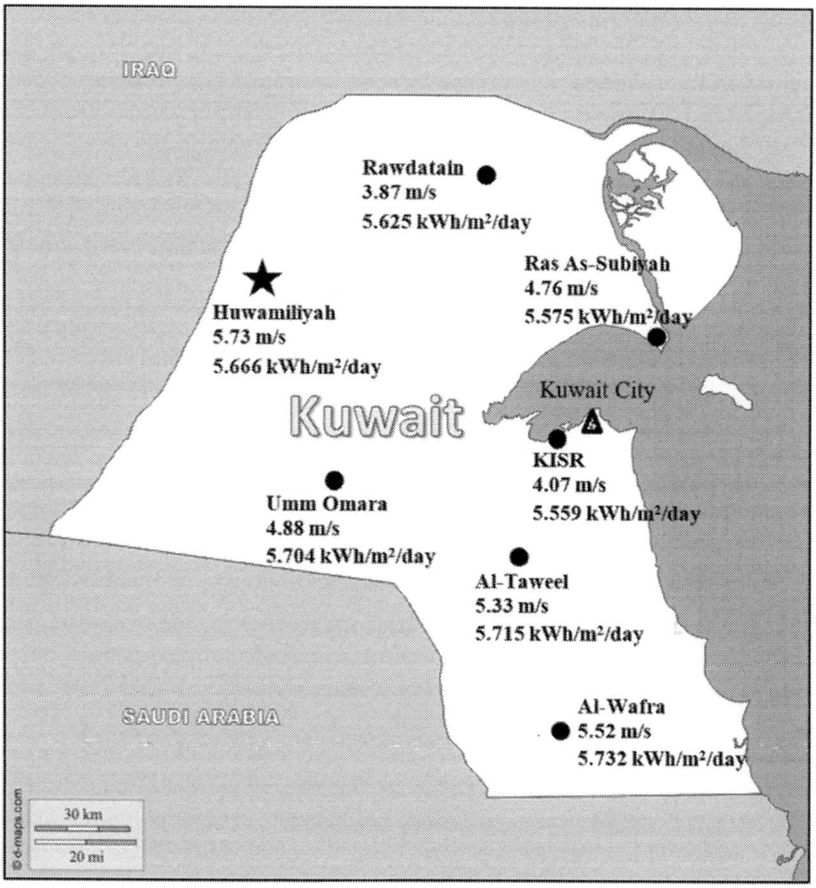

FIGURE 7.28 Kuwait map,(d-maps, 2016) showing locations of analyzed systems (in *circles*) by Al-Nassar (2005) and the location (in *star*) selected for the hybrid systems evaluated in this case study. Annual average wind speed and daily average global horizontal solar radiation are provided for all locations (Solargis, 2017).

value of 15.2% to a 12.2%–8.5% range. Their study has indicated also that installing building-integrated PV systems could save 15%–25% of the total energy consumption for the residential buildings located in the GCC region. However, these PV systems are not cost effective for the households due the current low electricity prices (Radhi, 2011). Shaahid and Elhadidy (2008) have evaluated the techno-economic feasibility of a standalone PV-diesel-battery hybrid system in meeting the electrical load of a residential building located in Dhahran, Saudi Arabia. Based on \$6900/kW capital cost for PV arrays and \$0.1/L diesel fuel price, the analysis indicated that the optimum cost of energy is about \$0.129/kWh using only diesel generators. When the renewable fraction from PV is set to be 22%, the cost of generating electricity increases to \$0.179/kWh.

Ramli et al. (2014) carried out a similar analysis for a standalone PV/diesel hybrid system in Makah, a city located in the western region of Saudi Arabia. The analysis results showed that a diesel generation system has the lowest LCOE of \$0.102/kWh. However, LCOE increases to \$0.117/kWh when 15% of the total electricity is generated from the PV system. Moreover, Elhadidy (2002) evaluated the performance of wind/solar/diesel standalone hybrid systems to meet 702,358 kWh/yr electrical load for 20 two-bedroom houses in Dhahran, Saudi Arabia. With installed costs of \$1000/kW for wind turbines and \$4000/kW for PV arrays, it was found that PV is not cost effective while wind could be competitive compared to the grid. In addition, Shawon (2013) carried out a comprehensive feasibility analysis of wind electrical power generation throughout the Middle East region. In particular, the analysis indicated that 20 GWh of electricity could be generated per year in Kuwait from a 10-MW wind farm at a cost of \$0.0702/kWh with a payback period of 8 years.

In summary, most of the reported studies for Kuwait did not consider utility-scale power plants and focused only on standalone systems. The case study outlined in this section evaluates the economic and environmental benefits for government-funded utility-scale hybrid power plants. In addition, the study considers the integration of high-energy performance designs of buildings as part of evaluating power generation systems suitable to meet the electrical loads of residential buildings. The general design approach outlined in Fig. 7.1 is used to determine optimal hybrid power generation systems for the residential communities considered in this case study. First, the solar and wind resources in Kuwait are presented. Then, the design specifications for the buildings within the communities are outlined (Ameer and Krarti, 2016). Then, select results obtained from a series of sensitivity and optimization analyses are presented to determine the optimal hybrid systems to meet the energy needs of the residential communities (Ameer and Krarti, 2017).

7.5.1 Renewable Energy Resources

Wind and solar resources for the selected site in Kuwait considered for the study are shown in Fig. 7.29. The site, located near Huwamiliayah (identified in Fig. 7.28 by a star), is selected to ensure that there is sufficient land available to install large hybrid power generation systems. In general, average monthly wind speeds are above 5 m/s throughout the year and are more than 6 m/s during the summer months. Moreover, the site has good solar resources with average daily global horizontal radiation above 6 kWh/(m^2·day) during the summer period ranging from May to September. In addition, the clearness index is consistently above 50% throughout the year. In addition, Kuwait has one of the hottest climates in the world with dry-bulb temperatures reaching more than 50°C and wet-bulb temperatures topping 34°C during summer (SAM, 2016). It should be noted that high wet-bulb temperatures affect the performance of water-cooling electrical power plants while high dry-bulb temperatures reduce the conversion efficiency of PV systems.

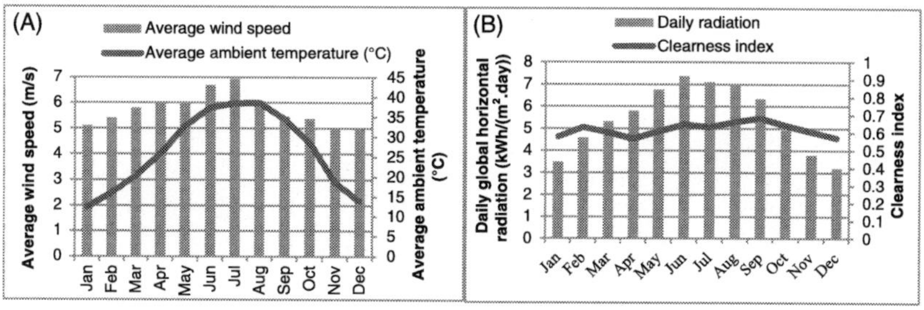

FIGURE 7.29 (A) Monthly wind speed at 10 m height. (B) Daily global horizontal radiation.

7.5.2 Building Design Specifications

As part of the case study outlined by Ameer and Krarti (2017), three community sizes are evaluated including 2000, 10,000, and 50,000 residential units. A prototypical villa is considered to represent all the residential buildings located within the community. The energy loads of the residential communities are estimated using the bottom–up approach by modeling the energy performance of the prototypical villa using various designs. In this study, a two-story villa with 500vm^2 total conditioned floor area is considered (Ameer and Krarti, 2016). The villa is first designed to meet the building energy efficiency code prevalent in Kuwait both in its 1983 and 2010 versions. Then, through optimization analysis, low-energy designs are identified for the villa using life-cycle cost (LCC) analysis approach over a 30-year period. It should be noted that Kuwaiti residential buildings use only electricity for all the end-uses including space heating and domestic hot water. Thus, only electrical loads are considered for the residential communities considered in this analysis. The details of the villa energy model, building energy analysis approach, and the optimization analysis results are described in detail by Ameer and Krarti (2016). Table 7.17 summarizes the specifications of the three design options considered for the prototypical villa. Fig. 7.30 compares energy and demand end-use distributions obtained for the three design alternatives of the prototypical villa. As indicated in Fig. 7.30, the air conditioning system (i.e., HVAC) is responsible for most of the annual electricity used as well as of the peak demand for the villa especially when it is designed according to code-1983 design. With an optimal design, the villa uses significantly less annual electrical energy and has lower peak demand. It should be noted that optimal villa design is characterized by improved building envelope components, higher efficiency air conditioning system, LED lighting fixtures, and high-performance refrigerators. Specifically, the optimal design saves about 56% and 42% of peak load compared to code-1983 design and code-2010 design, respectively, as noted in Fig. 7.30. Moreover, the optimal design alternative reduces annual electricity consumption by the villa by 51% and 37% relative to the previous and current Kuwaiti building energy efficiency codes.

TABLE 7.17 Main Specifications for Code-1983, Code-2010, Optimal Design Alternatives for a Prototypical Villa in Kuwait

Specifications	Code-1983	Code-2010	Optimal
Ext. walls	$U = 0.57$ W/m^2·°C (0.10 Btu/h·ft^2·°F)	$U = 0.57$ W/m^2·°C (0.10 Btu/h·ft^2·F)	$U = 0.47$ W/m^2·°C (0.08 Btu/h·ft^2·°F)
Ext. roof	$U = 0.40$ W/m^2·°C (0.07 Btu/h·ft^2·°F)	$U = 0.40$ W/m^2·°C (0.07 Btu/h·ft^2·°F)	$U = 0.31$ W/m^2·°C (0.055 Btu/h·ft^2·°F)
Window type	Double clear	Double reflective	Single reflective
Window over hang and fins	No overhang and fins	No overhang and fins	0.4 m overhang and fins
Window-to-wall ratio	13.0%	13.0%	7.8%
Cooling COP	2.0	2.3	3.5
Lighting power density	15.0 W/m^2 (1.4 W/ft^2)	9.7 W/m^2 (0.9 W/ft^2)	6.7 W/m^2 (0.62 W/ft^2)

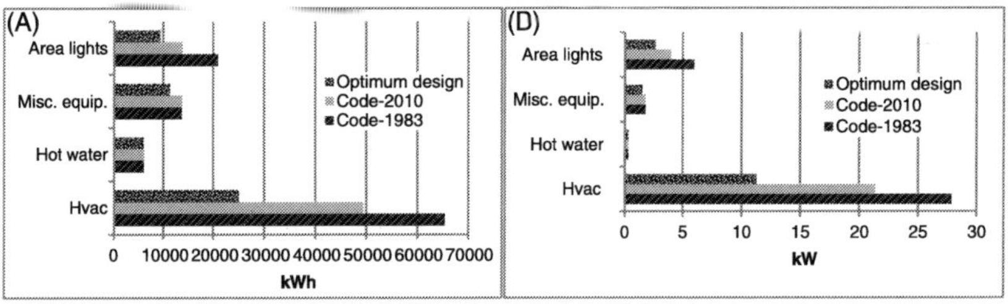

FIGURE 7.30 (A) Main end-use annual energy distribution and (B) peak load distribution for the three models.

7.5.3 Renewable Generation Systems

In addition to the grid and gas-fired generators, renewable energy systems are considered in this study to meet all or part of the electrical demand for the residential communities. Specifically, three renewable power generation systems are considered including PV arrays, CSP plants, and wind turbines. Table 7.18 lists the estimated installation, replacement, and maintenance costs for various components associated with PV, wind, and gas-fired generators considered in the analyses carried out in this study.

The CSP plant with TES outlined in Fig. 7.15 is considered in the analysis carried out for this case study (Turchi, 2010). Table 7.19 summarizes the basic design features and economic parameters of the CSP model. For the economic analysis presented in this chapter, a 30-year lifetime is assumed for all the evaluated system configurations with a 5% discount rate.

TABLE 7.18 Installation, Replacement, and Maintenance Costs Assumed for Various Systems and Technologies

Equipment	Initial cost	Repl. cost	O&M	Source
PV array	4183 ($/kW)	3973 ($/kW) Assuming (95% of initial cost)	27.8 ($/yr)	EIA (2013)
Wind turbine Enercon E-33 (330 kW capacity)	1850 ($/kW) or 610,500 ($/turbine)	390,720 ($/turbine) Assuming (64% of initial cost)	13,052 ($/yr)	EIA (2013), IRENA (2012)
Wind turbine Enercon E-115 (2500 kW capacity)	1850 ($/kW) or 4,625,000 ($/turbine)	2,960,000 ($/turbine) Assuming (64% of initial cost)	98,875 ($/yr)	EIA (2013), IRENA (2012)
Battery-Surrette 6CS25P	1250 ($/unit)	1050 ($/unit)	15 ($/yr)	Batterystuff (2015)
Inverter	600 ($/kW)	540 ($/kW)	0	NREL (2006)
Natural gas generator	973 ($/kW)	827 ($/kW) Assuming (85% of initial cost)	0.016 ($/h)	EIA (2013)

TABLE 7.19 Specifications and Economic Parameters for the CSP Plants Considered in This Study

Solar HTF fluid	Therminol VP-1
Storage HTF fluid	Hitec solar salt
Solar collector type	Siemens sunfield 6
Solar receiver type	Siemens UVAC 2010
Solar field design loop inlet temp.	295°C
Solar field design loop outlet temp.	390°C
Irradiation at design	900 W/m^2
Condenser type in the power plant	Evaporative cooling
Nominal discount rate	5%
Number of mirror washes per year	130 because of dusty climate (default is 62)

(Continued)

TABLE 7.19 Specifications and Economic Parameters for the CSP Plants Considered in This Study (*Conti*)

Direct capital cost	
Site improvement	\$30/m²
Solar field	\$270/m²
HTF system	\$80/m²
Storage	\$80/kWht
Power plant	\$830/kWe
Balance of plant	\$110/kWe
Contingency	7%
Total direct cost	\$879,593,506
Indirect capital cost	
EPC (engineer-procure-construct) cost	11% of the direct capital cost
Sales task	5% applied on 50% of direct capital cost
O&M (fixed cost per capacity)	\$65/kW·yr
O&M (variable cost by generation)	\$4/MWh
Total indirect cost	\$118,745,123

7.5.4 Optimization Analysis

In this section, select results of the optimization analysis are briefly presented.

7.5.4.1 Impact of Generation Technologies

Based on the economic parameters considered in the analysis using the available energy resources for the selected Kuwaiti site, wind turbines combined with the grid provide the best option to generate electricity for the residential communities independently of their size and their building design specifications and can ultimately achieve carbon-neutral conditions. Table 7.20 shows the components of the optimal power generation systems found for two scenarios of electricity prices from the grid for a 2000-unit residential community. The first case considers that the electricity price from the grid is 38 fils/kWh (\$0.133/kWh) to account for the current generation and distribution costs in Kuwait, while the second option assumes a grid electricity price of 29 fils/kWh (\$0.103/kWh) representing only the generation costs. LCOE values obtained for the optimal generation systems range from \$0.067/kWh to \$0.074/kWh significantly lower than the grid electricity production costs. As indicated in Table 7.20, the optimal systems in both cases include only wind turbines providing about 80% of the total load requirements for the 2000-unit residential community.

Fig. 7.31A illustrates the optimization analysis results for identifying the most economical hybrid generation system specific to the 2000-unit residential community. In particular, Fig. 7.31B illustrates the Pareto diagram presenting

TABLE 7.20 Optimal Design Features for the Grid-Connected Distributed Generation System for 2000-Unit Kuwaiti Residential Community

Category	Generation and distribution grid price 38 fils/kWh ($0.133/kWh)	Generation only grid price 29.5 fils/kWh ($0.103/kWh)
Peak load (MW)	32	32
Wind turbine, E-115 capacity (kW)	25,000 (10 units)	20,000 (8 units)
Wind turbine production (kWh/yr)	110,843,360	88,674,520
Grid purchases (kWh/yr)	24,646,948	33,276,154
Total production (kWh/yr)	135,490,304	121,276,154
Renewable fraction (%)	81.8	72.7
Carbon dioxide (kg/yr) of the system/only grid option	−3,292,823/66,760,052	10,717,756/66,760,052
Installation cost ($)	46,250,000	37,000,000
O&M cost ($/yr)	4,769,742	4,620,802
LCC ($)	119,572,624	108,033,056
LCOE ($/kWh)	0.074	0.067

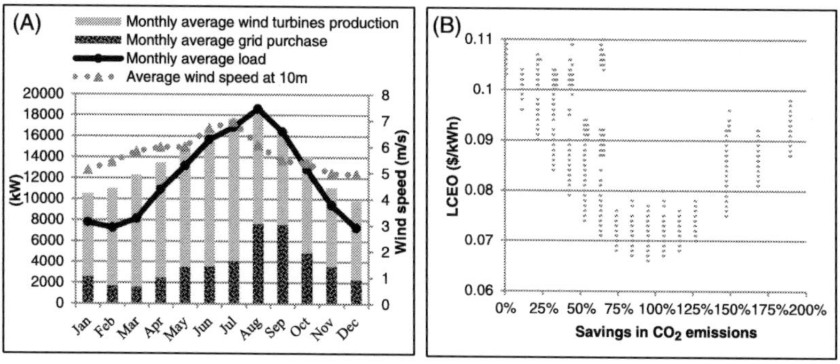

FIGURE 7.31 (A) Monthly electricity produced by the wind turbine, grid purchases, and average electrical load for an optimal DG system associated with 2000-unit residential community. (B) Variation of LCOE vs. percent savings in CO_2 emissions for various hybrid DG system options.

a wide range of power generation system configurations, with the variation of LCOE as a function of savings in carbon emissions relative to the grid-only option considered as the baseline for this analysis. When higher reduction level of carbon emissions has to be set, more renewable energy technologies have to be considered in the generation mix. It is clear from Fig. 7.31B that the power generation system with the lowest LCOE (optimal system) provides 94% reduction in CO_2 emissions reaching almost carbon-neutral conditions. In the analysis, the carbon emission factor for the grid is set to 632 g/kWh. As noted in Table 7.20, wind turbines constitute the optimal systems for the 2000-unit residential community in Kuwait. Fig. 7.31A shows the monthly electricity produced by the wind turbines and purchased from the grid compared to the electrical load required for the residential community. It should be noted that hourly analysis is conducted to compare electricity produced by the wind turbines and the load required by the community. When electricity produced by wind turbines exceeds the residential community load, then the differential generated power is provided to the grid. On the other hand, when electricity from the wind turbines are not sufficient to meet the residential community loads, the shortage power is purchased from the grid. During most months, there is an overall excess of electricity generated from the wind turbines as noted in Fig. 7.31A. However, for specific hours during almost all months, load has to be met by purchasing electricity from the grid due to the intermittent nature of wind power generation. The analysis assumes that the grid can be used as an electricity storage system by transferring the excess power. The negative carbon emissions listed in Table 7.20 indicate simply that the power generation system is transferring more electricity (generated from wind turbines) than power purchased to the grid and thus may exceed carbon-neutral conditions.

Fig. 7.32 provides a more extensive representation of the results of the simulation analysis to determine the optimal grid-connected hybrid generation system to serve the 2000-unit residential community. In particular, the performance

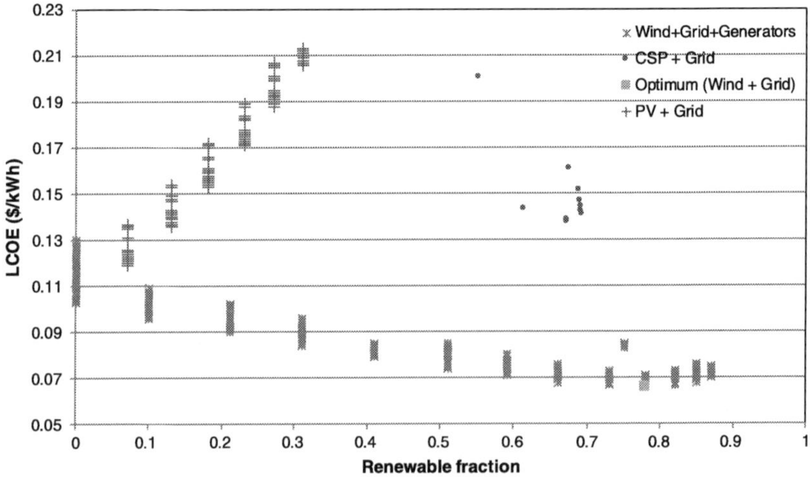

FIGURE 7.32 Variation of LCOE versus renewable fraction for different power generation systems.

indicators of all the generation technologies are shown including those of PV and CSP systems. As indicated in Fig. 7.31, LCOE values achieved by both PV and CSP systems are too high compared to those obtained using the wind turbines, the grid, and even the gas-fired generators. Additional analyses of the performance of PV and CSP plants are provided in later sections of this chapter to assess some of the conditions when these systems can become cost effective compared to the grid.

7.5.4.2 *Impact of Community Size and Design*

Table 7.21 summarizes the optimization results obtained for the power generation system to meet the electrical loads for the residential communities with three building energy efficiency designs as defined in Table 7.17 and three community sizes as 2000, 10,000, and 50,000 units. In particular, the basic characteristics and performance for the optimal systems corresponding to the various designs are provided in Table 7.21. It should be noted that the annual peak electricity demand for the three communities with the three building energy efficiency designs are provided. For each residential community design and size, the optimal power generation system is selected to minimize the LCOE. Fig. 7.33 compares the initial costs (IC) and LCC for power generation systems obtained when the buildings of the three residential communities are constructed using code-2010 or optimal building energy efficiency design rather than following the specifications of code-1983. Meanwhile, Fig. 7.34 compares the carbon emissions avoided due to constructing more energy efficient buildings (i.e., using code-2010 or optimal design) for the three residential community sizes instead of using energy conservation code-1983. As noted in Fig. 7.33, high-energy performance buildings result in lower electricity consumption and reduced optimal power generation system capacities and capital costs for all the residential community sizes. Specifically, the generation system for a residential community designed according to code-2010 can save up to 25% in the IC and up to 22% in LCCs compared with the optimal generation system obtained for a residential community designed based on code-1983. However, a residential community designed based on the optimal building energy efficiency design can save up to 60% of IC and 52% of LCC when compared to that designed using code-1983. In addition, Fig. 7.34 indicates that the optimal generation systems for residential communities constructed based on optimal building energy efficiency designs emits 54%–63% less CO_2 than power generation systems required for residential communities built based on code-1983 specifications.

7.5.4.3 *Impact of Grid Electricity Price*

A series of sensitivity analyses is carried out to assess the impact of various costs and design parameters on the selection of optimal grid-connected hybrid power generation systems to serve the three residential communities. In particular,

TABLE 7.21 Characteristics and Performance of Optimal Grid-Connect Hybrid DG Systems for Various Building Designs for a 2000-, 10,000-, and 50,000-Unit Residential Communities in Kuwait

Category	Size of the community (units)	Units designed using code-1983 Generation only grid price 29.5 fils/kWh ($0.103/kWh)	Units designed using to code-2010 Generation only grid price 29.5 fils/kWh ($0.103/kWh)	Units optimally designed Generation only grid price 29.5 fils/kWh ($0.103/kWh)
Peak load (MW)	2,000	72	55	32
	10,000	360	276	159
	50,000	1,802	1,378	794
Wind turbine, E-115 capacity (kW)	2,000	50,000 (20 units)	37,500 (15 units)	20,000 (8 units)
	10,000	237,500 (95 units)	187,500 (75 units)	112,500 (45 units)
	50,000	1,187,500 (475 units)	937,500 (375 units)	562,500 (225 units)
Wind turbine production (kWh/yr)	2,000	221,686,720	166,266,112	88,674,520
	10,000	1,053,006,208	831,321,920	498,795,904
	50,000	5,265,037,824	4,156,648,960	2,493,959,168
Grid purchases (kWh/yr)	2,000	54,123,264	44,277,764	33,276,154
	10,000	291,457,056	221,389,232	142,497,968
	50,000	1,457,284,608	1,106,946,176	712,489,472
Total production (kWh/yr)	2,000	275,809,984	210,543,872	121,276,154
	10,000	1,344,463,232	1,052,711,168	641,293,888
	50,000	6,722,322,432	5,263,595,008	3,206,448,640
Renewable fraction (%)	2,000	80.4	79	72.7
	10,000	78.3	79	77.8
	50,000	78.3	79	77.8
Carbon dioxide (kg/yr) of the system/only grid option	2,000	−4,635,248/ 135,470,592	1,161,212/ 106,240,584	10,717,756/ 66,760,052
	10,000	11,850,131/ 677,353,728	5,806,462/ 531,203,104	18,561,830/ 333,800,288
	50,000	59,249,688/ 3,386,761,728	29,032,280/ 2,656,015,360	92,808,824/ 1,668,999,040

TABLE 7.21 Characteristics and Performance of Optimal Grid-Connect Hybrid DG Systems for Various Building Designs for a 2000-, 10,000-, and 50,000-Unit Residential Communities in Kuwait (*cont.*)

		Units designed using code-1983	Units designed using to code-2010	Units optimally designed
Category	Size of the community (units)	Generation only grid price 29.5 fils/kWh ($0.103/kWh)	Generation only grid price 29.5 fils/kWh ($0.103/kWh)	Generation only grid price 29.5 fils/kWh ($0.103/kWh)
Installation cost ($)	2,000	92,500,000	69,375,000	37,000,000
	10,000	439,375,008	346,875,008	208,124,992
	50,000	2,196,875,008	1,734,375,040	1,040,625,024
O&M cost ($/yr)	2,000	8,558,092	6,798,157	4,620,802
	10,000	44,191,208	33,990,824	21,389,930
	50,000	220,955,984	169,954,144	106,949,624
LCC ($)	2,000	224,058,848	173,879,328	108,033,056
	10,000	1,118,702,208	869,397,312	536,940,672
	50,000	5,593,509,888	4,346,986,496	2,684,702,720
LCOE ($/kWh)	2,000	0.068	0.067	0.067
	10,000	0.068	0.067	0.066
	50,000	0.068	0.067	0.066

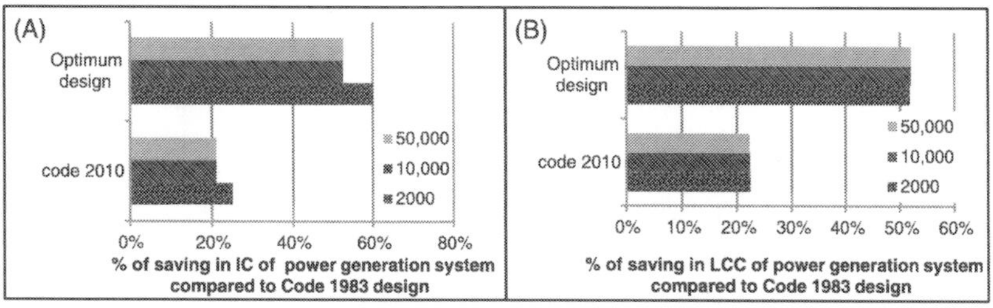

FIGURE 7.33 Comparative analysis in initial cost reduction of hybrid generation systems for buildings constructed using energy efficiency code-2010 and optimal design for three sizes of residential communities in Kuwait. The savings are estimated relative to buildings constructed using the original Kuwaiti energy efficiency code-1983.

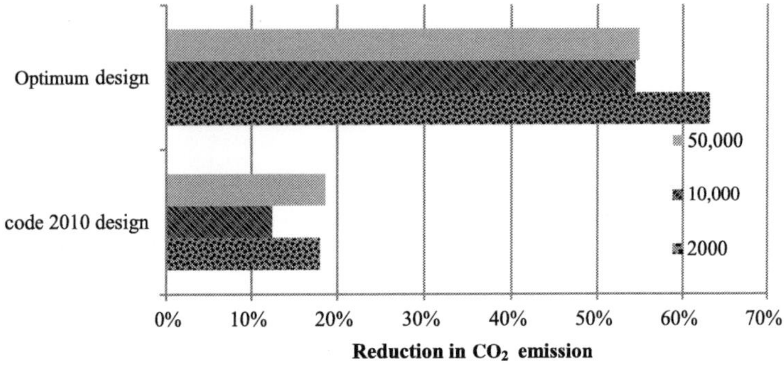

FIGURE 7.34 Savings in CO_2 emissions for power generation systems when residential community buildings are constructed using code-2010 and optimal design specifications instead of using code-1983.

Fig. 7.35 shows the effect of grid electricity prices and wind speed variations on the selection of an optimal power generation system for a 2000-unit residential community. Specifically, the analysis shows that when the grid electricity price increases above \$0.165/kWh while natural gas price remains less than \$0.40/ m^3, the best hybrid power generation system includes a mix of wind turbines and natural gas–powered generators with the grid providing complementary electricity needs for the residential community. Otherwise, wind turbines combined with the grid are the only systems that can optimally generate electricity for the residential community as long as the annual site wind speed is higher 3.9 m/s. That is, if a site with a lower wind speed than 3.9 m/s is considered, then wind turbines are no longer competitive with the grid in Kuwait.

7.5.4.4 *Impact of PV Installation Costs*

It should be noted from the results shown in Fig. 7.10 that PV and CSP are not selected to be part of the optimal power generation systems to serve the

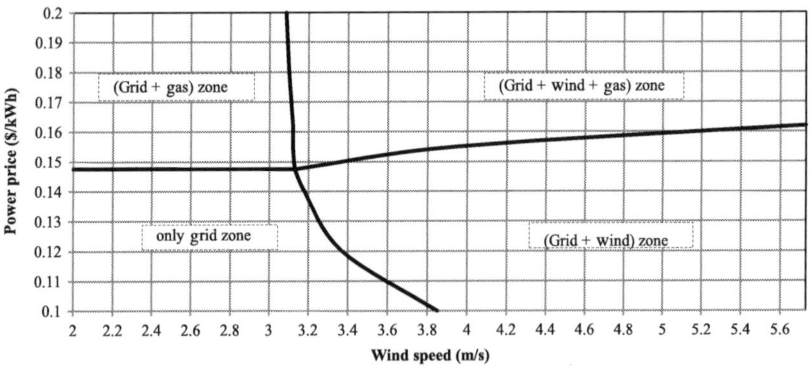

FIGURE 7.35 Impact of grid electricity price and average wind speed on the optimal DG system for a 2000-unit residential community in Kuwait.

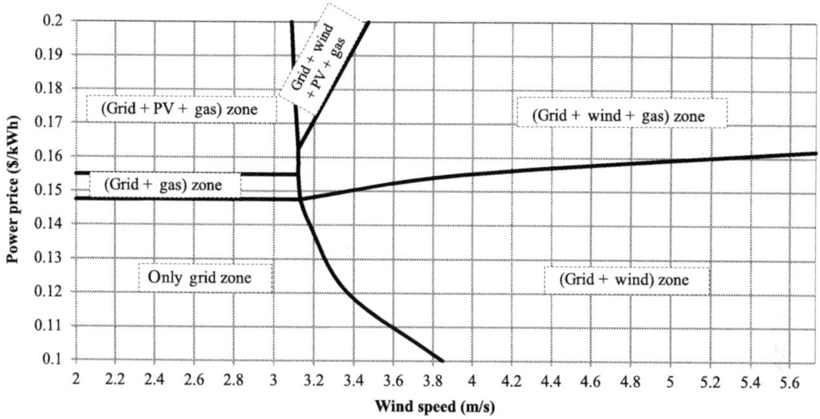

FIGURE 7.36 **Impact of grid electricity price and average wind speed on the optimal DG system for a 2000-unit residential community in Kuwait when the PV installed cost is reduced by 50% relative to the baseline.**

residential communities. As indicated earlier, the installation cost of PV systems considered in the analysis is $4183/kW regardless of the grid electricity prices or wind speed levels. However, when the installation costs are reduced by 50%, the cost-effectiveness of PV improves as indicated by the optimization results shown in Fig. 7.36. Specifically, Fig. 7.36 shows the impact of grid electricity prices and wind speed variations on the selection of the optimal power generation system for a 2000-unit residential community when the PV system installation costs are reduced to $2092/kW. According to the results of Fig. 7.35, the PV system starts to be cost-effective and part of the optimal hybrid power generation system when the annual average wind speed is less than 3.1 m/s and the grid electricity price is higher than $0.16/kWh.

7.5.4.5 Analysis of CSP Plants

While CSP is found not to be cost-effective compared to wind turbines for the Kuwaiti site considered for the study, a series of parametric analyses is carried out to determine the optimal specifications for the CSP plants that are designed to serve the three residential communities. In particular, the following CSP system parameters are varied: stow angle, deploy angle, solar multiple, full load storage hours, row spacing, number of subsections, and number of solar collector assemblies (SCAs) per loop are considered to optimize the cost and the performance of the CSP plants. Table 7.22 summarizes the design specifications, cost data, and performance indicators for the CSP plants required to meet the loads for the three residential communities considered in the analysis. In particular, Table 7.22 provides the electricity output of the power plants, the required land areas, the water usages, and the total estimated installation costs. The nominal LCOE values are estimated to be about $0.130/kWh for the

TABLE 7.22 Summary of the Specifications for the CSP Plants to Serve the Residential Communities

Size of residential community	2000-unit	10000-unit	50000-unit
Rated capacity	35 MW	160 MW	795 MW
Solar field substations number	4	6	10
Number of SCA per loop	7	7	9
Row spacing	24 m	24 m	20 m
Deploy angle	2 degrees	1 degree	3 degrees
Stow angle	176 degrees	176 degrees	175 degrees
Solar multiple	3.6	3.4	3
Full load storage hours	12 h	12 h	9.5 h
Parallel tank pairs	2	2	10
Tank height	14 m	14 m	14 m
Tank diameter	29 m	62 m	54.8 m
Annual energy output	161,021,344 kWh	698,695,296 kWh	2,885,747,968 kWh
Capacity factor	52.4%	49.8%	41.5%
Real LCOE	\$0.102/kWh	\$0.104/kWh	\$0.111/kWh
Nominal LCOE	\$0.130/kWh	\$0.133/kWh	\$0.141/kWh
Installation cost	\$423,807,816	\$1,871,056,594	\$8,056,916,822
Annual water usage	572,409 m^3	2,483,764 m^3	10,597,497 m^3
Required land area	3.3 km^2	14.1 km^2	52.45 km^2

2000-unit, \$0.133/kWh for the 10,000-unit, and \$0.141/kWh for the 50,000-unit residential communities.

Fig. 7.37 shows typical hourly performance of the 35-MW CSP plant. Fig. 7.38 compares electrical load profiles from the 2000-unit residential community obtained for different building energy efficiency designs against the power generated by the 35-MW CSP plant. Specifically, Fig. 7.37 shows the annual hourly profile for the absorbed thermal energy, thermal energy transferred to the power plant (either from the solar field or from the TES system), the thermal energy transferred to the TES system, and the electrical power output. It should be noted that the electrical power output is significantly lower than the thermal energy transferred to the power plant due to the rather low energy efficiency of the power plant. Indeed, the Rankine power cycle for the CSP plant has an estimated average efficiency of 36.6% and the thermal losses in the piping, heat exchangers, and TES tanks represent 7% of the total energy absorbed

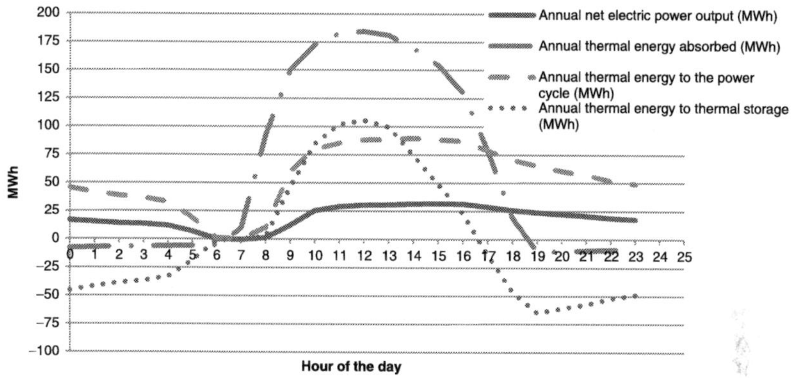

FIGURE 7.37 **Annual hourly variations of absorbed, transferred to power plant, stored energy, and electrical output of a 35-MW CSP plant in Kuwait.**

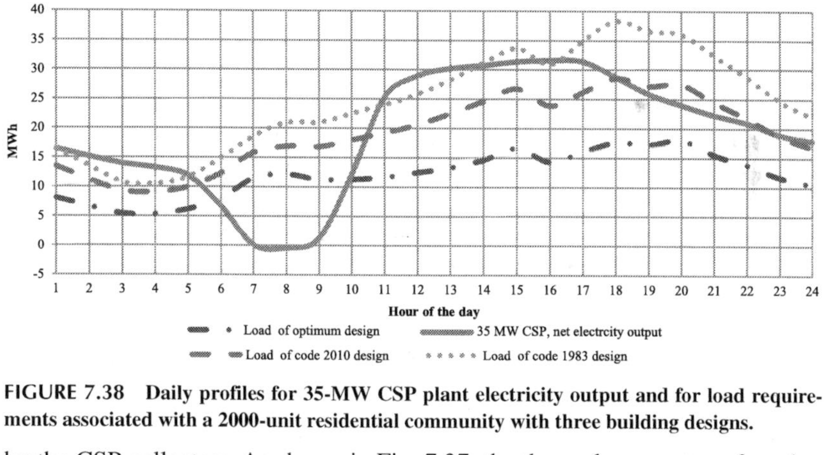

FIGURE 7.38 **Daily profiles for 35-MW CSP plant electricity output and for load requirements associated with a 2000-unit residential community with three building designs.**

by the CSP collectors. As shown in Fig. 7.37, the thermal energy transferred to the storage system is negative during some hours indicating that the TES system operates in a discharge mode when it transfers energy to the power plant. Moreover, the benefits of having a TES system for the CSP plant are clearly outlined in Fig. 7.37. In particular, the TES system allows the CSP plant to produce electricity during nighttime (except during 2 h). In addition, the TES system allows the power plant to operate even during cloudy periods.

From the results shown in Fig. 7.38, it is clear that a 35-MW CSP plant is not able to generate sufficient electricity during morning hours ranging from 6 a.m. to 10 a.m. to meet all the buildings' electrical loads even for the optimally designed residential community. Moreover, the same 35-MW CSP plant does not produce enough electricity during some night hours to meet the electrical load for residential communities designed using code-1983 and to a lesser extent using code-2010. Similar results are obtained for CSP plants serving larger residential communities.

A sensitivity analysis is conducted to evaluate the effect of several parameters on the performance and cost benefit of CSP plants. For each parameter, 20% variations in its optimal value are considered. The sensitivity analysis results are summarized in Table 7.23 for a 35-MW CSP plant suitable to serve the 2000-unit residential community. It can be noted that the solar field has the highest effect on LCOE since changing 20% of its cost results in 7% change in LCOE. The storage system cost also has a significant effect on LCOE with 20% change in its cost leading to 4.6% variation in LCOE value. Moreover, a 20% variation in the fixed O&M costs provides a 3% change in LCOE. Most of the other parameters such as costs associated with EPC, variable O&M, and site improvements do not affect significantly the LCOE of the CSP plant. Indeed, 20% variations in these parameters would result in

TABLE 7.23 Summary of the Parametric Analysis for a 35-MW CSP Plant in Kuwait

Sensitivity parameter	Sensitivity values	LCOE at the lower value (¢/kWh)	LCOE at the higher value (¢/kWh)	Installation cost at the lower value ($)	Installation cost at the higher value ($)
EPC cost (% of direct cost)	11% of direct cost (+/−20%)	12.77 (1.5% reduction)	13.17 (1.5% increasing)	415,593,039 (1.9% reduction)	432,022,593 (1.9% increasing)
Solar field cost	270 $/m^2 (+/−20%)	12.06 (7.0% reduction)	13.88 (7.0% increasing)	387,030,055 (8.7% reduction)	460,585,576 (8.7% increasing)
Power plant cost	830 $/kW (+/−20%)	12.77 (1.5% reduction)	13.17 (1.5% increasing)	415,945,466 (1.9% reduction)	431,670,165 (1.9% increasing)
Storage system cost	80 $/kW (+/−10%)	12.37 (4.6% reduction)	13.57 (4.6% increasing)	399,711,892 (5.7% reduction)	447,903,739 (5.7% increasing)
Site improvement	30 $/m^2 (+/−20%)	12.87 (0.8% reduction)	13.07 (0.8% increasing)	419,721,398 (1.0% reduction)	427,894,233 (1.0% increasing)
O&M (fixed cost/capacity)	65 $/kW (+/−20%)	12.59 (2.9% reduction)	13.35 (2.9% increasing)	–	–
O&M (variable cost by generation)	4 $/MWh (+/−20%)	12.86 (0.8% reduction)	13.08 (0.8% increasing)	–	–

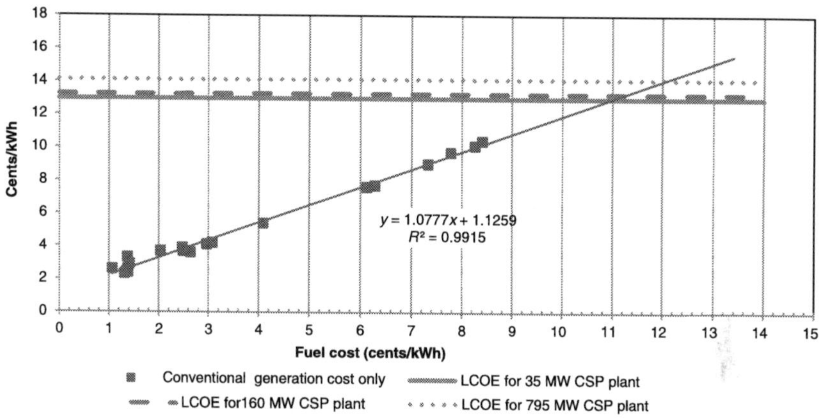

FIGURE 7.39 Variations of conventional generation cost and LCOE obtained from three CSP plants as functions of fuel cost in Kuwait.

less than 2% change in LCOE values as noted in Table 7.23. Therefore, the most important parameters that can affect the cost benefit of CSP plants in Kuwait include solar field, thermal storage system, and fixed O&M costs in that order.

To determine when CSP plants can be cost competitive compared to the grid in meeting the electrical loads for the residential communities, the fuel cost is varied and the grid production cost is determined. Fig. 7.39 shows the variation of the grid electricity production price with the fuel cost. The LCOE values estimated for three CSP plant sizes are also indicated in Fig. 7.39. The electricity from the CSP plants remains not competitive unless fuel cost per kWh becomes higher than $0.11/kWh. In other terms, CSP becomes cost competitive with the grid only when the fuel saved from not using the conventional power plant can be sold at a price of more than $0.11/kWh. This price is far higher than the current fuel prices in the international markets (with a $45/barrel price corresponding to $0.025/kWh). It should be noted that the results of Fig. 7.39 indicate that when the CSP plant capacity increases, LCOE also increases and the fuel price/kWh must be even higher to make CSP cost effective relative to the grid. For instance, the 795-MW CSP plant can only be cost effective when fuel prices exceed $0.12/kWh.

7.5.5 Recommendations

For the case study of residential communities in Kuwait, renewable energy technologies such as PV arrays, wind turbines, batteries, and CSP plants as well as natural gas–powered generators are considered to optimize grid-connected power generation systems to serve the electricity needs of residential communities in Kuwait. Specifically, the electrical energy requirements are estimated for three community sizes with 2000, 10,000, and 50,000 residential units designed according to different energy efficiency standards. In particular, residential buildings can be constructed according to the Kuwait conservation code of practice of 1983 and of 2010 as well as using optimized LCC analysis design. The results of the analysis outlined in this chapter show that the use of wind turbines can generate electricity at costs that are lower than the current grid production cost in Kuwait. A series of sensitivity analyses has been carried out to assess the impact of design parameters, capital costs, and grid generation costs on the optimal power generation system and its cost-effectiveness. In particular, it is found that the wind turbine is cost-effective for locations with average annual speed higher than 3.9 m/s and for grid-generated electricity prices higher than \$0.07/kWh. Natural gas–powered generators can be cost effective only when grid-generated electricity price is above \$0.15/kWh. Moreover, a PV system could be cost effective only when its capital costs are reduced by 50% of its current level and the grid electricity price is higher than \$0.16/kWh. CSP could be competitive with the grid only when the fuel prices become significantly high and exceed \$0.11/kWh.

Moreover, the analysis results have indicated that when the residential community is made up of buildings constructed with optimal design specifications, the power generation system capacity and installation cost can be reduced by 60% and 35% compared to cases where buildings are constructed using specifications of code-1983 and code-2010, respectively. Moreover, the high energy performance building designs allow to select power generation systems that can achieve carbon-neutral residential communities by reducing carbon emissions by up to 63% compared to buildings that meet code-1983 requirements.

In addition, the performance and the economic benefits of parabolic trough CSP plants are evaluated in this study. Specifically, three capacities for the CSP plants are considered including 35 MW to serve a 2000-unit community, 160 MW for a 10,000-unit community, and 795 MW for a 50,000-unit community. A series of parametric analyses is conducted to determine the impact of several parameters such as solar multiple, full load hours, row spacing, deploy and stow angles, number of SCA per loop, and number of subfield stations on the performance and the optimal design of the CSP plants. The results of the analyses have indicated that the lowest LCOE obtained for the CSP plants ranges from \$0.15/kWh to \$0.18/kWh. Therefore, CSP is not cost effective compared to the other considered generation systems including the grid.

Based on the results of the study presented for this case study, the following recommendations can be provided:

- Renewable power generation technologies can be cost-effective and should be considered by the Kuwaiti government to reduce its reliance on fossil fuel to produce electricity to preserve its main income source and decrease its carbon emissions.
- Wind technology is found to be highly cost-effective in Kuwait when considering the current grid production costs and should be considered to meet at least some of the electrical loads for newly constructed residential communities.

- The cost-effectiveness of the CSP plants is found to be highly dependent on the fuel costs. However, CSP remains not competitive with the current fuel prices in Kuwait.
- Before relying on any power generation systems, improving the current energy efficiency requirements for Kuwaiti buildings should be considered since this option provides the most cost-effective approach to save fuel use and reduce the need for new power plants in Kuwait.

7.6 CASE STUDY 2: DESIGN OF OPTIMAL HYBRID SYSTEMS

This case study outlined the design procedure for optimal hybrid distributed generation (DG) systems for three remote communities in Alaska. All three of these towns have isolated electrical grids and currently rely on diesel fuel for 100% of their electricity and heating requirements. This assessment included an analysis of each community's electrical and thermal load, a wind and solar resource evaluation, modeling and optimization of various DG systems using LCC analysis of these systems. For all three of the communities, hybrid wind-diesel systems have the potential to provide reductions in the cost of energy. However, the economic feasibility of these systems is extremely site-specific. In addition to providing possible decreases in the cost of energy, these hybrid systems can also provide significant environmental benefits, such as reductions in diesel fuel use and CO_2 emissions.

7.6.1 Introduction

In the state of Alaska, large transmission systems exist in the most populated areas such as the Railbelt region, which runs from Fairbanks through Anchorage to the Kenai Peninsula. Transmission systems are also present in some areas in Southeast Alaska; however, over 150 rural Alaskan communities have their own isolated electrical grids. About 70% of the electricity requirements for the Railbelt region are met with natural gas generators whereas in the majority of the remote areas, most of the electricity and heating requirements are met with diesel generators (AEA, 2011a). In 2010, residents in these rural areas had electricity rates between 40 and 60 cents per kilowatt hour (kWh), with some paying over $1.00/kWh (AEA, 2011b). Conversely, communities in the Railbelt region that get electricity from hydroelectric facilities or natural gas generators have electricity rates between 10 and 15 cents per kWh (Fay et al., 2011).

Due to rising electricity rates and diesel prices, many rural Alaskan communities are considering the addition of renewable energy sources and other DG technologies to form hybrid DG systems. Alaska's significant wind resource has led many coastal communities to invest in wind energy projects (AEA, 2011a). As of spring 2009, 10 hybrid wind–diesel systems were operating in rural communities, with six more systems in development (Baring-Gould and Dabo, 2009). Preliminary results are promising; hybrid DG systems have the potential to lower the cost of electricity, improve the reliability of electricity and heating systems, and lower the emissions from diesel generators. Based on their study in rural Alaska, Clark and Isherwood (2004) believe that hybrid DG systems can compete with conventional fossil fuel generation for communities with high electricity costs, available renewable resources, and no interconnection with a large grid. Indeed, the results indicated that a wind–diesel hybrid system could result in diesel fuel savings of over 50% and cost savings of over 30% compared to the base case (diesel only).

This section summarizes the results of an optimization analysis of hybrid DG systems for three rural communities in Alaska: Mountain Village, Deering, and Ambler conducted by Willman and Krarti (2013). The study includes an analysis of each community's electrical and thermal load; a wind and solar resource evaluation, modeling, and optimization of hybrid DG systems; an economic analysis of these systems; and an evaluation of the environmental benefits that they can provide. In this section, a brief description of the analysis approach and design recommendations is provided. More details for the other two sites can be found in the published study by Willman and Krarti (2013).

7.6.2 Optimization Design Approach

The analysis to determine the best hybrid distributed generation system to meet the energy requirements for a residential community considers the following steps:

Step 1: Determine the hourly electrical and thermal loads for the community. The estimation can be based on measured data or simulation analysis.
Step 2: Identify the main energy resources available within or close to the site where the community is located. These resources typically include solar, wind, geothermal, hydroelectric, fossil fuel, and biomass.
Step 3: Collect specific analysis data including economic parameters such as fuel costs and electricity prices from the grid as well as the capital and O&M costs for various power generating technologies.
Step 4: Carry out optimal analysis for the various technically feasible options for power generation technologies including fuel-based as well as renewable energy technologies. Simulation tools are typically used in the analysis.
Step 5: ank and select the most cost-effective hybrid system including the optimal capacities. The economic analysis using LCOE or LCC is typically considered to determine the cost-effectiveness of the various hybrid distributed systems considered in the evaluation. Chapter 9 discusses the various economic analysis methods suitable to assess the cost-effectiveness of energy projects.

A description of the steps and the results for the analysis is presented with a focus on one rural community in Alaska located in Mountain Village (Willman and Krarti, 2013). Note that all three communities have a stand-alone electrical grid and currently rely on diesel fuel for all of their electricity and heating requirements.

7.6.3 Electrical and Thermal Loads

The annual electricity generation, diesel fuel use for electricity generation, and CO_2 emissions from electricity generation for three Alaskan rural remote communities are shown in Table 7.24.

The total generating capacity for the community of Mountain Village is 2212 kW (AVEC, 2011). However, the type and capacity of the diesel generators are unknown. As a result, Mountain Village was modeled as having four 455-kW Caterpillar generators and a 410-kW Cummins generator for a total capacity of 2230 kW. Deering has four generators with a total capacity of 578 kW. These units include 100-kW and 138-kW John Deere generators and two 170-kW Cummins generators (Vaught, 2007). Ambler has three generators with a total capacity of 982 kW (AVEC, 2011) but the type and capacity of the generators are unknown. As a result, Ambler was modeled as having three Cummins generators with the following capacities: 270, 315, and 410 kW for a total generating capacity of 995 kW. Since hourly electricity and heating demand data were not available for any of the communities, a whole building simulation program was used to model the hourly electrical and thermal load for each village (Willman and Krarti, 2013). The results of the simulation including the peak and average electric and thermal loads are shown in Table 7.25.

TABLE 7.24 Electricity Generation, Diesel Fuel, and Carbon Emissions for Three Alaska Communities

Community	Electricity generation (kWh/yr)	Diesel fuel use (gal/yr)	CO_2 emissions (tons/yr)
Mountain Village	2,799,595	189,184	1,920
Deering	711,319	55,145	560
Ambler	1,249,161	89,892	912

TABLE 7.25 Average and Peak Electrical and Thermal Loads for Three Alaska Communities

Community	Average electric load (kW)	Average thermal load (kW)	Peak electric load (kW)	Peak thermal load (kW)
Mountain Village	303	2,277	512	7,846
Deering	74	775	132	2,232
Ambler	139	1,212	285	4,439

TABLE 7.26 Annual Average Wind Speed, Wind Power Density, and Solar Radiation for Three Alaska Communities

Community	Annual avg. wind speed (m/s)	Annual avg. wind power density (W/m^2)	Annual avg. solar insolation (kWh/m^2/d)
Mountain Village	7.28	523	2.80
Deering	5.22	322	2.61
Ambler	3.60	88	2.58

7.6.3.1 *Energy Resource Assessment*

Using the Alaska Energy Authority (AEA) extensive study on renewable energy resources (AEA, 2011a), wind and solar resource assessment is performed for each of the villages. Although Alaska's solar resource is minimal during the winter, a solar resource assessment was also completed. Hourly wind and solar resource data were obtained from the AEA and the National Solar Radiation Database. The annual average wind speed, wind power density, and solar radiation values for each community are listed in Table 7.26.

7.6.4 Electricity and Fuel Prices

Power Cost Equalization (PCE) program statistics for fiscal year 2010 were obtained for each community (AEA, 2011b). This program provides subsidies for residential customers in rural areas to decrease the cost of electricity for the first 500 kWh used by a customer each month. Community facilities are also eligible for a subsidized rate, but state and federal customers and commercial facilities are not. PCE subsidies are based on nonfuel and fuel costs (AEA, 2007). The PCE program statistics provide electricity price information for the electricity that is eligible for a subsidy and that which is not eligible. Diesel fuel costs for each community are \$3.15/gal for Mountain Village, \$4.69/gal for Deering, and \$3.75/gal for Ambler. The electricity sales and PCE rates are summarized in Table 7.27 for the three communities.

The results from a study performed by the Institute of Social and Economic Research at the University of Alaska Anchorage indicate that electricity rates and PCE subsidies decrease as wind penetration increases (Fay et al., 2010). As wind power supplies more electricity to a community, diesel fuel use decreases, which lowers the electricity rate. However, a reduction in the electricity rate also results in a lower subsidy from the PCE program. It should be noted that this has been viewed as a disincentive for communities to invest in wind–diesel systems. As diesel fuel prices increase, the PCE sub-

TABLE 7.27 Electricity Sales and Rates Based on PCE Program

Community	Mountain Village	Deering	Ambler
Total electricity sales (kWh)	2,650,142	647,749	1,216,727
PCE eligible sales (kWh)	1,094,841	Unknown	540,632
Non-PCE eligible sales (kWh)	1,555,301	Unknown	676,095
Electricity rate (¢/kWh)	50.62	77.37	53.80
PCE subsidy (¢/kWh)	29.20	30.74	32.22
PCE electricity rate (¢/kWh)	21.42	46.63	21.58

sidy also increases. However, communities with wind–diesel systems are not protected by the PCE program to the same extent as those with diesel systems. This effect has the most significant impact on communities whose wind–diesel systems operate at low wind penetration levels. For systems with high penetration levels, the effect is not as great, since an increase in wind penetration protects the community from high fuel prices. Despite the issues surrounding the PCE subsidy, wind systems continue to be an attractive option for remote communities with an abundant wind resource due to their potential to lower electricity rates, reduce CO_2 emissions, and protect these areas from diesel price fluctuations.

7.6.5 Optimization Analysis

DG systems were modeled and optimized using HOMER software (HOMER, 2016). Each town was modeled with its electrical and thermal load, a boiler, diesel generators with heat recovery, wind turbines, solar panels, batteries, and converters. For the sensitivity analyses, the diesel price was varied from \$1/L to \$3/L (about \$3.80/gal to \$11.40/gal). These values are consistent with the fuel price projections from the Institute of Social and Economic Research. Based on other wind–diesel system studies, a project lifetime of 25 years and a discount rate of 3% were chosen for the analysis.

The capital cost and operation and maintenance (O&M) costs that were used for the simulations are shown in Table 7.28. The capital cost for the solar panels was obtained through the examination of grant applications for the Alaska Renewable Energy Fund. Battery system costs were reported by Susitna Energy and Sandia National Laboratories and converter costs were obtained from ABS Alaskan. Based on the types of wind turbines currently installed in Alaska, the Entegrity 50 kW turbine was considered for each hybrid system.

TABLE 7.28 Economic Parameters for HOMER Simulations

System component	Capital cost ($/kW)	O&M ($/kWh)	References
Diesel generator	0	0.02	Vaught (2007)
Wind turbine	4000–15,000 [11]	0.04	Fay et al. (2010)
Solar panels	11,000	110 ($/yr)	Arvizu et al. (2011)
Battery (Surrette 6CS25P)	$1375 (total cost)	15 ($/yr)	
Converter	1100	50 ($/yr)	

7.6.6 Selected Results

The analysis results for the community's current diesel fuel cost of $3.16/gal are shown in Table 7.29. This table displays the capital cost, the net present cost, and the cost of energy (COE) for the optimal system for each level of wind turbine capital costs. The first wind–diesel–battery (W–D–B) hybrid system listed in the table includes eight wind turbines, a 455-kW John Deere generator, a 410-kW Cummins generator, 100 batteries, and a 200-kW converter system.

The monthly average electric production profile for the hybrid system with eight wind turbines is shown in Fig. 7.40. For this case, the annual electricity production from wind is 1,227,169 kWh and that from the diesel generators is 1,748,822 kWh. Wind accounts for 41% of electricity production. The diesel generator electricity production, diesel fuel use for electricity generation, and CO_2 emissions from electricity generation for this hybrid system and the baseline system are shown in Table 7.30. The W–D–B hybrid system results in a reduction of 1,050,773 kWh of diesel generator electricity production. This

TABLE 7.29 Mountain Village Simulation Results

Optimal system	Turbine capital cost ($/kW)	Initial capital ($)	Total NPC ($)	LCOE ($/kWh)
W–D–B	4,000	1,887,500	44,168,904	0.210
W–D–B	6,000	2,082,500	44,776,100	0.223
W–D–B	8,000	1,787,500	45,186,308	0.232
D–B	10,000–15,000	247,500	45,315,064	0.235

NPC, net present cost.

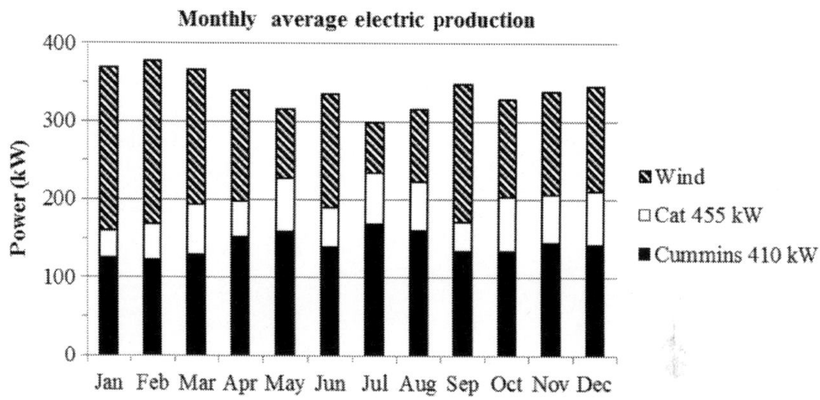

FIGURE 7.40 Electric production profile for the hybrid system.

TABLE 7.30 Comparison Between Baseline System and the Hybrid DG System

System	Electricity production from diesel (kWh/yr)	Diesel fuel use (gal/yr)	CO_2 emissions (t/yr)
Baseline	2,799,595	189,184	1,920
Hybrid	1,748,822	118,164	1,199

equates to a reduction in diesel fuel use of about 71,020 gal and a CO_2 emissions reduction of 721 metric tons per year.

The results from the sensitivity analysis are shown in Fig. 7.41. The sensitivity variables are the diesel price and the capital cost of the wind turbines, denoted by eW15 Capital Multiplier. The capital cost ranges from $4000 to $15,000 per kilowatt of installed capacity. The LCOE for each system is also shown in the figure. A W–D–B hybrid system is the optimal system, except for low diesel price and high capital cost values.

Even though PV is not included in the optimal system, it could still be economically feasible. According to the simulation results, the COE for solar hybrid systems at a diesel price of $3.16/gal ranges from $0.214/kWh to $0.265/kWh. Solar hybrid systems become part of the optimal system if renewable energy fraction (REF) constraints are imposed on the electric profile. The optimal system for REFs ranging from 0% to 50% and wind turbine capital costs ranging from $4000 to $15,000 per kilowatt are shown in Fig. 7.42. Specifically, Fig. 7.42 indicates the optimal systems for a fixed diesel price of $3.16/gal, the current cost of fuel. As shown in the figure, a W–D–B system is the predominant optimal system; however, wind–PV–diesel–battery systems are the optimal sys-

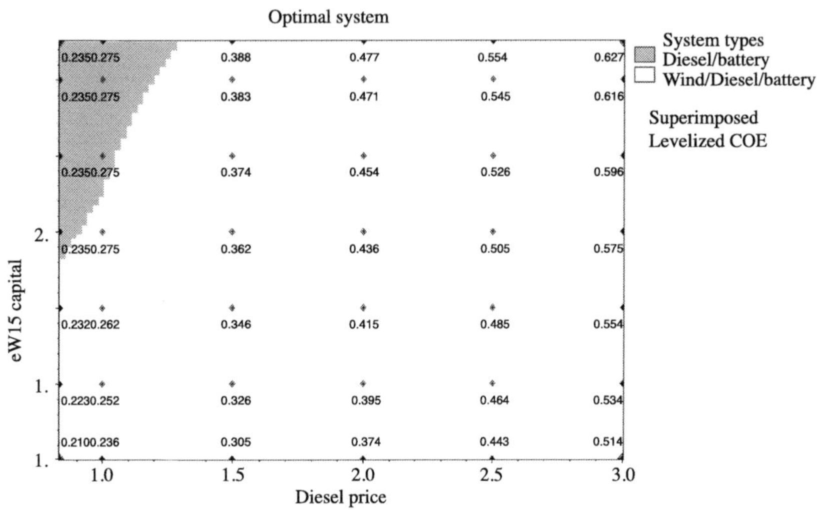

FIGURE 7.41 Mountain Village wind turbine sensitivity analysis.

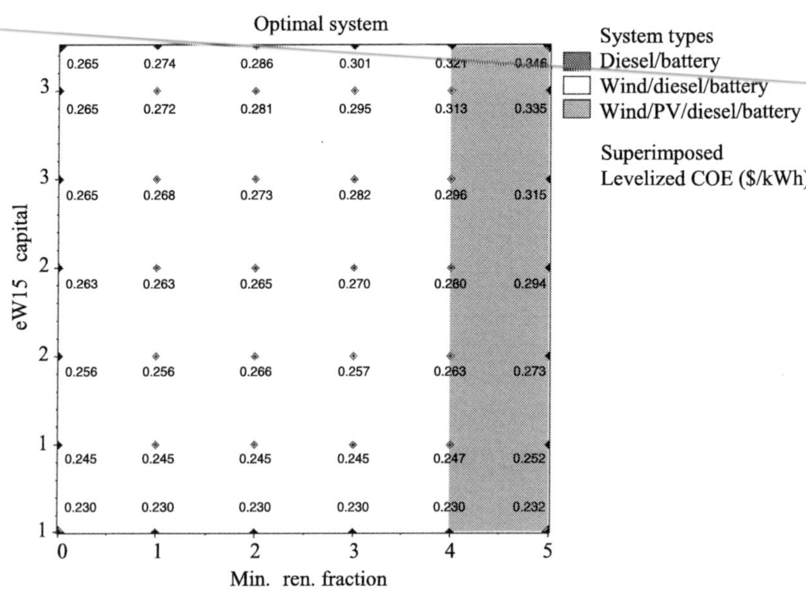

FIGURE 7.42 Mountain Village REF sensitivity analysis.

tem for high REF constraints. For all of these systems, the COE falls below Mountain Village's standard electricity rate of $0.51/kWh.

The LCOE as a function of the percent reduction in CO_2 emissions from electricity generation is shown in Fig. 7.43. As indicated in the chart, the optimal

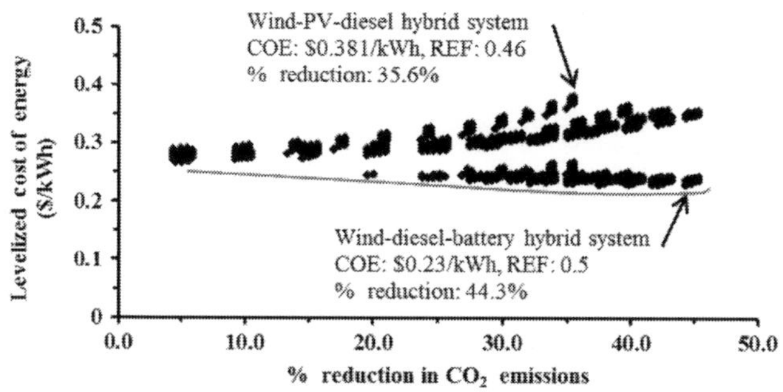

FIGURE 7.43 COE as a function of reduction in CO_2 emissions for the Mountain Village community.

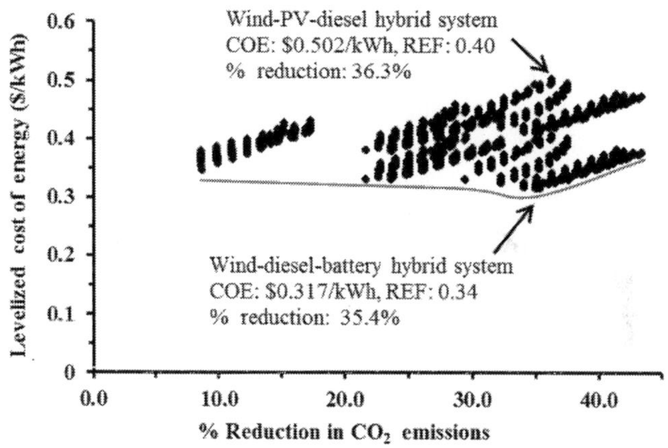

FIGURE 7.44 COE as a function of reduction in CO_2 emissions for the Deering community.

system in terms of the COE has a REF of 50%, a COE of $0.23/kWh, and a reduction in CO_2 emissions of 44.3%. These values are achieved with a W–D–B system with 10 turbines at a capital cost of $4000 per kilowatt, two generators, 200 batteries, and a 300-kW converter system. With a capital cost of $10,000 per kilowatt, the COE for this system increases to $0.292/kWh. The system with the highest COE includes 10 turbines at a capital cost of $15,000 per kilowatt, a 20-kW PV system, two generators, and a 400-kW converter system.

The LCOE as a function of the percent reduction in CO_2 emissions from electricity generation obtained for the rural community in Deering is shown in Fig. 7.44. As indicated in the chart, the optimal system in terms of the COE has a REF of 34%, a COE of $0.317/kWh, and a reduction in CO_2 emissions of 35.4%. These values are achieved with a W–D–B system with two turbines at a capital cost of $4000 per kilowatt, three generators, 50 batteries, and a 75-kW converter system. With a capital cost of $10,000 per kilowatt, the COE for this system increases to $0.369/kWh. The system with the highest COE has two turbines at a capital cost of $15,000 per kilowatt, a 70-kW PV system, three generators, and a 100-kW converter system.

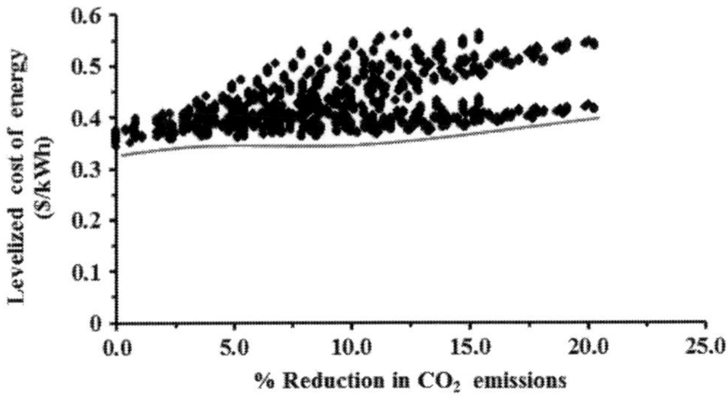

FIGURE 7.45 COE as a function of reduction in CO_2 emissions for the Ambler community.

For the Ambler community, the LCOE obtained as a function of the percent reduction in CO_2 emissions from electricity generation is outlined in Fig. 7.45. Unlike the other two communities, the COE increases with the percent reduction in CO_2 emissions for all systems. The optimal system in terms of the COE is a system with a REF of zero and therefore no reduction in CO_2 emissions.

7.6.7 Recommendations

To evaluate the feasibility of distributed generation systems for Mountain Village, Deering, and Ambler, HOMER software was used to model these systems. For Mountain Village and Deering, W–D–B systems are the optimal system for current diesel fuel prices, depending on the capital cost of the wind turbines. The levelized cost of electricity for these hybrid systems ranges from $0.210/kWh to $0.232/kWh for Mountain Village and from $0.288/kWh to $0.305/kWh for Deering. These ranges are well below the standard electricity prices for each of these communities.

The sensitivity analysis results indicate that the optimal system for Mountain Village is a W–D–B system for almost the entire range of diesel fuel prices and wind turbine capital costs that were considered. For Deering, when the diesel price is high (above about $2.3/L), a W–D–B system is the optimal system. When the diesel price is low, this hybrid system is only the optimal system for relatively low wind turbine capital costs. Solar systems could also be economically feasible for Mountain Village and Deering, but they are not part of the optimal system unless a REF constraint is imposed on the electric profile. Unlike Mountain Village and Deering, a W–D–B system is not the optimal system for Ambler. For this community, a diesel system with heat recovery is the optimal system for almost the entire range of diesel fuel prices and wind turbine capital costs that were considered in the sensitivity analysis. However, it should be noted that although a wind–diesel system is not the optimal system for Ambler, this type of system could still be economically feasible for this community. This is also true for systems with photovoltaics.

The range of COE results supports the findings from previous studies that have shown that the economic feasibility of wind–diesel systems is extremely site-specific. The hybrid systems for Mountain Village have a lower COE than those for Deering. It is the wind resource that has the largest effect on this result, since Mountain Village has a significantly better wind resource than Deering. The wind resource also has an effect on the optimal system for Ambler. This community has the worst wind resource; as a result, a wind–diesel system is not the optimal system. Additionally, solar hybrid systems may be more economically feasible than wind systems for Ambler, especially when the cost of the wind turbines is high.

Overall, the results from this assessment indicate that hybrid wind–diesel systems may be an economically beneficial option for rural Alaskan communities with an abundant wind resource. However, it should be noted that these results are based on simulated electricity and heating demand data since real hourly demand data were not available for any of the communities. The electricity and heating demand have a significant impact on the optimization, and any changes in the demand may alter the HOMER results for the optimal system type and the levelized cost of energy for each system. As a result, a more detailed analysis with real hourly electricity and heating demand data should be completed to more accurately evaluate the economic feasibility of these hybrid systems. Additionally, more research is needed to determine the impact that the PCE program may have on communities that develop these systems. Despite the issues surrounding the PCE subsidy, wind systems continue to be an attractive option for remote communities due to their potential to lower electricity rates and provide significant environmental benefits, such as reductions in diesel fuel use and CO_2 emissions. In some cases, these benefits may be achieved without an increase in the cost of energy.

7.7 SUMMARY

In this chapter, a design analysis approach for sustainable communities is presented with a focus on hybrid power systems that combine renewable and fuel generation technologies. The analysis relies on cost optimization to meet the energy needs of grid-connected or off-grid communities. An overview of the characteristics and performance metrics of hybrid power technologies is outlined along with some simplified analysis methods. Two case studies are presented to showcase the design approach for communities in different climates: the first case study considers the energy needs for residential communities in Kuwait while the second case study indicates the optimal hybrid systems for serving the microgrids of remote communities in Alaska. In both case studies, renewable power generation technologies were found to be cost-effective and able to meet significant portions of the energy needs of the communities.

PROBLEMS

7.1 Consider a 110-kW AC wind power turbine that is grid-connected to meet a local, constant base load of 100 kW. Make appropriate adjustments where needed to account for the following:

1. Hub height of 80 m (adjust for difference between the wind measurement level which is 10 m as provided in Table 7.1 and the turbine hub height).
2. Assume no daily or hourly variability in the loads.
3. The complete power curve is not available (as is often the case); therefore, use a 100-kW turbine power curve from HOMER database.
4. Account for cut in and cut out wind speed.

a. Calculate the annual energy delivered to the load and the cost per kilowatt for 25 years and 5% discount using HOMER. Also determine how much energy is delivered to the grid and its value if the average energy payment rate for the utility is \$0.09/kWh.

b. Determine the impact of daily and hourly variability of the load on the energy delivered by the turbine (both to the load and the grid).

c. Carry out a sensitivity analysis to assess the impact of the wind speed of the site and the utility rate on the cost-effectiveness of using the wind turbine to generate electricity.

Table P.1 provides the site monthly wind speed data at 10 m.

7.2 For your location, determine the monthly electricity produced by 100-kW PV arrays using fixed mount, one-axis tracking, and two-axis tracking. You can use any simulation tool such as SAM, PVWATTS, and HOMER. Plot the results as shown in Fig. 7.12.

7.3 A solar cell generates a maximum power of 2.8 W at regular solar conditions at 25°C. The open circuit voltage is measured at 0.50 V. Determine the maximum power of the solar cell if a concentrated PV system is considered with $C = 15$.

7.4 Determine the variation of the optimal concentration ratio for a solar cell in CPV system with the series resistance (varying from 0.0 to 0.050 Ω) if the cell's short-circuit current of 0.16 A at 25°C. Comment on the impact of using cells with high series resistance on the optimal concentration ration.

TABLE P.1 Site Monthly Wind Speed Data

Month	1	2	3	4	5	6	7	8	9	10	11	12
V (m/s)	7.5	8.0	8.2	9.5	9.7	8.9	7.7	7.5	8.5	9.0	8.2	7.8

7.5 Select a state or a country to design a 50-MW CSP plant. Ensure that the selected site has a good solar exposure and is connected to the grid. Using SAM simulation tool, determine the cost of implementing such plant using CSP technology. Estimate the LCC and the LCOE. Use SAM default values for the analysis. Perform a sensitivity analysis to determine the impact of various costs on the LCC and LCOE of the plant.
Optimize the design of the CSP plant.

7.6 Provide a simple payback analysis for implementing a cogeneration system in a hospital. Assume the following characteristics for the cogeneration system:

- Fuel input rate: 8000 Btu/kWh
- Heat recovery rate: 4800 Btu/kWh
- Maintenance cost: $0.02/kWh
- Maximum electrical output: 200, 300, 400, 500, 600, 700, or 800 kW
- Installed equipment cost: $700/kW, $1000/kW, or $1500/kW

Table P.2 summarizes the energy usage and cost of the hospital. Assume that the boiler(s) efficiency is 75%. The heating value of diesel fuel is 140,000 Btu/gal. Present of the results in one graph: the payback period vs. the equipment cost for various equipment sizes.

TABLE P.2 Monthly Utility Data for the Hospital Considered in Problem 7.6

Utility summary					
	Electricity			Fuel Oil	
Month	(kWh)	(kW)	($)	(Gallon)	($)
January	226,400	546	2,7020	20,659	14,911
February	273,600	572	28,949	20,555	12,639
March	280,800	564	31,048	16,713	9,670
April	228,000	526	25,251	10,235	4,742
May	246,000	692	28,755	12,193	5,347
June	301,200	884	36,604	12,352	9,001
July	346,800	1,040	45,031	20,604	3,122
August	403,200	944	46,374	17,276	5,711
September	303,600	860	36,541	10,457	3,762
October	276,000	872	33,559	10,890	3,726
November	272,400	662	28,042	13,478	5,255
December	276,000	524	25,041	17,661	7,808
Total	3,434,000		393,215	183,073	85,694

7.7 Using the same cogeneration system outlined for Problem 7.6 but assuming that the electric cost varies from \$0.05/kWh to \$0.15/kWh (including demand charge), determine the variation of the cogeneration system payback period vs. electricity cost. For this question, assume that the capacity and the cost of the cogeneration system are 300 kW and \$2000/kW, respectively.

7.8 For the same cogeneration system considered for Problem 7.6 but assuming that the life of the cogeneration system is 40 years, provide the optimal size of the cogeneration system for the hospital when the average interest rate is 8% and the inflation rate is 4%.

7.9 A 400-kW cogeneration system has an overall efficiency of 86%. The cogeneration system produces hot water at a thermal efficiency of 24%. Determine the cost-effectiveness of installing the cogeneration system in a facility at a cost of \$60,000. The cost of electricity, if purchased directly from the utility, is \$0.09/kWh and the cost of natural gas, used to operate the cogeneration system, is \$1.50/therm. Without the cogeneration system, a gas-fired boiler with an efficiency of 85% is utilized to produce hot water needed by the facility. Assume the following:

- The cogeneration system, as well as the facility, is operated 6500 h/yr.
- The maintenance cost of the cogeneration system is \$0.015/kWh.

Perform a life cycle analysis using a life cycle of 20 years and a discount rate of 5%.

7.10 Consider a facility located in Denver, Colorado with the hourly profiles shown in Table P.3. The owner would like to invest in a hybrid system including PV, wind, fuel cell, or microturbine. The facility is connected to the grid. Table P.4 provides basic efficiency and costs related to the hybrid system to be considered for the facility. A local utility serves the facility using the following time of use rate:

- The time of use rate has a customer charge of \$250 per month and a consumption charge of \$0.065/kWh, but it also includes a demand charge of \$7.50/kW except between 7 am to 9 pm Monday through Friday when the demand charge is \$12.50/kW. This demand charge is not ratcheted.

1. Find any relevant incentives and tax credit programs available for the client.
2. Use HOMER to find the best combination of DG systems (if any) without any incentives
3. Use HOMER to find the best combination of DG systems if the client wants at least 20% of the electricity to be generated from renewable energy systems with and without incentives.
4. Use HOMER to find the carbon emission reductions for optimal option found question (3). Estimate any benefits to the system economics if these emission reductions are valued at \$30/ton of carbon.

TABLE P.3 Average Monthly Building Electrical Load Profiles

Avg. kW	Month											
Hour	Jan	Feb	Mar	Apr	May	Jun	Jul	Aug	Sep	Oct	Nov	Dec
1	63.5	65.2	58.8	49.9	49.8	49.6	49.5	49.5	49.6	49.9	58.7	69.8
2	93.0	87.0	68.2	54.5	49.8	49.6	49.5	49.5	49.7	54.1	72.8	103.9
3	87.0	85.4	58.9	55.0	49.8	49.6	49.5	49.5	49.7	49.9	59.3	74.3
4	93.7	97.5	81.4	55.2	49.9	49.6	49.5	49.5	49.7	54.4	82.5	115.5
5	93.6	108.6	63.1	64.9	49.8	49.6	49.5	49.5	49.7	54.0	72.6	94.6
6	104.9	98.5	86.0	165.5	180.2	231.0	271.4	256.7	204.0	165.5	91.4	110.8
7	171.2	171.2	174.9	372.7	396.2	432.0	492.8	467.4	388.9	350.8	170.4	171.2
8	360.6	359.7	382.5	383.0	407.7	466.9	530.7	511.5	431.0	380.5	350.9	360.2
9	365.3	359.7	381.7	417.7	418.1	503.7	563.9	553.3	453.8	411.8	349.4	360.2
10	368.6	356.4	395.4	433.0	444.0	523.2	589.8	584.3	472.8	424.3	352.6	360.2
11	373.3	366.8	400.9	439.3	456.5	535.0	610.1	608.3	502.7	460.0	364.5	358.7
12	371.8	365.2	404.3	447.7	452.6	549.3	604.1	631.0	508.3	471.6	369.6	359.7
13	369.5	368.8	403.0	441.1	460.4	553.3	642.2	653.7	524.4	477.9	356.0	363.1
14	378.1	357.1	407.4	434.0	463.8	557.1	636.1	664.6	536.7	482.6	373.2	350.5
15	366.9	362.8	407.0	434.2	470.2	557.5	661.3	656.5	563.1	483.0	361.7	355.0
16	365.0	355.8	392.5	429.6	472.0	567.1	634.5	651.2	550.1	478.8	342.4	349.0
17	361.2	343.4	386.9	227.2	273.9	345.2	407.6	401.4	343.1	294.0	333.1	344.5
18	180.9	162.5	181.2	49.8	49.6	49.6	49.5	49.5	49.6	70.1	155.3	160.8
19	50.0	50.0	50.0	63.7	103.1	104.3	143.7	136.2	64.2	49.7	50.0	50.0
20	72.5	70.1	54.1	49.8	49.7	49.6	49.5	49.5	49.6	49.7	54.2	69.5
21	50.0	54.9	50.0	49.9	61.7	49.6	93.8	88.4	49.6	49.8	50.0	54.4
22	73.4	70.3	62.6	49.9	49.7	49.6	49.5	49.5	49.6	49.9	54.8	69.9
23	54.3	60.2	54.3	49.9	49.8	49.6	49.5	49.5	49.6	49.9	50.0	54.5
24	82.5	75.6	63.4	49.9	49.8	49.6	49.5	49.5	49.6	49.9	54.8	78.9

TABLE P.4 Summary of Basic Performance Characteristics of Distributed Generation Technologies

	Wind	Microturbine	PVs	Fuel cells
Efficiency	45.0%	38%	12%	48%
Capital cost ($/kW)	1200	800	6600	4250
O&M cost ($/kWh)	0.01	0.00575	0.0025	0.0017
Fuel	Wind	Natural gas	Sun	Hydrogen
Fuel cost	0	$1.0/CCF	0	$1.20/lb

5. Perform a sensitivity analysis to determine the impact on your findings for questions (3) of the following parameters:

- Initial costs of PV systems (what happens of the cost are reduced by half?).
- Cost of electricity from the utility provider (what happens if the energy charges increases to 0.13/kWh?).

For this analysis, you can select the specific characteristics of the hybrid system such as the tilt angle of the PV system and the power curve of the turbine. State these assumptions in your analysis.

REFERENCES

Adaramola, M.S., Paul, S.S., Oyewola, O.M., 2014. Assessment of decentralized hybrid PV solar-diesel power system for applications in Northern part of Nigeria. Energy Sustain. Dev. 19, 72–82.

AEA, 2007. Alaska Energy Authority, Power Cost Equalization Program Guide. Alaska Energy Authority, Aug. 2007.

AEA 2011a. Alaska Energy Authority, Renewable Energy Atlas of Alaska. Alaska Energy Authority, Aug. 2011.

AEA, 2011b. Alaska Energy Authority, Statistical Report of the Power Cost Equalization Program. Alaska Energy Authority, Mar. 2011.

Akinyele, D.O., Rayudu, R.K., 2016. Strategy for developing energy systems for remote communities: Insights to best practices and sustainability. Sustain. Energy Technol. Assess. 16, 106–127.

Al-Enezi, F.Q., Sykulski, J.K., Ahmed, N.A., 2011. Visibility and potential of solar energy on horizontal surface at Kuwait area. Energy Procedia 12, 862–872.

Al-Nassar, S., 2005. Potential wind power generation in the State of Kuwait. Renew. Energy 30, 2149–2161.

Al-Soud, M.S., Hrayshat, E.S., 2009. A 50-MW concentrating solar power plant for Jordan. J. Clean Prod. 17 (6), 625–635.

Ameer, B., Krarti, M., 2016. Impact of subsidization on high energy performance designs for Kuwaiti residential buildings. Energy Build. 116, 249–262, (research-article | March 2017).

Ameer, B., Krarti, M., 2017. Design of carbon-neutral residential communities in Kuwait. ASME J. Solar Energy Eng. 139, 4036054. doi: 10.1115/1.4036054.

American Gas, 1999. Distributing power in the future, an article written by Kabous. Am. Gas J. 16, July 16, 1999.

Ansari, M.J., 2013. Kuwait utility sector, Industry Research report, Capital Standards, Kuwait City, Kuwait. http://www.infomercatiesteri.it/public/images/paesi/107/files/Kuwait%20Utilities%20Sector%20Report_pdf%206_13.pdf.

Arvizu, D., Balaya, P., Cabeza, L., Hollands, T., Jäger-Waldau, A., Kondo, M., Konseibo, C., Meleshko, V., Stein, W., Tamaura, Y., Xu, H., Zilles, R., 2011. IPCC Special Report on Renewable Energy Sources and Climate Change Mitigation-Direct Solar Energy. Cambridge University Press, Cambridge, UK and New York, NY.

AVEC, 2011. Alaska Village Electric Cooperative, AVEC Facilities, AVEC Communities Served, 2011, http://www.avec.org/communities/.

Baring-Gould, I., Dabo, M., 2009. Technology, performance, and market report of wind-diesel applications for remote and island communities. Preprint for the European Wind Energy Conference. Marseille, France, Feb. 2009. National Renewable Energy Laboratory.

Batterystuff, 2015. Grants Pass, OR [Online]. Available from: http://www.batterystuff.com/batteries/pv-solar/6-cs-25ps.html.

Bhattacharyya, S.C., 2015. Mini-grid based electrification in Bangladesh: technical configuration and business analysis. Renew. Energy 75, 745–761.

Blanco, J., Palenzuela, P., Alarcón-Padilla, D., Zaragoza, G., Ibarra, M., 2013. Preliminary thermoeconomic analysis of combined parabolic trough solar power and desalination plant in port Safaga (Egypt). Desalin. Water Treat. 51 (7–9), 1887–1899.

Boukelia, T.E., Mecibah, M.S., 2013. Parabolic trough solar thermal power plant: potential, and projects development in Algeria. Renew. Sustain. Energy Rev. 21, 288–297.

Brand, B., Boudghene Stambouli, A., Zejli, D., 2012. The value of dispatchability of CSP plants in the electricity systems of Morocco and Algeria. Energy Policy 47, 321–331, D.

Clark, W., Isherwood, W., 2004. Distributed generation: remote power systems with advanced storage technologies. Energy Policy 32, 1573–1589, Sep 2004.

Corral, N., Anrique, N., Fernandes, D., Parrado, C., Cáceres, G., 2012. Power, placement and LEC evaluation to install CSP plants in northern Chile. Renew. Sustain. Energy Rev. 16 (9), 6678–6685.

CPV Consortium, 2017. Listing of Projects. Available from: http://cpvconsortium.org/projects, Sacramento, CA.

Domenech, B., Ferrer-Marti, L., Lillo, P., Pastor, R., Chiroque, J., 2014. A community electrification project: combination of microgrids and household systems fed by wind, PV or micro-hydro energies according to micro-scale resource evaluation and social constraints. Energy Sustain. Dev. 23, 275–285.

d-Maps, 2016. "d-Maps," Peynier, France. Available from: http://www.dmaps.com/pays.php?num_pay=93&lang=en.

EIA, 2013. Updated Capital Cost Estimates for Utility Scale Electricity Generating Plants, Washington D.C.

EIA, 2015. Levelized Cost and Levelized Avoided Cost of New Generation Resources in the Annual Energy Outlook 2015. Energy Information Agency, Department of Energy, Washington D.C. Available from: https://www.eia.gov/forecasts/archive/aeo15/pdf/electricity_generation_2015.pdf.

Elhadidy, M., 2002. Performance evaluation of hybrid (wind/solar/diesel) power systems. Renew. Energy 26 (3), 401–413.

Energy Nexus Group, 2002a. Technology Characterization: Steam Turbines, A report prepared for EPA, Climate Protection Partnership Division, Washington, D.C.

Energy Nexus Group, 2002b. Technology Characterization: Reciprocating Engines, A report prepared for EPA, Climate Protection Partnership Division, Washington, D.C.

Energy Nexus Group, 2002c. Technology Characterization: Gas Turbines, A report prepared for EPA, Climate Protection Partnership Division, Washington, D.C.

Energy Nexus Group, 2002d. Technology Characterization: Micro-Turbines, A report prepared for EPA, Climate Protection Partnership Division, Washington, D.C.

Energy Nexus Group, 2002e. Technology Characterization: Fuel Cells (A Report Prepared for EPA). Climate Protection Partnership Division, Washington, DC.

EnergyPlus, 2017. US Department of Energy, Energy Efficiency for Buildings, Building Technologies Program, Washington D.C. Available from: https://energyplus.net/.

eQUEST, 2015, DOE-2.2 Simulation Engine for eQUEST, James J. Hirsch and Associates, Berkeley, CA. http://www.doe2.com/.

Fay, G., Keith, K., Schwörer, T., 2010. Alaska Isolated Wind-Diesel Systems: Performance and Economic Analysis, report prepared for Alaska Energy Authority and Denali Commission, Institute of Social and Economic Research, University of Alaska Anchorage, Jun. 2010.

Fay, G., Meléndez, A.V., Converse, A., 2011. Alaska Energy Statistics 1960–2010, Preliminary Data, report prepared for Alaska Energy Authority, Institute of Social and Economic Research, University of Alaska Anchorage, Sep. 2011.

FERC, 1978. Public Utility Regulatory Policies Act of 1978 (PURPA). Federal Energy Regulatory Commission, Washington, DC.

Fluri, T.P., 2009. The potential of concentrating solar power in South Africa. Energy Policy 37 (12), 5075–5080.

Grande, G., Islas, J., Rios, M., 2015. Technical and economic analysis of domestic high consumption tariff niche market for photovoltaic systems in the Mexican household sector. Renew. Sustain. Energy Rev. 48, 738–748.

Hinkley, J.T., Hayward, J.A., Curtin, B., Wonhas, A., Boyd, R., Grima, C., et al., 2013. An analysis of the costs and opportunities for concentrating solar power in Australia. Renew. Energy 57, 653–661.

HOMER, 2016. HOMER ENERGY, [Online]. Available from: http://www.homerenergy.com/software.html.

HOMER, 2016. "Homerenergy", HOMER Energy, Boulder, CO. Available from: http://www.homerenergy.com/software.html.

Horn, M., Führing, H., Rheinländer, J., 2004. Economic analysis of integrated solar combined cycle power plants. Energy 29 (5-6), 935–945.

IEA, 2010. Technology Roadmap: Concentrating Solar Power. International Energy Agency (IEA), Paris, http://www.iea.org/papers/2010/csp_roadmap.pdf.

IRENA, 2012. Renewable Energy Technologies: Cost Analysis Series, Concentrated Solar Power, 1: Power Sector, Issue: 2/5, International Renewable Energy Agency, Abu Dhabi, UAE.

Jain, A., Vu, T., Mehta, R., Mittal, S.K., 2013. Optimizing the cost and performance of parabolic trough solar plants with thermal energy storage in India. Sustain. Energy 32 (3), 824–829.

Kanase-Patil, A.B., Saini, R.P., Sharma, M.P., 2010. Integrated renewable energy systems for off-grid rural electrification of remote area. Renew. Energy 35, 1342–1349.

Kaygusuz, K., 2011. Prospect of concentrating solar power in Turkey: the sustainable future. Renew. Sustain. Energy Rev. 15 (1), 808–814.

Kutscher, C., Mehos, M., Turchi, C., Glatzmaier, G., Moss, T., 2010. Line-Focus Solar Power Plant Cost Reduction Plan, Report NREL/TP-5500-48175, National Renewable Energy Laboratory, Golden, CO.

Li, C., Ge, X., Zheng, Y., Xu, C., Ren, Y., Song, C., 2013. Techno-economic feasibility study of autonomous hybrid wind/PV/battery power system for a household in Urumqi, China. Energy 55, 263–272.

Luque, A., 1989. Solar Cells and Optics for Photovoltaic Concentration. CRC Press, Bristol, England.

MEW, 2012. Summary of the cost of production and distribution of electric power, Ministry of Electricity and Water, Kuwait, 2010/2011.

Munawwar, S., Ghedira, H., 2014. A review of renewable energy and solar industry growth in the GCC region. Energy Procedia 57, 3191–3202.

NREL, 2006. A Review of PV Inverter Technology Cost and Performance Projections, National Renewable Energy Laboratory, Golden, CO.

NREL, 2016a. Map of US Annual Average Wind Speed (m/s) Off-Shore. http://maps.nrel.gov/re_atlas.

NREL, 2016b. Map of US Global Horizontal Solar Insolation. http://maps.nrel.gov/re_atlas.

NREL, 2016c. Map of US Direct Normal Solar Insolation. http://maps.nrel.gov/re_atlas.

Pavlović, T.M., Radonjić, I.S., Milosavljević, D.D., Pantić, L.S., 2012. A review of concentrating solar power plants in the world and their potential use in Serbia. Renew. Sustain. Energy Rev. 16 (6), 3891–3902.

Posadillo, R., Luque, R.L., 2008. Approaches for developing a sizing method for standalone PV systems with variable demand. Renew. Energy 33, 1037–1048.

Poullikkas, A., 2009. Economic analysis of power generation from parabolic trough solar thermal plants for the Mediterranean region—A case study for the island of Cyprus. Renew. Sustain. Energy Rev. 13 (9), 2474–2484.

Radhi, H., 2011. On the value of decentralised PV systems for the GCC residential sector. Energy Policy 39, 2020–2027.

Ramli, M.A., Hiendro, A., Bouchekara, H.H., 2014. Performance analysis of hybrid PV/diesel energy system in western region of Saudi Arabia. Int. J. Photoenergy, 240–251.

Ravindra, K., Iyer, P.P., 2014. Decentralized demand-supply matching using community microgrids and consumer demand response: a scenario analysis. Energy 76, 32–41.

Rezzouk, H., Mellit, A., 2015. Feasibility study and sensitivity analysis of a stand-alone photovoltaic–diesel–battery hybrid energy system in the north of Algeria. Renew. Sustain. Energy Rev. 43, 1134–1150.

SAM 2016. System Advisor Model, National Renewable Energy Laboratory, NREL, [Online]. Available from: https://sam.nrel.gov/.

SE4All, 2017. Sustainable Energy for All Initiative. Available from: http://www.se4all.org/.

Shaahid, S., Elhadidy, M., 2008. Economic analysis of hybrid photovoltaic–diesel–battery power systems for residential loads in hot regions-A step to clean future. Renew. Sustain. Energy Rev. 12 (2), 488–503.

Shawon, M.J., 2013. Overview of wind energy and its cost in the Middle East. Sustain. Energy Technol. Assess. 2, 1–11.

Solargis, 2017. solargis.info, Solargis s.r.o., [Online]. Available from: http://solargis.info/imaps/#tl=Google:hybrid&bm=satellite&loc=29.866847,47.320862&c=29.288794,47.831726&z=9.

Turchi, C., 2010. Parabolic Trough Reference Plant for Cost Modeling with the Solar Advisor Model (SAM), NREL Report, National Renewable Energy Laboratory, Golden, CO, July, 2010.

Turchi, C., Mehos, M., Ho, C.K., Kolb, G.J., 2010. Current and future costs for Parabolic trough and power tower systems in the US market, National Renewable Energy Laboratory (NREL), Golden, CO.

Ummadisingu, A., Soni, M.S., 2011. Concentrating solar power-technology, potential and policy in India. Renew. Sustain. Energy Rev. 15 (9), 5169–5175.

Vaught, D., 2007. Deering Wind-Diesel Hybrid Feasibility Study Report, V3 Energy, LLC, Aug. 26, 2011, 38 pp.

Weber, A.S., 2013. Review of sustainable and renewable energy activities in the state of Qatar. In: Proceedings for the International Renewable and Sustainable Energy Conference, 91–95.

Willman, L., Krarti, M., 2013. Optimization of hybrid distributed generation systems for rural communities in Alaska. Distrib. Gener. Altern. Energy J. 28 (4), 7–31.

Zhang, H.L., Baeyens, J., Degrève, J., Cacères, G., 2013. Concentrated solar power plants: review and design methodology. Renew. Sustain. Energy Rev. 22, 466–481.

Chapter 8

Integrated Design of Energy Efficient Cities

Outline

8.1 INTRODUCTION

Over 50% of the world's population currently lives in urban areas (UN-Habitat, 2016). In addition, cities are responsible for over 70% of the carbon emissions in the world and for almost 80% of the global economic output

Optimal Design and Retrofit of Energy Efficient Buildings, Communities, and Urban Centers.
http://dx.doi.org/10.1016/B978-0-12-849869-9.00008-9

(UN-Habitat, 2016). Throughout the world, energy consumption of cities varies widely and depends on several factors including primarily population density, building stock, and urban transportation mode. Indeed, several studies have shown that there is a correlation between the per capita energy consumption and population density in cities (Newman and Kenworthy, 1999). Using data reported for 2014 (CMS, 2016; Schulz, 2010), the population density and per capita energy consumption values are depicted for select cities around the world in Fig. 8.1. The densely populated Asian cities consume lower energy per person than North American and Australian urban centers, which have lower population densities. European cities have generally medium population densities with low energy consumption per capita. The dominance of apartment buildings in Asian and European cities compared to the prevalence of detached homes in North America and Australia is cited as the main reason for the lower per capita energy consumption.

A detailed analysis of data obtained for select US cities has indicated that urban per capita energy consumption is attributed mostly to three sectors: (1) transport, (2) residential buildings, and (3) commercial or industrial facilities as shown in Fig. 8.2 (ACEEE, 2017). For cities located typically in the East Coast (i.e., Boston, New York City, and Philadelphia, Washington, DC), with large transit transportation networks, housing consumes most of the energy. However, for other US cities located mostly in the south and the west (i.e., Atlanta, Austin, Denver, San Diego), transport utilizes most of the urban energy consumption due most likely to the limited transportation options.

Fig. 8.3 shows the trend between population density and per capita energy consumption associated, respectively, to total, transport, housing, and commercial/industrial facilities for select US cities considered in Fig. 8.2. As noted by

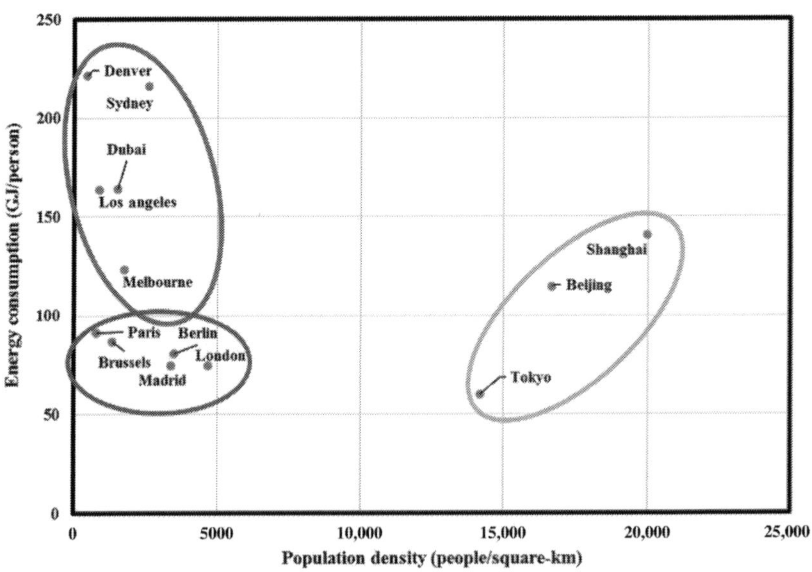

FIGURE 8.1 Electricity consumption per capita for various cities around the world.

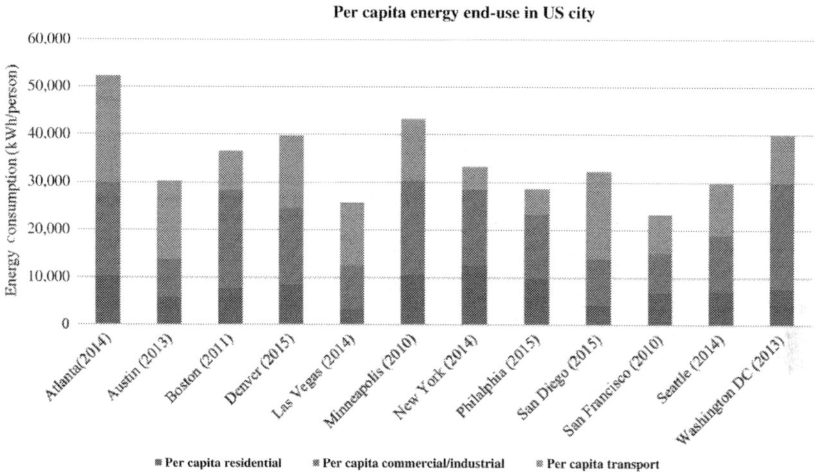

FIGURE 8.2 Per capita energy consumption and population density for select US cities.

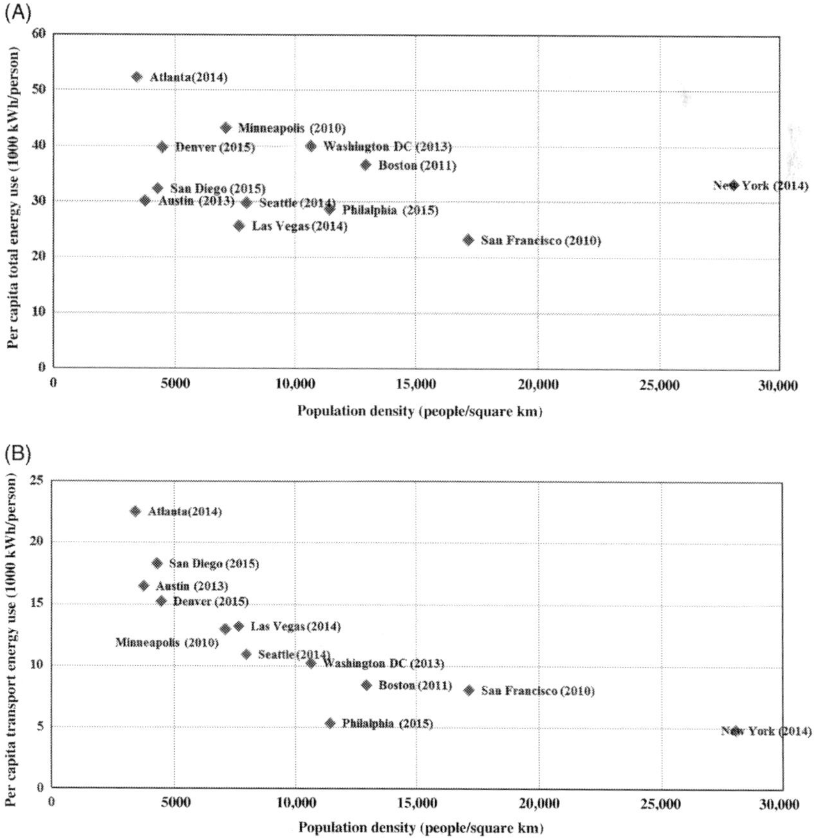

FIGURE 8.3 Energy use as a function of population density for US cities. (A) Total, (B) transport, (C) residential buildings, and (D) commercial and industrial.

(C)

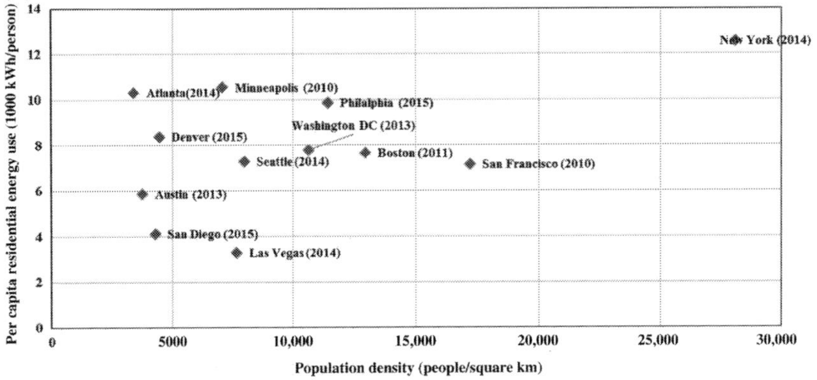

(D)

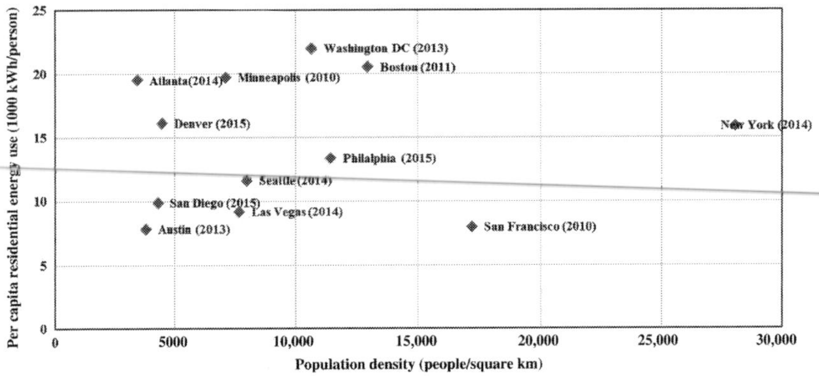

FIGURE 8.3 ***(Cont.)***

Newman and Kenworthy (1999), there is a clear correlation between the per capita energy consumption by the transport sector and population density for the US cities considered in Fig. 8.3. Typically, cities that are spread out have high energy consumption for transport. In these cities, most of the population live in large detached family homes typically located in the suburbs; the main mode of transportation is private vehicles. Such is the case for Atlanta, Austin, and Denver. The correlations between population density and per capita energy consumption for residential and commercial/industrial are similar albeit not as clear as that obtained for the transport sector.

Designing individual buildings within cities should account for its impacts on its environment including other buildings and outdoor spaces. Several accounts have been reported on some of the unintended effects of building forms and their specific features on the environment (Futcher et al., 2017). For instance, a 160-m-tall building in London with a curved façade was found to reflect down significant solar energy that was sufficient to melt a parked car. This effect has resulted in an expensive retrofit to add shading devices. Moreover, the form of the same building was found to be responsible for redirecting higher levels of winds to the lower level roads and pavements causing substantial discomfort to pedestrians (Futcher et al., 2017). Another tall building in London is planned to be constructed along an existing tall building equipped with PV panels along its entire south façade. This new building is expected to reduce the energy output of the PV system of the existing building by over 40% (Futcher et al., 2017). The use of energy modeling tools for designing urban buildings is rather limited. Indeed, urban building energy modeling has been developed and applied only in the last decade (Reinhart and Davila, 2016). Both top-down and bottom-up approaches can be considered for energy modeling of urban buildings as reviewed by Frayssinet et al. (2018). Moreover, several studies have shown that urban forms including buildings and other structures such as roads and pavements often create microclimates that affect outdoor air quality, thermal comfort, and energy use of buildings. For instance, urban heat island (UHI) effects are common in cities and are associated with warmer ambient air and surface temperatures. The magnitude of UHI is more prominent in highly dense urban centers with little green spaces than at the edges of cities with low building density and large green areas (Oikonomou et al., 2012).

Therefore, an integrated approach for urban developments has been promoted in the last decade to account for the interactions between buildings and their impacts on energy and natural resources accessibility as well as outdoor environmental quality. Currently, there are several sustainability rating systems for urban centers and communities including LEED-ND (USGC, 2009), BREEM Communities (BREEAM, 2012), STARS (Star Communities, 2012), and SITES (SSI, 2014). The rating systems identify most of the factors that need to be considered for the development of sustainable cities.

In the literature, there is a wide range of studies and analyses related to sustainable cities with different concepts and factors considered to design, operate, and rate urban centers (Larco, 2016). However, most of the studies consider that sustainable cities have reduced energy consumption and emission of greenhouse gases. A clear distinction between energy use for transport and non-transport (i.e., buildings and urban lighting) is often noted. Other areas that are considered to achieve sustainable urban design include equity and health, water access and quality, and ecological habitat preservation. Specific metrics have been proposed to provide direct or proxy indicators of the sustainability levels achieved by cities. Based on a review of the current rating systems for communities and urban centers, Fig. 8.4 provides the main pillars and performance categories for designing, retrofitting, and operating sustainable cities. Table 8.1 summarizes some of the primary and related indicators for various sustainability focus areas in cities as noted in Fig. 8.4 (Larco, 2016).

In this chapter, some technologies and strategies to design or retrofit cities to be more sustainable are presented. Only select performance categories outlined in Fig. 8.3 are addressed by the strategies outlined in this chapter. First, the basic proven strategies to improve the sustainability of the transport sector

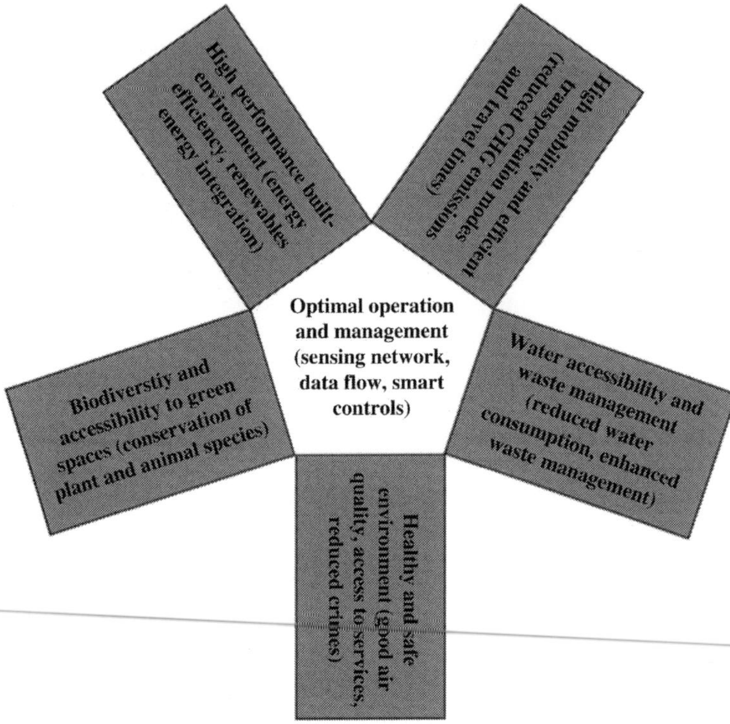

FIGURE 8.4 Basic pillars and performance for sustainable cities.

within urban centers are outlined. Then, technologies specific to district energy systems and cool roofs are presented with several illustrative examples to improve the energy efficiency of urban buildings and mitigate urban heat island effects. Finally, the integration of smart grids and intelligent controls are briefly discussed as significant strategies to operate sustainable cities.

8.2 SUSTAINABLE URBAN TRANSPORT

In the US, petroleum products provided about 92% of the total fuel consumed by the transport sector. The contribution of biofuels (i.e., ethanol and biodiesel), natural gas, and electricity account, respectively, for 5%, 3%, and less than 1% of the total energy used by the US transport sector (EIA, 2017). The shares of total US transportation energy use by modes of transportation 2016 are (EIA, 2017) as follows:

- Light-duty vehicles (cars, small trucks, vans, and sport utility vehicles) and motorcycles: 56%
- Large trucks: 23%
- Jets, planes, and other aircraft: 8%
- Boats, ships, and other watercraft: 4%
- Trains and buses: 3%

TABLE 8.1 Primary and Other Indicators for Common Sustainability Focus Areas in Urban Centers

Sustainability focus area	Primary indicators	Other indicators
Nontransport energy use	Energy use and energy use Intensity for residential and commercial buildings Energy use for street lighting	Heat island effect (for instance, ambient air temperature increase) Types of building stock
Transport energy use	Energy consumption by urban transportation Greenhouse gases emissions by urban transportation	Distance traveled by transportation mode Density (inhabitants per unit land area) Land use mix
Water accessibility and quality	Water consumption (potable, grey, and wastewater) Water recharge rate (rain and Water-storm run-off Water quality level (particulate concentration, temperature)	Number of water pipes Percent of permeable surface Water velocity
Ecology and habitat	Type and number of plant and animal species	Health levels for available species Green spaces and parks
Equity and health	Accessibility level to services (schools, hospitals, etc.) Affordability level for service use/access	Safety level (crime rate and accident rate) Availability of green and open spaces (limit exposure to pollutants and increase physical activities)

The remaining transport energy consumption are associated with mobility of military, pipelines, and lubricants.

While energy use for buildings depends on their form and function, energy consumption for urban transportation is related to the city compactness, parking and street design, and land use. In addition, buildings utilize energy generated typically remotely while transport use fuel directly and thus emit carbon and greenhouse gases within the city. Several studies have shown that compact cities equipped with transit transportation and dedicated walking and cycling paths are more energy efficient and safer. Cities that rely on individual vehicles for transportation often foster sprawls due to the need for larger roads and increased number of parking spaces resulting in expanded cities to accommodate these infrastructures. Subsequently, this expansion makes trips within cities longer and inefficient even for public transportation options. Therefore, the energy efficiency of transport within cities can be enhanced using an integrated urban design and planning.

Sustainable urban transport considers the efficiency at three levels including system (i.e., avoid or reduce travel need and time), travel (i.e., shift to more efficient modes of transportation), and vehicle (i.e., improve the efficiency of transportation modes). Some of the basic actions to improve each of these efficiencies are outlined (GIZ, 2012):

- *System efficiency*: Optimization of land use and urban planning is essential to reduce or even eliminate the need for motorized transportation. Typically, dense cities with mixed-use (that is, with shops located close to workplaces and residences) are crucial to increase urban transport system efficiency since people can easily walk and bike within cities.
- *Travel efficiency*: Use of more energy efficient modes of transportation is an important pillar of sustainable cities. Indeed, transit transport modes are generally more energy efficient and often more affordable and less time-consuming than private motorized cars. Table 8.2 compares the energy efficiency of common modes of transportation. Several strategies have been considered to encourage the shift to more efficient urban travel options. These strategies include (1) limiting available parking spaces within urban centers, (2) restricting cars with specific plate numbers to drive through city roads, (iii) charging fees to use city roads during peak travel periods, and (4) organizing events to reduce or eliminate the use of private cars within city centers (such as free cars days). Recently, car and bike sharing programs have been piloted by several cities to provide affordable and often energy efficient options to travel within urban centers. Moreover, car and biking sharing programs are found to increase the mobility of low-income communities and allow them to connect with difficult-to-access job centers to improve their economic status.
- *Vehicle efficiency*: Through technological and design improvements, fuel efficiency of vehicles can be significantly increased. In the US, the Corporate

TABLE 8.2 Travel Distance for Some Common Modes of Transport Within Cities

Mode of urban transport	Kilometers/passenger-liter	Miles/passenger-gallon
Walking	Large (∞)	Large (∞)
Biking	Large (∞)	Large (∞)
Bus (diesel)	310	729
Metro rail	160	376
Car (diesel)	70	165
Car (gasoline)	60	141
Scooter	45	106

Average Fuel Economy or CAFE program, established since 1975, has set regulations to improve fuel efficiency and reduce carbon emission of cars. Similarly, the European Union has introduced a set of standards for CO_2 emission requirements for all sold vehicles.

8.3 DISTRICT ENERGY SYSTEMS

A study reported by the United Nations Environment Programme has surveyed several low-carbon cities throughout the world to determine the main ingredients and factors to foster sustainable urban centers including reducing energy consumption and reducing greenhouse gas emissions (UNEP, 2015). One of the key findings of the report is that the application of district energy systems (DES) offers a cost-effective and sustainable approach to improve energy efficiency and integrate renewable energy within cities. In this section, the main components and technologies utilized for DES are presented. In addition, some of the DES benefits are discussed. Then, a simplified analysis of the cost-effectiveness analysis of DES systems is outlined with a specific example. In subsequent sections, more detailed descriptions and analysis approaches for various DES technologies are provided.

8.3.1 Overview of DES

A typical DES configuration and its main components are illustrated in Fig. 8.5. In particular, a DES has three main components:

1. Central plant where heating (i.e., hot water or steam) and/or cooling (i.e., chilled water) are/is produced. The various systems used to generate heating and cooling are first outlined in Sections 8.3.3 and 8.3.4. Then, detailed descriptions and analysis methods for heating systems, cooling systems, and thermal energy systems are provided in, respectively, Sections 8.4–8.6.
2. Distribution network consisting of pipes typically buried underground or located in concrete tunnel. An overview of the distribution network

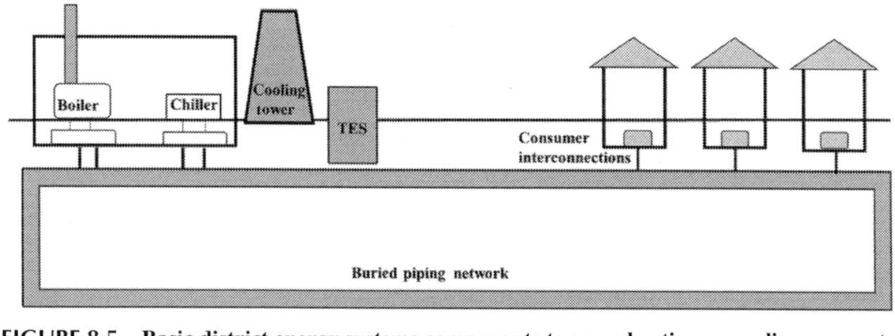

FIGURE 8.5 Basic district energy systems components to serve heating or cooling energy to buildings.

configurations is discussed in Section 8.3.5. Then, a thermal analysis of underground piping systems is presented in Section 8.7 of this chapter.

3. End-user systems that include the connections with the HVAC systems located within the buildings. Typically heat exchangers are utilized to connect the distribution network to the buildings. A brief outline of the end-user system types is given in Section 8.3.6.

For district energy systems to be cost-effective, heating/cooling load density, that is the heating and/or cooling requirements per unit area, should be relatively high. Indeed, it is uneconomical to distribute energy to sparsely populated areas where distribution piping costs and thermal losses are relatively high compared to conventional systems where heating and/or cooling is generated within each building. High total thermal loads improve thermal efficiencies of heating/cooling equipment (ASHRAE, 2016). A life-cycle cost analysis of DES is provided in Section 8.3.7.

8.3.2 Benefits of DES

DES has been utilized effectively to serve the heating and cooling needs for urban centers as well as for large buildings. Indeed, DES have been successfully applied to serve thermal loads of apartment complexes, hospitals, universities, groups of office buildings, and industrial facilities (UNEP, 2015). Fig. 8.6 illustrates typical building types that a district cooling system (DCS) serves in urban centers. Moreover, Fig. 8.6 shows that in addition to a set of modular chillers and distribution pumps, a DCS can benefit from thermal energy storage (TES) systems to reduce peak demand as discussed in detail in Section 8.7.

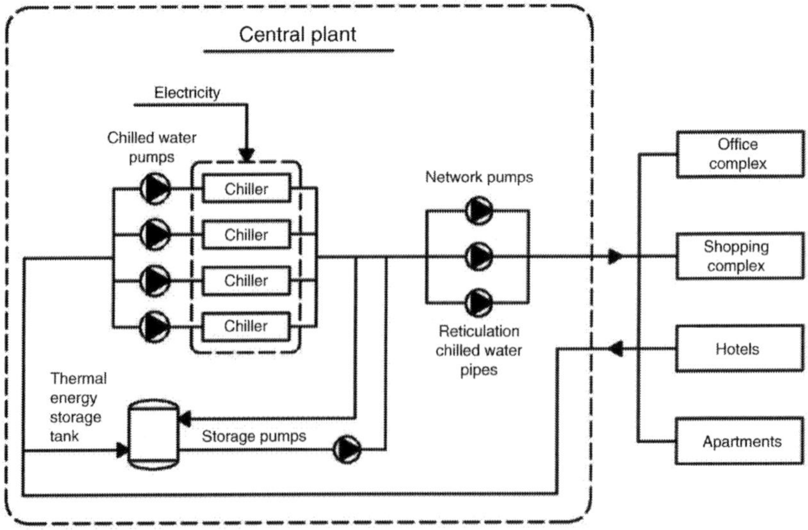

FIGURE 8.6 **Typical components of a district cooling system (DCS).**

Since they are serving a diverse group of customers with varying load requirements, DES have to accommodate a relatively large total thermal load with potentially wide variations from hour to hour and season to season. Since individual customers often experience peak loads at different times of the day, the peak demand exhibited by a DES is reduced when compared to the sum of all the individual peak loads. Thus, the installed equipment capacity in DES is typically less than that of conventional decentralized systems.

Figs. 8.7 and 8.8 show typical hourly cooling demand profiles for two large buildings, one for a commercial office building and the other for an apartment building. Both buildings demonstrate significant cooling demand variation with

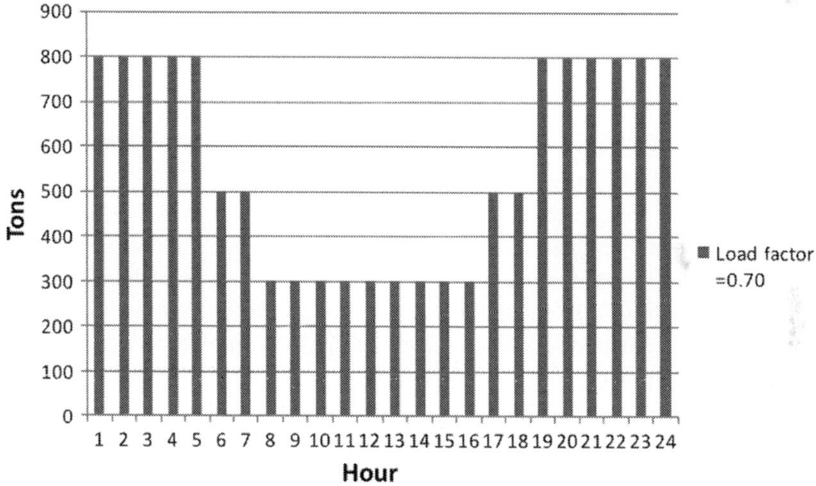

FIGURE 8.7 Cooling load profile for an apartment building.

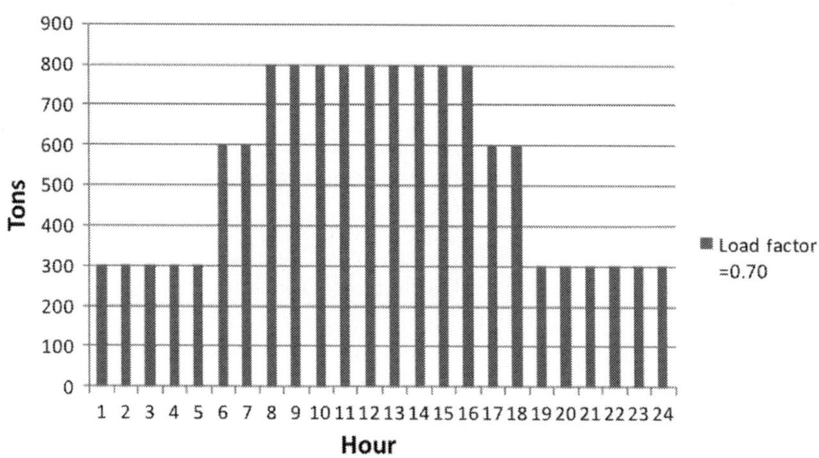

FIGURE 8.8 Cooling load profile for an office building.

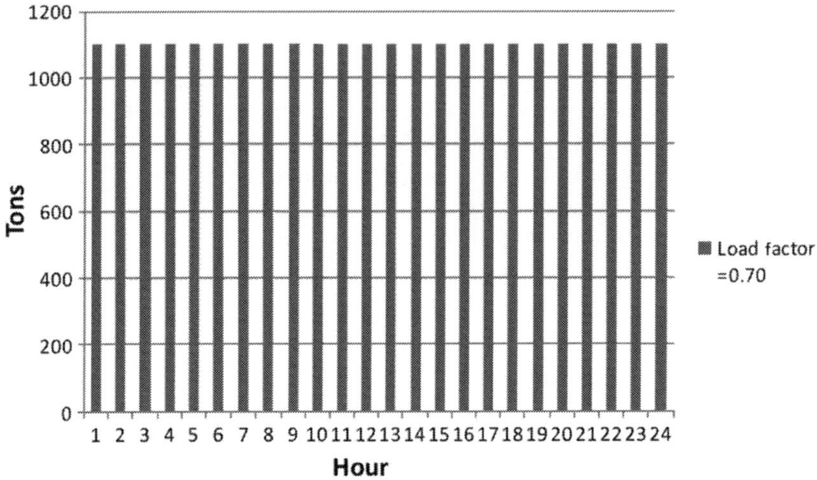

FIGURE 8.9 **Combined cooling load profile for both the apartment and the office buildings.**

similar load factor (i.e., average to peak load ratio). However, the peak occurs during daytime hours for the office building and during night hours for the apartment building. Fig. 8.9 shows the demand profile for both buildings and demonstrates the flattening effect on cooling demand for a DCS when used by a variety of customer types.

Depending on total system peak and average load requirements and the load variations from day to day and season to season, DES equipment can be designed. A relatively simple DES might utilize a single energy production facility, consisting for example of a natural gas-fired boiler or an electrically driven centrifugal chiller. For more complex DES, several thermal energy production units can be considered with different energy sources. The operation of these units can be optimized depending on fuel prices, availability of such alternatives, proximity of the load to such sources, environmental concerns, and other factors.

Based on their design specifications and operation principles, DES can have multiple benefits. Some of these benefits are summarized in the following categories:

- *Energy efficiency*: DES can improve the energy efficiency of meeting the heating and cooling needs for urban buildings. As illustrated in Figs. 8.7–8.9, the thermal loads related to heating and cooling are generally uniform and steady for DES throughout the year compared to individual buildings that are operated based on the specific demand and behavior of occupants. This uniform load improves equipment including boilers and chillers. Moreover, DES allow the use of free sources of heat and cold as well as cost-effective integration of renewable energy systems.

- *Economic benefits*: DES allow building owners and cities to reduce their energy expenses for heating and cooling services. Moreover, cities can reduce their needs for water usage associated with individual cooling systems often requiring individual water-cooling towers. In addition, cities can attract businesses due to the availability of low-cost utility services. Moreover, several skilled jobs can be created to operate and maintain DES and integrate renewable energy technologies.
- *Energy security and resiliency*: DES can generate power, heating, and cooling even under extreme events and thus increase the energy security and resiliency of cities. Moreover, the integration of local and renewable sources of energy allow cities to be less dependent on fuels and ultimately shielded against variations in availabilities and prices. Located typically in one centralized site, DES can be easily retrofitted to include the best available technologies as the equipment age.
- *Environmental benefits*: Due to high efficiency and integration of renewable energy sources, DES can reduce significantly carbon and greenhouse gas emissions including sulfur dioxide, nitrogen oxides, and particulates and ultimately improve air quality within cities. The improved air quality results in better health conditions (refer to Fig. 8.4 and Table 8.1) and thus more productive cities (refer to Chapter 9 on the impact of air quality on work productivity).

8.3.3 Technologies for Heating Systems

Several heating sources can be considered for DES and district heating systems (DHS) including the following options:

- Geothermal energy can be used as a heating source by installing brine heat exchangers to extract heat available naturally in underground reservoirs. The extracted free renewable energy can meet baseload demand for space heating of urban buildings. The main requirement for the cost-effectiveness of this option is that geothermal wells be close to the urban areas to be served. Moreover, deep geothermal reservoirs result in high costs for this technology to be cost-effective. Buildings in several areas in Paris are still heated by geothermal wells (UNEP, 2015).
- Waste-generated thermal energy can be used to supply heat needed by buildings in urban areas. Indeed, municipal and other combustible waste can be incinerated to produce large amounts of heat to meet the thermal and electrical needs of cities. The incinerators may have to be located away from the urban centers to avoid air pollution even though modern systems can reduce significantly the concentrations of undesired combustion products exhausted by the flue gases. Moreover, the bottom ash and slag, byproducts of the combustion process, can be used for construction applications. It is estimated that waste heat has displaced the use of over 1.4 million equivalent of barrels of oil annually in the city of Copenhagen (Thornton, 2009).

- Fossil fuel-fired boilers are common in generating heat for DHS throughout the world. Various fuels can be used including coal, oil, natural gas, biogas, and biomass to generate heat for peak or baseload demands. With available air pollution control technologies, the combustion byproducts especially particle matter, sulfur dioxide, and nitrogen oxide can be significantly reduced. Section 8.4 discusses in more detail the types and the thermal performance of fossil fuel-fired boilers.
- Combined heat and power (CHP) systems are often considered to provide heat and electricity needs for urban buildings. Typically, fossil fuels including coal, oil, natural gas, and biomass as well as municipal solid waste can be used for CHP applications. In addition to higher overall efficiency compared to the conventional boiler systems, the main advantage for CHPs is the ability to generate electricity independently of the utility and thus can act as emergency power generators in case of extreme events. Chapter 7 provides an overview of CHP systems and their performance.
- Waste heat recovery systems can also be considered for DHS when process thermal energy from industries or low-grade heat from sewage are available. Heat exchangers can be used to recover wasted heat to be used for space heating of urban buildings. For instance, Vancouver is using heat from raw wastewater to meet 70% of the heating need of 7000 residential units connected to a DHS (UNEP, 2015).
- Heat pumps can be used to recover heat from one source to provide space heating of urban buildings through DHS. Heat sources are generally free and can include ambient air, ground, and waste heat. With their high coefficient of performance (COP) especially at low temperature levels, heat pumps can be cost-effective in providing heating and in some cases cooling to buildings. Several cities in Europe currently utilize heat pumps to meet some of their heating needs. For instance, Helsinki has one of the largest heat pumps in the world to capture 165,000 GWh of heat from the city wastewater system (UNEP, 2015).
- Solar collectors can be used to capture thermal energy to heat buildings using DHS. The main challenges of these systems are the significant land requirement as well as the need for back-up heating systems (i.e., boilers) to meet peak demands especially in case of cloudy days. However, integration with DHS allow installation of large solar collectors to be cost-effective and avoid the need for thermal storage systems by individual buildings. For instance, the city of St. Paul, Minnesota in the US has integrated with its DHS solar collectors that can generate a thermal peak capacity of 1.2 MW (UNEP, 2015).

8.3.4 Technologies for Cooling Systems

Three options are generally considered to generate chilled water required for DES and DCS to meet the cooling needs for urban buildings:

1. Electrical chillers can operate using various types of mechanical compression systems as discussed in Section 8.5. The COP of these typically large

electric chillers is higher than that of small air-conditioners used for residential and small commercial buildings. Moreover, large chillers use refrigerants that have lower global warming potential than that of small air conditioning units. Most of the cities with DCS utilize electric chillers to produce chiller water for cooling. For instance, Paris uses electric chillers for its DCS to cool most of its buildings. Pearl, a new city center in Doha, has a DCS with electric chillers that have a total cooling capacity of 130,000 tons or 456 MW (UNEP, 2015).

2. Free cooling can be considered for DCS using cold water from oceans, seas, lakes, rivers, aquifers, and waste cooling energy sources (such as in processing of liquefied natural gas). Heat exchangers are typically installed to capture free cooling and deliver it to urban buildings through the DCS distribution network. The cooling plant needs to be close to the source and the buildings where the cold water is delivered. Moreover, backup cooling systems are needed especially when thermal demands are high. The city of Toronto has a DCS that utilizes a water pipeline network to extract cooling energy from the Ontario Lake through a set of heat exchangers and pumps to cool buildings.
3. Absorption chillers are also considered for DCS to produce cooling energy when free heat is available such as from waste incinerators, industrial processes, and power generation. Heat from CHPs can be also used to produce cooling for DCS. In this case, CHPs, referred to as trigeneration systems, generate three utilities: electricity, heat, and cooling. While the COP of absorption chillers are low compared to electric chillers, the free thermal energy can make them cost-effective in producing cooling energy. Moreover, absorption cooling increases heat demand during the summer period and thus makes CHPs more cost-effective as discussed in Chapter 7. The Olympic park in London utilizes a 4-MW absorption chiller plant to generate cooling needs for most of the buildings within the site during the summer months when the heat demands are low (UNEP, 2015).

In addition to systems for generating heating and/or cooling, DES typically integrate TES systems to ensure continuous supply of hot or chilled water especially during peak demands. For DCS operating with electric chillers, TES systems allow cooling shifting strategies to reduce or avoid the use of electricity during peak periods with higher energy prices. Section 8.6 describes in more detail the types and operation strategies of TES systems for cooling applications.

8.3.5 Types of Distribution Systems

DES thermal energy, whether in the form of chilled water (for cooling applications) or in the form of hot water or steam (for heating applications), is delivered to customers via a network of supply pipelines. In addition, a return pipeline network is used to return the chilled water or hot water back to the DES production plant. The piping for both supply and return distribution network is typically buried in the ground as noted in Fig. 8.5. For industrial applications, above-ground piping systems are also considered. Typically, the piping distribution systems are located within building basements and are often insulated to minimize thermal losses as discussed in Section 8.6.

Depending on the applications, four types of distribution piping systems can be suitable for DES:

1. Single-pipe systems are used when steam is supplied without condensate return. These systems are simple to design and relatively inexpensive to install. However, the overall efficiency of the DES may be low resulting in higher operating costs and carbon emissions.
2. Two-pipe systems are considered to allow for both supply and return of water and/or steam. The two-pipe systems can be used to deliver either cooling or heating energy but do not allow both simultaneously. A distinct switchover set-up has to be scheduled between heating and cooling seasons to use the same two-pipe systems to deliver hot water/steam and chilled water from the production plant. Two-pipe systems are often used for DES that supply residential and apartment complexes or similar buildings where heating and cooling are not required during the same season. It should be noted that steam can be used to meet heating and cooling requirements when absorption chillers are installed in the customers' site. Moreover, when the two-pipe systems are utilized to deliver chilled water during the cooling season, customers need to have other alternatives for generating domestic hot water within their buildings.
3. Three-pipe systems are specified to enhance the capabilities of the two-pipe systems since they provide an additional piping line. In particular, this line can be utilized to supply the needs of DHW heating during summer and winter seasons. As in the case of two-pipe systems, the two other lines provide either chilled water (i.e., for space cooling during the summer) or hot water/steam (i.e., for space heating during the winter).
4. Four-pipe systems provide the greatest flexibility for DES and can supply simultaneous heating and cooling needs within buildings and among buildings. Indeed, two pipes can be solely dedicated to chilled water to supply cooling while the other two pipes can carry hot water or steam to meet any heating needs. These systems are the most expensive among the four types of distribution configurations but may be required to serve a diverse group of buildings that have simultaneously heating and cooling needs.

The piping distribution network is often the most expensive component of the DES and the part that requires the most time and effort to install. Indeed, the required excavation for burying or repairing piping systems often result in significant disturbances to traffic and inconveniences to pedestrians and motorists especially in urban centers. Underground tunnels are often considered as alternatives to buried pipes especially when such infrastructures are already available or needed for other purposes (i.e., transportation or sewage system).

8.3.6 End-User Systems for DES

Two types of systems are generally specified at the end-user side to receive heating and/or cooling energy from the DES production plant via the distribution network:

1. Direct systems that allow the supplied chilled water, hot water, or steam to be directly used by the HVAC equipment present within the buildings such as coils and radiators. Finned tubes are common coils within air handling units to receive DES-supplied hot water or chilled water.
2. Indirect systems that utilize heat exchangers to transmit energy from the DES distribution network to the building HVAC equipment. Unlike the direct systems, the indirect systems isolate the building from the DES production and distribution network.

In the early DES installations, direct systems were prevalent. However, indirect systems are now more common and are preferred due to their inherent advantages including the following:

- Better equipment protection especially for the end-users. Indeed, damages to radiators and coils may occur due to pressure surges from the distribution system.
- Lower maintenance and treatment costs for DES. Indeed, customers' equipment (such as radiators and coils) may leak due to lack of maintenance and/or due to their age. These issues can result in a significant operation problem for DES with direct systems.
- Higher flexibility for load diversity. Indeed, indirect systems can easily accommodate customers with various thermal needs and pressure requirements.
- Better control options. Indeed, DES production and utilization are essentially isolated when using indirect systems and thus both DES operations can be controlled independently.
- Higher flexibility to contractual set-ups. Indeed, ownership of equipment for DES production, distribution, and utilization can be clearly defined when indirect systems are used. In particular, ownership of various equipment can be easily separated if desired.

DES can serve new and existing buildings that are located close to the distribution network. For existing buildings, retrofit requirements from standalone HVAC systems to DES-connected systems are generally straightforward as long as the existing systems are compatible with DES-supplied chilled water for cooling and hot water or steam for heating. However, existing HVAC systems that call for DX cooling or electric heating may require significant modifications to be connected to the DES distribution network. For buildings with air handling units and fan coils served by boilers and chillers, the connections to DES can be made relatively easily. Most of the changes occur in the mechanical rooms where boilers and chillers are located. Specifically, the existing boilers and chillers are replaced with heat exchangers that are connected to the DES distribution network. The same or different heat exchangers can be used to provide hot water/steam and chilled water. Moreover, existing cooling towers can be eliminated with a conversion to DES. It should be noted that spaces for circulation pumps, control valves, water treatment systems, and metering devices are required for DES as well as conventional HVAC systems.

8.3.7 Cost Analysis of DES

The cost-effectiveness of DES depend on several factors including the profile of thermal loads, the costs of production, distribution, and end-use equipment as well as the operation costs. To perform a cost analysis of DES, detailed analysis approaches are required and may include the use of hourly simulation tools. However, a simplified analysis can be considered to assess the feasibility of DES compared to conventional heating and cooling systems. Example 8.1 illustrates a simplified cost analysis of DES using the life-cycle cost (LCC) method as discussed in Chapter 5.

Example 8.1

For a building with a peak load of 2000 tons and equivalent full load hours of 2500 h/yr, determine the most economical option to deliver cooling needs between the two design alternatives:

- Conventional cooling plant with an electric chiller (sized to be 2000 tons) with an efficiency of 0.70 kW/ton. The cost of the cooling plant is estimated to be \$2000/ton with an annual operation and maintenance cost of \$90,000/yr. Electricity rate include an energy charge of \$0.10/kWh and a monthly demand charge of \$25/kW. The peak demand of 2000 tons occur every month of the year.
- District cooling provider with a pipe connection 50 m in length. The cost of the connection including piping is estimated to be \$3000/m. The rate for the chilled water is \$0.12/ton·hr with a capacity charge of \$300/ton. The operation and maintenance costs are included in the provider's rate.

Use LCC analysis with a life cycle of 20 years and a discount rate of 5%.

Solution

First, the initial costs for both options are estimated. For the conventional cooling plant with the electric chiller, the initial cost, IC_{conv} is:

$$IC_{conv} = (2000\text{ tons}) * \left(\frac{\$2500}{\text{ton}}\right) = \$5,000,000$$

For the DCS, the cost for the connection, IC_{DCS}, is

$$IC_{DCS} = (50\text{ m}) * \left(\frac{\$300}{\text{m}}\right) = \$150,000$$

Then, the annual costs related are determined. For the conventional cooling plant, the annual costs, AC_{conv}, include the energy charges, EC_{conv}, the monthly demand charges, DC_{conv}, and the operation and maintenance costs, OM_{conv}:

$$EC_{conv} = (2000\text{ tons}) * \left(2{,}500\frac{\text{hrs}}{\text{yr}}\right) * \left(0.7\frac{\text{kW}}{\text{ton}}\right) * \left(\frac{\$0.10}{\text{kWh}}\right) = \$350{,}000/\text{yr}$$

$$DC_{conv} = (2000\text{ tons}) * \left(0.7\frac{\text{kW}}{\text{ton}}\right) * \left(\frac{\$25}{\text{kW}}\right) * (12\text{months}) = \$420{,}000/\text{yr}$$

$$OM_{conv} = \$90{,}000/\text{yr}$$

Therefore,

$$AC_{conv} = \frac{\$350{,}000}{\text{yr}} + \frac{\$420{,}000}{\text{yr}} + \frac{\$90{,}000}{\text{yr}} = \$860{,}000/\text{yr}$$

For the district cooling system, the annual costs, AC_{DCS}, include the capacity charge and the chiller water costs and is determined to be

$$AC_{DCS} = (2000\text{ tons}) * \left(\frac{\$300}{\text{ton}}\right) + (2000\text{ tons}) * \left(2500\frac{\text{hrs}}{\text{yr}}\right) * \left(\frac{\$0.12}{\text{ton}-\text{hr}}\right) = \$1{,}200{,}000/\text{yr}$$

The LCC values for each option, LCC_{conv} and LCC_{DCS}, can be estimated using the following expression (refer to Chapter 5):

$$LCC_{conv} = IC_{conv} + USPW.\ AC_{conv}$$

$$LCC_{DCS} = IC_{DCS} + USPW.\ AC_{DCS}$$

with USPW estimated for a lifetime of 20 years and a discount rate of 5% (refer to Chapter 5):

$$USPW = 12.46\text{ years}$$

Thus,

$$LCC_{conv} = \$5{,}000{,}000 + (12.46\text{ years}) * \left(\frac{\$860{,}000}{\text{yr}}\right) = \$15{,}717{,}500$$

$$LCC_{DCS} = \$150{,}000 + (12.46\text{ years}) * \left(\frac{\$1{,}200{,}000}{\text{yr}}\right) = \$15{,}106{,}500$$

The DCS option of Example 8.1 is cost-effective when compared to the conventional configuration with the building supplying its cooling needs using a cooling plant. However, the cost-effectiveness of the DCS depends on several factors including the cost of the chiller for the conventional cooling plant and the chilled water from the DCS provider. Table 8.3 and Fig. 8.10 illustrate the impact of both of these factors on the net present value (i.e., $NPV = LCC_{conv} - LCC_{DCS}$) of the district cooling system option relative to the conventional option with the cooling plant. As noted in Table 8.3 and Fig. 8.12, when the chiller cost is high and the chilled water rate is low, the district cooling

TABLE 8.3 Impact of Chiller Cost and Chilled Water Rate on NPV for the DCS Relative to the Cooling Plant

NPV ($)	Cost of chiller		
Chilled water cost	$2500/ton	$2000/ton	$3000/ton
0.8	3,105,291	2,105,291	4,105,291
0.1	1,859,070	859,069.5	2,859,070
0.12	612,848.5	−387,152	1,612,848
0.14	−633,373	−1,633,373	366,627.4
0.16	−1,879,594	−2,879,594	−879,594

DCS, District cooling system; *NPV*, net present value.

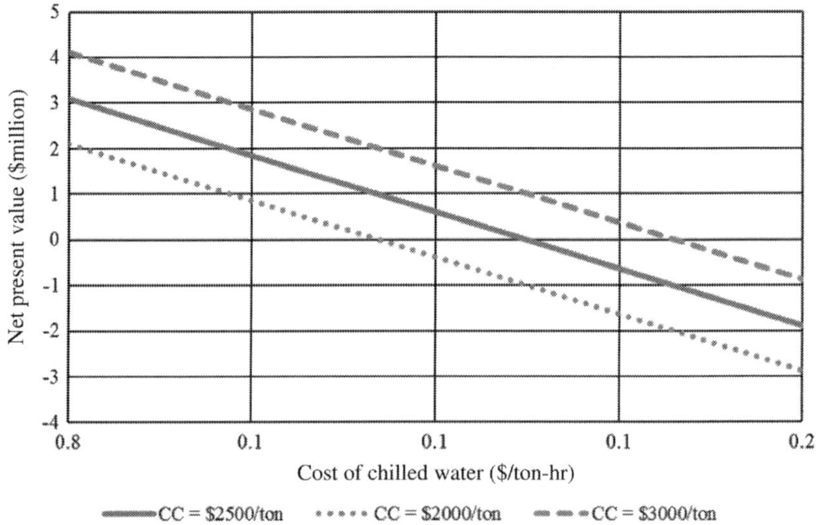

FIGURE 8.10 Variation of net present value for the DCS relative to the cooling plant with chilled water rate for three chiller cost levels.

system is more cost-effective than specifying a cooling plant for the building considered in Example 8.1.

8.4 FOSSIL FUEL-FIRED BOILERS

Fossil fuel-fired boilers are commonly used to generate hot water or steam that is circulated through a DES distribution network to heat buildings. Boilers have several parts including an insulated jacket, a burner, a mechanical draft system, tubes and chambers for combustion gas, tubes and chambers for water or steam circulation, and controls (ASHRAE, 2013). The types of fuel-fired boilers are briefly presented in this section. In addition, simplified analysis methods are outlined to estimate the potential energy use savings associated with common energy efficiency measures applied to boilers.

8.4.1 Types of Boilers

Boilers commonly used in DES can be divided into two basic groups: (1) fire-tube or water-tube depending on the relative location of the hot combustion gases, and (2) the fluid being heated within the boiler. In the following sections, common boilers and burner types are briefly described.

Fire-tube boilers: The hot combustion products flow through tubes submerged within the water contained in these boilers. To increase the contact surface area between the hot gases and the water, two to four passes are used for the tubes. The multipass tubes increase the efficiency of the boiler but require greater fan power. Due to economics, the highest capacity of fire-tube boilers is currently in the 10,000 kg of steam per hour with an operating pressure of 16 atm (250 psi). The fire-tube boilers are generally simple to install and maintain. Moreover, they have the ability to meet sudden and wide load fluctuations with only small pressure changes.

Water-tube boilers: In these boilers, the water flows inside tubes surrounded by flue combustion gases. The water flow is generally maintained by the density variation between cold feed water and the hot water/steam mixture in the riser. The water-tube boilers are classified in several groups depending on the shape, and drum location, capacity, and number. The size of water-tube boilers can be as small as 400 kg of steam per hour and as large as 1000 MW unit. The largest industrial boilers are generally about 250,000 kg of steam per hour.

The firing systems of boilers depend on the fuel used. For gas-fired boilers, natural gas is introduced to the burner through orifices that provide jets that allow the fuel to mix rapidly with the combustion air supply. There is a wide range of burner designs depending on the orientation, the number, and the location of the orifices. As part of routine tune-up and maintenance of gas-fired boilers, it is important to inspect gas injection orifices to check that all the passages are unobstructed. In addition, it is important to identify and replace any burned off or missing burner parts. For oil-fired boilers, fuel is first prepared and treated before final delivery to the burner following these general steps:

- Use of strainers and filters to clean the oil fuel and remove any deposit or solid foreign material.
- Add flow line preheaters to deliver the fuel oil with a proper viscosity.
- Use atomizers to deliver the fuel oil in small droplets before mixing with the combustion air supply. A gun fitted with a tip that has several orifices to produce fine spray can carry out the atomization of the fuel oil.

8.4.2 Simplified Analysis of Energy Performance of Boilers

There are several factors that influence the energy performance of boilers including fuel characteristics, firing method, steam pressure, and heating capacity. The main challenge to ensure optimal operating conditions for boilers is to provide the proper excess air for the fuel combustion. It is generally agreed that 10% excess air provides the optimum air-to-fuel ratio for complete combustion. Too much excess air causes higher stack losses and requires more fuel to heat ambient air to stack temperatures. On the other hand, if insufficient air is supplied, incomplete combustion occurs and the flame temperature is reduced.

The general definition of overall boiler thermal efficiency is the ratio of the heat output, E_{out}, over the heat input, E_{in} (Krarti, 2011):

$$\eta_b = \frac{E_{out}}{E_{in}} \tag{8.1}$$

The overall boiler efficiency accounts for combustion efficiency, stack heat loss, and radiation and convection losses from the outer surfaces of the boiler. The combustion efficiency refers to the effectiveness of the burner in providing the optimum fuel/air ratio for complete fuel combustion. To determine the overall boiler thermal efficiency, some measurements are required. The most common test used for boilers is the flue gas analysis using an Orsat apparatus to determine the percentage by volume the amount of CO_2, CO, O_2, and N_2 in the combustion gas leaving the stack. Based on the flue gas composition and temperature, some adjustments can be made to tune-up the boiler and to determine the best air-to-fuel ratio to improve the boiler efficiency. Monographs are available to determine the overall boiler efficiency based on measurement of flue gas composition and temperature. One of these monographs applies to gas-fired and oil-fired boilers and is reproduced in Fig. 8.11 (for IP units) and Fig. 8.12 (for SI units), and Example 8.2 illustrates how the monograph can be used to determine the boiler efficiency.

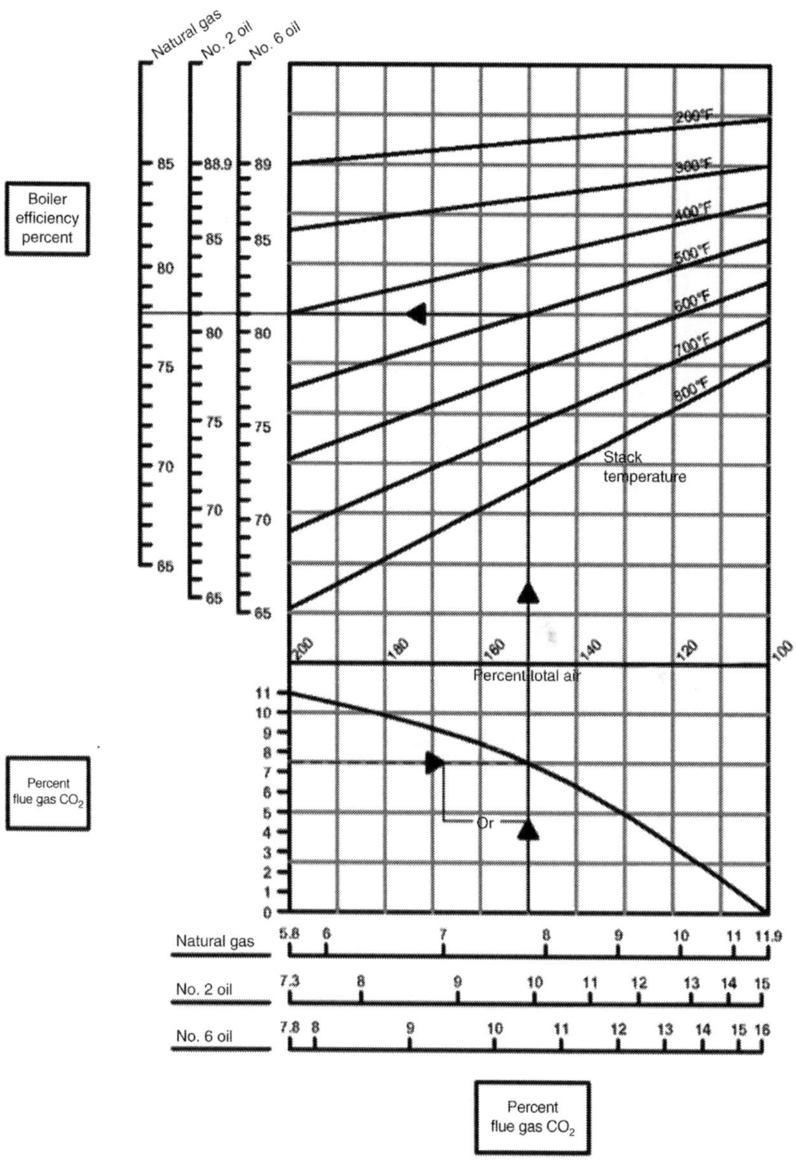

FIGURE 8.11 Monograph for boiler efficiency estimation using IP units.

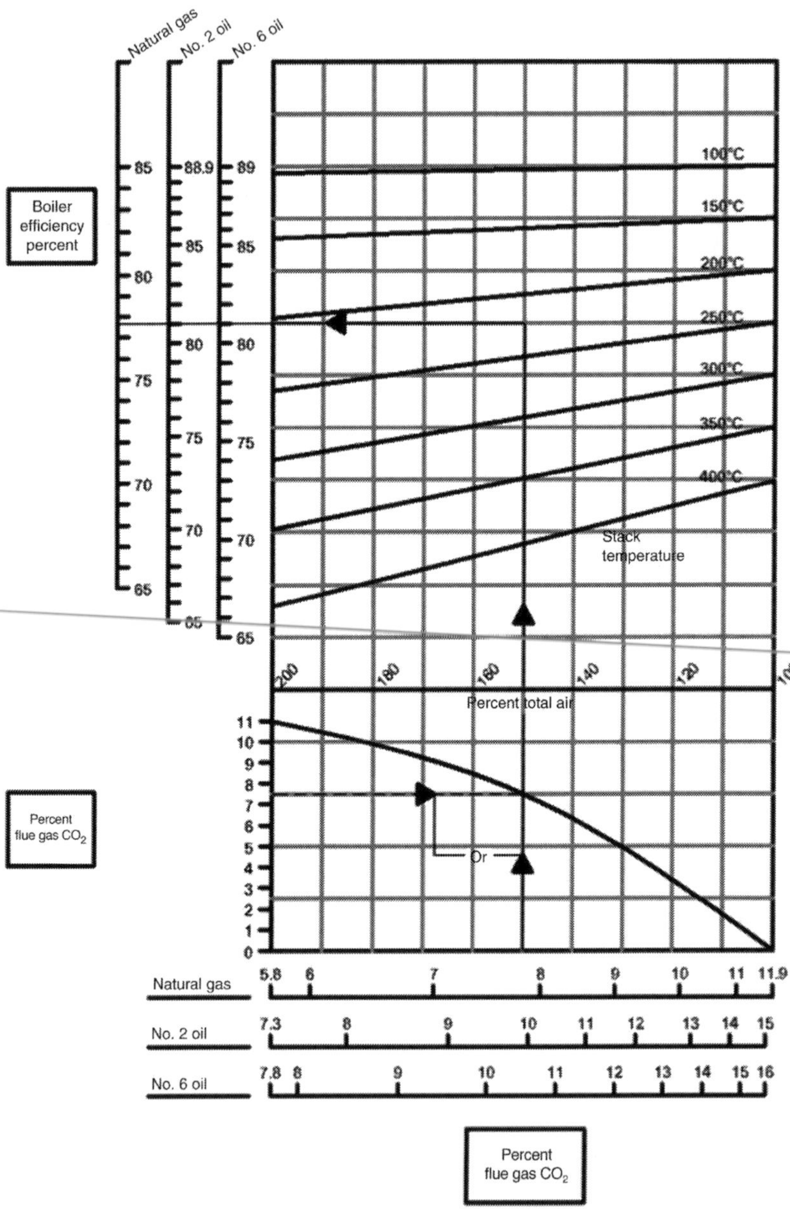

FIGURE 8.12 Monograph for boiler efficiency estimation using SI units.

Example 8.2

A flue gas analysis of a gas-fired boiler indicates that the CO_2 content is 8% with a gas flue temperature of 204°C (400°F). Determine the overall thermal efficiency of the boiler.

Solution

By reading the monograph of Fig. 8.12, the combustion occurs with an excess air of 48% and excess O_2 of 7%. The overall boiler thermal efficiency is about 81%.

The long-term performance of boilers is expressed in terms of annual fuel utilization efficiency or AFUE and accounts for part-load operation. There are several measures by which the boiler efficiency of a heating plant can be improved such as the installation of condensing boilers, the periodic tune-up of the boilers, and control operations especially for modular boilers. The net effect of any energy efficiency measure is some savings in the fuel use by the heating plant. To calculate the savings in fuel use, ΔFU, related to the change in the boiler efficiency, the following equation can be used:

$$\Delta FU = \frac{AFUE_{eff} - AFUE_{std}}{AFUE_{eff}} FU_{std} \tag{8.2}$$

where $AFUE_{std}, AFUE_{eff}$ are, respectively, the old and the new seasonal thermal efficiency of the boiler, and FU_{std} is the fuel consumption before any retrofit of the boiler system.

Example 8.2 illustrates how the cost-effectiveness of a boiler tune-up kit can be evaluated.

Example 8.3

The boiler of Example 8.2 uses 1,500,000 L of fuel oil per year. An instrumentation kit is purchased at a cost of $20,000 and is used to adjust the boiler operation so its excess of O_2 is only 3%. Determine the payback of the instrumentation if the cost of fuel oil is $0.20/L.

Solution

From Example 8.2, the existing boiler has an excess O_2 of 6% and an overall boiler thermal efficiency of about 78% (i.e., $\eta_{std} = 0.78$).

After the boiler tune-up, the excess O_2 is 3%. Using the monograph of Fig. 8.12, the new boiler efficiency can be determined from the flue gas temperature (343°C or 650°F) and the excess O_2 (3%). It is found to be $\eta_{eff} = 84\% = 0.84$. Using Eq. (8.4), the fuel savings can be calculated:

$$\Delta FU = \frac{0.84 - 0.78}{0.84} .1{,}500{,}000 = 107{,}140 \quad \mathrm{L/yr}$$

Therefore, the simple payback period for the instrumentation is

$$SPP = \frac{\$20{,}000}{10{,}714 L/\mathrm{yr} * \$0.20/L} . \approx 1.0 \ \text{year}$$

Other measures that can be considered to increase the overall efficiency of boilers include (Krarti, 2011) the following:

- Install turbulators in the fire-tubes to create more turbulence and thus increase the heat transfer between the hot combustion gas and the water. The improvement in boiler efficiency can be determined by measuring the stack flue gas temperature. The stack gas temperature should decrease when the turbulators are installed. As a rule of thumb, a 2.5% increase in the boiler efficiency is expected for each 50°C (90°F) decrease in the stack flue gas temperature.
- Insulate the jacket of the boiler to reduce heat losses. The improvement of the boiler efficiency depends on the surface temperature.
- Install soot-blowers to remove boiler tube deposits that reduce heat transfer between the hot combustion gas and the water. The improvement in the boiler efficiency depends on the flue gas temperature.
- Use economizers to transfer energy from stack flue gases to incoming feedwater. As a rule of thumb, a 1% increase in the boiler efficiency is expected for each 5°C (9°F) increase in the feedwater temperature.
- Use of air preheaters to transfer energy from stack flue gases to combustion air.

The stack flue gas heat recovery equipment (i.e., air preheaters and economizers) are typically the most cost-effective auxiliary equipment that can be added to improve the overall thermal efficiency of the boiler system.

8.5 COOLING SYSTEMS

In this section, the common types of cooling equipment used in DES are first described. The benefits of operating multiple chiller to meeting variable cooling loads are then outlined. Finally, simplified analysis methods for evaluating the energy performance of these equipment are presented with some illustrative calculation examples.

8.5.1 Types of Chillers

For DES as well as large building complexes, large chillers are used to produce chiller water that can be supplied to meet requirements for space air conditioning of various buildings. Central chillers are generally powered by electric motors, fossil fuel engines, or turbines. Some chillers use hot water or steam to generate chilled water. A description of various types of central chillers is provided next.

- *Electric chillers*: There are currently three major types of electric chillers available in the market using centrifugal, reciprocating, or rotary compressors. All these chillers use mechanical vapor compression cycle. A brief outline of the typical features of each of the electric chiller types is provided next.
  - *Centrifugal compressors* use rotating impellers to increase refrigerant gas pressure and temperature. Chillers with centrifugal compressors have capacities in the range of 300–25,000 kW (or 85–7000 tons). For

capacities above 4500 kW (or 1250 tons), the centrifugal compressors are typically field erected.

- *Reciprocating compressors* use pistons to raise the pressure and the temperature of refrigerant gases. Two or more compressors can be used under part-load conditions to achieve higher operating efficiencies. Capacities of 35–700 kW (or 10–200 tons) are typical for chillers with reciprocating compressors.
- *Rotary compressors* use revolving motions to increase refrigerant gas pressure. One of the most ingenious rotary compressors is the scroll compressor. The most conventional rotary compressors are the screw compressors that can have several configurations. The capacity of the rotary chillers can range from 3 kW to 1750 kW (1–500 tons).

- *Absorption chillers*: Absorption chillers operate using a concentration–dilution cycle to change the energy level of the refrigerant (water) by using lithium bromide to alternately absorb heat at low temperatures and reject heat at high temperatures. The absorption chillers can be direct fired (using natural gas or oil fuel), or indirect fired. Indirect fired units may use as heat source steam or hot water (from a boiler, a district heating network, an industrial process, or a waste heat). A brief description of both types of absorption chiller is provided in the following sections.
  - Direct-fired absorption chillers can be cost-effective when the price of natural gas is favorable. Some of the direct-fired chillers can be used to produce chilled and hot water. Thus, they can provide cooling and heating and are sometimes referred to as chillers/heaters. These chillers/heaters can be cost-effective especially when heating needs exist during the cooling season (for instance, buildings with large service hot water requirements). Two types of absorption chillers are available in the market: single-effect and double-effect chillers. Some prototypes of tripe-effect absorption chillers have been built in the US. Capacities ranging from 100 kW to 5000 kW (or 30–1500 tons) are available for direct-fired chillers.
  - *Indirect-fired absorption chillers* operate with steam (with pressures as low as 15 psig) or hot water (with temperatures as low as 140°C). Some small absorption chillers using solar energy (to generate hot water) have been proposed and some prototypes have been developed and evaluated. Cooling capacities from 15 kW to 425 kW (4–120 tons) are available even though typical sizes range from 200 kW to 5000 kW (55–1450 tons). Double-effect chillers can be considered only for high-temperature hot water and steam or with hot industrial waste gases.
- *Engine-driven chillers*: Like electrically driven chillers, the engine-driven chillers can use reciprocating, rotary, or centrifugal compressors to provide mechanical refrigeration. The compressors can be powered by turbines or gas-fired engines. The engine-driven chillers can have large capacities up to 15,000 kW (4250 tons) but have usually high first costs.

TABLE 8.4 Typical COP for Different Chiller Types

Type of chillers	Range for COP
Small electric chillers	
Air cooled	2.2–3.2
Large electric chillers	
Air cooled	3.7–4.1
Water cooled	4.6–5.3
Absorption chillers	
Single-effect	0.4–0.6
Double-effect	0.8–1.1
Engine-driven chillers	1.2–2.0

COP, Coefficient of performance.

Table 8.4 provides typical energy efficiency indicators expressed by the range of COP values of chiller types described in this section.

8.5.2 Operation of Multiple Chillers

Instead of a single chiller, two or more smaller chillers may be more cost-effective to meet varying load requirements especially for DES applications. Indeed, parallel staging of multiple chillers is a common method of meeting peak load in larger installations. Multiple chillers also provide redundancy for routine maintenance and equipment failure. For some applications, sizing one chiller at one-third and another chiller at two-thirds of the peak load enables the system to meet most cooling conditions at relatively high chiller part-load efficiencies (Krarti, 2011). These staged units can also be sized optimally for different conditions. For example, one chiller can be operated for peak efficiency at summer conditions and the other chiller can be optimized for winter conditions. Furthermore, proper sequencing helps to maintain the flow rate through each evaporator within the range recommended by the chiller manufacturer. As the system flow nears the maximum limit for the operating chiller(s), another machine must be brought online. Similarly, as the system load and flow decrease, chillers must be shut down to reduce the need for bypass water flow.

The energy savings attributed to switching from a single-chiller plant to a multichiller cooling plant depend on the load profile and climatic conditions. Table 8.5 shows the number of hours of operating one-chiller system at various part-load ratios for a load profile associated to office buildings located in four US climates. The potential savings in total electrical energy use as a function of the number of chillers in a multichiller cooling plant is provided in Fig. 8.13.

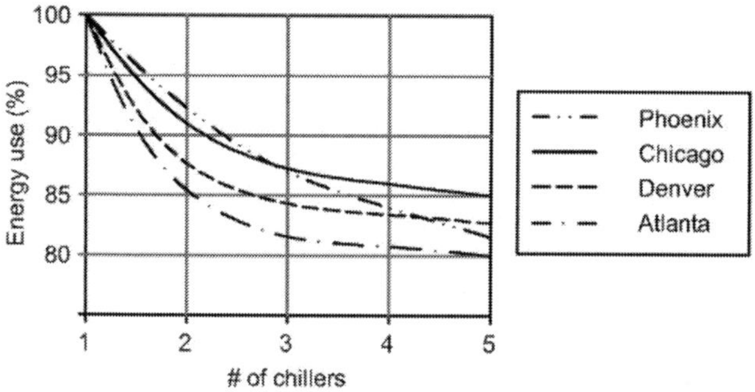

FIGURE 8.13 **Energy use reduction as a function of number of chillers for office buildings located in four US climates.**

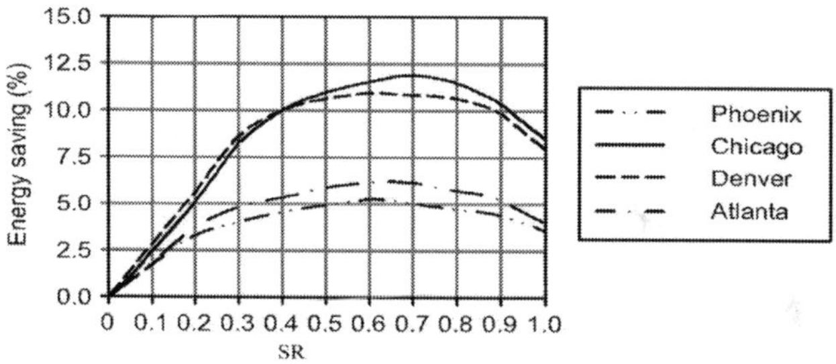

FIGURE 8.14 **Energy use savings as a function of chiller size ratio in a two-chiller central cooling plant for office buildings located in four US locations.**

Fig. 8.14 illustrates the potential savings associated with the annual building electrical energy savings when using a two-chiller central cooling plant with various chiller sizing ratios. The chiller size ratio, *SR*, is defined as the ratio between the sizes of the small chiller to that of the larger chiller in a two-chiller cooling plant. The analysis was carried out by letting the *SR* vary from 0 (single chiller configuration) to 1 (two equal chiller).

8.5.3 Simplified Analysis of Energy Performance of Chillers

To reduce the energy use of cooling systems, the energy efficiency of the equipment has to be improved under full load and partial load conditions. In general, the improvement of the energy efficiency of cooling systems can be achieved by using high-performance chillers and optimizing the controls of the cooling equipment. Indeed, significant improvements in the overall efficiency of DES cooling plants have been achieved by the introduction of two-compressor reciprocating and centrifugal chillers, variable-speed centrifugal chillers, and scroll compressor chillers. A brief description of each of these

TABLE 8.5 Hours at Percent Part-Load Ratio Using One Chiller

Climate	0–10	10–20	20–30	30–40	40–50	50–60	60–70	70–80	80–90	90–100	100+
Denver	539	601	1214	1182	703	140	25	3	2	1	0
Chicago	644	523	980	1290	657	139	19	1	0	0	0
Phoenix	350	456	790	1270	1349	600	55	25	16	8	1
Atlanta	589	595	888	1167	633	635	45	2	2	0	0

chiller configurations is presented next with some estimation of their energy efficiency:

- Multiple compressor chillers can be reciprocating, screw, or centrifugal with capacities in the range of 100–7000 kW (i.e., 30–2000 tons). They are energy efficient to operate especially under part load conditions. Some studies indicate that chillers equipped with multiple compressors can save up to 25% of the cooling energy use compared to single-compressor chillers (Tuluca, 1997).
- Variable-speed compressor chillers are in general centrifugal and operate with variable head pressure using variable speed motors. Therefore, the variable-speed compressor chillers work best when their cooling load is most of the time below the peak. The typical capacity of a variable-speed compressor chiller is in the range of 500–2500 kW (i.e., 150–700 tons). It is reported that chillers with a variable speed compressor can reduce the cooling energy use by almost 50% (Tuluca, 1997).
- The scroll compressor is a rotary compression device with two primary components, a fixed scroll and an orbiting scroll, both needed to compress and increase the pressure of the refrigerant. The scroll compressors are more energy efficient than the centrifugal compressor since the heat loss between the discharge and the suction gases is reduced. Manufacturers of scroll compressors report that the COP of the scroll chillers exceeds 3.2.

The energy savings calculation from increased energy efficiency of cooling systems can be estimated using the simplified but general expression provided by Eq. (8.3):

$$\Delta E_C = \left(\frac{\dot{Q}_C \cdot N_{h,C} \cdot LF_C}{SEER}\right)_S - \left(\frac{\dot{Q}_C \cdot N_{h,C} \cdot LF_C}{SEER}\right)_E \tag{8.3}$$

where the indices E and S refer to the values of the parameters, respectively, for energy efficient and standard cooling system. Moreover, the following parameters are used in Eq. (8.3):

- SEER is the seasonal efficiency ratio of the cooling unit. When available, the average seasonal COP can be used instead of the SEER.

- $\dot{Q}_C$ is the rated capacity of the cooling system.
- $N_{h,c}$ is the number of equivalent cooling full-load cooling hours.
- LF_C is the rated load factor and is defined as the ratio of the peak cooling load experienced by the building over the rated capacity of the cooling equipment. This load factor compensates for over-sizing of the cooling unit.

It should be noted that the units for Q_C and *SEER* have to be consistent, that is, if *SEER* has no dimension (using the European definition of EER), Q_C has to be expressed in kilowatts.

Example 8.4 illustrates a sample of calculation to determine the cost-effectiveness of replacing an existing chiller with a high energy efficiency chiller.

Example 8.4

An existing chiller with a capacity of 800 kW and with an average seasonal COP of 3.5 is to be replaced by a new chiller with the same capacity but with an average seasonal COP of 4.5. Determine the simple payback period of the chiller replacement if the cost of electricity is $0.07/kWh and the cost differential of the new chiller is $15,000. Assume that the number of equivalent full-load hours for the chiller is 1000/yr before and after the replacement.

Solution

In this example, the energy use savings can be calculated using Eq. (9.12) with $SEER_e = 3.5$; $SEER_r = 4.5$; $N_{h,C} = 1000$; and $Q_C = 800$ kW; $LF_C = 1.0$ (it is assumed that the chiller is sized correctly):

$$\Delta E_C = 800\ \text{kW} * 1000\ \text{h/yr} * 1.0 * \left(\frac{1}{3.5} - \frac{1}{4.5}\right) = 50{,}800\ \text{kWh/yr}$$

Therefore, the simple payback period for investing in a high-efficiency chiller rather than a standard chiller can be estimated as follows:

$$SPP = \frac{\$15{,}000}{50{,}800\ \text{kWh/yr} * \$0.07/\text{kWh}} = 4.2\ \text{years}$$

An LCC analysis may be required to determine if the investment in a high energy efficiency chiller is really warranted.

8.6 THERMAL ENERGY SYSTEMS

Thermal energy storage (TES) is generally defined as the temporary storage of energy for later use when heating or cooling is needed. For heating applications, heat storage systems are used with energy stored at high temperatures (above 20°C [68°F]). For cooling applications, energy is stored at low temperatures (below 20°C [68°F]). The concept of TES is not new and has been used few centuries ago to cool churches using block of ice that were stored in the cellar. TES can be effective in reducing operating cost of cooling plant equipment for DES applications. By operating chillers during off-peak hours to recharge the storage system and discharge the storage during on-peak hours, a significant fraction of the on-peak electrical demand and energy consumption is shifted to off-peak periods. Time-of-use rates, as outlined in Chapter 3, constitute an incentive to reduce or avoid operation of the cooling plant during on-peak periods by the cool storage system. Large differential between on- and off-peak energy and peak consumption rates generally make cool storage systems economically feasible.

Moreover, TES systems can be also used to reduce the size and thus the initial cost of chillers especially in applications where peak loads occur only for limited periods (Krarti, 2011). While cooling plants with TES systems may actually consume more energy that cooling plants without TES systems, TES systems if designed and controlled properly can reduce the overall operating costs of cooling plants. To determine the potential of operating cost reduction, it is important to carefully account for various factors that affect the design and operation of TES systems. This section provides description of basic principles for designing and operating thermal energy storage (TES) systems typically considered to reduce operating costs of cooling plants in DES applications. First, an overview of the various types of cooling thermal energy storage systems is presented. Then various control strategies for TES systems are discussed to minimize the operating costs of cooling plants. A simplified calculation procedure is provided to determine the energy cost savings incurred from installing TES systems. Finally, typical operating cost savings incurred by implementing optimal controls to operate TES systems are presented.

8.6.1 Types of TES Systems

Thermal energy storage can be achieved by two mechanisms:

1. Sensible energy storage by increasing (for heating applications) or decreasing (for cooling applications) the temperature of the storage medium (water for instance)
2. Latent energy storage by changing the phase of the storage medium (Phase Change Materials – PCM, eutectic salt solutions, or ice-water mixtures).

For cooling applications, there are several types of TES systems that have been installed in various commercial buildings and industrial applications. Among these TES systems are the following:

1. *Chilled water storage systems*: These systems consist typically of tanks where chilled water (temperature above freezing point) is stored before it is used during off-peak periods. There is no change of phase for the water in these systems and thus can store a limited density of energy. Water is selected since it has the highest specific heat of all common materials (4.18 kJ/kg°C). Typically, a tank volume varying from 0.09 to 0.17 m^3 is required to store 1 kWh of energy using chilled water.

2. *Eutectic salts*: In these systems, a solution of salts is used to store energy at low temperatures. The advantage of these systems is that a temperature below 0°C can be achieved before the solution is frozen. In addition, some salts have heat of fusion comparable to that of ice. It should be noted that the solution of salts needs to be mixed in a controlled ratio to ensure that the mixture melts completely and have the same composition in liquid and solid phases. For eutectic salts, the volume requirement for the storage tank is estimated to be 0.05 m^3/kWh.
3. *Ice storage systems*: In these systems, the water is transformed into ice, which is stored in tanks. Therefore, the water can be present in the form of two phases (liquid and solid) inside the tank. Typically, the ice is made during the off-peak periods (charging) and is melted during on-peak periods (discharging). Ice storage systems have higher energy density compared to chilled water systems. Thus, the volume of the storage tank required for ice systems is significantly less than that for chilled water systems (almost one-fourth). In addition, ice storage systems allow for innovative HVAC system design such as cold air distribution systems, which have lower initial costs compared to conventional distribution systems. Common ice storage systems include the following:
   - Ice harvesters (Fig. 8.15A). In these systems, thin ice layers are formed around vertical plates (evaporator) that are sprayed with water that is pumped from the tank. The ice layers are harvested to the storage tank by circulating hot gases through the evaporator. The ice mixed with water is stored in the tank to obtain what is often referred to as ice slurry. The volume requirement for the storage tank used in ice harvesters is about 0.025 m^3/kWh.
   - Internal melt ice-on-soil storage systems (see Fig. 8.15B). In these systems, direct expansion coils are fitted inside the storage tank that is filled with water. Brine solution (mixture of water and ethylene glycol) is typically circulated through the coils with a temperature in the range of −6°C to −3°C. In the charging mode, ice layers are formed around the coils. In the discharging mode, the ice is melted by circulating warm brine solution in the coils to be cooled to provide space cooling. The volume of the storage tank required for internal melt ice-on-coil systems varies from 0.019 to 0.023 m^3/kWh.
   - External melt ice-on-coil storage systems. They are similar to the internal melt ice-on-coil system in that the ice is made around coils filled with brine solution. However, the water that results from melting ice in the storage tank is directly used to provide space cooling. Typically, a volume of 0.023 m^3/kWh is used to size storage tanks for external melt ice-on-coil systems.
   - Containerized ice storage systems (see Fig. 8.15C). In these systems, small containers of various shapes (typically spherical) filled with water are used inside a tank to store energy. The water inside the containers is

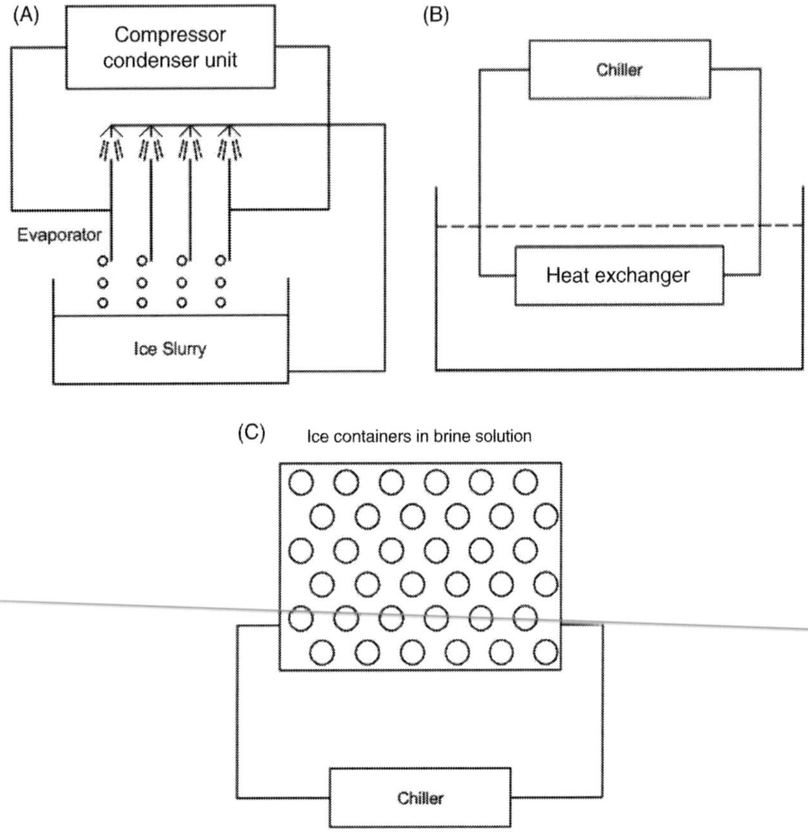

FIGURE 8.15 Commonly installed ice storage systems. (A) Ice harvester storage system. (B) Ice-on-coil with internal melt storage system. (C) Containerized ice storage system.

frozen by directly cooling the solution inside the tank (which acts as the evaporator). The typical volume requirement for containerized ice storage systems is 0.048 m^3/kWh.

8.6.2 Principles of TES Systems

Fig. 8.16 illustrates a typical configuration for a cooling plant with a TES system. Instead of one chiller, some cooling plants may have a base-load chiller that provides cooling up to a threshold load (determined by the capacity of the base chiller). Any additional cooling loads are either met directly by a second chiller (TES chiller) or the storage system. This second chiller is used to charge the TES system during unoccupied periods (or off-peak hours).

Typically, chillers operate at low efficiencies when they are used to charge ice storage tanks rather than to directly meet space cooling load. In addition, there are energy losses from the storage tanks when ice is kept for long periods

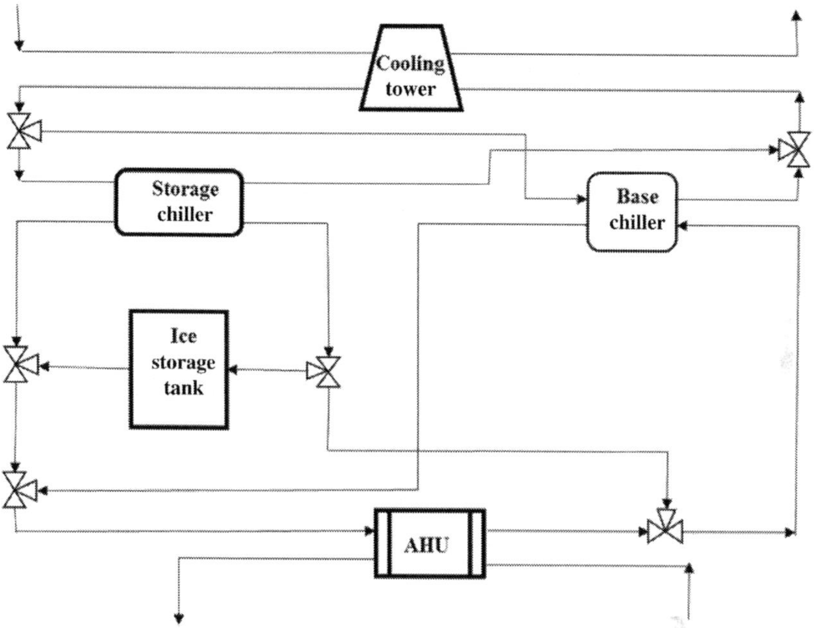

FIGURE 8.16 **Typical configuration for a cooling plant with a thermal energy storage system.**

of time without being utilized. Therefore, the energy used to meet a cooling load through TES systems may be actually higher than that consumed by chillers used to provide cooling directly. It is therefore important to pinpoint that the main advantage in the use of cooling plants with TES systems is not primarily to save energy but rather to reduce the electrical demand during on-peak periods and/or to provide additional cooling beyond the chiller capacity. In most applications including, DES, TES systems are specified primarily to reduce the cost of operating a cooling plant. The factors that affect the operation of cooling plants with TES systems include the following:

- TES and cooling plant performance during charging/discharging
- Control strategies used to operate the TES system
- Utility rate structures (real-time-pricing and time-of-use rates including ratchet clauses)
- Cooling load profile and noncooling electrical load profile

Some of the effects of these factors on operating cost savings of cooling plants are discussed in the following sections.

Charging/Discharging of TES Systems

The operation of the TES systems may lead to partial—rather than full—charging and discharging of the storage tank. The performance of TES systems under partial charging/discharging can affect the overall performance of the cooling plant especially for ice storage systems characterized by ice breakage effects for low charging levels. Typically, charging is performed during night time (off-peak period) and discharging occurs during day time (on-peak period). In some instances, the ice storage tank is not fully charged and/or not fully discharged. Partial charging and discharging cycles may occur when the peak load is not high or when the charging/discharging time is limited as is the case for some real-time pricing rates (Henze and Krarti, 1999; Krarti et al., 1999). In other cases, the chiller may be too small to fully charge the entire tank during one charging period (i.e., one day). In these cases, the storage tank operates almost exclusively under partial charging and discharging cycles.

During partial charging and discharging sequences, thin ice layers can be trapped between water layers within an internal melt ice-on-coil storage tank. In particular, the coils are surrounded by ice, which can be trapped by a water layer during a charging cycle. Fig. 12.3 shows a cross-section view of one coil during the charging cycle when the ice layers for adjacent coils do not overlap. When the tank is discharged again, the heat transfer between the brine circulating within the coils and the surrounding ice is relatively high. Similarly, Fig. 12.4 illustrates a cross-section view of the coil during the discharging cycle.

Partial charging/discharging cycles affect the TES performance, which is generally difficult to model. Some of the effects that characterize ice storage systems especially under partial charging/discharging operation include the following:

- Effects of water flow within the tank during charging or discharging cycles.
- Gravitational effects that deform the ice formations around the coils.
- Effects of ice breakage at the end of discharging cycles.

8.6.3 TES Control Strategies

Once installed, the TES system needs to be properly operated and controlled to achieve the desired cost savings. Several control strategies have been proposed and actually used to operate existing TES installations. The control of a TES system depends on the operating mode. Two TES operating modes are commonly considered: full and partial storage.

Full Storage

This operating strategy is also called load shifting and consists of generating the entire on-peak cooling load during off-peak periods when no significant cooling load exists. Therefore, the TES system operates at full capacity and the chillers do not operate at all during the on-peak hours. Thus, to implement a full storage operating mode, the TES system has to be sized properly so it can hold enough energy to meet the cooling load for the entire design day on-peak hours. The full storage strategy is best suited for applications where the length of the on-peak cooling period is short compared to the off-peak period when the TES system can be charged. It can be an effective operating strategy when the on-peak demand charges are high. Moreover, the control under a full storage operating strategy is simple since all that is needed is a timer clock to define charging and discharging periods. However, the full storage strategy requires large chiller and storage capacities and thus high initial costs.

Partial Storage

It can be defined as an operating strategy when the TES system meets only part of the on-peak cooling load. The remainder of the load is provided directly by the chiller. Partial storage requires lower initial costs than full storage since the chiller and the storage for partial storage are of smaller size. However, partial storage operating strategies require more complex controls than full storage systems. Some of these controls are discussed next.

- *Chiller-priority control*: The simplest of TES control strategies used is the *chiller-priority control*. For this strategy, the chiller runs continuously under conventional chiller control (direct cooling) possibly subject to a demand-limit while the storage provides the remaining cooling capacity if required. The simplicity lies in the fact that the conventional chiller control is not altered yielding high average part loads and a smooth demand curve, however, the meltdown of the ice is not well controlled to allow for maximal demand reduction. In addition, this control strategy does not account for the time-of-day dependent energy rate structure.
- *Constant-proportion control*: This control strategy considers that the storage meets a constant fraction of the cooling load under all conditions. Thus, neither the chiller nor the storage has priority in providing cooling. This simple control strategy provides a greater demand reduction than chiller-priority control since the chiller capacity fraction used for a particular month will track the fraction of the annual cooling design load the building experiences during that month. For example, in a month in which 50% of the annual design load occurs, only 50% of the chiller capacity will be requested. Constant-proportion control is rather easy to implement in practice by assigning a fixed fraction of the total temperature difference between brine supply and return flow to be realized by the storage and the remainder by the chiller. Finding the best load fraction for each application is a matter of trial and error.
- *Storage-priority control*: As the name reflects, *storage-priority control* requires melting as much ice as possible during the on-peak period. It is generally defined as a control strategy that aims at fully discharging the available storage capacity over the next available on-peak period. Thus, both the simultaneous operation of the chiller plus the storage and the terminal state-of-charge are specified. The following implementation of storage priority has been suggested by Henze et al. (1997): The chiller is base loaded during off-peak hours to recharge the ice storage for the next on-peak period. The chiller operates in one of two modes during the on-peak period. In the first mode, the chiller operates at a reduced capacity in parallel with the storage during on-peak hours so that at the end of the on-peak period the storage inventory is just depleted. If this is not possible without prematurely depleting storage, the control switches to the second mode in which the storage provides a constant proportion of the load in each on-peak hour similar to constant-proportion control.
- *Optimal controls*: Several optimal control strategies have been proposed to minimize energy use and operating cost of active TES systems while

maintaining occupant comfort. The main requirement for an optimal control strategy is typically the forecasting of cooling load and weather. Typically, an optimal control strategy provides a sequence of control actions that minimizes the total operating cost (i.e., including demand and energy charges) of the cooling plant over the simulation period. One optimization technique evaluated and tested is a predictive model-based control based on a "closed-loop" optimization approach. A more detailed description and discussion of the predictive optimal controller is provided in Chapter 5.

8.6.4 Simplified Analysis of TES Performance

To assess the cost effectiveness of TES systems, a detailed analysis is generally needed. Indeed, the utility rates that provide the major incentives for TES systems are set on an hourly and/or seasonal basis. Therefore, the use of a dynamic energy analysis tool is typically warranted to determine the overall operating cost of cooling plants with TES systems. However, simplified calculation methods can be sometimes used to especially assess whether or not a TES system is economically viable.

In this section, a simplified analysis method is presented to evaluate the cost-effectiveness of specifying and operating a TES system in a cooling plant. Specifically, the electrical power demand reduction due to the use of TES systems can be estimated as follows (Krarti, 2011):

$$\Delta kW_{TES} = \left(\frac{\dot{Q}_C}{SEER_{CHW}} \right)_S - \left(\frac{(1-X).\dot{Q}_C}{SEER_{CHW}} \right)_E \tag{8.4}$$

With the electric demand reduction calculated, savings in the demand charges can be estimated.

Similarly, the energy cost savings calculation incurred from the use of the TES system can be calculated as follows:

$$\Delta EC_{TES} = \dot{Q}_C . N_{h,c}^{TES} . \left(\frac{C_{on-pk}}{SEER_{CHW}} - \frac{C_{off-pk}}{SEER_{ICE}} \right) \tag{8.5}$$

where the indices *e* and *r* indicate the values of the parameters, respectively, before and after retrofitting the cooling unit (i.e., adding the TES system). SEER is the seasonal efficiency ratio of the cooling unit. When available, the average seasonal COP can be used instead of the SEER. Typically, the $SEER_{CHW}$ for producing chilled water (to directly cool the space) is higher than $SEER_{ICE}$ for making ice (to charge the TES system). $\dot{Q}_C$ is the rated capacity of the cooling system. X is the fraction of the on-peak cooling load (occurring during the hour when maximum electrical power demand is obtained) shifted to off-peak period. $N_{h,C}^{TES}$ is the number of equivalent on-peak cooling full-load cooling hours, which have been shifted during off-peak periods using the TES system.

Example 8.5 illustrates a sample of calculation to determine the cost-effectiveness of installing a TES system for a commercial building.

Example 8.5

Consider the cooling load profile for an office building shown in Fig. 8.17. The cooling is provided by a chiller having a capacity of 1000 kW and with an average seasonal COP of 3.5. The noncooling profile experienced by the same office building is illustrated in Fig. 8.18. It is proposed to install an ice storage system. When making ice the chiller has an average COP of 3.0. Determine the simple payback period of installing an ice storage system if the cost of electricity is as follows:

- Energy charges: \$0.07/kWh for on-peak hours (between 10:00 and 15:00 during week-days) and only \$0.02/kWh during other hours.
- Demand charges: \$15/kW during on-peak hours and \$0/kW during off-peak hours. The demand charges are assessed on a monthly basis.

The installed cost of the TES system is \$100/kW. Assume that the number of typical cooling days during the entire year before and after the installation of the TES system is 250 days. The TES system is operated with demand-leveling control so that the power demand during on-peak hours never exceeds 500 kW (which is the maximum noncooling load).

Solution

In this example, the entire cooling load during on-peak hours have to be shifted and thus in Eq. (8.4) the fraction $X = 1$. Therefore, the savings in

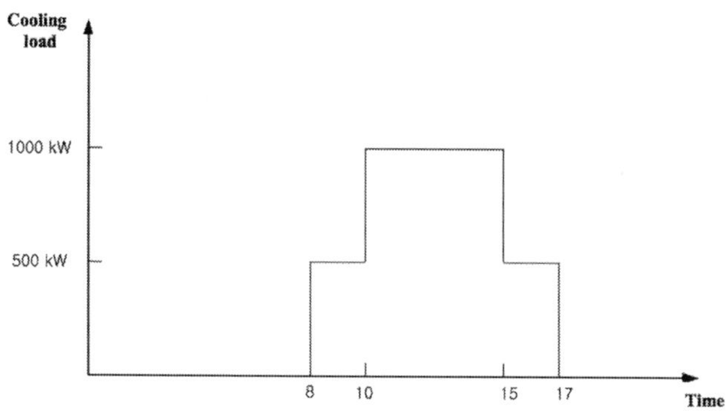

FIGURE 8.17 Cooling load profile used in Example 8.3.

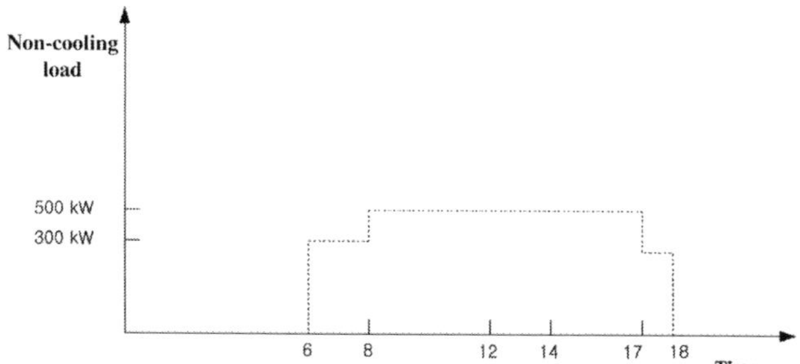

FIGURE 8.18 Electrical noncooling load-profile used in Example 8.3.

the electric power demand can be calculated using $SEER_{CHW} = 3.5$ and $Q_C = 1000$ kW:

$$\Delta kW_{TES} = 1000\text{ kW} * \left(\frac{1}{3.5}\right) = 286\text{ kW}$$

Thus, the cost savings due to demand charges are: 286 kW*$15/kW*12 months/yr = $51,480/yr.

The energy cost savings of using the TES can be calculated using Eq. (8.5) with the assumption that there are 250 typical cooling days per year:

$$\Delta EC_{TES} = 1000\text{ kW} * 5\text{ h/day} * 250\text{ days/yr}\left(\frac{\$0.07/\text{kWh}}{3.5} - \frac{\$0.02/\text{kWh}}{3.0}\right) = \$16{,}667/\text{yr}$$

The size of the ice storage tank is such that it can hold the cooling energy required during on-peak period, that is, 1000 kW*5 h = 5000 kWh. Therefore, the simple payback period for installing TES is estimated as follows:

$$SPB = \frac{\$100/\text{kWh} * 5000\text{ kWh}}{(\$51{,}480 + \$16{,}667)} = 7.3\text{ years}$$

An LCC analysis may be required to determine if the investment in a TES system is really warranted.

It should be noted that more savings can be obtained if the chiller is replaced. Indeed, when the TES is installed the chiller capacity can be reduced. For the case of this example, a chiller with 500-kW cooling capacity is sufficient (instead of a 1000-kW chiller) to charge the TES system and cool the office building during off-peak periods.

8.7 THERMAL ANALYSIS OF UNDERGROUND PIPING SYSTEMS

As noted in Section 8.3.6, distribution systems for DES consist of piping networks commonly buried underground. Therefore, it is important to determine the main factors that affect the thermal performance of underground pipes used to distribute chilled water, hot water, and steam for production locations to end-use consumers. An overview of the basic thermal analysis of heat loss/gain for buried pipes is provided in this section with some illustrative examples.

Using steady-state analysis, the rate of heat transfer from pipe to the soil, Q_{pipe}, can be estimated (ASHRAE, 2016):

$$Q_{pipe} = \frac{T_f - T_s}{R_{tot\quad s}} L_{pipe} \tag{8.6}$$

where T_f is the fluid temperature flowing in the pipe (i.e., hot water, steam, or chilled water), T_s is the soil temperature at the depth where the pipe is buried, R_{tot} is the total thermal resistance between the pipe, and L_{pipe} is the length of the pipe buried in the soil medium.

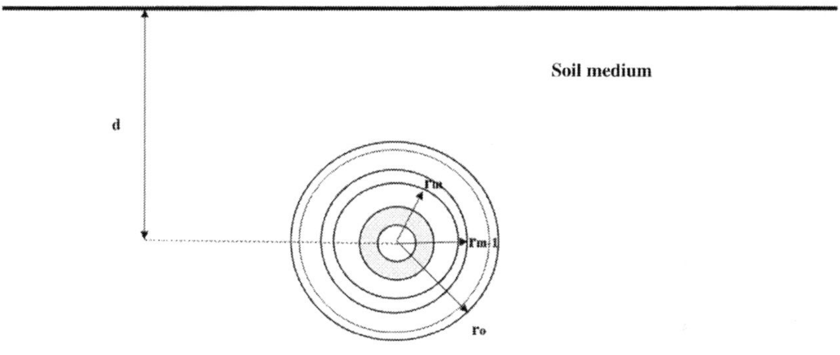

FIGURE 8.19 **Buried pipe in a soil medium with different layered materials.**

The total thermal resistance of the buried pipe, illustrated in Fig. 8.19, can be calculated using the following expression:

$$R_{tot} = R_s + \sum_{m=1}^{N} R_m \tag{8.7}$$

where R_s is the soil thermal resistance for a pipe of outer radius, r_o, buried at a depth d. Eqs. (8.8) and (8.9) provide simplified expressions to estimate R_s. R_m is the thermal resistance of the layer m covering the pipe including insulation, air layer, or pipe material. Eqs. (8.10) and (8.11) show the calculation procedures for R_m depending on the layer's properties.

When d/r_o is more than 2 but less than 4 (i.e., $2 < d/r_o < 4$), R_s can be estimated using the following equation:

$$R_s = \frac{1}{2\pi k_s} \cdot Ln\left[\frac{d}{r_o} + \sqrt{\left(\frac{d}{r_o}\right)^2 - 1}\right]\left\{\frac{d}{r_o} + \sqrt{\left(\frac{d}{r_o}\right)^2 - 1}\right\} \tag{8.8}$$

When d/r_o is more than 4 (i.e., $d/r_o > 4$), R_s is determined as follows:

$$R_s = \frac{1}{2\pi k_s} \cdot Ln\left(\frac{2d}{r_o}\right)\left\{\frac{2d}{r_o}\right\} \tag{8.9}$$

In Eqs. (8.8) and (8.9), k_s is the soil thermal conductivity and depends on the type of soil and its moisture content.

The thermal resistance, R_m, for layer m with outer radius, r_{m-1} and inner radius r_m (i.e., $r_{m-1} > r_m$) with a solid material having k_m as a thermal conductivity can be calculated using Eq. (8.10):

$$R_m = \frac{1}{2\pi k_m} \cdot Ln\left(\frac{r_{m-1}}{r_m}\right)\left\{\frac{r_{m-1}}{r_m}\right\} \tag{8.10}$$

When the layer m is made of air, the thermal resistance, R_m, can be estimated as follows:

$$R_m = \frac{1}{2\pi r_{m-1} h_m} \tag{8.11}$$

With the convective heat transfer $h_m = 17.0\ \text{W/m}^2{}^\circ\text{C} = 3.0\ \text{Btu/h·ft}^2{}^\circ\text{F}$.

Table 8.6 lists the thermal properties as per ASTM standards of some common insulation and pipe materials used for the distribution network of district heating and cooling systems (ASHRAE, 2016). Examples 8.6 and 8.7 illustrate the calculation of heat loss and gain through the pipe buried underground for DES applications.

TABLE 8.6 Thermal Properties of Common Insulation Materials Used for DES Piping System

Material	Thermal conductivity		Density kg/m³ (lb/ft³)	Maximum temperature allowed °C (°F)
	(W/m°C)	(Btu/h ft°F)		
Urethane foam				
At 38°C (100°F)	0.022	0.013		
At 93°C (200°F)	0.024	0.014	30–51 (1.9–3.2)	121°C (250°F)
Calcium silicate				
At 38°C (100°F)	0.048	0.028		
At 93°C (200°F)	0.066–0.078	0.038–0.045		
At 149°C (300°F)	0.073–0.083	0.042–0.048		
At 204°C (400°F)	0.080–0.088	0.046–0.051	240–352 (15–22)	649°C (1200°F)
Cellular glass				
At 38°C (100°F)	0.052	0.030		
At 93°C (200°F)	0.064	0.037		
At 149°C (300°F)	0.083	0.045		
At 204°C (400°F)	0.094	0.054	107–147 (6.7–9.2)	427°C (800°F)
Mineral fiber				
At 38°C (100°F)	0.036	0.021		
At 93°C (200°F)	0.045	0.026		
At 149°C (300°F)	0.057	0.033		
At 204°C (400°F)	0.074	0.043	128–176 (8–11)	454°C (850°F)

DES, District energy systems.

Example 8.6

Consider an insulated buried pipe for DCS with the following characteristics:

- The pipe has an inner radius of $r_i = 0.05$ m, a thickness of $\Delta r_p = 0.010$ m and is made of PVC material with a thermal conductivity of $k_p = 0.173$ W/m°C.
- The thickness of the foam urethane insulation is $\Delta r_{ins} = 0.025$ m with a thermal conductivity $k_{ins} = 0.022$ W/m°C.
- The protective jacket for the insulation has a thickness of $\Delta r_j = 0.005$ m and is made of PVC material with a thermal conductivity of $k_p = 0.173$ W/m°C.
- Determine the pipe heat transfer per unit length if the chilled water temperature is $T_f = 7$°C and soil medium temperature is $T_s = 14$°C. The depth of the pipe is $d = 1$ m and the soil thermal conductivity is $k_s = 1.73$ W/m°C.

Solution

The pipe has three layers: the pipe with a thickness $\Delta r_p = 0.010$ m, the insulation with a thickness $\Delta r_{ins} = 0.025$ m, and the protection jacket with a thickness, $\Delta r_j = 0.005$ m. Therefore, the outdoor radius of the pipe, r_o, is

$$r_o = r_i + \Delta r_p + \Delta r_{ins} + \Delta r_j = 0.05 + 0.01 + 0.025 + 0.05 = 0.09 \text{ m}$$

Since $d/r_o = 1/0.09 = 11.1 > 4$, Eq. (8.9) can be used to estimate the soil resistance, R_s:

$$R_s = \frac{1}{2\pi k_s}.Ln\left(\frac{2d}{r_o}\right)\left\{\frac{2d}{r_o}\right\} = \frac{1}{2\pi(1.73)}Ln(22.2) = 0.285\,\text{m°C / W}$$

The thermal resistances for, respectively, the pipe, R_p, the insulation layer, R_{ins}, and the protective jacket, R_j, can be estimated using Eq. (8.10):

$$R_p = \frac{1}{2\pi k_p}.Ln\left(\frac{r_i + \Delta r_p}{r_i}\right)\left\{\frac{r_i + \Delta r_p}{r_i}\right\} = \frac{1}{2\pi(0.173)}Ln\left(\frac{0.05+0.01}{0.05}\right)\left[\frac{0.05+0.01}{0.05}\right] = 0.168 \text{ m°C / W}$$

$$R_{ins} = \frac{1}{2\pi k_{ins}}.Ln\left(\frac{r_i + \Delta r_p + \Delta r_{ins}}{r_i + \Delta r_p}\right)\left\{\frac{r_i + \Delta r_p + \Delta r_{ins}}{r_i + \Delta r_p}\right\} = \frac{1}{2\pi(0.022)}$$

$$Ln\left(\frac{0.05+0.01+0.025}{0.05+0.01}\right)\left[\frac{0.05+0.01+0.025}{0.05+0.01}\right] = 2.519 \text{ m°C / W}$$

$$R_j = \frac{1}{2\pi k_j}.Ln\left(\frac{r_i + \Delta r_p + \Delta r_{ins} + \Delta r_j}{r_i + \Delta r_p + \Delta r_{ins}}\right)\left\{\frac{r_i + \Delta r_p + \Delta r_{ins} + \Delta r_j}{r_i + \Delta r_p + \Delta r_{ins}}\right\} = \frac{1}{2\pi(0.173)}$$

$$Ln\left(\frac{0.05+0.01+0.025+0.05}{0.05+0.01+0.025}\right)\left[\frac{0.05+0.01+0.025+0.05}{0.05+0.01+0.025}\right]$$

Or,

$$R_j = 0.053 \text{ m°C / W}$$

Thus, the total thermal resistance for the buried pipe according to Eq. (8.7) is

$$R_{tot} = 0.285 + 0.168 + 2.519 + 0.053 = 3.025 \text{ m°C / W}$$

The heat transfer per unit pipe length can be estimated using Eq. (8.6) :

$$\frac{Q_{pipe}}{L_{pipe}} = \frac{T_f - T_s}{R_{tot \ s}} = \frac{7-14}{3.025} = -2.314 \text{ W / m}$$

The negative sign indicates that the heat is gained from the ground medium to the chilled water carried by the buried pipe.

Example 8.7

Consider a buried steel pipe for DHS insulation with two layers as follows:

- The pipe has an inner radius of $r_i = 0.20$ m and is made of steel.
- The thickness of the mineral fiber insulation is $\Delta r_{ins,1} = 0.025$ m with a thermal conductivity $k_{ins} = 0.045$ W/m°C.
- A layer of air of a thickness of $\Delta r_{air} = 0.025$ m is then placed between the two insulation layers. In fact, the pipe and the mineral fiber insulation is housed inside a steel conduit that has an air gap.
- The thickness of a foam urethane insulation around the conduit is $\Delta r_{ins,2} = 0.025$ m with a thermal conductivity $k_{ins} = 0.022$ W/m°C.

Determine the pipe heat transfer per unit length if the hot water temperature is $T_f = 175$°C and soil medium temperature is $T_s = 14$°C. The depth of the pipe is $d = 3$ m and the soil thermal conductivity is $k_s = 1.73$ W/m°C. Neglect the thermal resistances of the pipe material, the conduit, and any protective jacket.

Solution

Thermally, the pipe has three layers: the mineral fiber insulation with a thickness $\Delta r_{ins,1} = 0.025$ m, the air layer with a thickness of $\Delta r_{air} = 0.025$ m, and the urethane foam insulation with a thickness $\Delta r_{ins,2} = 0.025$ m. Therefore, the outdoor radius of the pipe, r_o, is:

$$r_o = r_i + \Delta r_{ins,1} + \Delta r_{air} + \Delta r_{ins,2} = 0.2 + 0.025 + 0.025 + 0.025 = 0.275 \text{ m}$$

Since $d/r_o = 3/0.275 = 10.9 > 4$, Eq. (8.9) can be used to estimate the soil resistance, R_s:

$$R_s = \frac{1}{2\pi k_s}.Ln\left(\frac{2d}{r_o}\right)\left\{\frac{2d}{r_o}\right\} = \frac{1}{2\pi(1.73)}Ln(21.8) = 0.284 \text{ m°C / W}$$

The thermal resistance for the mineral fiber insulation, $R_{ins,1}$, can be estimated using Eq. (8.10):

$$R_{ins,1} = \frac{1}{2\pi k_{ins,1}}.Ln\left(\frac{r_i + \Delta r_{ins,1}}{r_i}\right)\left\{\frac{r_i + \Delta r_{ins,1}}{r_i}\right\} = \frac{1}{2\pi(0.045)}$$

$$Ln\left(\frac{0.20 + 0.025}{0.2}\right)\left[\frac{0.20 + 0.025}{0.2}\right] = 0.417 \text{ m°C / W}$$

The thermal resistance for air layer, R_{air}, is determined using Eq. (8.11):

$$R_{air} = \frac{1}{2\pi h_{air}\left(r_i + \Delta r_{ins,1}\right)} = \frac{1}{2\pi(17)(0.2+0.025)} = 0.042\ \text{m°C / W}$$

Finally, the thermal resistance for the urethane foam insulation, $R_{ins,2}$, is calculated using Eq. (8.10):

$$R_{ins,2} = \frac{1}{2\pi k_{ins,2}}.Ln\left(\frac{r_i + \Delta r_{ins,1} + \Delta r_{air} + \Delta r_{ins,2}}{r_i + \Delta r_{ins,1} + \Delta r_{air}}\right)\left\{\frac{r_i + \Delta r_{ins,1} + \Delta r_{air} + \Delta r_{ins,2}}{r_i + \Delta r_{ins,1} + \Delta r_{air}}\right\}$$

That is,

$$R_{ins,1} = \frac{1}{2\pi(0.022)} Ln\left(\frac{0.2+0.025+0.025+0.025}{0.2+0.025+0.025}\right)\left[\frac{0.2+0.025+0.025+0.025}{0.2+0.025+0.025}\right]$$
$$= 0.689\ \text{m°C / W}$$

Thus, the total thermal resistance for the buried pipe according to Eq. (8.7) is

$$R_{tot} = 0.284 + 0.417 + 0.042 + 0.689 = 1.431\ \text{m°C / W}$$

The heat transfer per unit pipe length can be estimated using Eq. (8.6):

$$\frac{Q_{pipe}}{L_{pipe}} = \frac{T_f - T_s}{R_{tot}} = \frac{175-14}{1.431} = 112.5\ \text{W / m}$$

The positive sign indicates that the heat is lost from the hot water carried by the pipe to the ground medium.

It should be noted that the calculations can be made more accurate since the thermal conductivity of thermal insulation depends on their temperature as noted in Table 8.6. Thus, the temperatures at various layers can be determined based on the first iteration to assess the thermal conductivity values listed in Table 8.6 are adequate. For instance, the temperature of the outer surface of the mineral fiber layer, $T_{o,ins,1}$ can be estimated as

$$T_{o,ins,1} = T_f - \left({Q_{pipe}}/{L_{pipe}}\right) * \left(R_s + R_{ins,1}\right)$$

Therefore,

$$T_{o,ins,1} = 175 - 117.2 * (0.284 + 0.417) = 96.2°\text{C}$$

Similarly, the inner surface temperature of the mineral fiber layer, $T_{i,ins,1}$ can estimated to be

$$T_{i,ins,1} = 175 - 117.2 * (0.284) = 143.1°\text{C}$$

The mean temperature of the mineral fiber is 119.7°C and is close to the 93°C for which the thermal conductivity is listed to be 0.045 W/m°C in Table 8.6. No additional iteration is needed to estimate the thermal resistance for the mineral fiber layer. Similar calculations should be carried out to estimate the medium temperature and ultimately the thermal resistance for the urethane foam.

Other configurations for buried pipes can be considered such as placing the pipes in tunnels or setting parallel pipes for return and supply. For these configurations, numerical or semianalytical analyses may be required to estimate the heat transfer between the pipe and the ground medium. For instance, Krarti and Kreider (1996) provided a semianalytical solution for heat transfer from a buried tunnel.

It should be noted that the insulation level for the piping system can be optimized based on the LCC analysis using the energy cost and insulation cost. Example 8.8 illustrates the selection of the optimal insulation thickness for a distribution piping system for a DHS application.

Example 8.8

Consider the insulated buried steel pipe for DHS described in Example 8.7. Determine the optimal insulation thickness for the urethane foam layer using the LCC analysis with a lifetime of 20 years and a discount rate of 5%. The cost for the insulation varies with thickness as noted in Table 8.7 while the hot water is generated for 8760 h/yr using a gas-fired boiler with an average efficiency of 0.85 and a natural gas cost of $0.025/kWh.

Solution

The LCC can be estimated from the insulation cost, IC_{ins}, and the annual energy cost, EC, as follows:

$$LCC = IC_{ins} + USPW.\,EC$$

The energy cost depends on the heat loss per unit pipe length, Q_{pipe}/L_{pipe}, as well as the cost of natural gas, C_{ng}, and the boiler efficiency, η_B:

$$EC = \frac{1}{\eta_B}\left({}^{Q_{pipe}}\!\big/_{L_{pipe}} \right) * C_{ng} * N_h$$

The calculations for Q_{pipe}/L_{pipe} depend on the urethane foam insulation thickness as noted in Example 8.7. The results for the calculations as well as the insulation costs, annual energy costs, and the LCC values for various insulation thicknesses are summarized in Table 8.7 and illustrated in Fig. 8.20. It is clear that the optimal insulation thickness for the urethane foam is 0.10 m since the LCC is at its minimum at this insulation level.

TABLE 8.7 Analysis Results for Example 8.8

Urethane foam thickness (m)	Q_{pipe}/L_{pipe} (W/m)	Cost of insulation ($/m)	Energy cost ($/m·yr^{-1})	LCC ($/m)
0	214.54	0	55.27	779.04
0.025	112.50	7	28.99	415.52
0.05	78.44	17	20.21	301.84
0.075	61.35	32	15.81	254.79
0.1	51.06	50	13.15	235.40
0.125	44.16	76	11.38	236.35
0.15	39.20	108	10.10	250.35
0.175	35.46349	150	9.14	278.78

LCC, Life-cycle cost.

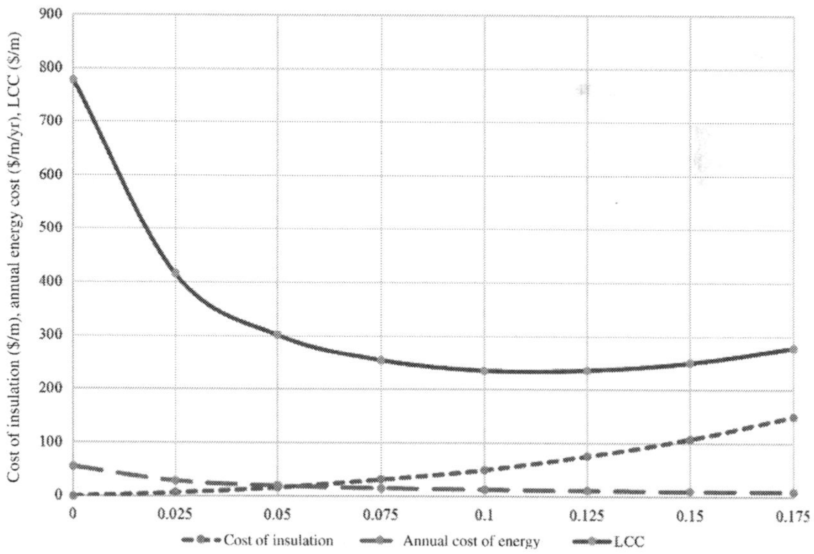

FIGURE 8.20 **Variation of the cost of insulation, annual energy cost, and life-cycle cost as a function of insulation thickness for Example 8.8.**

8.8 COOL ROOFS

In urban centers, higher ambient air temperatures have been observed when compared to neighboring rural areas or lightly populated suburbs (Gonzalez and Krarti, 2017). This phenomenon occurs especially during the summer seasons and is referred to as urban heat island (UHI) effects. The UHI magnitude depends on several factors including the albedo of exposed urban surfaces (i.e.,

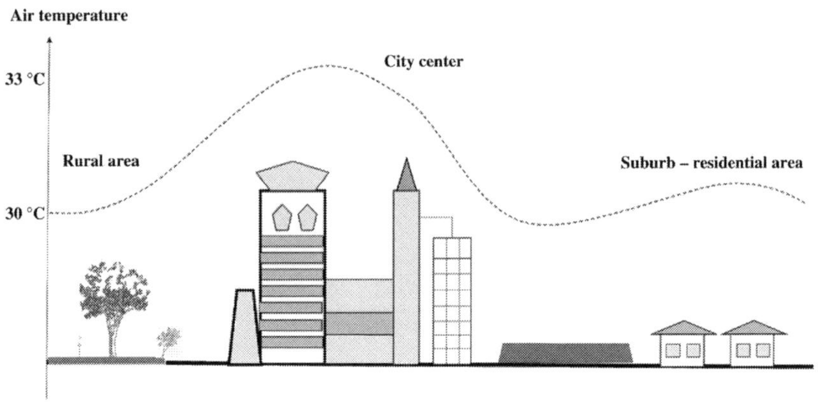

FIGURE 8.21 **Air temperature differences between city centers and neighboring regions.**

roofs and pavements), density of vegetation, thermal mass of buildings, and weather conditions (i.e., temperature, humidity, wind, solar radiation). Fig. 8.21 illustrates typical spatial variation of ambient air temperature between urban center and its neighboring regions. The magnitude of the UHI intensity has been measured in several cities throughout the world and has found to vary significantly depending on the city location and the measurement period. For instance, the UHI intensity of 12°C has been measured in Phoenix during summertime (Hawkins et al., 2004). The impacts of UHI include poor air quality and increase in building energy consumption. A wide range of mitigation techniques has been proposed to reduce the magnitude and consequently the impacts UHI in cities (Aleksandrowicz et al., 2017; Giguere, 2012). These mitigation techniques can be classified into four main categories:

- Building surface treatments such as cool roofs, green roofs, and green facades to reduce heat absorbed by the building envelope components.
- Road infrastructure treatments such as cool pavements to reduce the heat absorbed and then reflected by urban streets and pavements.
- Vegetation and tree enhancements by increasing the density of green spaces and parks within cities to create microclimates with lower temperatures.
- Bioclimatic urban planning and design to consider better shading strategies and street orientation to reduce solar heat gains and enhance the breeze cooling effects.

In this section, the benefits of the applications of static and dynamic reflective surfaces on the energy consumption of building are discussed in detail. By increasing the solar reflectance of surfaces (including roofs and pavements) in urban centers, the outflow of short-wave solar radiation increases, less solar heat energy is then absorbed leading to lower surface temperatures and reduced out-flow of thermal radiation into the atmosphere. This process of "negative radiative forcing" effectively counters global warming. Indeed, Akbari and

Matthews (2012) estimated that increasing the albedo of urban roofs and pavement surfaces worldwide would induce a negative radiative forcing equivalent to offsetting at least 40–160 Gt of emitted CO_2.

Reflective coatings are typically applied to roofs of residential and commercial buildings to constitute what commonly are referred to as cool roofs. Indeed, heat gain through a roof can be a dominant component of a building's total cooling load since it is highly susceptible to solar radiation (Konopacki et al., 1997). One common solution to reducing the cooling thermal loads due to heat gains through roofs is to increase the solar reflectance of the rooftop surfaces by installing highly reflective coatings or commonly referred to as cool roofs. Due to their low solar absorptivity and high thermal emissivity, cool roofs maintain lower surface temperatures and reduce the heat flow into buildings (Levinson and Akbari, 2010). Indeed, the low solar absorption increases the solar reflectance and the high thermal emittance enhances the ability of the roof surface to radiate any absorbed solar energy (Berdahl and Bretz, 1997; Levinson et al., 1996). Based on the existing literature, an average standard roof reflectance ranges from 0.20 to 0.30, an average initial low-sloped cool roof white material ranges from 0.70 to 0.80, and an average aged low-sloped cool roof white material ranges from 0.55 to 0.70 (Akbari, 1998; Gartland et al., 1996; Konopacki et al., 1997). The typical thermal emittance of most roofing materials, regardless of color or reflectance is typically between 0.80 and 0.90 (Gartland et al., 1996; Levinson and Akbari, 2010).

In addition to white cool roofing materials, there has been recent developments in colored cool roofing materials specifically for steep-sloped roofs. A cool colored surface absorbs the visible part of the spectrum to appear a specific color, but is highly reflecting in the near-infrared (NIR) part of the spectrum, which accounts for nearly 50% of solar radiation (Akbari and Levinson, 2008). Synnefa and Santamouris (2007) have developed prototypes of cool colored coatings using NIR reflective pigments added to acryl-based coatings, which can reduce solar reflectance by 0.06 (light blue) and 0.22 (black). Levinson et al. (2007) have developed coating systems for various types of substrates and found that coated steel and glazed clay-tile roofing achieved NIR reflectance as high as 0.50 and 0.75, respectively, using a one-coat system. Pisello et al. (2014) have developed a new coating system for colored clay tiles, which includes a white englobe above the substrate and a sodium silicate–based binder with pigment added for color. The study found that the new coatings are able to reflect NIR by 75%, which is 10% more than traditional tiles with an equivalent visible appearance.

8.8.1 Cool Roof Properties

Cool roofs have exterior surfaces or coatings that minimize solar absorption and maximize thermal emittance. They maintain lower surface temperatures and reduce heat flows into buildings (Levinson and Akbari, 2010; Levinson et al., 1996). In particular, solar absorption is minimized by increasing roof solar reflectance, defined as the fraction of the solar radiation that is diffusely reflected away by the surface. Lighter surfaces have higher reflectance values than dark surfaces. Specifically, the solar reflectance for black surfaces is close to 0 and for white surfaces are close to 1. In addition to color, surface roughness and presence of impurities can also affect roof solar reflectance (Berdahl and Bretz, 1997). Roof thermal emittance measures the surface's ability to radiate any absorbed solar energy. The higher their thermal emittance is, the more rapidly roofs can cool (ASHRAE, 2011). Cool roofs can also increase their thermal resistance since the thermal conductivity values of insulation layers typically increase with temperature (Gartland et al., 1996).

The majority of buildings in the US have relatively dark roofs. For instance, aerial photos of Chicago, Philadelphia, and Washington, DC have indicated that building roofs in these cities have an average solar reflectance of 0.25 (Konopacki et al., 1997). According to a survey conducted for ASHRAE in 1998, it is found that standard shingles have a solar reflectance ranging from 0.03 to 0.26, with most values between 0.10 and 0.15. Moreover, nonwhite roofing membranes, such as single-ply roofing materials, smooth bitumen, and granule-surface bitumen have solar reflectance values that range from 0.06 to 0.26. Similarly, gravel roofs have low solar reflectance values varying from 0.12 to 0.34.

Commercial roof products that qualify as cool roofs fall into three categories: single ply, liquid applied coating, and white finished metal panels (ASHRAE, 2011). Freshly applied white elastomeric coatings have solar reflectance values that range from 0.60 to 0.85. Solar reflectance of new white single-ply roofing membranes typically exceeds 0.70 (Akbari, 1998). Thermal emittance values of nonmetallic surfaces are generally between 0.80 and 0.90. Shiny metal roofs can have a relatively high solar reflectance, upwards of 0.60, but they tend to have low thermal emittances, which can make them as hot as dark roofs. Therefore, they would not be good materials for cool roofs. One method to quantify the properties of cool roofs is the estimation of their Solar Reflectance Index (SRI). SRI is calculated by using solar reflectance and thermal emittance values adjusted with wind coefficients. The higher the SRI value of a roof is, the lower is its surface temperature and thus the lower is the heat gains into the building. SRI is defined to be zero for a clean black roof (with a solar reflectance of 0.05 and a thermal emittance of 0.90) and 100 for a clean white roof (with a solar reflectance of 0.80 and a thermal emittance of 0.90) (Akbari and Levinson, 2008). The SRI of a roof surface can be determined by the following equations:

$$SRI = 123.97 - 141.35(x) + 9.655(x^2) \tag{8.12}$$

with

$$x = \frac{20.979 * \alpha - 0.603 * \varepsilon}{9.5205 * \varepsilon + 12.0} \tag{8.13}$$

TABLE 8.8 Typical Roofing Material Properties

Product	Solar reflectance	Infrared emittance	Temperature rise	SRI
Smooth bitumen	0.06	0.86	83	−1
Generic black shingle	0.05	0.91	82	1
Grey EPDM	0.23	0.87	68	21
Shasta white shingle	0.26	0.91	64	27
Light gravel	0.34	0.90	57	37
Aluminum	0.61	0.25	48	56
White EPDM	0.69	0.87	25	84
White coating on shingle	0.71	0.91	23	87
White PVC (Sarnafil)	0.83	0.92	11	104

SRI, Solar reflectance index.

where α is the absorption and ε is the emissivity of the roof surface.

To be considered cool roofs, surfaces should have their initial SRI exceed 78 (ASHRAE, 2011). Table 8.8 shows typical roofing materials with their solar reflectance, infrared emittance, temperature rise, and SRI values.

The properties provided in Table 8.8 are specific for newly installed roofs. Over time, solar reflectance values of cool roofs are likely to decrease because of surface dirt accumulations and material degradations. The emittance, however, would not decrease significantly (Bretz and Akbari, 1997). Indeed, soiling and accumulation of soot particles can reduce roof surface solar reflectance by about 0.15 with most of the degradation occurring during the first year (Berdahl et al., 2001). Standard maintenance and washing of roof surfaces can restore some of their original solar reflectance properties depending on the washing method. Akbari et al. found in a study on aging that solar reflectance values of cool roofs aged 5–8 years dropped from 0.80 to 0.50. However, they found that simply washing the roof surfaces with water could restore their solar reflectance to 70%–100% of the original values (Akbari et al., 2005).

8.8.2 Energy Standards for Cool Roofs

Over the last two decades, several national and state energy standards have added cool roof requirements or credits for specific climate zones. Some standards require the use of cool roofs by prescribing a specific roof reflectance and emittance to save energy consumption, reduce peak power demand, and improve air quality. Cool roof credits in standards do not require the use of cool roofs, but promote their use by allowing more flexibility in the design approach and permitting the use of less energy-efficient systems for buildings that integrate cool roofs. Such credits for cool roofs can be energy neutral, but may still reduce peak power demand and improve air quality. They may also reduce the initial cost of buildings (Akbari and Levinson, 2008).

Current energy standards that incorporate cool roofs within the US include ASHRAE 90.1, ASHRAE 90.2, and California Title 24. In particular, ASHRAE 90.1-2013 prescribes the use of cool materials for low-sloped roofs on nonresidential buildings in climate zones 1–3. The cool roofs must have a minimum 3-year aged reflectance of 0.55 and a thermal emittance of at least 0.75, or a minimum aged SRI of 64 (ASHRAE, 2013). If cool roofs are not specified, then roof insulation levels must be increased to a value set by ASHRAE. ASHRAE 90.2 2007 uses a credit method that allows the reduction of the roof insulation level for residential buildings if cool roofs are prescribed in climate zones 1, 2, or 3. The standard 90.2 requires cool roofs to have a minimum initial solar reflectance of 0.65 and a minimum thermal emittance of 0.75, or have a minimum SRI of 75 (ASHRAE, 2007).

California Title 24-2013 also prescribes the use of cool roofs within the state of California for low and steep-sloped roofs for nonresidential buildings in all California climate zones. A minimum aged reflectance of 0.63 and thermal emittance of 0.75 or an aged SRI of 75 is required for low-sloped roofs, and a minimum aged reflectance of 0.20 and thermal emittance of 0.75 of an aged SRI of 16 is required for steep-sloped roofs. In addition, Title 24 prescribes cool roofs for high-rise residential buildings with low-sloped roofs in climate zones 9–15, and for high-rise residential buildings with steep-sloped roofs in California climate zones 2–15. A minimum aged reflectance of 0.55 and thermal emittance of 0.75 or a minimum aged SRI of 64 is required for low-sloped roofs, and a minimum aged reflectance of 0.20 and thermal emittance of 0.75 of an aged SRI of 16 is required for steep-sloped roofs (CEC, 2013). The roof U-factor is then prescribed according to aged solar reflectance.

In addition to energy standards, there are also voluntary energy credit programs that have adopted cool roof credits. The US EPA Energy Star rewards points for the use of cool roofs. It requires that low-sloped roofs have an initial and aged reflectance of at least 0.65 and 0.50, respectively, and that steep-sloped roofs have an initial and aged reflectance of at least 0.25 and 0.15, respectively (Akbari and Levinson, 2008). LEED also rewards 2 points for the use of cool roofs that has an initial and aged SRI of 82 and 64, respectively, for low-sloped roofs, and an initial and aged SRI of 39 and 32, respectively, for steep-sloped roofs (USGBC, 2016).

8.8.3 Performance of Static Cool Roofs

In this section, static cool roofs are those roofs that have coatings with optical properties including reflectivity that remain constant throughout the year. Several studies have tested, simulated, and documented cooling energy savings and cooling peak demand reductions from static cool roofs (SCRs) in a wide range of climates. Testa and Krarti (2017a) conducted a detailed literature review of the reported performance of static cool roofs. One of the first study of SCRs was conducted by Konopacki et al. (1997) and Akbari et al. (1999). This study has documented the potential energy use savings from increased roof solar reflectance for residential and commercial buildings in 11 cities throughout the US including Atlanta, Chicago, Los Angeles, Dallas/Fort Worth, Houston, Miami/Fort Lauderdale, New Orleans, New York City, Philadelphia, Phoenix, and DC/Baltimore. The building prototypes modeled in each city consisted of a new/old residential building, new/old office building, new/old retail building, primary and secondary school, hospital, nursing home, and a grocery store. All building prototypes were single-story, since the energy use savings associated with SCRs are assumed to only affect the top floors. For all old vintage building prototypes, schools, health care buildings, and grocery stores, R-11 roofs and R-7 walls were considered. For all new vintage building prototypes, R-19 roofs and R-11 walls were used. Whole-building energy simulations were completed for each building type and location for two cases: (1) standard roof reflectance and (2) high roof reflectance. A standard roof reflectance of 0.25 was used for residential and commercial prototypes, and a high roof reflectance of 0.55 and 0.70 were used for residential and commercial prototypes, respectively. The roofs were modeled using a single-layered construction with a thermal emittance of 0.91. The study results indicated that SCRs achieved annual cooling electricity savings per square meter of flat roof area ranging from 1.1 to 5.2 kWh (7%–17%) for old residential buildings, 0.4 to 2.8 kWh (6%–15%) for new residences, 1.8 to 6.0 kWh (5%–9%) for old offices, 1.1 to 3.3 kWh (3%–7%) for new offices, 2.6 to 7.8 kWh (5%–10%) for old retail stores, and 1.4 to 4.5 kWh (3%–7%) for new retail stores. However, SCRs also resulted in annual heating energy use penalties per square meter of flat roof area ranging from 0.002 to 0.17 therm (25%–5%) for old residences, 0 to 0.13 therm (0%–5%) for new residences, 0 to 0.19 therm (05%–14%) for old offices, 0 to 0.11 therm (05%–12%) for new offices, 0 to 0.10 therm (0%–12%) for old retail stores, and 0 to 0.04 therm (05%–7%) for new retail stores. The relative energy use savings were found to be most significant for shell-dominated residential building prototypes. Moreover, buildings with longer hours of occupancy or operation (such as hospitals and nursing homes) showed larger savings per roof area than those with shortened occupancy schedules. In addition, old residential, office, and retail store prototypes provided larger electricity savings and gas penalties than those of the new vintage. This result is associated with lower roof *R*-values specific to the old vintage buildings. Additionally, prototypes with gas heating systems provided larger net cost savings than prototypes with electric heat pumps, because of the higher cost of electric heat compared to gas (Akbari et al., 1999). Building location was found to have a significant effect on energy use savings and penalties associated with SCRs. For example, new offices with electric heat pumps in Chicago and Philadelphia, new residences with electric heat pumps in New York City, and new residences with gas furnaces in Philadelphia all displayed limited impact of SCRs with net energy deficits or savings very close to zero. The total impact of using SCRs for all 11 metropolitan areas were estimated to include electricity savings of 2.6 TWh (average of 2.15 kWh/m^2), natural gas deficit of 6.9 TBtu (average of 0.05 therms/m^2), average energy bill net savings of \$194 M (average \$0.16/m^2), and a peak demand savings of 1.7 GW (average of 1.45 W/m^2).

As noted by Testa and Krarti (2017a), the reported literature demonstrates that several factors including building location, building construction, building type, and roof solar reflectance impact the potential energy use savings and peak demand reductions that SCRs can achieve. In warmer climates with abundant sunshine, cooling energy use and cost savings greatly outweigh heating energy use and cost penalties. In colder climates, or climates with equal heating and cooling seasons, SCRs may cause significant heating need increases. Thus, the best climates for cool roofs would have a long cooling season and a short heating season. In addition to outdoor temperature variations, global horizontal solar irradiance that the building can experience also affect cooling energy savings and heating thermal penalties. Building construction also can impact significantly the benefits and the penalties of SCRs. Older vintage buildings with low insulation levels would have higher cooling energy use savings and higher heating thermal penalties. Solar roof reflectance has a direct effect on the performance of SCRs. When roof solar reflectance is doubled, cooling energy use savings and heating thermal penalty would roughly double. Finally, building type and occupant levels can impact the potential savings from SCRs. Residential buildings tend to exhibit more relative cooling energy use savings and heating thermal penalty than commercial buildings. Commercial buildings, however, provide more relative cooling energy use savings when compared to heating thermal penalties. Moreover, Additionally, buildings with longer operation schedules provide larger savings per roof area than those with shorter schedules.

8.8.4 Performance of Dynamic Cool Roofs

It is clear from the reported literature that while SCRs have several benefits, they have some limitations and disadvantages. The largest disadvantage being higher potential heating energy use, which may reduce the cost-effectiveness of SCRs to building owners and to utilities. Indeed, buildings with SCRs experience energy penalties as they transition from cooling to heating modes due to the increase in reflected solar energy that is no longer transferred into heat gains at the roof surfaces (Gray, 2015). An ideal alternative of SCRs would be roofing systems that can be highly reflective when buildings are in a cooling mode and highly absorptive when buildings transition to a heating mode; in other terms, switchable cool roofs with coatings that have the ability to change reflectance throughout the year. These types of cool roofs are referred to as dynamic cool roofs or DCRs.

For building applications, two types of switchable materials have been used for windows: thermochromic (TC) coatings and electrochromic (EC) coatings. Both materials contain a specific category of polymers known as chromogenic polymers, which react to external stimuli by changing their visible optical properties (Erdem and Saffa, 2015; Kamalisarvestani et al., 2013). TC coatings contain a chromogenic polymer that reacts to temperature while EC glass contains a chromogenic polymer that reacts to an electric field (Gray, 2015). Both switchable coating types have advantages and disadvantages. The major difference is that EC coatings can be manually controlled or potentially linked to a thermostat, while TC coatings are only controlled based on the temperature of the material itself from a low reflectance value at low temperatures to a high reflectance value at high temperatures (Gao, 2012). TC coatings are currently less expensive and require less maintenance than EC coatings (Lee et al., 2013). Currently, very limited literature exists on applying switchable coatings to building roofs. There has been no reported research to date on the application of EC coatings or any other manually controlled switchable roofing systems, but Azari and Bierman (2008) discussed the potential for applying these coatings to be used for roofs and be linked to a thermostat without any specific analysis and quantification of the benefits of such controls. Recently, Gray (2015) has applied TC coatings to cool roofs so that their color changes from black to clear to reveal the actual light color of the roofs. It was found that the low temperature reflectance potential by applying a TC coating to a cool roof is 0.3 and the high temperature reflectance potential is about 0.55. The difference between the low and high temperatures when the coating changes its optical properties was found to be 11°C. The emissivity of the switchable system was found to be about 0.95. However, Gray (2015) found that there were several challenges encountered with the application of the TC coatings as switchable materials including adhesion to nonmetallic surfaces, nonuniformity, and the need for higher and lower transparency temperatures to further increase cool roof potential of heating and cooling energy savings. Furthermore, the applied TC coating needed a long and unpractical period to completely shift from the low reflectance to the high reflectance values requiring an 11°C temperature difference to occur. Ideally, a more instantaneous shift in reflectance would be needed for energy use savings of DCRs to materialize (Gray, 2015). A recent simulation-based analysis evaluated the impacts of ideal switchable reflectance coatings on the potential energy use savings associated with DCRs for office buildings located in four US climates (Park and Krarti, 2016). The benefits of DCRs compared to conventional SCRs with constant optical properties have been quantified using ideal coatings that can instantaneously switch reflectance between a low value of 0.30 to a high value of 0.55 (Park and Krarti, 2016; Testa and Krarti, 2017a).

As an alternative to switchable reflectance roofs, Hooshangi (2015) explored the performance of heterogenous directional reflective materials (DRMs) with corrugated tilted surfaces with reflective sides facing toward the sky and absorptive sides facing toward the ground. The tilt angle of the corrugation surfaces can be set depending on sun location in the summer and winter. The reflective sides reflect sunlight during the summer season when the sun is high in the sky and the absorptive sides absorb sunlight during the winter season when the sun is low in the sky. Akbari and Touchaei (2014) have developed a model to calculate the hourly reflectance of DRMs as a function of zenith and azimuth angles of the suns position. No analysis has been reported to compare potential energy use savings from DRM roofs compared to static cool roofs.

The main challenge for the application of DCRs is identifying the best control strategies to optimize their reflectance settings. Indeed, heat transfer through building roofs is quite complex with several different mechanisms contributing to the transient heat transfer including solar radiation at the interior and exterior, net longwave radiation at the exterior and interior, surface convection at the interior and exterior, and heat conduction within the roof assembly (Gray, 2015). Therefore, it is difficult to draw a direct relationship between roof reflectance and a control setting. Any change in roof reflectance can, of course, affect solar heat gains from the roof. However, building thermal loads are not only dependent on solar gains. In a very cold but sunny climate, a building may still require heating even if solar heat gains are high; thus, a lower reflectance roof would be desirable to increase heat gains. The opposite may be true in a warm climate that temporarily experiences low solar heat gains; a high reflectance roof would still be desirable to reduce further any solar heat gains a building may experience from the roof. The building insulation level and thermal mass can also have important effects on the magnitude of the heat gains transferred from the roof to the conditioned building. There is currently very little research on optimizing control strategies for DCRs. An ideal control system would lead to a DCR that consumes the same amount of cooling energy as an SCR and the same amount of heating energy as a standard roof without a highly reflective coating. Some evaluation of potential control strategies for switchable roofing systems have been reported using simulation analyses (Gao, 2012; Gray, 2015). It was found that several parameters should be accounted to determine the best strategies to change the reflectance levels of roofs including climate, building type, occupancy schedule, insulation level, and thermal mass, with little correlation between location (Gray, 2015).

8.8.5 Energy Savings Potential of Cool Roofs

As noted earlier, limited reported studies have been carried out to determine the benefits of DCRs compared to SCRs (Park and Krarti, 2016; Testa and Krarti, 2017a). Selected results from these studies are briefly outlined in the following sections.

Impact of Building Insulation Level

The performance of DCRs was found to be highly dependent on the overall building insulation level. The ranges of source energy savings reported for

TABLE 8.9 Ranges of Source Annual Energy Savings From DCRs Compared to SCRs for Three Roof Insulation Levels

Insulation level	Building prototype			
(*R*-wall/*R*-roof)	Residential building		Office building	
m²·K/W	MJ/m²	%	MJ/m²	%
Low (0.9/1.85)	4.3–19.4	1.6–4.9	1.2–18.0	0.3–3.9
Medium (2.12/3.46)	2.3–10.2	1.3–4.1	0.4–9.4	0.1–2.6
High (3.42/4.85)	1.5–7.3	0.9–3.5	0.2–6.5	0.1–1.9

DCRs, Dynamic cool roofs; *SCRs*, static cool roofs.

prototypical residential and commercial buildings for three insulation levels are summarized in Table 8.9.

The results of Table 8.9 show that the amount of insulation of the building is an important factor that affects the energy savings resulting from DCRs when compared to SCRs. In particular, that annual energy use is lower for DCRs, regardless of the insulation level. However, when the *R*-value is large (the building is well insulated), heat transfer between the surface of the roof and the interior of the building is small and the impact on the energy use of using DCRs is not as significant compared to the case for buildings with low insulation levels.

Impact of Building Mass

The impact of building mass level on the performance of DCRs was also evaluated with switchable roof reflectance of 0.30/0.55 and SCRs with a set reflectance value of 0.55. The ranges of source energy savings found for various building types and mass levels are summarized in Table 8.10.

TABLE 8.10 Ranges of Source Energy Savings From DCRs Compared to SCRS for of Three Roof Mass Levels

Mass level	Building prototype			
(Wall/roof mass)	Residential building		Office building	
kg/m²	MJ/m²	%	MJ/m²	%
Light	1.5–16.9	0.9–3.9	0.2–15.6	0.1–3.2
Medium	1.7–19.4	1.1–4.9	0.3–18.0	0.1–3.9
Heavy	1.7–18.3	1.1–4.8	0.3–17.4	0.1–3.8

DCRs, Dynamic cool roofs; *SCRs*, static cool roofs.

Table 8.10 indicates that energy savings associated with DCRs relative to SCRs do not steadily increase with the thermal mass level. Rather the energy savings increase to a maximum level as mass increases, then start to decrease after a threshold roof mass level.

Impact of Roof Reflectance

The impact of roof reflectance settings has been evaluated on the performance of DCRs. Indeed, this parameter is particularly of interest should the technology improve and allows large reflectance differential with lower low reflectance or higher high reflectance. The analysis carried out by Testa and Krarti (2017b) has indicated that there is a linear relationship between savings and the difference in roof high–low reflectance values. Moreover, the analysis has found that as the high reflectance increases, the annual energy use savings from DCRs increases at a faster rate than those obtained from SCRs. This result is due to the fact that as the reflectance value for SCRs increases, both the cooling savings and the heating penalties increase. Whereas for DRCs with switchable reflectance values, only the cooling savings increase while the heating penalties remain minimal. This behavior is more pronounced for warmer climates than for colder climates. Indeed, the source energy savings associated with cool roofs applied in warm climates to include space heating and cooling using SRC and DRC systems can be expressed as a function of the roof reflectivity, ρ_{roof}, by Eqs. (8.14) and (8.15), respectively:

$$\Delta E_{source,SRC}\left(\frac{MJ}{m^2}\right) = 40.0\rho_{roof,SRC} - 13.0 \tag{8.14}$$

$$\Delta E_{source,DRC}\left(\frac{MJ}{m^2}\right) = 66.0\rho_{roof,DRC} - 22.2 \tag{8.15}$$

For cold climates, the heating and cooling source energy savings associated with cool roofs using SRC and DRC systems can be expressed as a function of the roof reflectivity, ρ_{roof}, by Eqs. (8.16) and (8.17), respectively:

$$\Delta E_{source,SRC}\left(\frac{MJ}{m^2}\right) = -5.0\rho_{roof,SRC} + 2.8 \tag{8.16}$$

$$\Delta E_{source,DRC}\left(\frac{MJ}{m^2}\right) = 37.0\rho_{roof,DRC} - 11.9 \tag{8.17}$$

Problem 8.9

An office building has a roof area of 1000 m^2 and is located in Atlanta. Estimate the potential savings of using (i) SRC and (ii) DRC when the reflectivity of the SCR and DCR coatings is 0.80.

Solution

Atlanta has a warm climate. Thus, the energy use savings from SRC and DRC can be estimated using Eqs. (8.14) and (8.15), respectively, multiplied by the roof area (1000 m^2):

$$\Delta E_{source,SRC} = 1000 * \left[40.0 * (0.80) - 13.0\right] = 19{,}000 \text{ MJ / year}$$

$$\Delta E_{source,DRC} = 1000 * \left[66.0 * (0.80) - 22.2\right] = 30{,}600 \text{ Mj / year}$$

Thus, DRC provides more than 80% of the source energy saved by the SRC system for this building.

Impact of Climate

As part of a comprehensive analysis, the DRC performance has been evaluated for residential and commercial buildings located in various US climates. Table 8.11 summarizes the results of the analysis using the heating degree-days (HDDs) estimated using a base temperature of 18°C. Specifically, Table 8.11 lists the ranges of annual source energy savings obtained for DCRs with a switchable low/high reflectance of 0.30/0.55 compared to SCRs with a set reflectance of 0.55 for various US locations considered in the analysis. Since the additional energy savings achieved by a switchable roof come from a decrease in heating energy use, the HDDs were used to represent the climatic characteristics for all US locations.

TABLE 8.11 Ranges of Annual Source Energy Savings From DCRs Relative to SCRs for Residential and Commercial Buildings in Various US Locations

Location	HDD	Building prototype			
		Residential building		Office building	
		MJ/m^2	%	MJ/m^2	%
New Orleans	787	1.5–5.5	0.9–2.2	0.2–2.0	0.1–0.6
Las Vegas	1264	3.5–12.6	1.6–3.4	1.2–6.8	0.4–2.3
Atlanta	1571	3.5–11.4	2.0–3.8	1.9–7.1	0.8–2.3
Portland	2426	4.5–13.8	3.1–4.9	2.7–9.4	1.7–3.9
Baltimore	2574	5.2–16.4	2.4–4.1	3.9–12.8	1.5–3.4
Chicago	3607	6.2–17.6	2.1–3.3	4.7–14.9	1.4–3.0
St. Paul	4379	6.9–19.4	1.2–3.0	6.1–18.0	15–2.9

DCRs, Dynamic cool roofs; *HDD*, heating degree-day; *SCRs*, static cool roofs.

It is clear from Table 8.11 that there is a strong relationship between HDDs and source energy savings incurred from DCRs relative to SCRs. Figs. 8.22 and 8.23 illustrate the variation of source energy savings as a function of HDD for, respectively, a light weight residential building and an office building with three insulation levels. A similar trend was found for all building configurations. In particular, the results of Table 8.11 and Figs. 8.22 and 8.23 show that the annual energy use savings associated with the application of DCRs instead of SCRs increase as HDD increases. Therefore, implementing DCRs would be more beneficial in colder climates.

8.8.6 Economic Analysis of Cool Roofs

The cost-effectiveness of DCRs using break-even costs has been investigated for residential and commercial buildings in various US sites (Testa and Krarti, 2017b). While the life span of variable reflective roofing systems is uncertain from the literature, the average traditional commercial roof life span is approximately 22 years according to Coffelt and Hendrickson (2010). The economic analysis is based on the LCC method using estimated DRC annual energy savings combined with electricity and natural gas rates reported by EIA for all US locations (EIA, 2016a; EIA, 2016b). Table 8.12

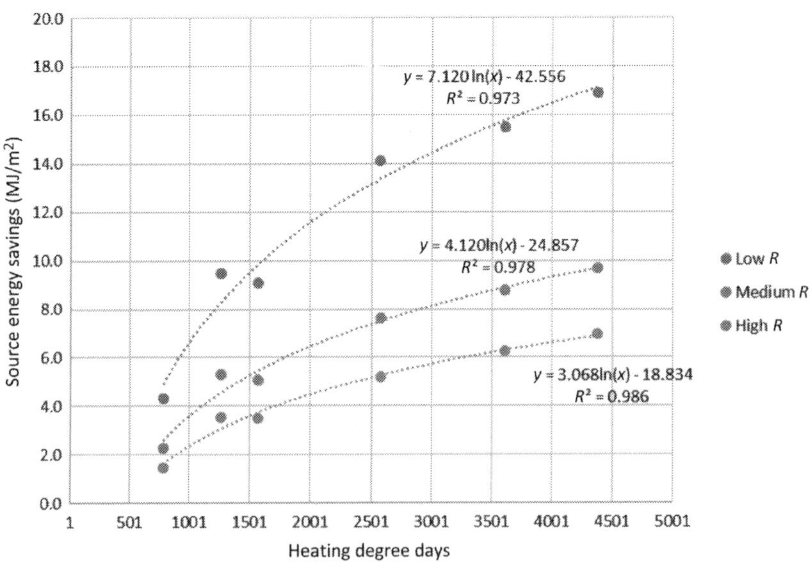

FIGURE 8.22 Effect of heating degree-days (HDDs) on annual source energy savings obtained from dynamic cool roofs (DCRs) relative to static cool roofs (SCRs) for light mass residential buildings.

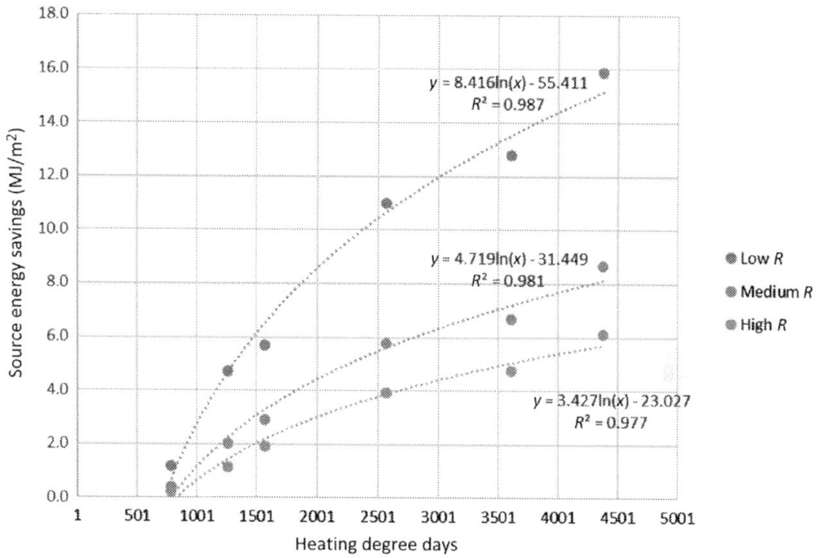

FIGURE 8.23 Effect of HDDs on annual source energy savings obtained from DCRs relative to SCRs for light mass office buildings.

shows the break-even costs for DCRs relative to standard roofs (i.e., with a low reflectance value of 0.30) and Table 8.13 shows the break-even costs required when switching from SCRs to DCRs for residential and commercial buildings.

It is clear from Tables 2.7 and 2.8 that when compared to standard roofs, office buildings with DCRs save more energy cost per square meter than residential buildings. The opposite, however, is true when comparing the cost savings from SCRs to those obtained from DCRs. This result is most likely associated to the fact that an office building has higher cooling loads due to high internal loads. Therefore, when compared to a residential building, more savings are seen when the cooling load decreases and less savings are seen when the heating load decreases.

Another, more obvious general trend is that the highest break-even costs and cost savings are achieved for low insulated building configurations. In particular, the results show that the break-even cost for DCRs for low insulated buildings is more than 1.5 times greater than for high insulated buildings. Therefore, older buildings would be a better target for the application of DCRs. As noted in the literature review, while some TC coatings have been developed and tested, their cost is still high. However, the use of DRM roofing system or a mechanically controlled system, which only switches reflectance seasonally, could be economically feasible based on the benchmark costs listed in Tables 8.12 and 8.13.

TABLE 8.12 Initial Cost ($\$/m^2$) Required to Break Even if Switching From a Standard Roof With a Reflectance of 0.30 to a Switchable Cool Roof With a Reflectance of 0.3/0.55

Location	Residential			Office		
	Low *R*	Medium *R*	High *R*	Low *R*	Medium *R*	High *R*
New Orleans, 2A						
Light mass	2.96	1.76	1.28	2.83	1.65	1.19
Medium mass	3.27	1.82	1.33	3.00	1.66	1.23
Heavy mass	3.08	1.77	1.30	2.78	1.60	1.21
Las Vegas, 2B						
Light mass	4.46	2.60	1.88	4.67	2.76	1.99
Medium mass	4.84	2.67	1.94	4.92	2.80	2.04
Heavy mass	4.59	2.60	1.90	4.57	2.69	1.99
Atlanta, 3A						
Light mass	2.94	1.76	1.27	2.89	1.78	1.30
Medium mass	3.28	1.83	1.33	3.11	1.80	1.34
Heavy mass	3.10	1.77	1.30	2.91	1.73	1.32
Portland, 3C						
Light mass	1.47	0.94	0.72	1.82	1.19	0.89
Medium mass	1.64	0.99	0.75	1.84	1.19	0.91
Heavy mass	1.53	0.96	0.73	1.71	1.14	0.89
Baltimore, 4A						
Light mass	3.06	1.78	1.28	3.17	1.92	1.40
Medium mass	3.33	1.81	1.31	3.31	1.95	1.45
Heavy mass	3.13	1.76	1.28	3.07	1.88	1.43
Chicago, 5A						
Light mass	2.02	1.20	0.88	2.08	1.25	0.91
Medium mass	2.16	1.24	0.92	2.16	1.25	0.93
Heavy mass	2.04	1.20	0.90	1.99	1.20	0.91
St. Paul, 6A						
Light mass	1.81	1.12	0.84	1.98	1.19	0.86
Medium mass	1.98	1.11	0.82	2.07	1.21	0.89
Heavy mass	1.87	1.08	0.80	1.93	1.16	0.87

TABLE 8.13 Initial Cost ($\$/m^2$) Required to Break Even if Switching From SCRs with a Reflectance of 0.55 to DCRs With a Low/High Reflectance of 0.3/0.55

Location	Residential			Office		
	Low *R*	Medium *R*	High *R*	Low *R*	Medium *R*	High *R*
New Orleans, 2A						
Light mass	0.54	0.29	0.18	0.17	0.06	0.03
Medium mass	0.68	0.31	0.21	0.28	0.08	0.04
Heavy mass	0.67	0.32	0.21	0.28	0.07	0.04
Las Vegas, 2B						
Light mass	1.36	0.76	0.51	0.68	0.30	0.17
Medium mass	1.79	0.87	0.57	0.97	0.38	0.22
Heavy mass	1.79	0.88	0.58	0.96	0.39	0.22
Atlanta, 3A						
Light mass	1.74	0.97	0.67	1.08	0.56	0.37
Medium mass	2.13	1.07	0.73	1.37	0.61	0.41
Heavy mass	2.08	1.06	0.73	1.33	0.59	0.40
Portland, 3C						
Light mass	1.68	0.92	0.64	1.15	0.58	0.39
Medium mass	1.96	1.00	0.69	1.35	0.64	0.43
Heavy mass	1.91	1.00	0.69	1.29	0.63	0.43
Baltimore, 4A						
Light mass	2.00	1.08	0.74	1.58	0.83	0.56
Medium mass	2.35	1.16	0.80	1.84	0.92	0.63
Heavy mass	2.28	1.15	0.80	1.76	0.90	0.63
Chicago, 5A						
Light mass	1.44	0.82	0.58	1.14	0.60	0.43
Medium mass	1.64	0.87	0.61	1.35	0.65	0.46
Heavy mass	1.59	0.86	0.61	1.31	0.64	0.45
St. Paul, 6A						
Light mass	1.73	1.02	0.73	1.66	0.90	0.64
Medium mass	2.02	1.06	0.75	1.89	0.98	0.68
Heavy mass	1.97	1.06	0.75	1.84	0.96	0.67

DCRs, Dynamic cool roofs; *SCRs*, static cool roofs.

8.8.7 Selection Guidelines and Benefits of Cool Roofs

It is clear from the reported analyses that SCRs provide cooling energy use savings during hot seasons, but may also cause significant heating penalties during cold seasons. For this reason, SCRs are most beneficial in climates with a long cooling season and a short heating season. Due to higher prices of electricity compared to those of gas, SCRs in several US climates can provide an overall cost savings. As outlined in this chapter, the specific levels of savings and penalties would be affected by several factors including the roof insulation level, building type, and cool roof reflectance value. Typical reflectance values for standard commercial roofing materials range from 0.05 to 0.36, while typical aged reflectance values of cool roofing materials range from 0.55 to 0.65. The thermal emittance for most reported cool roofing materials is typically 0.90. Typical electricity savings seen from static cool roofs with reflectance values of 0.3–0.4 for US climates range from 0.1 to 8.6 kWh/m^2 for residential buildings, 1.1 to 8.2 kWh/m^2 for office buildings, and 1.4 to 10.9 kWh/m^2 for retail buildings. Typical reported fuel use increases due to application of SCRs range from 0 to 0.22 therm/m^2 for residential buildings, 0 to 0.19 therm/m^2 for commercial buildings, and 0 to 10 therm/m^2 for retail buildings. Additionally, overall annual energy cost savings from SCRs range from \$0.16/m^2 to \$0.36/m^2.

Though SCRs in several climates have been shown to provide overall source energy use savings, there is potential to provide even more cost and energy use savings by applying switchable reflectance coatings to roof surfaces to achieve what is referred to as DCRs. A switchable reflectance coating would especially be useful in climates with heating and cooling seasons with a similar length, or in heating-dominated climates with a large swing in temperatures between the heating and cooling seasons. Though there are currently switchable technologies used for windows (EC and TC coatings), there has been limited reported research on switchable roof coatings. The reviewed available literature indicates that a possible technology would be to apply switchable TC coatings over cool roofs so their overall reflectance changes from 0.3 to 0.55. The major challenge identified for switchable roofing systems is determining their best reflectance settings to optimize their performance. Currently, the only applicable switchable technologies have two different control types: manual controlled (EC coating) and temperature controlled (TC coating).

Based on the limited reported analysis results for the performance of DCRs, it is found that seasonal switchable reflectance roofs can avoid heating energy use penalties and achieve additional source energy savings and cost savings when compared to SCRs for residential and commercial buildings. In particular, the analysis has indicated that DCRs can provide the following benefits:

- *Residential buildings*: For low-insulation buildings, the source energy savings range from 4.33 to 19.44 MJ/m^2 (1.6%–4.9%) and the cost savings range from 0.04 to \$0.18/m^2 (1.9%–5.0%); for medium insulation buildings, the source energy savings range from 2.30 to 10.22 MJ/m^2 (1.3%–4.0%)

and the cost savings range from \$0.02/m^2 to \$0.09/m^2 (1.5%–4.2%); and for high insulation residential buildings, the source energy savings range from 1.46 to 7.25 MJ/m^2 (0.9%–3.5%) and the cost savings range from \$0.01/m^2 to \$0.06/m^2 (1.1%–3.6%).

- *Commercial buildings*: For low-insulation office buildings, the source energy savings range from 1.17 to 18.00 MJ/m^2 (0.3%–3.9%) and the cost savings range from \$0.01/m^2 to \$0.14/m^2 (0.5%–4.3%); for medium insulation office buildings, the source energy savings range from 0.42 to 9.35 MJ/m^2 (0.14%–2.6%) and the cost savings range from \$0.0/m^2 to \$0.07/m^2 (0.2%–2.8%); and for high insulation office buildings, the source energy savings range from 0.21 to 6.49 MJ/m^2 (0.1%–1.9%) and the cost savings range from \$0.0/m^2 to \$0.05/m^2 (0.1%–2.1%).

Moreover, the reported analysis results have indicated that energy savings of DCRs depend largely on the climate, insulation level, and reflectance of the roof. The results also show that energy savings generally increase as thermal insulation level decreases for all building types. Making older buildings with low insulation levels a good target to retrofit with DCRs rather than well-insulated new constructions. In addition, the results of an LCC analysis show that for residential buildings the break-even cost using a 22-year life span ranges depending on the climate from \$0.80/m^2 to \$4.84/m^2 for switching to DCRs from standard roofs, and from \$0.18/m^2 to \$2.35/m^2 for switching to DCRs from SCRs. For commercial buildings, the break-even cost for a 22-year life span ranges from \$0.86/m^2 to \$4.92/m^2 for switching to DCRs from standard roofs, and from \$0.03/m^2 to \$1.89 \$/m^2 for switching to DCRs from SCRs.

While more research is needed to improve the performance and lower the cost of switchable reflectance materials like TC coatings, the limited reported analyses indicates that seasonally controlled DCRs such as mechanically controlled roofs or DRM roofing systems have the potential to improve the energy efficiency of residential and commercial buildings and should be further explored and investigated. Moreover, additional savings could be obtained from variable thermal emittance coatings. Indeed, a lower thermal emittance provides additional heating savings since it slows the release of long-wave radiation. The lower the thermal emittance, the slower the roof would cool. This technology could be particularly beneficial in providing additional heating savings during the heating period or even at night during swing months.

8.9 INTELLIGENT CONTROLS AND MANAGEMENT

As illustrated in Fig. 8.4, optimal operation and management of various utilities and urban infrastructures is an important pillar of sustainable cities. In this section, an overview is outlined for sensing systems typically utilized to gather a wide range of data for operating smart urban centers. Moreover, the main components and operation strategies for smart microgrids are presented to introduce some integration mechanisms of renewable energy technologies to meet energy needs of cities.

8.9.1 Data Collection Systems for Smart Cities

Over 178 cities are currently trying to install a set of sensors, cameras, and devices to acquire flow of data to control several urban systems including traffic lights, streetlights, pedestrian displays, and power distribution networks. These data are then sent to the cloud where algorithms extract the required information, perform a series of analyses, and send signals to make the cities safer and more efficient. For instance, Singapore has built increasingly smart infrastructures equipped with high-speed internet and several Internet of Things technologies (Kosowatz, 2017). For instance, in a Singapore neighborhood, thousands of sensors have been installed in individual apartments to monitor in real time energy demand, water use, and waste production. San Diego officials have allocated $30 million to make their city smarter. In particular, the plan is to install over 14,000 LED streetlights and 3200 sensors to gather a wide range of information including vehicular traffic, air quality, and pedestrian movement. The data can be available to several stakeholders. In particular, entrepreneurs can use the pedestrian traffic data to determine the best location for new retail stores or the rental value for residential and commercial properties.

In most smart cities, light poles are typically used to install energy-efficient LED lights but also sensors, wireless transmitters, and microprocessors to develop a digital network within the urban neighborhoods. The network of sensors create a significant amount of data that is becoming increasingly difficult to analyze and use. Data analytics are often combined with knowledge from experts in urban traffic, services, and power distribution to make data-driven intelligent decisions. For instance, machine-learning algorithms have been used to improve the value add of LED streetlights and cut the payback period for lighting retrofit projects in cities. These algorithms develop models that can optimize the operation of individual streetlights based on historic data of pedestrian and vehicular traffic. The algorithms decide when to dim or turn-off streetlights. Similar machine-learning algorithms have been used in the city of Belford, France to optimize the public transportation flow. Data about the number of passengers, tickets sold, and bus speed have been collected on the 100 buses that the city operates along five routes. The algorithms are then used to determine the best bus schedules to avoid crowding and reduce operating costs (Kosowatz, 2017).

Security of the data flow remains one of the challenges of developing open platforms for smart cities. Indeed, experiences have indicated that hackers can tamper with urban infrastructures through intrusion into the digital networks. New security standards for smart networks are being developed and enforced including ISO/IEC 18598 specific to the security of the physical elements of automated infrastructure management (AIM) systems. The AIM elements include sensors, cables, computer rooms, and other hardware. The ISO/IEC 18598 standard mandates stricter security but higher interoperability for software interfaces from different vendors so new technologies and devices can be easily integrated into the existing digital infrastructure. Moreover, the standard requires that AIM systems be able to locate all AIM elements (i.e., sensors, devices, and cables) to identify and block intrusions and viruses at the component levels.

The significant capital needed to build a digital infrastructure remains an important hurdle for developing smart cities. Moreover, the life cycle of most elements of a digital infrastructure is rather short (i.e., 3–5 years) making the recurrent need for investing and upgrading smart cities even more significant. However, the benefits of greater connectivity and more intelligent infrastructure outweigh the costs of making cities smarter especially when advances in digital technologies are moving fast with autonomous cars, flying drones, and wearable communication devices becoming realities. Cities have to use real-time data to be able to optimally provide their citizens with optimal sets of ever changing services and utilities.

8.9.2 Overview of Smart Grids

One of the first steps to develop smart cities is to modernize and optimize the power grid operation. Indeed, smart grid allows the integration of sustainable technologies including clean energy sources, reduced air pollution, and accessibility to city services (Angelakis et al., 2017). In Chapter 4, microgrids were briefly described including their components and their benefits to serve the energy needs of communities and cities. In particular, it has been noted that microgrids can increase the reliability and the energy efficiency of generating and distributing power needs to communities and cities. Moreover, microgrids allow the integration of distributed generation technologies including renewable energy systems into the grid as showcased by case studies discussed in Chapter 7. Smart grids go a step further by integrating communications and computer-based technologies to monitor, control, and optimize the entire power systems.

Smart grids are generally used to refer to sets of advanced control systems and technologies to operate efficiently power systems to generate, transmit, and distribute electricity to various end-users including buildings. Typically, smart grids utilize digital data streams (DDS) combined with information technologies to enhance the integration of various components of power systems including matching and controlling electricity demand and supply. In particular, electrical loads of buildings can be controlled and varied through communications with smart grids to optimize the operation of the utility generating power systems. The interactions between smart grids and buildings can occur through smart meters to transmit data about the operation settings and energy use levels of building energy systems. With these data sets of DDS, smart grids and buildings when equipped with smart controls can learn the buildings' loads and their patterns to make better decisions in electricity demand and supply.

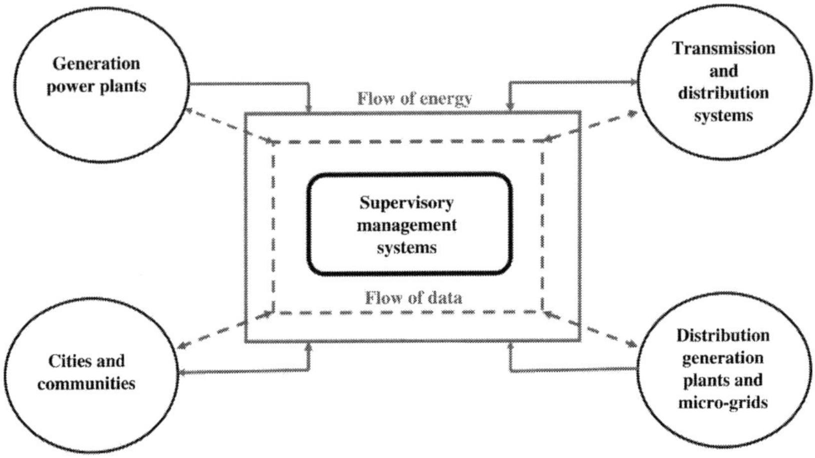

FIGURE 8.24 Schematic electricity and information flows for a smart grid.

The main components of a smart grid that allow for two-way communication interactions between the grid and the end-users include the following, as illustrated in Fig. 8.24:

- Smart meters to measure end-uses and other key drivers of electricity consumption.
- Management tools to account for demand and supply of electricity.
- Information technology systems.
- Building energy management systems.
- Intelligent or smart electricity-consuming equipment such as appliances and air conditioners.

Some of the advantages of smart grids compared to the traditional grids are the following:

- Optimization of the power generation operation;
- Ability to defer electricity generation, transmission, and distribution capacity;
- Minimization of ancillary service costs;
- Reduction of congestion costs;
- Decrease in equipment failures and wide-scale blackouts;
- Minimization of maintenance costs;
- Reduction of meter reading costs;
- Decrease in electricity theft; and
- Reduction in greenhouse gas emissions.

The optimized operation of energy systems can be carried out at different scales starting from small systems to large systems. For electrical grids, the proposed scales include the following:

1. Individual equipment or systems in a building
2. Individual buildings

3. Nanogrids (communities and campuses) or microgrids (cities)
4. Regions and States
5. Countries (country-wide coordinated system)
6. Continents

While a traditional grid attempts only to meet specific loads from end-users using the available generating capacity, a smart grid tries also to manage electricity usage to shape the demand to match the supply. The loads can be shifted using demand response controls from on-peak periods to off-peak periods where electricity is cheaper to generate. The ability to manage the demand as well as the supply allows smart grids to reduce energy use as well as to lower the required generation capacity. This reduced capacity and increased efficiency are often considered the main benefits of smart grids. The main differences between the traditional and smart grid are summarized in Table 8.14.

8.10 SUMMARY

In this chapter, the main pillars and categories and associated metrics for designing sustainable urban centers are described. Specific technologies and strategies for improving the efficiency and flexibility of urban transport modes are then discussed. To provide heating and cooling required for urban buildings, DES are presented to overview their component design configurations and analysis approaches. In particular, several examples are provided throughout the chapter to illustrate the applications of simplified analysis methods for various DES and operation strategies. To mitigate the urban heat island effects, the performance of static and dynamic cool roofs is presented. Finally, the design elements and operation guidelines of smart cities are introduced to highlight the importance of data collection and intelligent controls of various infrastructures and utility resources to optimally operate sustainable urban centers.

TABLE 8.14 Comparison of Features and Characteristics of Smart Again Traditional Grids

Features and characteristics	Conventional grid	Smart grid
Information flow	One-way	Two-way
Electricity generation	Centralized	Distributed
Grid architecture	Radial	Network
Integration of renewables	Low	High
Sensor need	Low	High
Optimized controls	Low	High
Resiliency level	Low	High
Energy efficiency	Low	High
Environmental impact	High	Low

PROBLEMS

8.1 An efficiency test of a boiler fired by fuel No. 2 indicated a flue gas temperature of 350°C and excess air of 35%. The annual fuel consumption of the boiler is 300,000 L/yr. The cost of fuel No. 2 is $0.60/L.

(a) Estimate the efficiency of the boiler as well as the percent in CO_2 and O_2 in the flue gas.

(b) Determine the new boiler efficiency and the annual cost savings in fuel use if the percent O_2 in the flue gas is reduced to 3% and the stack temperature is set to 250°C.

(c) Determine the payback period of installing an automatic control system—at a cost of $9500—to maintain the same efficiency found in (b) throughout the life of the boiler.

8.2 A boiler has an annual fuel oil no. 6 consumption of 1,000,000 L. The cost of fuel is $0.65/L. A combustion efficiency test done on the boiler indicated that the stack temperature is 325°F and the CO_2 content is 11%.

(a) Determine the excess combustion air and the boiler efficiency.

(b) If by adjusting the fuel–air ratio so that the boiler efficiency is increased to 84%, determine the payback period, the net present worth, and the rate of return (discount rate is 5% over 10 years) of installing a full metering kit with O_2 trim control that cost $55,000. Comment on the cost effectiveness of the kit.

8.3 A combustion efficiency test done on a gas-fired boiler indicated that the stack temperature is 700°F and the CO_2 content is 7%.

(a) Determine the excess combustion air and the boiler efficiency.

(b) The boiler efficiency is increased to 83% by adjusting the fuel–air ratio so that the excess air is limited to 30%. Estimate the new stack temperature and the CO_2 content.

(c) Estimate the annual natural gas consumption (in therms) if a combustion gas analyzer kit, installed at a cost of $12,500, is paid back after 4 years. The kit helps maintain the boiler efficiency at 83%. The annual cost of natural gas is $1.75/therm.

8.4 A combustion efficiency test done on a boiler indicated that the stack temperature is 400°C and the CO_2 content is 20%.

(a) Determine the excess combustion air and the boiler efficiency if the boiler uses fuel No. 2.

(b) Determine the excess combustion air and the boiler efficiency if the boiler uses fuel No. 6.

(c) Determine the excess combustion air and the boiler efficiency if the boiler uses natural gas.

8.3 Consider switching from one large chiller (2000 ton) to either (i) two chillers (each rated1000 ton) or (ii) four chillers (each rated 500 ton). Using the normalized savings shown in Fig. 8.13, estimate the annual energy savings for using multichiller systems for options (i) and (ii) in a building located in Chicago, IL, if the large chiller operates for 5500 equivalent full load hours per year and has a rated COP of 4.0.

8.4 (a) Repeat Problem No. 8.3 when the cooling plant is located in Atlanta. (b) Estimate the sizing ratio for the two-chiller cooling plant configuration and the potential additional savings relative to the case of both chillers are equally sized (refer to the chart shown in Fig. 8.14)

8.5 A 400-ton chiller with 1.0 kW/ton average seasonal efficiency is operated 4500 h/yr with an average load factor of 70%. This chiller needs to be replaced by either a chiller A or chiller B. The manufacturers indicated that the COP for chiller A is 4.0 and for chiller B is 4.5.

(a) Estimate the energy cost savings due to replacing the existing chiller with chiller A or with chiller B. The electricity cost is $0.12/kWh.

(b) The cost differential between chiller B and A is $35,000. Determine if it is cost-effective to replace the existing chiller with chiller B rather than chiller A. For this question, a simple payback analysis can be used.

(c) If the discount rate is 5%, determine the electricity price for which it is more cost-effective to replace the existing chiller with chiller A rather than chiller B. Assume a life cycle of 10 years for both chillers.

8.6 Consider DES with 10,000 tons peak cooling load. In the peak-cooling day, the cooling load occurs during 12 h with a load factor of 75%. During the entire year, there are 5000 equivalent full-load hours. Cooling is required only during the on-peak period of the utility rate structure.

The utility rate structure includes (i) demand charge: $20/kW with 12 months ratchet period, (ii) energy charge: the energy charge differential between on-peak and off-peak is $0.05/kWh. The installed cost of a chiller is $500/ton. The cost of a storage system is $50/ton·hr. Consider that the efficiency of the chiller is on average 1.0 kW/ton.

(1) Determine the cost-effectiveness (using a simple payback analysis) of installing (a) a full storage system and (b) a partial storage system with 50% undersized chiller. For the partial storage system, assume that only 2000 of equivalent full-load hours (relative to the new chiller size) have to be supplied by the chiller during the on-peak period.

(2) Determine the effect of the number of hours that the chiller has to be used during the on-peak period on the cost-effectiveness of the partial storage system.

8.7 Consider a cooling plant with 5000 tons peak cooling load. In the peak-cooling day, the cooling load occurs during 10 h with a load factor of 60%. During the entire year, there are 3500 equivalent full-load hours. Cooling is required only during the on-peak period of the utility rate structure. The utility rate structure includes (i) demand charge: $15/kW with 12 months ratchet period, (ii) energy charge: the energy charge differential between on-peak and off-peak is $0.075/kWh. The installed cost of a chiller is $600/ton. The cost of a storage system is $70/ton·h. Consider that the efficiency of the chiller is 0.8 kW/ton.

(i) Determine the cost-effectiveness (using a simple payback analysis) of installing (a) a full storage system and (b) a partial storage system with a 50% undersized chiller. For the partial storage system, assume that only 400 of equivalent full-load hours (relative to the new chiller size) have to be supplied by the chiller during on-peak period.

(ii) Determine the effect of the number of hours that the chiller has to be used during the on-peak period on the cost-effectiveness of the partial storage system.

8.8 An office building has a roof area of 4000 m^2 and is located in Chicago, IL. Estimate the potential savings of using (i) SCR and DCR. Assume a coating reflectivity in both cases of 0.70.

8.9 Repeat Problem No. 8.5 when the building is located in Atlanta, GA. Determine the simple payback periods for SCR and DCR if the cost of source energy is $0.05/MJ.

REFERENCES

ACEEE, 2017. Community-Wide Energy Consumption by City and Year. American Council for Energy Efficient Economy, Washington, DC, Report accessible online: http://www.database.aceee.org/sites/default/files/docs/local-energy-data.pdf.

Akbari, H., 1998. Cool roofs save energy. ASHRAE Trans. 104 (1B), 783–788.

Akbari, H., Levinson, R., 2008. Evolution of cool-roof standards in the US. Adv. Build. Energy Res. 2, 1–32.

Akbari, H., Matthews, H., 2012. Global cooling updates: reflective roof sand pavements. Energy Build. 55, 2–6.

Akbari, H., Touchaei, A., 2014. Modeling and labeling heterogeneous directional reflective roofing materials. Solar Energy Mater. Solar Cells 124, 192–210.

Akbari, H., Konopacki, S., Pomerantz, M., 1999. Cooling energy savings potential of reflective roofs for residential and commercial buildings in the United States. Energy 24, 391–407.

Akbari, H., Berhe, A., Levinson, R., Graveline, S., Foley, K., Delgado, A., et al., 2005. Aging and weathering of cool roofing membranes, Report. Lawrence Berkeley National Laboratory, Berkeley, CA.

Aleksandrowicz, O., Vuckovic, M., Kiesel, K., Mahdavi, A., 2017. Current trends in urban heat island mitigation research: observations based on a comprehensive research repository. Urban Clim. 21, 1–26.

Angelakis, V., Tragos, E., Pöhls, H.C., Kapovits, A., Bassi, A., 2017. Designing, Developing, and Facilitating Smart Cities, Urban Design to IoT Solutions. Springer International Publishing, Switzerland.

ASHRAE, 2007. ANSI/ASHRAE Standard 90.2-2007. Energy Standard for Low-Rise Residential Buildings. American Society of Heating, Refrigerating, and Air-Conditioning Engineers, Inc., Atlanta, GA.

ASHRAE, 2011. ANSI/ASHRAE. Advanced Energy Design Guide for Small to Medium Office Buildings. American Society of Heating, Refrigerating, and Air-Conditioning Engineers, Inc., Atlanta, GA.

ASHRAE, 2013. ANSI/ASHRAE Standard 90.1-2013. Energy Standard for Buildings Except Low-Rise Residential Buildings. American Society of Heating, Refrigerating, and Air-Conditioning Engineers, Inc., Atlanta, GA.

ASHRAE, 2016. ASHRAE Handbook: HVAC Systems and Equipment. American Society for Heating, Refrigerating, and Air Conditioning Engineers Inc., Atlanta, GA.

Azari, S., Bierman, J., 2008. System and Method for Energy-Conserving Roofing. US Patent 7,335,419 B2 filed September 2002 and issued February 2008.

Berdahl, P., Bretz, S., 1997. Preliminary survey of the solar reflectance of cool roofing materials. Energy Build. 25, 149–158.

Berdahl, P., Akbari, H., Rose, L., 2001. Aging of reflective roofs: soot deposition. Appl. Opt. 41 (2), 2335–2358.

BREEAM, 2012. BREEAM Communities – Technical Manual. BREEAM – BRE Global Limited, Watford, England.

Bretz, S., Akbari, H., 1997. Long-term performance of high-albedo roof coatings. Energy Build. 25, 159–167.

CEC, 2013. California Energy Code, California Code of Regulations. Title 24, Part 6. California Building Standards Commission, Sacramento, CA.

CMS, 2016. City Mayors Statistics. Available from http://www.citymayors.com/statistics/largest-cities-mayors-ad.html.

Coffelt, D., Hendrickson, C., 2010. Life-cycle costs of commercial roof systems. J. Arch. Eng. 16 (1), 29–36.

EIA, 2016a. Electricity Usage Tables. Available from: https://www.eia.gov/electricity/monthly/epm_table_grapher.php?t=epmt_5_6_a.

EIA, 2016b. Natural Gas Usage Tables. Accessed online: https://www.eia.gov/dnav/ng/ng_pri_sum_a_EPG0_PRS_DMcf_a.htm.

EIA, 2017. Annual Energy Review. Department of Energy, Energy Information Administration. Available from: http://www.doe.eia.gov.

Erdem, C., Saffa, R., 2015. A state-of-the-art review on innovative glazing technologies. Renew. Sustain. Energy Rev. 41, 695–714.

Frayssinet L, Merliera L, Kuznika F, Huberta JL, Millieza M, Rouxa JJ. Modeling the heating and cooling energy demand of urban buildings at city scale. Renew. Sustain. Energy Rev. 2018;81(2):2318–27.

Futcher, J., Mills, G., Emmanuel, R., Korolija, I., 2017. Creating sustainable cities one building at a time: towards an integrated urban design framework. Cities 66, 63–71.

Gao, Y., 2012. Nanoceramic VO_2 thermochromic smart glass: a review on progress in solution processing. Nano Energy 1 (2), 221–246.

Gartland, L.M., Konopacki, S.J., Akbari, H., 1996. Modeling the effects of reflective roofing. In: Proceedings of the 1996 ACEEE Summer Study on Energy Efficiency in Buildings. American Council for an Energy-Efficient Economy, 4, Pacific Grove, CA, pp. 117–124.

Giguere, M., 2012. Urban Heat Island Mitigation Strategies, report for the Government of Quebec., https://www.inspq.qc.ca/pdf/publications/1513_UrbanHeatIslandMitigationStrategies.pdf.

GIZ, 2012. Urban Transport and Energy Efficiency. a sourcebook published by the German Corporation for International Cooperation GmbH, Bonn, Germany, http://www.sutp.org.

Gonzalez, J., Krarti, M., 2017. ASME Handbook of Integrated and Sustainable Buildings Equipment and Systems, Volume I: Energy Systems. American Society for Mechanical Engineers, New York, NY, 520 pp.

Gray, C., 2015. Application of Adaptive Albedo Roofing Coatings in the Southeastern United States. A Dissertation. The University of Alabama, Birmingham, AL.

Hawkins, T., Brazel, A., Stefanov, W., Bigler, W., Saffel, E., 2004. The role of rural variability in urban heat island determination for Phoenix, Arizona. J. Appl. Meteorol. 43, 476–486.

Henze GP, Dodier RH, Krarti M. Development of a predictive optimal controller for thermal energy storage systems. HVAC&R Res. 1997;3(3):233–64.

Henze, G.P., Krarti, M., 1999. The impact of forecasting uncertainty on the performance of a predictive optimal controller for thermal energy storage systems. ASHRAE Trans. 105 (1), 553–561.

Hooshangi, H., 2015. Energy Performance Modeling of Buildings With Directional Reflective Roofs. A Thesis. Concordia University.

Kamalisarvestani, M., Saidur, M., Mekhilef, R., Javadi, F.S., 2013. Performance, materials and coating technologies of thermochromic thin films on smart windows. Renew. Sustain. Energy Rev. 26, 353–364.

Konopacki, S., Akbari, H., Pomerantz, M., Gabersek, S., Gartland, L., 1997. Cooling Energy Savings Potential of Light-Colored Roofs for Residential and Commercial Buildings in 11 U.S. Metropolitan Areas, Lawrence Berkeley National Laboratory, Berkeley, CA.

Kosowatz, J., 2017. Building a smart city. ASME Mech. Eng. J. 139 (8), 32–37.

Krarti, M., 2011. Energy Audit for Building Systems: An Engineering Approach, 2nd ed. CRC Press, Francis and Taylor, Roca Baton, FL, 600 pp.

Krarti, M., Henze, G.P., Bell, D., 1999. Planning horizon for a predictive optimal controller for thermal energy storage systems. ASHRAE Trans. 105 (1), 543–552.

Larco, N., 2016. Sustainable urban design—a (draft) framework. J. Urban Des. 21 (1), 1–29.

Lee, E., Fernandes, L., Goudey, C., Jonsson, C., Curcija, D., Pant, X., et al., 2013. A Pilot Demonstration of Electrochromic and Thermochromic Windows in the Denver Federal Center, Building 41, Denver, CO. Lawrence Berkeley National Laboratory, Berkeley, CA.

Levinson, R., Akbari, H., 2010. Potential benefits of cool roofs on commercial buildings: conserving energy, saving money, and reducing emission of greenhouse gases and air pollutants. Energy Effic. 3, 53–109.

Levinson, R., Akbari, H., Gartland, L.M., 1996. Impact of the temperature dependency of fiberglass insulation R-value on cooling energy use in buildings. In: Proceedings of the 1996 ACEEE Summer Study on Energy Efficiency in Buildings, vol. 10. American Council for an Energy-Efficient Economy, Pacific Grove, CA, pp. 85–94.

Levinson, R., Berdahl, P., Akbari, H., Millar, W., Joedicke, I., Reilly, J., et al., 2007. Methods of creating solar-reflective nonwhite surfaces and their application to residential roofing materials. Solar Energy Mater. Solar Cells 91, 304–314.

Newman, P., Kenworthy, J., 1999. Sustainability and Cities, Overcoming Automobile Dependence. Island Press, Washington, DC, 464 pp.

Oikonomou, E., Davies, M., Mavrogianni, A., Biddulph, P., Wilkinson, P., Kolokotroni, M., 2012. Modelling the relative importance of the urban heat island and the thermal quality of dwellings for overheating in London. Build. Environ. 57, 223–238.

Park, B., Krarti, M., 2016. Energy performance analysis of variable reflectivity envelope systems for commercial buildings. Energy Build. 124, 88–98.

Pisello, A., Cotana, F., Brinchi, L., 2014. On a cool coating for roof clay tiles: development of the prototype and thermal-energy assessment. Energy Proc. 45, 453–462.

Reinhart, C.F., Davila, C.C., 2016. Urban building energy modeling: a review of a nascent field. Build. Environ. 97, 196–202.

Schulz, N., 2010. Urban Energy Consumption Database and Estimations of Urban Energy Intensities (1.0). International Institute for Applied Systems Analysis (IIASA), Available from: http://

www.iiasa.ac.at/web/home/research/Flagship-Projects/Global-Energy-Assessment/KM18_City_energy_DB.pdf.

SSI, 2014. SITES V2 Rating System for Sustainable Land Design and Development. Sustainable Sites Initiative, Austin, TX.

Star Communities, 2012. STAR Community Rating System, Version 1.0. Star Communities, Washington, DC.

Synnefa, A., Santamouris, M., 2007. Cool-colored coatings fight the urban heat-island effect. SPIE – Int. Soc. Opt. Eng. 2007, 1–3.

Testa, J., Krarti, M., 2017a. A review of benefits and limitations of static and switchable cool roof systems. Renew. Sustain. Energy Rev. 77, 451–460.

Testa, J., Krarti, M., 2017b. Evaluation of energy savings potential of variable reflective roofing systems for US buildings. Sustain. Cities Soc. 31, 62–73.

Thornton, R.P., 2009. Copenhagen's District Heating System – Recycling Waste Heat Reduces Carbon Emissions and Delivers Energy Security. International District Energy Association, Westborough, MA.

Tuluca A. Energy Efficient Design and Construction for Commercial Buildings. New York, NY: Steven Winter and Associates, Mc-Graw Hills; 1997.

UNEP, 2015. District Energy in Cities: Unlocking the potential energy efficiency and renewable energy, report by United Nations Environment Programme, Paris, France. Accessed online: http://www.unep.org/energy/des.

UN-Habitat, 2016. Urbanization and Development, Emerging Futures, World Cities Report 2016, Nairobi, Kenya. Accessed online: http://www.unhabitat.org.

USGBC, 2009. LEED Reference Guide for Green Neighborhood Development LEED-ND. US Green Building Council, Washington, DC.

USGBC, 2016. US Green Building Council's Leadership in Energy and Environmental Design (LEED). Building Design and Construction Rating System. Accessed online: http://coolroofs.org/resources/leed.

Chapter 9

Analysis of Large-Scale Energy Efficiency Programs

Chapter Outline

9.1 INTRODUCTION

Due to economic and technological trends, it is expected that world energy productivity will improve by 40%–70% by 2035 through savings in annual production costs ranging from \$290 billion to \$390 billion mostly due to a reduction in oil consumption estimated at \$170–\$220 \$billion/yr (Winters, 2017). Indeed, the costs of extraction both oil and mining companies have increased. For instance, the average extraction or lift cost of oil jumped from \$8/barrel in 2004 to \$28/barrel in 2014 (Winters, 2017). In addition, advances in various technologies are making economies less dependent on resources. For instance, automation is minimizing energy and material use in manufacturing. Similarly, the adoption of hybrid and electric vehicles is reducing oil consumption. Moreover,

Optimal Design and Retrofit of Energy Efficient Buildings, Communities, and Urban Centers
http://dx.doi.org/10.1016/B978-0-12-849869-9.00009-0

renewable energy resources including solar and wind are becoming increasingly cost-competitive compared to fossil fuels (Winters, 2017).

In this chapter, tools and analysis methods are presented to perform assessments of the impacts of large-scale implementation of energy efficiency or renewable energy programs. First, an overview of energy building stock modeling techniques is provided. Then, examples of reference energy models are described for US residential and commercial buildings. These reference energy models provide prototypes suitable for performing bottom–up analyses of large-scale energy efficiency and renewable energy programs on building stocks. The impact of indoor environment on the well-being and productivity of occupants is presented using simplified analysis techniques. Finally, a case study is presented to evaluate the benefits of improving the existing residential building stock within the Gulf Cooperation Council (GCC) region.

9.2 ENERGY BUILDING STOCK MODELING

To predict the impact of various energy efficiency programs and policies on energy demand for large building stocks, specific modeling techniques can be considered. Two basic modeling methods are typically proposed:

- *Top–down modeling methods*: In these methods, aggregated energy demands for desired regions and/or periods are first collected to include the entire building stock. Then, an analysis is carried out to assess any correlations between these energy demands and other variables, typically related to econometric factors such as energy prices and income levels. In some cases, the aggregated energy demands can be divided by building types and/or locations.
- *Bottom–up modeling methods*: For these methods, models for individual buildings are generally considered to estimate energy consumption or energy end-uses. These individual energy demands and end-uses are then added using statistical or engineering calculation approaches to estimate the aggregated value for the entire building stock. Within the engineering calculation approaches, there are two main subcategories using either stochastic or deterministic analysis algorithms.

Fig. 9.1 outlines the basic building stock modeling methods commonly used to evaluate energy consumption and power demand for large cities, regions, and countries. In the following sections, a more detailed description is provided to highlight the common applications of various methods and approaches used for modeling energy demand for building stocks.

9.2.1 Top–Down Modeling Approaches

The top–down approaches are common in macroeconomic studies to assess the relationships between energy consumption and economic output. These methods

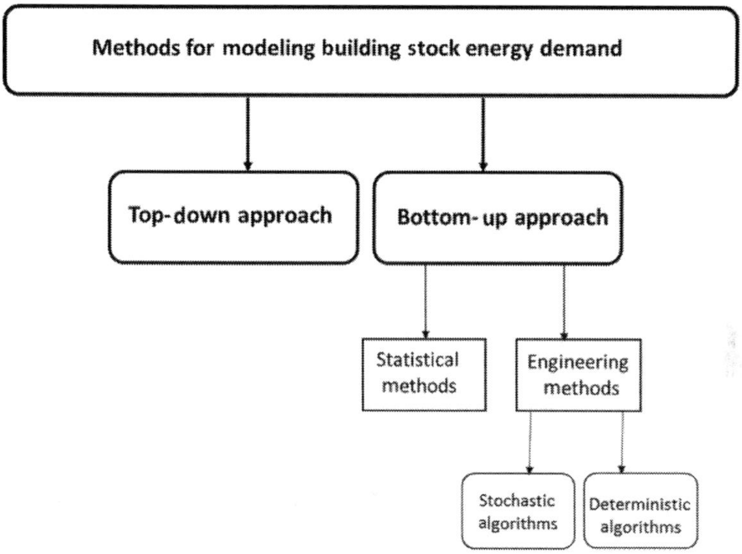

FIGURE 9.1 Methods used for modeling energy demand for building stocks.

are generally based on statistical and regression analysis methods to determine the correlations between the historical energy demand of the building stock energy demands and some socio-economic variables including population levels, climatic conditions, and gross domestic product values.

The National Energy Modeling System (NEMS) used by the US Energy Information Administration to forecast energy, economic, and environmental impacts of various energy policies is based on a top–down method (EIA, 2009). Similarly, several integrated energy system models used for macroeconomic analysis of the building sector are based on top–down methods including MARKAL (Fishbone and Abilock, 1981), the building energy model of MARKAL, GCAM (Edmonds et al., 1994; Kyle et al., 2010), and TIMES (Loulou et al., 2005). Most of these models account only for aggregate energy consumption for the entire building stock and do not consider finer discretization based on building types or locations. Typically, energy demand on an annual basis, national level, and sometimes end-uses can be predicted by the top–down methods. Some attempts have been made to use these methods to predict monthly and even hourly aggregate load profiles (Sailor and Lu, 2004). However, these predictions are generally not accurate and present large uncertainties.

Energy demand models obtained from the top–down approaches are valuable for policymakers to evaluate trends, setting benchmarks, and assess impacts of socio-economic strategies. However, their applications remain limited. Indeed, one of the main disadvantages of the top–down methods is its reliance on historical data with specific and known set of conditions; they are not suitable for predicting accurately the impacts of new technologies or energy efficiency policies including changes in building energy systems or operational strategies.

9.2.2 Bottom–Up Statistical Modeling Methods

Bottom–up methods are able to estimate the energy demand of individual buildings or groups of buildings including end-uses (i.e., lighting, heating, and cooling). Representative energy models for buildings as well as weighting analysis approaches are generally used by these methods to predict energy demand for the entire building stock for a region or a country (Swan and Ugursal, 2009). The statistical bottom–up methods rely on historical data to determine any correlations between known building characteristics and energy use data. Energy utility billing data as well as survey data for building characteristics are generally used along with some regression analysis approaches common for inverse modeling of building energy performance.

One of the first regression analysis of building energy use is the Princeton scorekeeping method or PRISM (Fels, 1986). Specifically, PRISM correlates the building energy use to the heating or cooling degree-days to assess the energy efficiency level as well as evaluate the impact of simple energy efficiency measures for a large building stock (Kavgic et al., 2010). However, PRISM and similar statistical models cannot be used for the analysis of all building types (i.e., large commercial buildings) and for all energy efficiency measures (i.e., improvements of lighting or cooling systems).

Recently, data from geographic information systems (GIS) have been used to improve the accuracy of bottom–up statistical models for cities or urban centers. For instance, Girardin et al. (2010) developed a GIS-based model to predict annual energy end-uses for Geneva, Switzerland, including heating and cooling requirements. Similarly, Kolter and Ferreira (2011) developed a regression-based model for energy consumption of residential and commercial buildings in Cambridge, MA, using utility data with and publicly available tax assessors as well as GIS data. Finally, Howard et al. (2012) used linear regression analysis to determine annual energy use intensity for several building functions in New Your City using GIS data as well as the RECS and CBECS databases (EIA, 2003 and 2005).

As noted, the bottom–up statistical models are based on regression analysis techniques and are easy to develop as long as sufficient data of both energy demand and building features are available. However, these models are not generally suitable for predicting the interactive impacts of several energy efficiency measures.

9.2.3 Bottom–Up Deterministic Engineering Modeling Methods

These bottom–up engineering models are based on algorithms that describe the physical and thermal interactions between various components of buildings including envelope, lighting, appliances, and heating, ventilating, and air conditioning systems. Well established thermodynamic and heat transfer principles are used to develop the deterministic engineering models. Most of the detailed simulation tools for modeling building energy performance are based on deterministic engineering models.

Representative or prototypical building energy models are commonly used for the bottom–up deterministic engineering methods to predict energy demand for building stocks. A general procedure to develop building prototypes includes the following:

1. *Step 1*: Classify building stock into categories based on desired attributes such as building geometric properties, function types, and climate conditions.
2. *Step 2*: Establish building energy models for archetypes with characteristics that represent specific types of building stock's category.
3. *Step 3*: Predict energy consumption and or end-uses for each archetype using one of well-established and verified building energy simulation tools.
4. *Step 4*: Determine aggregated energy consumption based on individual energy demands obtained for the individual building architypes using proper weighting techniques considering typically the number of units and the total floor area of the building stock.

The analysis of the aggregated energy consumption for the housing stock in several countries is based on the bottom–up deterministic engineering methods using several building prototypes (Kavgic et al., 2010; Oladokun and Odesola, 2015). The application of the bottom–up engineering building stock models to the commercial buildings is becoming more common even if the number of architypes is typically high when compared to those needed for the residential building stock. For instance, Yamaguchi et al. (2007) has developed 612 archetypes to represent the commercial building stock in Osaka City. Caputo et al. (2013) has established 56 building archetypes to model the aggregated energy consumption for the city of Milan, Italy. Recently, Davila et al. (2016) has developed an hourly urban building energy model for Boston, Massachusetts, using 76 building archetypes.

Due to the high number of individual archetypes, the bottom–up engineering methods need significant processing and computational efforts to estimate the energy consumption for a building stock especially if hourly and end-use data are desired. However, these methods have some significant advantages compared to the top–down and statistical bottom–up methods. Indeed, and since bottom–up engineering models are based on specific characteristics of individual buildings, they can be used to determine the impacts of several physical and behavior variables of each of the individual buildings. In particular, these models can determine the impacts of implementing new technologies on the energy demand of the building stock.

However, the bottom–up engineering models require a detailed building description to calculate the energy consumption. Some of the building data may not be known or available, thus some assumptions may need to be made, which can result in significant uncertainties in the energy demand predictions. Furthermore, a single archetype model may not be sufficient to adequately represent the energy performance of an entire building stock of the same type. In addition, deterministic engineering models do not typically consider the impacts of occupant behavior on building energy consumption.

9.2.4 Bottom–Up Stochastic Engineering Modeling Methods

As noted earlier, deterministic bottom–up engineering building stock energy models cannot handle adequately uncertainties inherent to the unknown input data or to the limited number of energy models specific to individual building archetypes. As an alternative, the stochastic approach has been proposed to consider uncertainties associated with deterministic building energy models (Eisenhower et al., 2012; Hopfe and Hensen, 2011; Reddy et al., 2007). In this approach, values of input parameters are randomly selected within desired ranges to develop sets of building energy models. The uncertainty levels are determined using variations of the input and output parameters. The Monte Carlo method has been used in several studies to represent the distribution of various building types within a building stock (Korolija et al., 2013; Nishio and Asano, 2006; Smith, 2009). Calibration with the measured data is an important step to validate the predictions from the stochastic models.

Only few attempts have been made to calibrate energy demand for building stock obtained from bottom–up stochastic engineering models. Three sources can cause uncertainties in building stock models (Booth et al., 2012):

- Accuracy of the energy modeling of building thermal performance.
- Variation of characteristics within the same archetype of buildings.
- Homogeneity with the building stock due to unknown or random variation in building features including occupant behaviors.

The lack of actual energy use data is one significant challenge to calibrate building stock energy models. Typically, only aggregated energy use data are available (i.e., city, region, or national level). Moreover, stochastic methods require high computing efforts especially when calibration analysis is considered. Some options to reduce the input parameters through screening approaches and to develop simplified energy simulation algorithms through the development of meta-models have been proposed to lower the computational costs needed to calibrate the building energy models (Kang and Krarti, 2016; Kim, 2016; Menberg et al., 2016; Pianosi et al., 2016). While bottom–up stochastic engineering models have some potential to predict energy demand for existing building stock, further research is needed to consider the noted challenges.

9.3 PROTOTYPES OF US RESIDENTIAL BUILDINGS

Energy use for residential housing across the US is about 8.4% of the total US energy consumption. In recent years, the new housing sector has made notable improvements in energy efficiency due to more stringent codes and standards.

Moreover, several programs have been developed to increase the energy efficiency of new residential homes such as the US Department of Energy (DOE) Building America program (DOE, 2009) and the Energy Star program (Energy Star, 2009) for homes. However, new homes represent only 8% of the US total housing stock; homes built before 2000 represent 92% (DOE, 2005). To reduce energy use in the residential building sector, older homes need to be retrofitted so its energy performance can be improved.

Retrofit efforts have so far been limited for the residential buildings. The largest effort for retrofitting existing housing includes the Energy Star Home Performance program developed by the US EPA and the US DOE. The Home Performance program offers a whole house approach to improve energy efficiency and comfort (Energy Star, 2009). This program provides training and technical support to local contractors to perform energy audits and retrofits for interested homeowners. Nonprofit organizations and energy utility companies from several states are currently participating in this program. For example, Xcel Energy in Colorado offers home energy audits and incentive programs for energy efficiency. Similarly, the Energy Trust of Oregon (2009) offers home energy audits and cash incentives for implementing retrofit projects. To assess the potential impact of regional and national retrofit programs, identification of the characteristics of existing housing stock is needed. Useful characteristics for homes include building envelope features (i.e., insulation level, glazing type, and air-leakage), lighting system, plug-load, heating and cooling equipment, age (vintage), location, and annual energy use.

Early efforts to identify the US home characteristics include studies by Andersson et al. (1985), Huang and Gu (2002), Ritschard et al. (1992), and Sherman and Dickerhoff (1998). Most of these studies utilize two common databases for the US housing stock including the US Census Bureau's American Housing Survey (AHS) and the DOE Residential Energy Consumption Survey (RECS). The two databases are briefly described next:

AHS Database: Every 2 years, the US Bureau of the Census collects data via surveys and interviews to develop a US housing stock inventory of the United States. The main purpose of the AHS database is used to measure specific changes in the housing inventory and assess progress toward the goal of providing sustainable living conditions to all Americans.

RECS Database: This database is a collection of statistical information covering the characteristics of the US housing units, with emphasis on the consumption of and expenditures for energy in these housing units (DOE, 2005). The RECS survey has been typically conducted every 3 to 6 years over the past few decades. The last RECS survey was conducted during 2015, 6 years after the 2009 survey.

Both AHS and RECS databases are based on surveys of a small fraction of existing residential buildings. Indeed, there are over 100 million housing units in the US ranging from single-family homes to large apartment buildings. These housing units are distributed throughout the US and are located in a wide range of climatic zones and regions as illustrated by Figs. 9.2 and 9.3 for single-family homes (DOE, 2005). Fig. 9.4 describes the Census US regions while Fig. 9.5 shows DOE climate zones defined based on heating and cooling degree days. Consequently, characteristics of these homes vary significantly especially in terms of building envelope features and types of heating and cooling systems. Moreover, the age of the housing units can span over 200 years. In the US, most of the homes were built between 1960 and 2000 as indicated in Fig. 9.6.

Fig. 9.7 shows the distribution of the houses per location and vintage. As expected, older houses are more predominant in the Northeast and Midwest where colonization began and the older settlements are located. As indicated in Fig. 9.7, most of the more recent construction activities took place in the South. The West had a slight construction peak during the 1970s and 1980s. Fig. 9.8 shows the distribution of the existing US house stock per type and vintage. Single-family residences represent about two-thirds of the US households. As indicated in Fig. 9.8, it was only after the 1960s that mobile homes started to be built in a significant number.

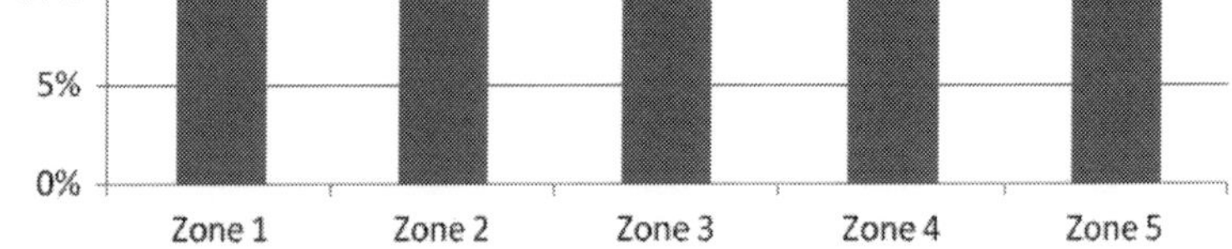

FIGURE 9.2 Distribution of single family detached houses among climate zones.

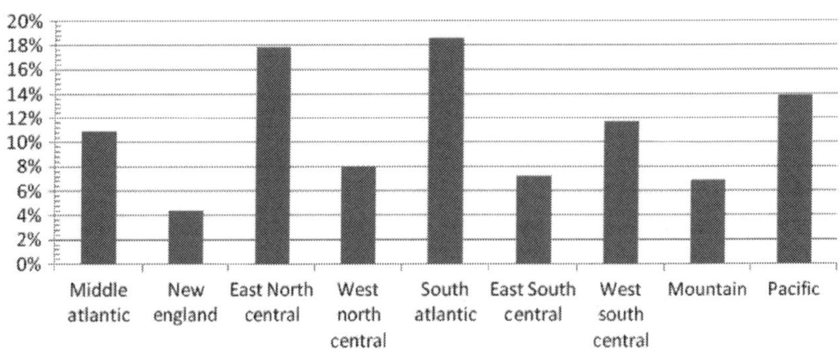

FIGURE 9.3 Distribution of single family detached houses by region.

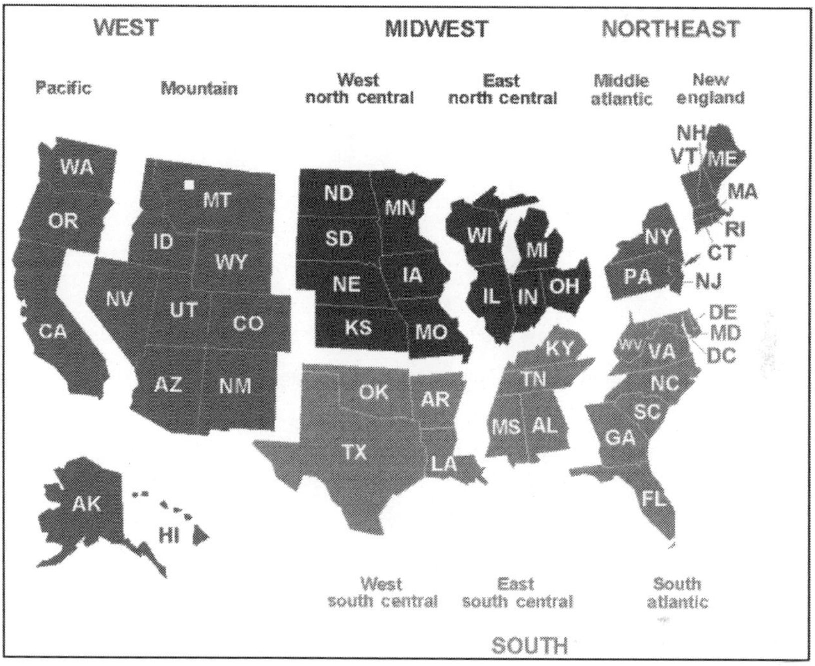

FIGURE 9.4 Department of Energy US regions.

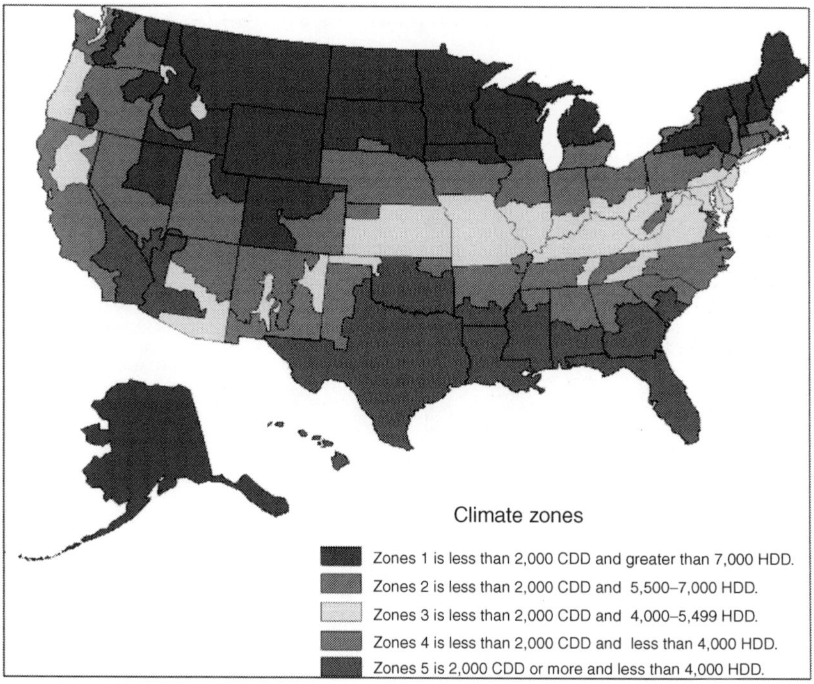

FIGURE 9.5 Department of Energy US climate zones.

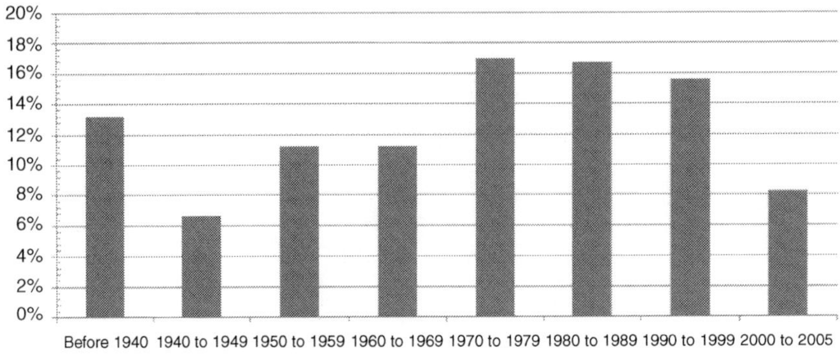

FIGURE 9.6 Distribution of all US housing units by vintage.

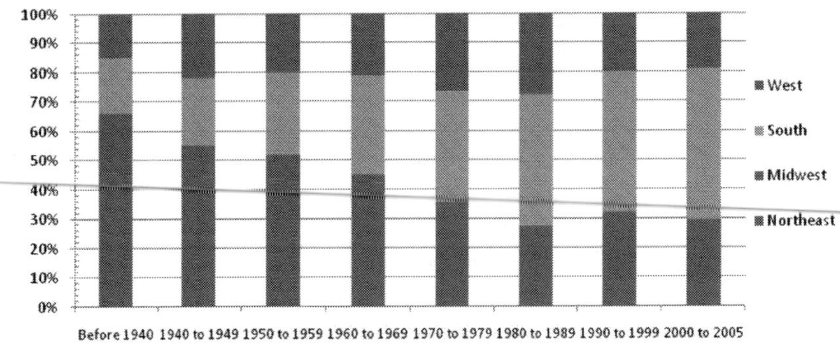

FIGURE 9.7 Distribution of US housing stock by vintage.

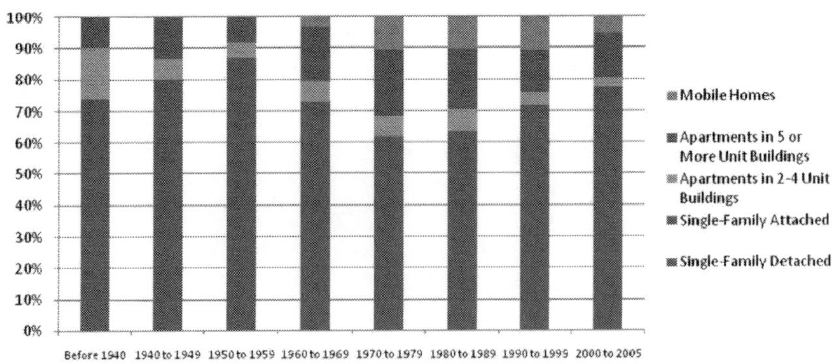

FIGURE 9.8 Distribution of US housing stock by type and vintage.

Tables 9.1 summarizes typical housing characteristics that are considered in both the AHS and the RECS databases. Based on the characteristics, 80 prototypes for homes have been developed to represent regions, climate zones, and vintages (Albertsen et al., 2011). Specifically, 16 cities have been considered

TABLE 9.1 Basic Characteristics Surveyed for Each Housing Type

Variable	Variable range or level
Single family (attached and detached)	
Forced-air distribution	Yes No
Floor area, m^2 (ft^2)	<149 (1600) 149 to 223 (1600 to 2399) >223 (2400)
Year built	<1940 1940–1969 1970–1989 1990–1997
Garage or carport	Yes No
Foundations type	Basement Crawl space Slab
Number of stories	1 2 ≥3
Apartments	
# of units in building	2–4 5–9 10–19 20–39 >40
Floor area, m^2 (ft^2)	<93 (1000) >93 (1000)
Year built	<1940 1940–1969 1970–1989 >1990
Central heating system	Yes No
Manufactured homes	
Floor area, m^2 (ft^2)	<149 (1600) >149 (1600)
Year built	<1940 1940–1969 1970–1989 >1990
Central forced-air system	Yes No

TABLE 9.2 Representative Cities and Climate Zones

Census region	City	Heating degree days	Cooling degree days	Climate zone
New England	Boston	5822	888	2
Mid-Atlantic	New York	4658	1597	3
East North Central	Chicago	5906	1166	2
West North Central	Minneapolis	6911	1003	1
	Kansas City	4638	1635	3
South Atlantic	Washington	4607	1343	3
	Atlanta	2678	1843	4
	Miami	131	4439	5
West South Central	Fort Worth	1921	3119	4
	New Orleans	1150	3204	5
Mountain	Denver	5582	925	2
	Albuquerque	3826	1562	3
	Phoenix	755	4714	5
Pacific North	Seattle	4446	164	3
Pacific South	San Francisco	2295	71	4
	Los Angeles	1182	526	4

to represent both the census regions as illustrated in Fig. 9.4 and DOE climate zones as outlined in Fig. 9.5. Table 9.2 lists the 16 cities representing US census regions and climate zones. For each city, five prototypes have been defined depending on the vintage including houses built. Tables 9.3 through 9.18 illustrate the features of housing prototypes covering US climate zones and five vintages.

The characteristics listed for US homes related to various vintages and regions can be used with an energy stock model to assess various large-scale energy efficiency and renewable energy programs including the cost-effectiveness of various retrofit programs at the regional as well as the national levels.

9.4 US COMMERCIAL REFERENCE BUILDING MODELS

The US DOE has developed commercial reference building models, also known as commercial building benchmark models (Deru et al., 2011). These models represent new and existing buildings in the US and are intended to be used in a wide range of research studies and policy assessments including evaluating new technologies, optimizing building designs, analyzing advanced controls, and developing energy codes and standards. The models have 16 building types that represent approximately 70% of the commercial buildings in the US (Deru et al., 2011). The models provide a set of specifications and data suitable for

TABLE 9.3 US Housing Archetypes by Census Division: New England (Boston)

	Prototype A	Prototype A1	Prototype B	Prototype B1	Prototype C
Year built	Pre-1950s	Pre-1950s	1950–1979	1950–1979	1980–2000
# Bedrooms	3	3	3	3	3
# Bathrooms	1	1	1	1	2
# Stories	2	2	2	2	2
Floor conditioned area (ft^2)	2090	2090	1626	1626	2092
Window area (ft^2)	233	233	212	212	242
Window-to-wall ratio	0.12	0.12	0.12	0.12	0.12
Glazing layers	1	2	1	2	2
Windows	Wood	Wood	Wood	Wood	Wood
Wall siding type	Wood	Wood	Wood	Wood	Wood
Wall type	Plaster	Plaster	Drywall	Drywall	Drywall
R-values					
Wall	0	7	0	7	13
Ceiling	0	22	22	22	27
Floor	0	0	0	0	0
Foundation type	Basement	Basement	Basement	Basement	Basement
Foundation insulation	None	None	None	None	None
Garage	Detached	Detached	Attached	Attached	Attached
Heating equipment	Natural gas	Natural gas	Natural gas	Natural gas	Natural gas
Furnace efficiency AFUE	80%	80%	80%	80%	80%
Cooling equipment	Window	Window	Window	Window	Window
A/C efficiency SEER	7	7	8	8	10
Cooling set points	76	76	76	76	76
Heating set points	67	67	67	67	67
Infiltration ACH 50Pa	0.7	0.7	0.7	0.7	0.4
Appliances	Non-ES[a]	Non-ES	Non-ES	Non-ES	Non-ES
Lighting (fluorescent)	14%	14%	14%	14%	14%

[a] *Non-Energy Star.*

TABLE 9.4 US Housing Archetypes by Census Division: Middle Atlantic (New York)

	Prototype A	Prototype A1	Prototype B	Prototype B1	Prototype C
Year built	Pre-1950s	Pre-1950s	1950–1979	1950–1979	1980–2000
# Bedrooms	3	3	3	3	3
# Bathrooms	1	1	1	1	2
# Stories	2	2	2	2	2
Floor conditioned area (ft^2)	1789	1789	1804	1804	2111
Window area (ft^2)	239	239	196	196	226
Window-to-wall ratio	0.12	0.12	0.12	0.12	0.12
Glazing layers	1	2	1	2	2
Windows	Wood	Wood	Wood	Wood	Wood
Wall siding type	Wood	Wood	Wood	Wood	Wood
Wall type	Plaster	Plaster	Drywall	Drywall	Drywall
R-values					
Wall	0	7	0	7	13
Ceiling	0	7	7	11	27
Floor	0	0	0	0	19
Foundation type	Basement	Basement	Basement	Basement	Basement
Foundation insulation	None	None	None	None	None
Garage	Detached	Detached	Attached	Attached	Attached
Heating equipment	Natural gas	Natural gas	Natural gas	Natural gas	Natural gas
Furnace efficiency AFUE	80%	80%	80%	80%	80%
Cooling equipment	Window	Window	Window	Window	Window
A/C efficiency SEER	7	7	8	8	10
Cooling set points	78	78	78	78	78
Heating set points	67	67	67	67	67
Infiltration ACH 50PA	0.7	0.7	0.7	0.7	0.4
Appliances	Non-ES[a]	Non-ES	Non-ES	Non-ES	Non-ES
Lighting (fluorescent)	14%	14%	14%	14%	14%

[a]*Non-Energy Star.*

TABLE 9.5 US Housing Archetypes by Census Division: East North (Chicago)

	Prototype A	Prototype A1	Prototype B	Prototype B1	Prototype C
Year built	Pre-1950s	Pre-1950s	1950–1979	1950–1979	1980–2000
# Bedrooms	3	3	3	3	3
# Bathrooms	1	1	1	1	2
# Stories	2	2	1	1	2
Floor conditioned area (ft^2)	1646	1646	1557	1557	2223
Window area (ft^2)	199	199	167	167	190
Window-to-wall ratio	0.12	0.12	0.12	0.12	0.12
Glazing layers	1	2	1	2	2
Windows	Wood	Wood	Wood	Wood	Wood
Wall siding type	Wood	Wood	Brick	Brick	Aluminum
Wall type	Plaster	Plaster	Drywall	Drywall	Drywall
R-values					
Wall	0	7	0	7	13
Ceiling	0	11	11	19	25
Floor	0	0	0	0	0
Foundation type	Basement	Basement	Basement	Basement	Basement
Foundation insulation	None	None	None	None	None
Garage	Detached	Detached	Attached	Attached	Attached
Heating equipment	Natural gas	Natural gas	Natural gas	Natural gas	Natural gas
Furnace efficiency AFUE	80%	80%	80%	80%	80%
Cooling equipment	Window	Window	Central	Central	Central
A/C efficiency SEER	7	7	8	8	10
Cooling set points	74	74	74	74	74
Heating set points	61	67	61	67	68
Infiltration ACH 50Pa	0.7	0.7	0.7	0.7	0.4
Appliances	Non-ES[a]	Non-ES	Non-ES	Non-ES	Non-ES
Lighting (fluorescent)	14%	14%	14%	14%	14%

[a]*Non-Energy Star.*

TABLE 9.6 US Housing Archetypes by Census Division: West North Central (Minneapolis)

	Prototype A	Prototype A1	Prototype B	Prototype B1	Prototype C
Year built	Pre-1950s	Pre-1950s	1950–1979	1950–1979	1980–2000
# Bedrooms	3	3	3	3	3
# Bathrooms	1	1	1	1	2
# Stories	2	2	1	1	2
Floor conditioned area (ft^2)	1621	1621	1487	1487	2048
Window area (ft^2)	209	209	174	174	231
Window-to-wall ratio	0.12	0.12	0.12	0.12	0.12
Glazing layers	1	1	1	2	2
Windows	Wood	Wood	Wood	Wood	Wood
Wall siding type	Wood	Wood	Wood	Wood	Wood
Wall type	Plaster	Plaster	Drywall	Drywall	Drywall
R-values					
Wall	0	7	0	7	19
Ceiling	0	11	11	22	27
Floor	0	0	0	0	0
Foundation type	Basement	Basement	Basement	Basement	Basement
Foundation insulation	None	None	None	None	R-5, 4-ft basement wall
Garage	Detached	Detached	Attached	Attached	Attached
Heating equipment	Natural gas	Natural gas	Natural gas	Natural gas	Natural gas
Furnace efficiency AFUE	80%	80%	80%	80%	80%
Cooling equipment	Window	Window	Central	Central	Central
A/C efficiency SEER	7	7	8	8	10
Cooling set points	74	74	72	72	72
Heating set points	59	66	59	66	68
Infiltration ACH 50Pa	0.3	0.3	0.3	0.3	0.1
Appliances	Non-ES[a]	Non-ES	Non-ES	Non-ES	Non-ES
Lighting (fluorescent)	14%	14%	14%	14%	14%

[a] *Non-Energy Star.*

TABLE 9.7 US Housing Archetypes by Census Division: West North Central (Kansas City)

	Prototype A	Prototype A1	Prototype B	Prototype B1	Prototype C
Year built	Pre-1950s	Pre-1950s	1950–1979	1950–1979	1980–2000
# Bedrooms	3	3	3	3	3
# Bathrooms	1	1	1	1	2
# Stories	2	2	1	1	2
Floor conditioned area (ft^2)	1621	1621	1487	1487	2048
Window area (ft^2)	209	209	174	174	231
Window-to-wall ratio	0.12	0.12	0.12	0.12	0.12
Glazing layers	1	1	2	2	2
Windows	Wood	Wood	Aluminum	Aluminum	Wood
Wall siding type	Wood	Wood	Wood	Wood	Wood
Wall type	Plaster	Plaster	Drywall	Drywall	Drywall
R-values					
Wall	0	7	0	7	19
Ceiling	0	11	11	22	27
Floor	0	0	0	0	0
Foundation type	Basement	Basement	Basement	Basement	Basement
Foundation insulation	None	None	None	None	None
Garage	Detached	Detached	Attached	Attached	Attached
Heating equipment	Natural gas	Natural gas	Natural gas	Natural gas	Natural gas
Furnace efficiency AFUE	80%	80%	80%	80%	80%
Cooling equipment	Window	Window	Central	Central	Central
A/C efficiency SEER	7	7	8	8	10
Cooling set points	74	74	72	72	72
Heating set points	62	67	62	67	69
Infiltration ACH 50Pa	0.7	0.7	0.7	0.7	0.4
Appliances	Non-ES[a]	Non-ES	Non-ES	Non-ES	Non-ES
Lighting (fluorescent)	14%	14%	14%	14%	14%

[a]*Non-Energy Star.*

TABLE 9.8 US Housing Archetypes by Census Division: South Atlantic: Washington, DC

	Prototype A	Prototype A1	Prototype B	Prototype B1	Prototype C
Year built	Pre-1950s	Pre-1950s	1950–1979	1950–1979	1980–2000
# Bedrooms	3	3	3	3	3
# Bathrooms	1	1	1	1	2
# Stories	1	1	1	1	2
Floor conditioned area (ft^2)	1831	1831	1643	1643	2211
Window area (ft^2)	163	163	165	165	160
Window-to-wall ratio	0.12	0.12	0.12	0.12	0.12
Glazing layers	1	1	1	1	2
Windows	Wood	Wood	Aluminum	Aluminum	Wood
Wall siding type	Wood	Wood	Brick	Brick	Aluminum
Wall type	Plaster	Plaster	Drywall	Drywall	Drywall
R-values					
Wall	0	7	0	7	11
Ceiling	0	7	7	22	23
Floor	0	0	0	0	19
Foundation type	Crawl-space	Crawl-space	Crawl-space	Crawl-space	Basement
Foundation insulation	None	None	None	None	None
Garage	Detached	Detached	Attached	Attached	Attached
Heating equipment	Natural gas	Natural gas	Electric furnace	Electric furnace	Electric furnace
Furnace efficiency AFUE	80%	80%	99%	99%	99%
Cooling equipment	Window	Window	Central	Central	Central
A/C efficiency SEER	7	7	8	8	10
Cooling set points	77	77	77	77	77
Heating set points	75	75	75	75	75
Infiltration ACH 50Pa	0.3	0.3	0.3	0.3	0.1
Appliances	Non-ES[a]	Non-ES	Non-ES	Non-ES	Non-ES
Lighting (fluorescent)	14%	14%	14%	14%	14%

[a]*Non-Energy Star.*

TABLE 9.9 US Housing Archetypes by Census Division: South Atlantic (Atlanta)

	Prototype A	Prototype A1	Prototype B	Prototype B1	Prototype C
Year built	Pre-1950s	Pre-1950s	1950–1979	1950–1979	1980–2000
# Bedrooms	3	3	3	3	3
# Bathrooms	1	1	1	1	2
# Stories	1	1	1	1	2
Floor conditioned area (ft^2)	1831	1831	1643	1643	2211
Window area (ft^2)	170	170	161	161	158
Window-to-wall ratio	0.12	0.12	0.12	0.12	0.12
Glazing layers	1	1	1	1	2
Windows	Wood	Wood	Aluminum	Aluminum	Aluminum
Wall siding type	Wood	Wood	Brick	Brick	Wood
Wall type	Plaster	Plaster	Drywall	Drywall	Drywall
R-values					
Wall	0	7	0	7	11
Ceiling	0	7	7	22	24
Floor	0	0	0	0	19
Foundation type	Crawl-space	Crawl-space	Crawl-space	Crawl-space	Basement
Foundation insulation	None	None	None	None	None
Garage	Detached	Detached	Attached	Attached	Attached
Heating equipment	Natural gas	Natural gas	Electric furnace	Electric furnace	Electric furnace
Furnace efficiency AFUE	80%	80%	99%	99%	99%
Cooling equipment	Window	Window	Central	Central	Central
A/C efficiency SEER	7	7	8	8	10
Cooling set points	77	77	77	77	77
Heating set points	75	75	75	75	75
Infiltration ACH 50Pa	0.3	0.3	0.3	0.3	0.1
Appliances	Non-ES[a]	Non-ES	Non-ES	Non-ES	Non-ES
Lighting (fluorescent)	14%	14%	14%	14%	14%

[a]*Non-Energy Star.*

TABLE 9.10 US Housing Archetypes by Census Division: South Atlantic (Miami)

	Prototype A	Prototype A1	Prototype B	Prototype B1	Prototype C
Year built	Pre-1950s	Pre-1950s	1950–1979	1950–1979	1980–2000
# Bedrooms	3	3	3	3	3
# Bathrooms	1	1	1	1	2
# Stories	1	1	1	1	1
Floor conditioned area (ft^2)	1831	1831	1643	1643	2211
Window area (ft^2)	163	163	165	165	160
Window-to-wall ratio	0.12	0.12	0.12	0.12	0.12
Glazing layers	1	1	1	1	1
Windows	Wood	Wood	Aluminum	Aluminum	Aluminum
Wall siding type	Wood	Wood	Brick	Brick	Stucco
Wall type	Plaster	Plaster	Drywall	Drywall	Drywall
R-values					
Wall	0	7	0	7	11
Ceiling	0	7	7	22	23
Floor	0	0	0	0	0
Foundation type	Crawl-space	Crawl-space	Crawl-space	Crawl-space	Slab
Foundation insulation	None	None	None	None	None
Garage	Detached	Detached	Attached	Attached	Attached
Heating equipment	Natural gas	Natural gas	Electric furnace	Electric furnace	Electric furnace
Furnace efficiency AFUE	80%	80%	99%	99%	99%
Cooling equipment	Window	Window	Central	Central	Central
A/C efficiency SEER	7	7	8	8	10
Cooling set points	77	77	77	77	77
Heating set points	75	75	75	75	75
Infiltration ACH 50Pa	1.0	1.1	1.1	1.1	0.7
Appliances	Non-ES[a]	Non-ES	Non-ES	Non-ES	Non-ES
Lighting (fluorescent)	14%	14%	14%	14%	14%

[a]*Non-Energy Star.*

TABLE 9.11 US Housing Archetypes by Census Division: West South Central (Fort Worth)

	Prototype A	Prototype A1	Prototype B	Prototype B1	Prototype C
Year built	Pre-1950s	Pre-1950s	1950–1979	1950–1979	1980–2000
# Bedrooms	3	3	3	3	3
# Bathrooms	1	1	1	1	2
# Stories	1	1	1	1	1
Floor conditioned area (ft^2)	1247	1247	1493	1493	2055
Window area (ft^2)	180	180	150	150	168
Window-to-wall ratio	0.12	0.12	0.12	0.12	0.12
Glazing layers	1	1	1	1	2
Windows	Wood	Wood	Aluminum	Aluminum	Aluminum
Wall siding type	Wood	Wood	Brick	Brick	Wood
Wall type	Plaster	Plaster	Drywall	Drywall	Drywall
R-values					
Wall	0	7	0	7	11
Ceiling	0	7	7	19	24
Floor	0	0	0	0	0
Foundation type	Slab	Slab	Slab	Slab	Slab
Foundation insulation	None	None	None	None	R-5, 2-ft slab edge
Garage	Detached	Detached	Attached	Attached	Attached
Heating equipment	Natural gas	Natural gas	Natural gas	Natural gas	Natural gas
Furnace efficiency AFUE	80%	80%	80%	80%	80%
Cooling equipment	Window	Window	Central	Central	Central
A/C efficiency SEER	7	7	8	8	10
Cooling set points	78	78	78	78	78
Heating set points	72	72	72	72	72
Infiltration ACH 50Pa	1.1	1.1	1.1	1.1	0.7
Appliances	Non-ES[a]	Non-ES	Non-ES	Non-ES	Non-ES
Lighting (fluorescent)	14%	14%	14%	14%	14%

[a]*Non-Energy Star.*

TABLE 9.12 US Housing Archetypes by Census Division: West South Central (New Orleans)

	Prototype A	Prototype A1	Prototype B	Prototype B1	Prototype C
Year built	Pre-1950s	Pre-1950s	1950–1979	1950–1979	1980–2000
# Bedrooms	3	3	3	3	3
# Bathrooms	1	1	1	1	2
# Stories	1	1	1	1	1
Floor conditioned area (ft^2)	1247	1247	1493	1493	2055
Window area (ft^2)	180	180	150	150	168
Window-to-wall ratio	0.12	0.12	0.12	0.12	0.12
Glazing layers	1	1	1	1	2
Windows	Wood	Wood	Aluminum	Aluminum	Aluminum
Wall siding type	Wood	Wood	Brick	Brick	Brick
Wall type	Plaster	Plaster	Drywall	Drywall	Drywall
R values					
Wall	0	7	0	7	11
Ceiling	0	7	7	19	24
Floor	0	0	0	0	0
Foundation type	Slab	Slab	Slab	Slab	Slab
Foundation insulation	None	None	None	None	None
Garage	Detached	Detached	Attached	Attached	Attached
Heating equipment	Natural gas	Natural gas	Electric furnace	Electric furnace	Electric furnace
Furnace efficiency AFUE	80%	80%	99%	99%	99%
Cooling equipment	Window	Window	Central	Central	Central
A/C efficiency SEER	7	7	8	8	10
Cooling set points	78	78	78	78	78
Heating set points	72	72	72	72	72
Infiltration ACH 50Pa	1.1	1.1	1.1	1.1	0.7
Appliances	Non-ES[a]	Non-ES	Non-ES	Non-ES	Non-ES
Lighting (fluorescent)	14%	14%	14%	14%	14%

[a]*Non-Energy Star.*

TABLE 9.13 US Housing Archetypes by Census Division: Mountain (Denver)

	Prototype A	Prototype A1	Prototype B	Prototype B1	Prototype C
Year built	Pre-1950s	Pre-1950s	1950–1979	1950–1979	1980–2000
# Bedrooms	3	3	3	3	3
# Bathrooms	1	1	1	1	2
# Stories	1	1	1	1	2
Floor conditioned area (ft^2)	1823	1823	1734	1734	2017
Window area (ft^2)	159	159	141	141	238
Window-to-wall ratio	0.12	0.12	0.12	0.12	0.12
Glazing layers	1	1	1	1	2
Windows	Wood	Wood	Wood	Wood	Wood
Wall siding type	Wood	Wood	Brick	Brick	Wood
Wall type	Plaster	Plaster	Drywall	Drywall	Drywall
R-values					
Wall	0	7	0	7	13
Ceiling	0	11	11	11	30
Floor	0	0	0	0	11
Foundation type	Basement	Basement	Slab	Slab	Basement
Foundation insulation	None	None	None	None	None
Garage	Detached	Detached	Attached	Attached	Attached
Heating equipment	Natural gas	Natural gas	Natural gas	Natural gas	Natural gas
Furnace efficiency AFUE	80%	80%	80%	80%	80%
Cooling equipment	Window	Window	Central	Central	Central
A/C efficiency SEER	7	7	8	8	10
Cooling set points	77	77	77	77	77
Heating set points	70	70	70	70	70
Infiltration ACH 50Pa	1.1	1.1	1.1	1.1	0.7
Appliances	Non-ES[a]	Non-ES	Non-ES	Non-ES	Non-ES
Lighting (fluorescent)	14%	14%	14%	14%	14%

[a]*Non-Energy Star.*

TABLE 9.14 US Housing Archetypes by Census Division: Mountain (Albuquerque)

	Prototype A	Prototype A1	Prototype B	Prototype B1	Prototype C
Year built	Pre-1950s	Pre-1950s	1950–1979	1950–1979	1980–2000
# Bedrooms	3	3	3	3	3
# Bathrooms	1	1	1	1	2
# Stories	1	1	1	1	1
Floor conditioned area (ft^2)	1823	1823	1734	1734	2017
Window area (ft^2)	159	159	141	141	238
Window-to-wall ratio	0.12	0.12	0.12	0.12	0.12
Glazing layers	1	2	1	2	2
Windows	Wood	Wood	Aluminum	Aluminum	Wood
Wall siding type	Wood	Wood	Brick	Brick	Stucco
Wall type	Plaster	Plaster	Drywall	Drywall	Drywall
R-values					
Wall	0	7	0	7	13
Ceiling	0	11	11	11	30
Floor	0	0	0	0	0
Foundation type	Basement	Basement	Slab	Slab	Slab
Foundation insulation	None	None	None	None	R-5, 2-ft slab edge
Garage	Detached	Detached	Attached	Attached	Attached
Heating equipment	Natural gas	Natural gas	Natural gas	Natural gas	Natural gas
Furnace efficiency AFUE	80%	80%	80%	80%	80%
Cooling equipment	Window	Window	Central	Central	Central
A/C efficiency SEER	7	7	8	8	10
Cooling set points	77	77	77	77	77
Heating set points	70	70	70	70	70
Infiltration ACH 50Pa	1.1	1.1	1.1	1.1	0.7
Appliances	Non-ES[a]	Non-ES	Non-ES	Non-ES	Non-ES
Lighting (fluorescent)	14%	14%	14%	14%	14%

[a]*Non-Energy Star.*

TABLE 9.15 US Housing Archetypes by Census Division: Mountain (Phoenix)

	Prototype A	Prototype A1	Prototype B	Prototype B1	Prototype C
Year built	Pre-1950s	Pre-1950s	1950–1979	1950–1979	1980–2000
# Bedrooms	3	3	3	3	3
# Bathrooms	1	1	1	1	2
# Stories	1	1	1	1	1
Floor conditioned area (ft^2)	1823	1823	1734	1734	2017
Window area (ft^2)	159	159	141	141	238
Window-to-wall ratio	0.12	0.12	0.12	0.12	0.12
Glazing layers	1	1	1	1	1
Windows	Wood	Wood	Aluminum	Aluminum	Aluminum
Wall siding type	Wood	Wood	Brick	Brick	Stucco
Wall type	Plaster	Plaster	Drywall	Drywall	Drywall
R-values					
Wall	0	7	0	7	13
Ceiling	0	11	11	11	30
Floor	0	0	0	0	0
Foundation type	Basement	Basement	Slab	Slab	Slab
Foundation insulation	None	None	None	None	None
Garage	Detached	Detached	Attached	Attached	Attached
Heating equipment	Natural gas	Natural gas	Natural gas	Natural gas	Natural gas
Furnace efficiency AFUE	80%	80%	80%	80%	80%
Cooling equipment	Window	Window	Central	Central	Central
A/C efficiency SEER	7	7	8	8	10
Cooling set points	80	80	78	78	78
Heating set points	70	70	70	70	70
Infiltration ACH 50Pa	0.4	0.3	0.4	0.3	0.1
Appliances	Non-ES[a]	Non-ES	Non-ES	Non-ES	Non-ES
Lighting (fluorescent)	14%	14%	14%	14%	14%

[a]*Non-Energy Star.*

TABLE 9.16 US Housing Archetypes by Census Division: Pacific North (Seattle)

	Prototype A	Prototype A1	Prototype B	Prototype B1	Prototype C
Year built	Pre-1950s	Pre-1950s	1950–1979	1950–1979	1980–2000
# Bedrooms	3	3	3	3	3
# Bathrooms	1	1	1	1	2
# Stories	1	1	1	1	2
Floor conditioned area (ft^2)	1786	1786	1722	1722	1978
Window area (ft^2)	180	180	142	142	194
Window-to-wall ratio	0.12	0.12	0.12	0.12	0.12
Glazing layers	1	1	1	1	1
Windows	Wood	Wood	Wood	Wood	Wood
Wall siding type	Wood	Wood	Wood	Wood	Wood
Wall type	Plaster	Plaster	Drywall	Drywall	Drywall
R-values					
Wall	0	7	0	7	11
Ceiling	0	11	11	19	23
Floor	0	0	0	0	19
Foundation type	Crawl-space	Crawl-space	Crawl-space	Crawl-space	Crawl-space
Foundation insulation	None	None	None	None	None
Garage	Detached	Detached	Attached	Attached	Attached
Heating equipment	Natural gas	Natural gas	Natural gas	Natural gas	Natural gas
Furnace efficiency AFUE	80%	80%	80%	80%	80%
Cooling equipment	Window	Window	Central	Central	Central
A/C efficiency SEER	7	7	8	8	10
Cooling set points	75	75	73	73	73
Heating set points	70	70	70	70	70
Infiltration ACH 50Pa	0.4	0.4	0.4	0.4	0.3
Appliances	Non-ES[a]	Non-ES	Non-ES	Non-ES	Non-ES
Lighting (fluorescent)	14%	14%	14%	14%	14%

[a] *Non-Energy Star.*

TABLE 9.17 US Housing Archetypes by Census Division: Pacific South (San Francisco)

	Prototype A	Prototype A1	Prototype B	Prototype B1	Prototype C
Year built	Pre-1950s	Pre-1950s	1950–1979	1950–1979	1980–2000
# Bedrooms	3	3	3	3	3
# Bathrooms	1	1	1	1	2
# Stories	1	1	1	1	2
Floor conditioned area (ft^2)	1786	1786	1722	1722	1978
Window area (ft^2)	180	180	142	142	194
Window-to-wall ratio	0.12	0.12	0.12	0.12	0.12
Glazing layers	1	1	1	1	2
Windows	Wood	Wood	Aluminum	Aluminum	Aluminum
Wall siding type	Wood	Wood	Stucco	Stucco	Stucco
Wall type	Plaster	Plaster	Drywall	Drywall	Drywall
R-values					
Wall	0	7	0	7	11
Ceiling	0	11	11	19	23
Floor	0	0	0	0	19
Foundation type	Crawl-space	Crawl-space	Crawl-space	Crawl-space	Crawl-space
Foundation insulation	None	None	None	None	None
Garage	Detached	Detached	Attached	Attached	Attached
Heating equipment	Natural gas	Natural gas	Natural gas	Natural gas	Natural gas
Furnace efficiency AFUE	80%	80%	80%	80%	80%
Cooling equipment	Window	Window	Central	Central	Central
A/C efficiency SEER	7	7	8	8	10
Cooling set points	75	75	73	73	73
Heating set points	70	70	70	70	70
Infiltration ACH 50Pa	1.1	1.1	1.1	1.1	0.4
Appliances	Non-ES[a]	Non-ES	Non-ES	Non-ES	Non-ES
Lighting (fluorescent)	14%	14%	14%	14%	14%

[a]*Non-Energy Star.*

TABLE 9.18 US Housing Archetypes by Census Division: Pacific South (Los Angeles)

	Prototype A	Prototype A1	Prototype B	Prototype B1	Prototype C
Year built	Pre-1950s	Pre-1950s	1950–1979	1950–1979	1980–2000
# Bedrooms	3	3	3	3	3
# Bathrooms	1	1	1	1	2
# Stories	1	1	1	1	2
Floor conditioned area (ft^2)	1786	1786	1722	1722	1978
Window area (ft^2)	180	180	142	142	194
Window-to-wall ratio	0.12	0.12	0.12	0.12	0.12
Glazing layers	1	1	1	1	2
Windows	Wood	Wood	Aluminum	Aluminum	Aluminum
Wall siding type	Wood	Wood	Stucco	Stucco	Stucco
Wall type	Plaster	Plaster	Drywall	Drywall	Drywall
R-values					
Wall	0	7	0	7	11
Ceiling	0	11	11	19	23
Floor	0	0	0	0	19
Foundation type	Crawl-space	Crawl-space	Crawl-space	Crawl-space	Slab
Foundation insulation	None	None	None	None	None
Garage	Detached	Detached	Attached	Attached	Attached
Heating equipment	Natural gas	Natural gas	Natural gas	Natural gas	Natural gas
Furnace efficiency AFUE	80%	80%	80%	80%	80%
Cooling equipment	Window	Window	Central	Central	Central
A/C efficiency SEER	7	7	8	8	10
Cooling set points	75	75	73	73	73
Heating set points	70	70	70	70	70
Infiltration ACH 50Pa	1.1	1.1	1.1	1.1	0.4
Appliances	Non-ES[a]	Non-ES	Non-ES	Non-ES	Non-ES
Lighting (fluorescent)	14%	14%	14%	14%	14%

[a]*Non-Energy Star.*

TABLE 9.19 List of US Commercial Reference Building Models and Their Floor Area

Building type	Number of floors	Floor area	
		m^2	ft^2
Large office	12	46,320	498,588
Medium office	3	4982	53,628
Small office	1	511	5500
Warehouse	1	4835	52,045
Stand-alone retail	1	2319	24,962
Strip mall	1	2090	22,500
Primary school	1	6871	73,960
Secondary school	2	19,592	210,887
Supermarket	1	4181	45,000
Quick service restaurant	1	232	2500
Full service restaurant	1	511	5500
Hospital	5	22,422	241,351
Outpatient health care	3	3804	40,946
Small hotel	4	4013	43,200
Large hotel	6	11,345	122,120
Midrise apartment	4	3135	33,740

computer energy simulations. In fact, the models are formatted as input files readily for use with the detailed whole-building simulation tool, EnergyPlus. Table 9.19 lists the types of the reference building models as well as their floor areas while Fig. 9.9 illustrates renderings of these models based on their EnergyPlus input files (Deru et al., 2011). Specifically, the commercial reference building models are available for the following categories:

- new construction
- existing buildings constructed in or after 1980 ("post-1980")
- existing buildings constructed before 1980 ("pre-1980")

The models are available for 16 locations representing the various climate zones within the US. Several sources have been considered to define the input parameters for the building models (Deru et al., 2011). Some of the resources include ASHRAE Standards 90.1-2004, 62.1-2004, and 62-1999 for new construction and Standard 90.1-1989 for post-1980 construction. Moreover, standard building practices are considered for some input parameters (Field et al., 2010).

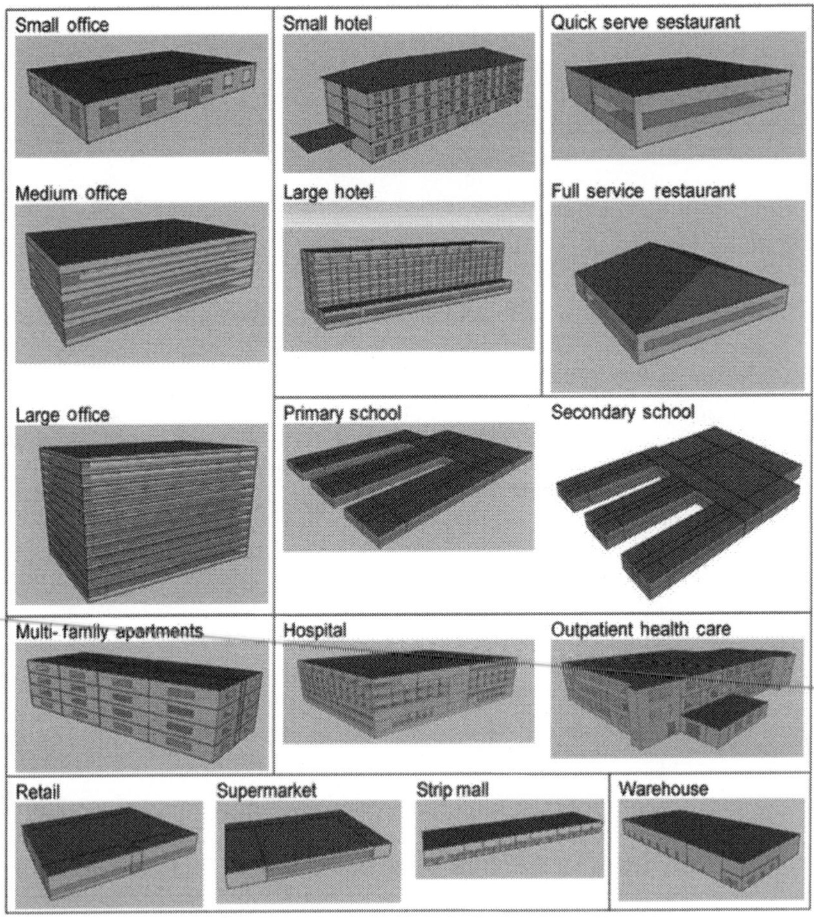

FIGURE 9.9 Renderings for US commercial reference building models.

Using these energy models, the DOE is currently developing a tool for estimating the energy and carbon impacts of various energy conservation measures (ECMs) on the US building sector (DOE, 2017). The tool, called Scout, has the ability to also model residential buildings using prototypes similar to those presented in Section 9.3. According to the DOE, Scout identifies ECMs using their relative or absolute performance characteristics such as installed cost, service lifetime, and year of introduction into the market. In addition, Scout can estimate the ECM energy performance using whole-building energy simulation with EnergyPlus using the prototypical energy building models. This approach allows fair comparison and assessment of ECMs using identical assumption. It also produces savings estimates disaggregated by end uses, climate regions, and building types. In particular, Scout ECMs are applied to a baseline US building stock as defined by the Energy Information Administration (EIA) based on various characteristics include building type (e.g., office, hospital, single family home), building vintage (e.g., new, existing), climate zone (e.g., hot, humid, cold), end use (e.g., heating, lighting), and fuel type (e.g., electricity, natural gas).

In Scout, baseline buildings stock is used to evaluate the performance of ECMs under two adoption scenarios (DOE, 2017):

(i) *Technical potential adoption scenario*: The ECMs are assumed to capture all their full baseline market potential upon introduction and to maintain full market saturation in later years.
(ii) *Maximum adoption scenario*: The ECMs are assumed to only capture the segments of their baseline markets associated to new, existing retrofit, or replacement applications based on realistic rates of new construction, retrofits, and turnover in baseline equipment.

The tool, Scout, can assess a set of combined ECMs to ensure that each ECM savings impacts are not double-counted. The ECMs compete for the same market segment based on their cost-effectiveness measured by the simple payback period, internal rate of return, cost of conserved energy (CCE), or cost of conserved carbon Typically, ECMs with lower incremental capital and operational costs (i.e., lower life-cycle costs) are assumed to capture larger portions of the market segments (DOE, 2017).

9.5 IMPACT OF ENERGY EFFICIENCY ON OCCUPANT PRODUCTIVITY

The effects of indoor environmental quality (IEQ) on the performance of occupants have been widely investigated. In particular, the impacts of thermal comfort, visual comfort, noise level, and air quality have been evaluated through a series of experimental analyses and surveys. A wide range of sometimes inconsistent results has been reported depending on the tested subjects, performance tasks, and indoor conditions (Al Horr et al., 2016; Clements-Croome, 2015; Tarantini et al., 2017). However, most of the findings concur that improved IEQ can optimize worker productivity, reduce absenteeism and lower staff turnover.

The economic advantages of maintaining optimal IEQ conditions can be significant. For instance, in office buildings, the salary cost of staff can exceed energy costs by 100 times especially in developed countries. Thus, a mere 1% improvement in work productivity can cover the energy cost of operating the building. Some studies have compared the value of the work productivity to the energy and maintenance costs as summarized in Table 9.20. As indicated in Table 9.20, the ratio of the staff to energy costs ranges from 87 to 218 for all the office buildings evaluated by various studies indicating that only 0.5%–1.2% work productivity improvement can cover the entire energy costs. The equivalent time required to cover the energy costs represents only 2–5 min/day per person for the office buildings listed in Table 9.20.

TABLE 9.20 Comparison of Staff and Energy Costs for Select Office Buildings

Building location	US	US	US	US	Japan
Reference	Rosenfeld	Lorsch and Abdou (1994)	EPA	BOMA	Nishihara and Tanabe (2010)
Staff cost ($\$ m^{-2} yr^{-1}$)	3000	2180	2000	1300	370
Energy cost ($\$ m^{-2} yr^{-1}$)	15	15–20	20	15	36
Staff to energy cost ratio	200	114–218	100	87	103
Productivity Improvement required to cover energy costs (%)	0.5	0.4–0.9	0.9	1.2	1.0
Daily working time to cover energy costs (min/day/person)	2.25	2.00–3.75	4.33	5.0	4.67

Some studies have attempted to assess the contribution of various environmental factors in occupant's perception and performance. For instance, Kawamura et al. (2007) has conducted a series of surveys using laboratory-controllable experimental testing and found by order of significance that improving thermal comfort, acoustic conditions, and illuminance levels affect occupants' perception. Through a questionnaire targeting workers in Hong Kong office buildings, Wong et al. (2008) ranked thermal comfort as the most important in worker's perception of indoor environment quality followed by indoor air quality, noise conditions, and lighting levels. Other studies have confirmed that thermal comfort and air quality are the most important indoor environmental factors that can affect the well-being and performance of occupants in buildings (Seppänen and Fisk, 2006; Seppänen et al., 2005; Wargocki, 2007; Wargocki and Wyon, 2017).

Specific results and models related to thermal comfort and air quality are discussed in the following sections.

9.5.1 Impact of Indoor Thermal Comfort

The effects of indoor thermal conditions on occupant's performance and productivity have been evaluated by several studies. However, a wide range of indicators has been considered to measure the thermal environment and performance level. Indicators of indoor thermal conditions include air temperature, relative humidity, operative temperature, thermal sensation, and predicted mean value (*PMV*). Productivity indicators considered in the experimental testing range from learning tasks in schools to thinking and typing speeds in offices. It is difficult to compare the results from various studies due to the large variance in measuring the indoor environment conditions and productivity metrics.

In general, it is well established that extreme indoor temperatures (i.e., too cold or tool hot) result in loss of performance. In particular, Wyon and Wargocki (2005) found that extreme air temperatures beyond specific ranges lower mental concentration, reduce manual dexterity, and affect the thinking and learning performance of occupants. Depending on the experiment and subjects considered in the tests and surveys, various ranges of temperatures and *PMV* values are recommended for indoor thermal conditions to optimize performance and productivity. For instance, Kosonen and Tan (2004) found that optimal productivity occurs at a temperature of 24°C and a *PMV* of 0.21. Seppänen et al. (2006) recommended 22°C as the optimal temperature. Initially, Tham (2004) indicated that a worker's performance is maximized at an indoor air temperature of 24.5°C for a call center in Singapore but later concluded through a further simulation analysis that air temperature around 20°C can improve mental performance in activities requiring mental attention (Tham and Willem, 2010). Tanabe et al. (2007) reported an increase of fatigue and a decrease in performance at an operative temperature of 33°C for college-age subjects.

Several authors have proposed models to predict the productivity loss, *PL*, due to thermal conditions. One of the simplified model is that proposed by Seppänen et al. (2005) correlating *PL* expressed in percentage to only the indoor air temperature, T_a, expressed in °C as noted in Eq. (9.1):

$$PL = 2T_a - 50 \tag{9.1}$$

Note that the Sappenen et al. model considers the impact of high indoor temperature (above 25°C and up to 33°C) on the loss of productivity. Fig. 9.10 illustrates the results of various experimental studies indicating the loss of productivity as a function of indoor air temperature relative to the simplified Sappenen et al. model. Generally, the simplified model predictions are consistent with the results obtained from other studies when the indoor temperatures exceed 25°C.

Seppänen and Fisk (2006) has proposed a more elaborate correlation between *PL* and T_a as indicated by Eq. (9.2):

$$PL = 100 * (1 + 0.468533 - 0.164752\ T_a + 0.005827\ T_a^2 - 0.0000623\ T_a^3 \tag{9.2}$$

Fig. 9.11 illustrates the variation of PL as a function of T_a for the applicable range of the correlation provided by Eq. (9.2). It is clear that when the indoor air temperature is 19–25°C, the loss of productivity remains small and is below 2%. However, when the air temperature exceeds 25°C, *PL* increases significantly.

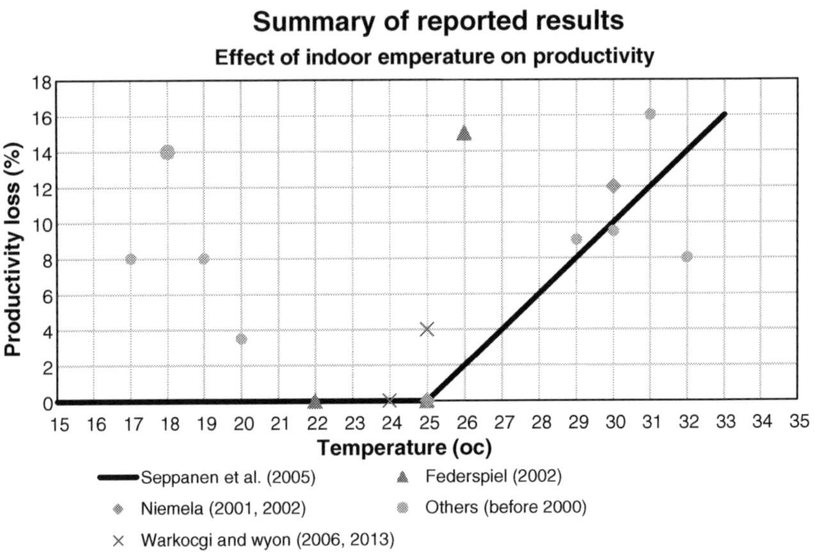

FIGURE 9.10 Impact of indoor air temperature on the productivity loss.

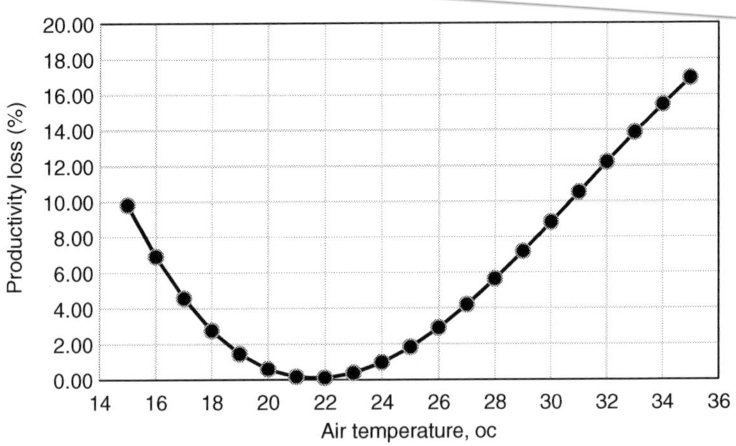

FIGURE 9.11 Impact of indoor air temperature on the productivity loss using Eq. (9.2).

For instance, when the indoor air temperature is 30°C, *PL* becomes close to 9% using Eq. (9.2) and 10% using Eq. (9.1).

Other models consider the effects of other indoor thermal comfort factors on the occupant performance such as relative humidity, operative temperature, and thermal sensation (Jensen et al., 2009; Lan et al., 2011; Loveday et al., 1995). Models correlating *PMV* to *PL* have also been proposed. In particular,

Roelofsen (2001) suggested a polynomial correlation between *PL* and PMV for office workers as indicated in Eq. (9.3):

$$PL = \left|1.2802 + 15.996\ PMV + 31.507\ PMV^2 + 11.755\ PMV^3 + 1.474\ PMV^4\right| \quad for\ PMV < 0 \tag{9.3}$$

Fig. 9.12 illustrates the variation of the productivity loss as a function of the thermal comfort indicator (*PMV*). As noted in Fig. 9.12, the worst is thermal comfort relative to the neural conditions (i.e., $PMV = 0$), the higher is productivity loss.

Kosonen and Tan (2004) suggested models of *PL* as a function of *PMV* depending on office activity type. In particular, they proposed models for typing and thinking activities as indicated by Eqs. (9.4) and (9.5), respectively:

For typing

$$\begin{aligned} PL &= -0.154 + 3.882\ PMV + 25.176\ PMV^2 - 26.641\ PMV^3 \\ &\quad + 13.110\ PMV^4 - 3.130\ PMV^5 + 0.293\ PMV^6 \quad for\ PMV > 0 \\ PL &= 4.999 + 32.123\ PMV + 50.240\ PMV^2 - 8.712\ PMV^3 \\ &\quad - 183.750\ PMV^4 + 198.410\ PMV^5 - 60.543\ PMV^6 \\ &\quad for - 0.21 < PMV < 1.25 \end{aligned} \tag{9.4}$$

For thinking

$$\begin{aligned} PL &= 1.876 + 13.389\ PMV + 19.226\ PMV^2 - 10.401\ PMV^3 \\ &\quad - 1.553\ PMV^4 + 1.593\ PMV^5 \quad for - 0.21 < PMV < 1.25 \end{aligned} \tag{9.5}$$

Fig. 9.13 illustrates the variation of the productivity loss as a function of *PMV* using Eqs. (9.4) and (9.5). It is clear that comfort level affects typing more than thinking activities in office settings. Indeed, a discomfort level character-

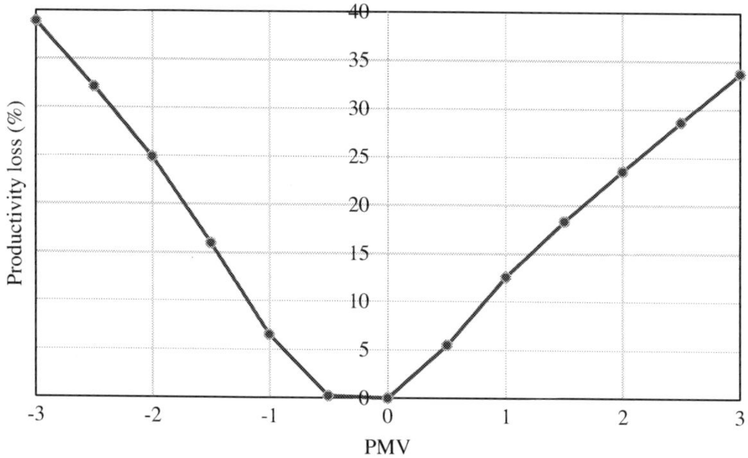

FIGURE 9.12 Variation of productivity loss vs. predicted mean value (*PMV*) using Eq. (9.3).

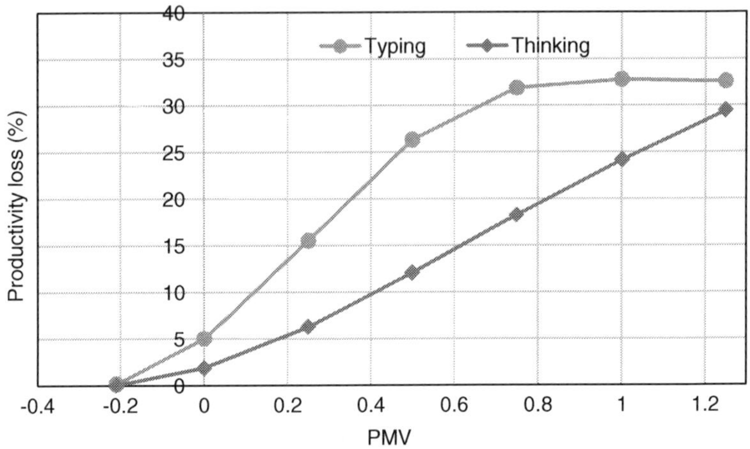

FIGURE 9.13 **Variation of productivity loss vs. *PMV* using Eqs. (9.4) and (9.5).**

ized by a *PMV* of 1.0 results in a productivity loss of 24% in thinking-based activities and of 32% in typing tasks.

9.5.2 Impact of Indoor Air Quality

In addition to the thermal conditions, the air quality of the indoor environment has been found to have a significant impact on occupant's well-being and performance. In particular, adequate ventilation of the indoor environment can have significant impacts on health, well-being, and ultimately productivity within buildings. For instance, Wargocki et al. (2000) found an improvement of 1.7% in typing and in creative thinking at a ventilation rate of 10 L/s per person compared to 3 L/s per person. Based on a study of eight primary schools in the UK, Bakó-Biró et al. (2008, 2012) have reported that children's concentration is affected by poor ventilation (1 L/s/person) with CO_2 levels reaching up to 5000 ppm. They recommended that adequate ventilation rates of 8 L/s per person be provided so that CO_2 level remains below 1000 ppm. Using a series of experimental test subjects under controllable conditions for an office space, Satish et al. (2012) indicated that CO_2 can affect occupants' decision-making and thus their performance at levels as low as 600 ppm, well below the commonly accepted level of 1000 ppm.

Fisk et al. (2012) compared the benefits and the costs of various outdoor air ventilation rates in the US offices and their effect on sick building syndrome symptoms, work performance, short-term absence, and building energy consumption. In particular, they reported that the additional productivity benefits exceed significantly the additional ventilation energy costs. The estimated economic annual benefits of $13 billion and $38 billion can be achieved by increasing minimum ventilation rates, respectively, from 8 to 10 L/s per person and from 8 to 15 L/s per person. Roelofsen (2001) has indicated based on surveys of over 7000 office workers in the Netherland that 2.5 days/yr representing 25% of total average absenteeism are directly attributed to inadequate indoor conditions. This represented a quarter of the total average absenteeism.

Fanger (1989) has proposed a simplified model to estimate the ventilation required to maintain acceptable indoor air quality. The model considers that the level of dissatisfaction with indoor air quality can be attributed to an equivalent number of occupants as the actual pollutant sources. In particular, the Fanger model provides the required ventilation in L/s, Q_V, as a function of the total pollutant generated indoors, G_i, the ventilation effectiveness, ε_V, as well as the perceived air quality levels indoors, C_i, and outdoors, C_o. Both C_i and C_o are expressed in decipols while G_i is expressed in olf.

$$Q_V = 10 \bullet \frac{G_i}{C_i - C_o} \bullet \frac{1}{\varepsilon_V} \tag{9.6}$$

The perceived air quality level expressed in decipols can be determined using the percent of dissatisfied PD:

$$C = 112 \bullet \left[Ln(PD) - 5.98 \right]^{-4} \tag{9.7}$$

The loss in productivity for occupants has been correlated to the percent dissatisfied with indoor air quality for office buildings as noted in Eq. (9.8) indicating that a 10% increase in PD leads to a 1.1% loss of productivity (Wargocki et al., 2000):

$$PL = 0.11\ PD \tag{9.8}$$

Fig. 9.14 illustrates the variation of C as a function of PD. Fig. 9.15 shows the variation of PL as a function of PD. It is clear as the PD increases, the perceived air quality as well as the loss of productivity increase.

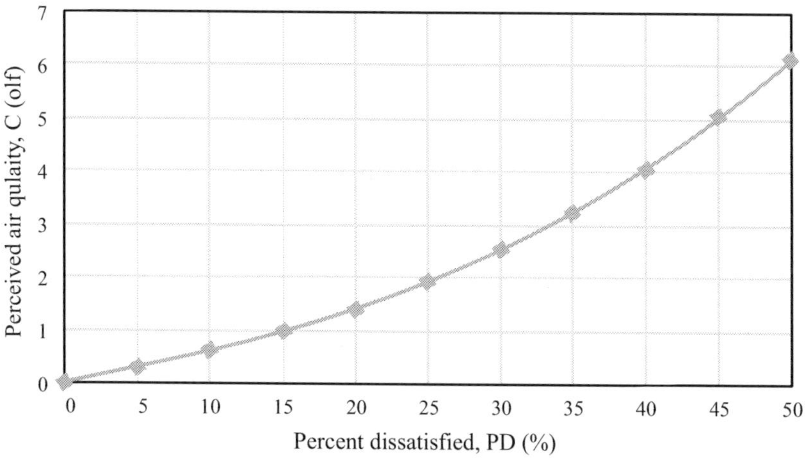

FIGURE 9.14 Variation of perceived air quality level as a function of percent dissatisfied using Eq. (9.7).

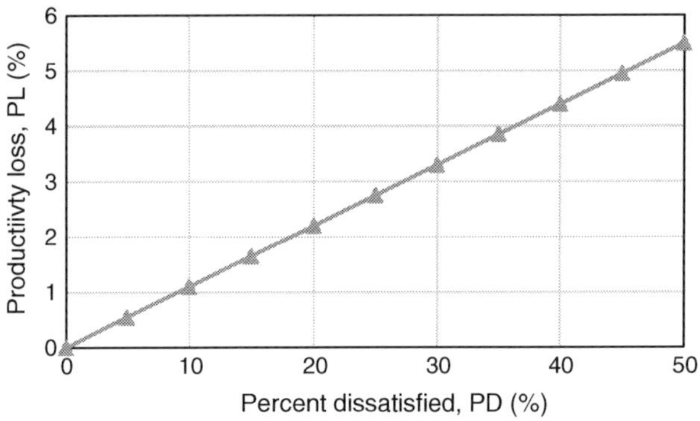

FIGURE 9.15 **Variation of productivity loss as a function of percent dissatisfied using Eq. (9.8).**

9.6 LARGE-SCALE EE PROGRAMS FOR RESIDENTIAL BUILDINGS

In this section, large-scale analysis using the concept of energy productivity (refer to Chapter 5), energy building stock (refer to Section 9.2), as well the impact of work productivity (refer to Section 9.5) to evaluate the benefits of energy efficiency and renewable energy programs for a set of countries. Due to the lack of significant energy efficiency policies for the building sector, the GCC countries are considered in this section based on a study by Krarti and Dubey (2017).

9.6.1 Housing Energy Consumption in GCC Region

In the last decade, the total final energy consumption for households in the GCC has increased significantly outpacing the increase in the population. Indeed, household energy use per capita has increased in all GCC countries except Kuwait during the 2005–2015 period as illustrated in Fig. 9.16. For instance, Qatar had an increase of 25% between 2005 and 2015 in the per capita energy consumption for residential buildings. Oman has the lowest energy use per capita among the GCC countries with 0.3 TOE/person. During the same period, the UK and to a lesser extent the US have seen their energy use per capita decrease for the residential building sector. However, the US remains to had the highest household energy use per capita of 0.8 TOE/person in 2015, almost double than the values reported for the UK and UAE.

Households in the GCC utilize almost exclusively electricity as final energy source even for domestic hot water and rarely needed space heating. Fig. 9.17 compares per capita electricity use for households among GCC countries and some other developed and developing countries. It is clear that residential electricity consumption per capita is much higher in the GCC region than in most other countries.

In the context of the building sector, this paper examines the potential benefits of energy efficiency retrofit programs in the GCC that reduce energy demand

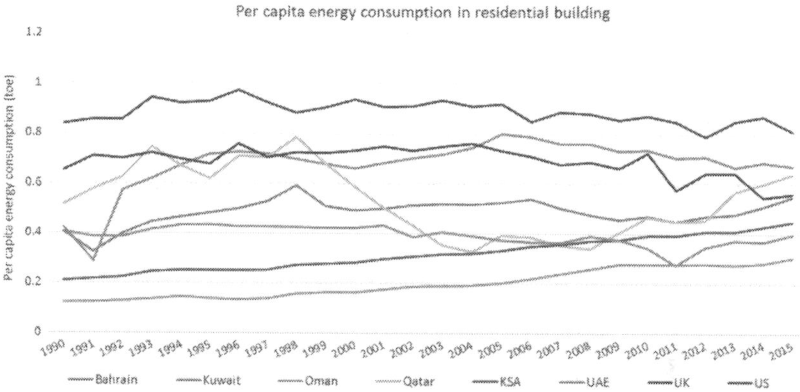

FIGURE 9.16 **Variation of annual residential building energy use per capita for Gulf Cooperation Council (GCC) countries, the US, and the UK between 1990 and 2015.**

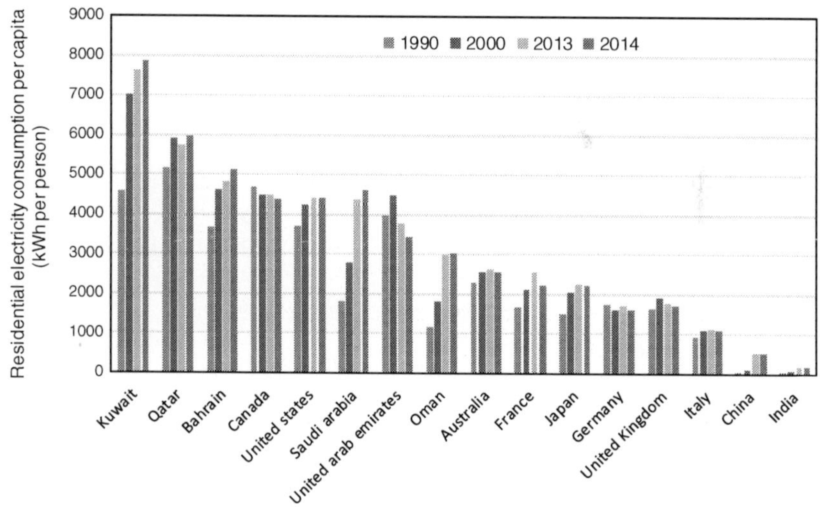

FIGURE 9.17 **Electricity consumption per capita for residential buildings in GCC and G-7 countries.**

of the existing residential building stock. In particular, the analysis presented in this paper incorporates findings from a growing body of work that examines the value, whether economic, environmental, or social, associated with improved services of energy efficient buildings. This paper demonstrates that an analysis based on energy productivity can be used to assess investment alternatives and assist decision-making for energy policies.

First, an overview is provided to discuss the cost-effectiveness of individual energy efficiency strategies in reducing energy consumption for households using the concept of cost of avoided or conserved energy use. Then, the potential benefits of large-scale energy retrofit programs are evaluated for the existing GCC housing stock by applying the energy productivity analysis approach. The energy productivity analysis is also applied to assess the benefits of improving indoor quality environment that result in healthier occupants and lower absenteeism and ultimately higher work productivity. The impact on energy productivity of increased property asset value of buildings with improved energy efficiency is also quantified. Finally, a comparative analysis of renewable and energy efficiency programs applied to GCC households is presented using the energy productivity framework.

9.6.2 Impact of Individual Energy Efficiency Strategies

Using the limited available data for the total floor area of residential buildings compiled for 2010, the energy use intensity (EUI) for KSA households has been estimated and compared to the energy use per capita for select European countries and the US as shown in Fig. 9.18 (Moura et al., 2015). The general trend is that countries with cold climates consume more energy per unit floor area and per capita than countries with warm climates including KSA due most likely to higher space heating needs. Moreover, the results of Fig. 9.18 indicate that energy use per capita can be a good proxy for energy use intensity and thus energy efficiency of households at least for countries considered in the analysis of Fig. 9.18. The analysis of the energy mix used in the residential buildings, as illustrated in Fig. 9.19A, confirms that electricity is widely used in countries with warm climates (i.e., KSA and Spain) while fossil fuels including processed heat

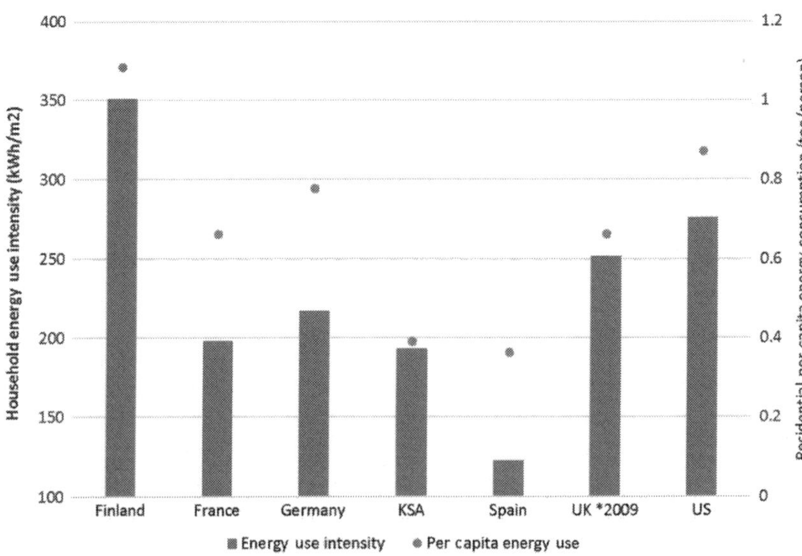

FIGURE 9.18 Energy consumption per floor area and per capita for residential buildings during 2010.

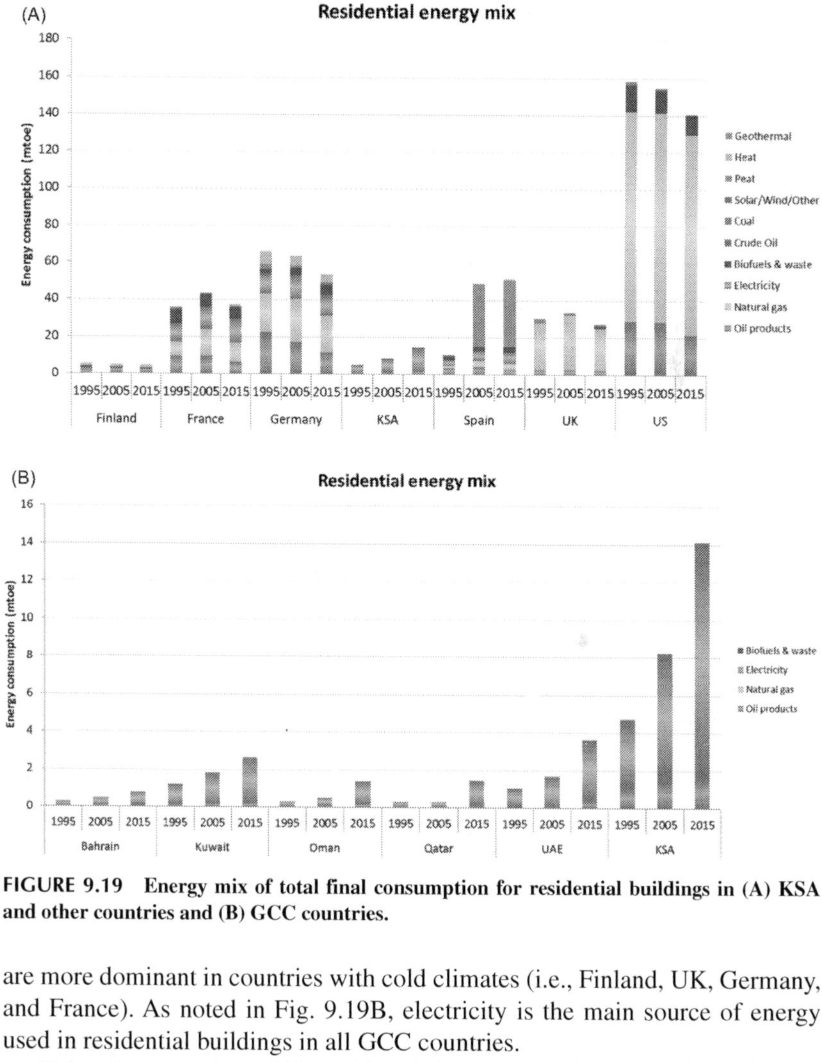

FIGURE 9.19 Energy mix of total final consumption for residential buildings in (A) KSA and other countries and (B) GCC countries.

are more dominant in countries with cold climates (i.e., Finland, UK, Germany, and France). As noted in Fig. 9.19B, electricity is the main source of energy used in residential buildings in all GCC countries.

Using the same data of Fig. 9.18 and the concept of energy affordability, as an indicator of energy productivity for buildings outlined in Chapter 4, Fig. 9.20 indicates that an energy block of 10,000 kWh/yr can serve more household floor area in warm climates that in cold climates. As discussed in Chapter 4, improvement in energy efficiency boosts household energy affordability and thus energy productivity resulting in more floor area to be served by the same energy block. Fig. 9.21 shows specific variations of EUI and energy affordability (EA) for KSA households due to the application of proven energy efficiency strategies. As expected, installing an energy efficient air conditioning system has the most impact among the measures considered in the analysis summarized in Fig. 9.21 with the potential to reduce EUI and increase EA by 28%.

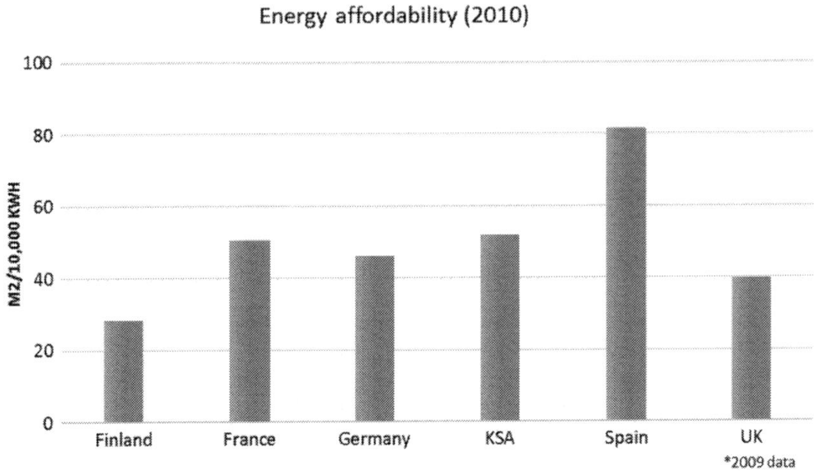

FIGURE 9.20 Floor area served by 100,000 kWh/yr for residential buildings during 2010.

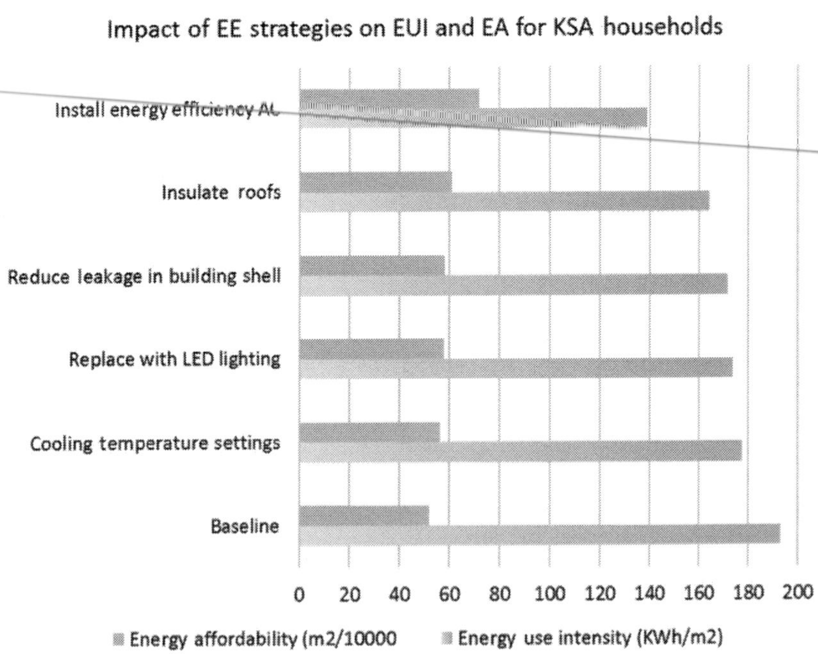

FIGURE 9.21 Effect of energy efficiency measures on energy use intensity and energy affordability for households in KSA.

The cost-effectiveness of the energy efficiency measures depends on their cost of implementation as well as the electricity rate charged to homeowners. Table 9.21 lists the cost for avoiding one 1 kWh/yr of electricity consumption as well as the simple payback periods for the energy efficiency measures under two scenarios for KSA electricity rates (Krarti et al., 2017):

TABLE 9.21 Cost of Avoided Energy Consumption and Period Pack of Energy Efficiency Measures for KSA Households for Two Electricity Price Scenarios

Energy efficiency strategy for households	Cost of avoided energy use ($/kWh)	Payback ($0.0479/kWh)	Payback ($0.1678/kWh)
Cooling temperature settings	0.00	0.00	0.00
Replace with LED lighting	0.0513	1.07	0.31
Reduce leakage in building shell	0.1220	2.55	0.73
Insulate roofs	0.4578	9.56	2.73
Install energy efficiency AC	0.2480	5.18	1.48

1. The cost of electricity at $0.0479/kWh (i.e., electricity cost of production in KSA).
2. Cost of electricity at $0.1678/kWh (i.e., opportunity electricity prices).

It should be noted that the current KSA electricity rate follows a tiered structure depending on the block of energy consumed monthly by the households as discussed by Krarti et al. (2017). Currently, KSA is undergoing an energy price reform. Within a few years, it is expected that electricity price would be closer to the second scenario than the first scenario, which relies on heavy energy subsidies from the KSA government. Typically, efficiency measures are recommended when their simple payback period is less than 5 years (Krarti, 2012). Based on the results outlined in Table 9.4, it is clear that under electricity price scenario 1 (i.e., $0.0479/kWh), the installation of high-efficiency air conditioning systems and the addition of thermal insulation in the roofs are not cost-effective. However, under electricity rate scenario 2 (i.e., $0.1678/kWh), all the listed energy efficiency measures listed in Table 9.21 become cost-effective and are highly recommended to be implemented by KSA households. In fact, when the average electricity price is set above the threshold rate of $0.092/kWh, all the measures become cost-effective for KSA households.

9.6.3 Analysis of Retrofit Programs for Existing Buildings

To assess the potential for the existing residential building stock to contribute to energy productivity in the GCC, three levels of energy efficiency retrofits (Krarti et al., 2017) are considered in this study:

- *Level 1* applies low-cost EEMs such as installing a programmable thermostat, use of LED lighting and weatherization of building shell to reduce air infiltration.
- *Level 2* improves the building envelope components to meet any energy efficiency code requirements for new buildings, including use of energy efficient cooling systems and appliances.

- *Level 3* applies a wide range of EEMs, including replacing windows and/or cooling systems, using variable speed drives and installing daylighting control systems. To minimize implementation costs, this type of energy retrofit is typically linked with architectural refits.

Based on a series of detailed analyses, the economic, environmental, and social benefits of the retrofit programs have been quantified for the existing housing stock of all the GCC countries (Krarti, 2015; Krarti and Dubey, 2017; Krarti et al., 2017). Moreover, the analysis estimated the capital investments needed to implement the retrofit programs. Specifically, the analysis evaluated the following benefits:

- Annual energy use savings and associated reduction in fuel use to generate electricity.
- Reduction in electricity peak demand representing avoided future demand in electrical power plants associated with growth in population and building stock.
- Reduction in CO_2 emissions associated with lowering annual fuel use for generating electricity.
- Annual creation of jobs to implement the energy retrofit programs. A 10-year implementation period is considered to retrofit the existing residential building stock.

Fig. 9.22 summarizes the results specific to the four benefits listed above for all six GCC countries and three levels of energy retrofit programs applied to the existing residential building stock. Due to its large population and building size

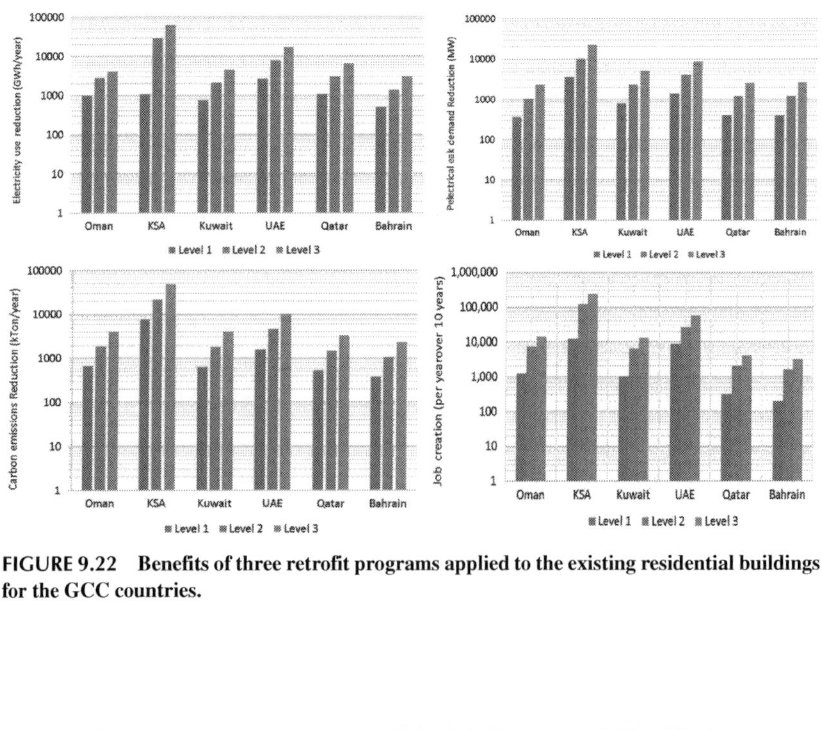

FIGURE 9.22 Benefits of three retrofit programs applied to the existing residential buildings for the GCC countries.

compared to other GCC countries, KSA has the highest potential in reducing energy use, peak demand, and carbon emissions as well as in creating jobs.

Table 9.22 summarizes results for each of the GCC countries, reflecting retrofit of the existing residential building stock. The value added of the benefits outlined in Table 9.22 can be quantified by estimating avoided oil consumption to generate electricity as well as reduced future needs to build power plants. In the value-added estimation, the following assumptions are made:

- oil prices of $45/BOE paired against oil production costs in each GCC country, as summarized in Table 9.23 (Knoema, 2016),
- electricity power plant building cost of $1700/kW (ECRA, 2015), and
- average power plant efficiency as noted in Table 9.23 (IEA, 2016).

As expect, the magnitude of the benefits increases with the level of retrofit program and with the size of the economy. For the entire GCC region, the highest impacts are achieved by level 3 retrofit program with potential reductions on annual electricity use of 97,830 GWh/yr, peak electrical demand of 23,673 MW, and annual carbon emissions of 71.123 metric tons/yr.

Using the results provided in Table 9.22 and the data listed in Table 9.23, the value added arising from various retrofit programs can be estimated for all GCC countries (Krarti and Dubey, 2017; Krarti et al., 2017). In particular, Table 9.24 provides the required investments, potential primary energy savings, and economic value released to the national governments due the conserved oil consumption from large-scale energy retrofit programs applied to the existing housing stock in the GCC region. For the entire GCC region, level 3 retrofits of the existing residential buildings require over $60 billion/yr of investments and would result in annual savings of 170.5 million BOE in national oil consumption that result in the availability of about $6.7 billion/yr in expenses related to generating electricity. The cost-effectiveness of the retrofit programs can be evaluated from the perspective of both national governments (if they provide the capital required for implementing these programs) and households (if they decide to invest in retrofitting their housing units). Table 9.25 provides the simple payback periods when only the electricity savings are considered and when all the value added from all the benefits listed in Table 9.2 are accounted for. As part of a preliminary analysis, the governments can utilize the simple payback periods to assess the cost-effectiveness of the energy efficiency programs. Moreover, Table 9.25 presents the CCE expressed per unit of saved electricity. Households can compare the CCE value to the average electricity price that households have to pay as part of their utility bills. It is clear that a level 1 retrofit in all GCC countries is highly cost-effective for the national governments and households. However, level 2 and level 3 are cost-effective for the national governments but not for households except in UAE and Qatar. When the energy price reforms are executed, it is expected that electricity prices in all GCC countries would be higher than any CCE value listed in Table 9.25 and thus all the retrofit programs would be cost-effective for households.

TABLE 9.22 Benefits From Energy Retrofit Programs Applied to Existing GCC Residential Buildings

Country	Energy retrofit level for existing building stock								
	Level 1			Level 2			Level 3		
	Reduced electrical power generation capacity (MW)	Electrical energy savings (GWh/yr)	Annual carbon emissions reduction (Mton/yr)	Reduced electrical power generation capacity (MW)	Electrical energy savings (GWh/yr)	Annual carbon emissions reduction (Mton/yr)	Reduced electrical power generation capacity (MW)	Electrical energy savings (GWh/yr)	Annual carbon emissions reduction (Mton/yr)
Bahrain	116	495	0.377	335	1422	1.083	727	3091	2.053
Kuwait	422	738	0.642	1213	2123	1.847	2637	4614	4.014
Qatar	248	1066	0.526	714	3064	1.513	1553	6660	3.290
Oman	214	957	0.660	616	2751	1.898	1339	3980	4.126
KSA	2290	10,054	7.611	6583	28,906	21.882	14,312	62,839	47.569
UAE	497	2663	1.611	1428	7657	4.633	3105	16,646	10.071
GCC	3787	15,973	11.427	10,889	45,923	32.856	23,673	97,830	71.123

GCC, Gulf Cooperation Council.

TABLE 9.23 Summary of Oil Production Costs and Thermal Efficiency of Power Plants in GCC

GCC countries	Oil production costs ($/BOE)[a]	Average electrical power plant efficiency[b] (%)	Building sector value added for 2013[c] ($billion/yr)	Building sector final energy consumption for 2013[b] (MTOE/yr)
Bahrain	8.4	27	1.83	1.06
Kuwait	4.4	34	4.50	3.61
Oman	5.3	36	4.10	2.32
Qatar	6.8	41	8.54	1.86
KSA	5.0	32	47.23	18.48
UAE	5.9	34	32.20	7.39

GCC, Gulf Cooperation Council.
[a]*From Knoema (2016).*
[b]*From IEA (2016).*
[c]*From IMF (2016).*

Energy productivity gains in the building sector resulting from large-scale energy retrofit programs in all GCC countries are shown when targeted for the residential building stock as illustrated in Fig. 9.23. The results of Fig. 9.23 for the energy productivity are normalized relative to the baseline of business-as-usual (i.e., no energy retrofit is implemented). Moreover, the data of Table 9.24 and oil price of $45/BOE as well as carbon emissions of $10/ton are used in the energy productivity-based analysis (Russell et al., 2015; Skumatz, 2014). In addition, data from 2013 for building sector value added and total final energy consumption, listed in Table 9.24, are considered to estimate the baseline energy productivity indicators since the energy efficiency analysis is carried out based on 2013 data (Krarti et al., 2017).

All three levels of retrofit for the residential building stock show some positive impact on energy productivity in each GCC country (see Fig. 9.23). As summarized in Tables 9.22 and 9.23, the benefits and impacts are evaluated and quantified for the three retrofit programs when different benefits are considered in the analysis (Krarti, 2015; Krarti and Dubey, 2017; Krarti et al., 2017). When accounting for avoided energy consumption only, in which lower domestic consumption holds the potential to make available to the government more oil that can be exported at a higher value, the energy productivity can be increased slightly (typically 5%) by applying a level 1 retrofit program to the entire building stock.

Level 3 retrofit programs, applied across all existing buildings, provide the best option to increase the building sector's energy productivity for all GCC countries (refer to Fig. 9.23). In the case of KSA, retrofitting the residential

TABLE 9.24 Required Investments, Potential Primary Energy Savings, and Economic Value Released to the GCC Governments From Large-Scale Energy Retrofit Programs for Existing Housing Stock

	Energy retrofit level for existing building stock								
	Level 1			Level 2			Level 3		
Country	Capital required ($ Billion)	Potential primary energy savings (MBOE/yr)	Economic value released to gov't ($ Billion/yr)	Capital required ($ Billion)	Potential primary energy savings (MBOE/yr)	Economic value released to gov't ($ Billion/yr	Capital required ($ Billion)	Potential primary energy savings (MBOE/yr)	Economic value released to gov't ($ Billion/yr
Bahrain	0.103	1.047	0.038	1.027	3.007	0.110	2.053	6.536	0.239
Kuwait	0.195	1.239	0.050	1.951	3.565	0.145	3.902	7.748	0.315
Qatar	0.071	1.691	0.067	0.707	4.859	0.193	1.413	10.562	0.419
Oman	0.174	1.333	0.051	1.739	3.831	0.146	3.477	5.542	0.212
KSA	2.836	17.938	0.718	28.365	51.573	2.063	47.569	112.116	4.485
UAE	0.126	4.472	0.175	1.255	12.358	0.503	2.511	27.952	1.093
GCC	4	27.719	1.099	35	79.693	3.160	61	170.457	6.762

GCC, Gulf Cooperation Council.

TABLE 9.25 Cost-Effectiveness Analysis of Large-Scale Energy Retrofit Programs for GCC Existing Housing Stock

Country	Energy retrofit level for existing building stock								
	Level 1			Level 2			Level 3		
	Simple payback period Only electrical energy savings (years)	Simple payback period All benefits (years)	Cost of conserved electricity ($/kWh)	Simple payback period Only electrical energy savings (years)	Simple payback period All benefits (years)	Cost of conserved electricity ($/kWh)	Simple payback period Only electrical energy savings (years)	Simple payback period All benefits (years)	Cost of conserved electricity ($/kWh)
Bahrain	2.7	0.0	0.014	11.0	5.3	0.055	8.6	3.1	0.043
Kuwait	3.9	0.0	0.017	13.5	0.0	0.060	12.4	0.0	0.055
Qatar	1.1	0.0	0.004	3.7	0.0	0.015	3.4	0.0	0.014
Oman	3.4	0.0	0.012	11.9	4.2	0.041	16.4	4.7	0.057
KSA	4.0	0.0	0.018	13.7	7.5	0.064	10.6	4.7	0.049
UAE	0.7	0.0	0.003	2.5	0.0	0.011	2.3	0.0	0.010
GCC	3.2	0.0	0.014	11.1	0.0	0.050	9.0	0.0	0.041

GCC, Gulf Cooperation Council.

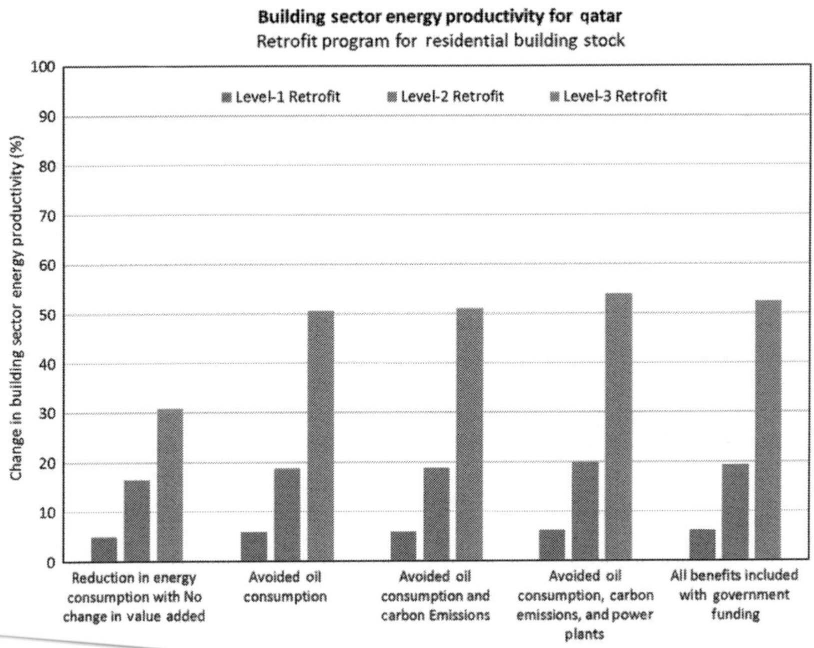

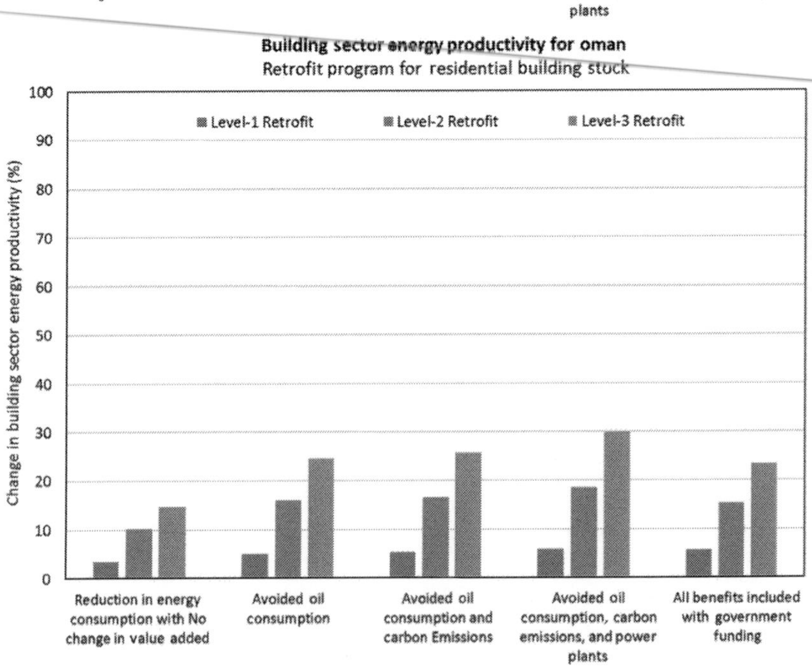

FIGURE 9.23 **Impact of energy retrofit programs for all existing buildings in GCC countries.**

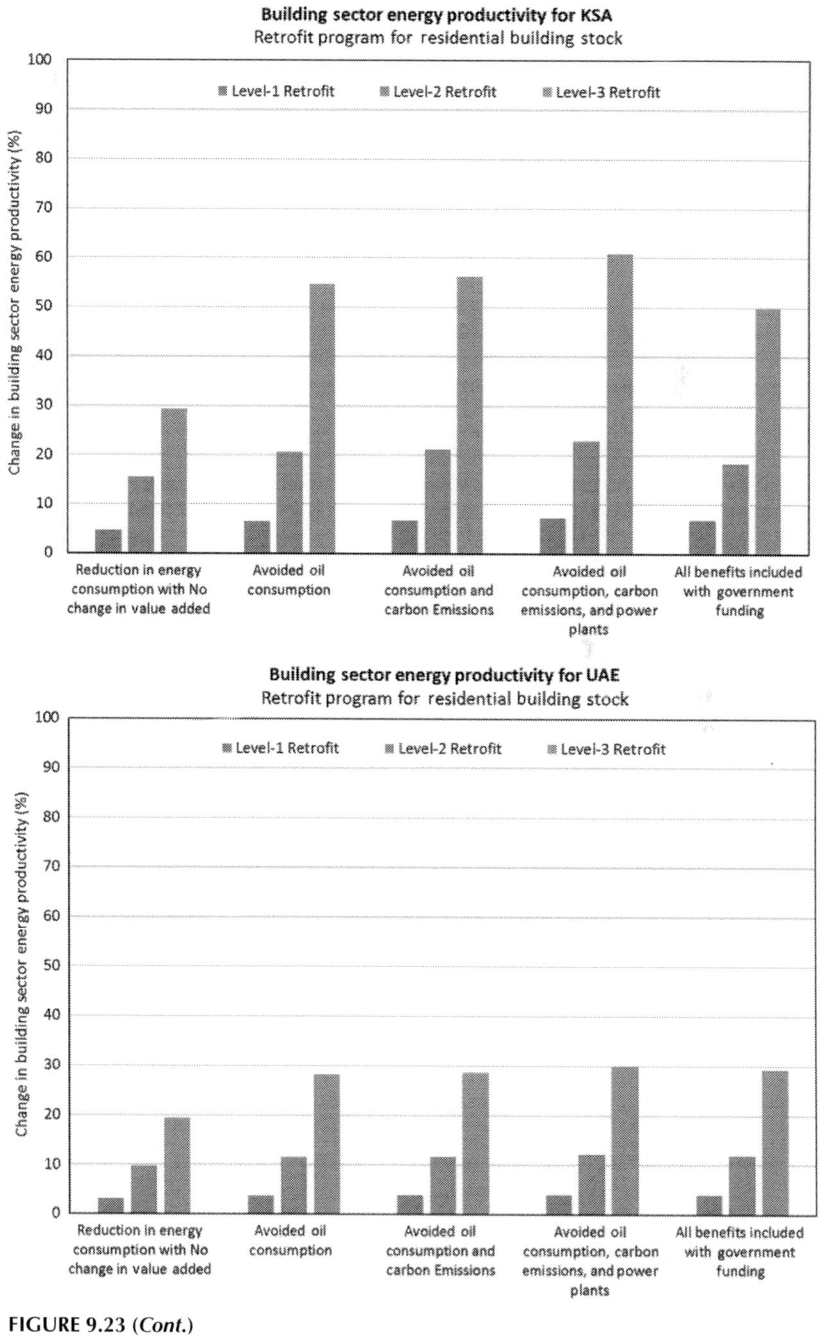

FIGURE 9.23 (*Cont.*)

building stock can increase energy productivity in the building sector by 50% when the government finances the entire retrofit programs. The increase is 60% when the private sector or the households pay for the retrofits.

9.6.4 Estimation of Nonenergy Benefits

Two types of nonenergy benefits (NEBs) of high-performance buildings are outlined in this section. The NEBs are mainly associated with enhancing work productivity and increase in property asset values. The work productivity is due to improvements of thermal comfort and indoor air quality. Higher asset values are related to the increase in rental and sale values.

9.6.4.1 Impact of Work Productivity

Improving the energy efficiency of buildings can provide higher quality services such as comfortable and healthy buildings and ultimately improve the work productivity of occupants. Indeed, a number of studies have linked energy efficient buildings to improvements in work productivity through reduced absenteeism and improved retention (DOE, 2013a). For instance, it is estimated that improved ventilation can increase work productivity by 11% and better lighting design can increase workforce economic output by 23% (WGBC, 2013). Moreover, comfortable and healthy buildings reduce absenteeism from schools and work places. Several energy retrofit projects can improve indoor air quality and health conditions for occupants. It is estimated that there is a 15% reduction in days off schools among children living in homes that benefitted from energy efficiency upgrades. A similar absenteeism rate reduction from the workplace has been noted for adults (EIA, 2014). These benefits have been monetized using daily costs of hiring caregivers for sick children and loss of wage earnings for employees. The net present value for the health benefits of the retrofit projects is found to increase by 11.5% (IEA, 2014).

Moreover, improvement in energy efficiency can directly affect economic growth. Based on statistical analysis of historical data obtained for 28 OECD countries (Vivid Economics, 2013), it is found that a 1% improvement in energy efficiency leads to 0.1% increase in the annual growth rate of GDP per capita. Other studies have found similar findings. For instance, doubling the US energy productivity is estimated to result in $25 billion or 2% in real GDP gains by 2030 or an annual growth rate of about 0.13% (Rhodium Group, 2013).

In this section, some NEBs associated with energy efficiency retrofit programs applied to residential buildings in the GCC countries are quantified. Specifically, improvements in indoor built environment and their impacts on enhancing work productivity and economic output are considered in the energy productivity analysis of retrofitting the housing stock. For this analysis, the energy efficiency retrofit is set to affect only the economic output of the country's service sector since indoor environmental quality of households would mostly affect workers in commercial and public services attributed mainly to improved

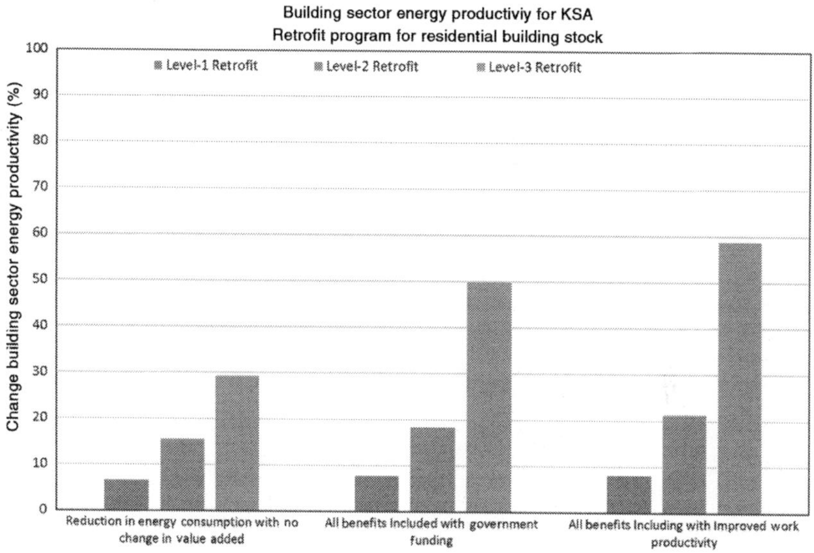

FIGURE 9.24 Impact of improved work productivity on the KSA building sector's energy productivity.

health and lower absenteeism. Using findings reported in the literature, it is assumed that 1% reduction in building energy consumption would lead to 0.1% in the value added for the service sector to account for higher work productivity associated with better indoor built environment resulting from the energy efficiency retrofit programs. Fig. 9.24 summarizes the analysis results obtained for KSA households and three retrofit levels. Similar findings are obtained for other GCC countries. In particular, Fig. 9.24 indicates that when the benefits of improved work productivity are accounted for in the analysis, the value added for the retrofit programs increases resulting in higher energy productivity of the building sector. Specifically, the improved work productivity increases by household's energy productivity relative to the case when only the quantifiable benefits are considered (i.e., reductions in energy use, peak demand, and carbon emissions) by 1%, 2%, and 5% for, respectively, level 1, level 2, and level 3 retrofit programs applied to the existing residential building stock.

9.6.4.2 *Impact of Real Estate Asset Value*

A number of studies have indicated that improvements in energy efficiency can increase the asset value of buildings through higher rentals and sales. Indeed, energy efficient buildings have higher sale and rental premiums due to higher performance and perceived benefits including lower operating costs and higher indoor environmental quality. In fact, higher sale and rental premiums of energy efficient buildings relative to the overall market are increasingly stimulating investments in sustainable design and energy retrofit of buildings (DOE, 2013b).

Few studies mostly in the US and UK have quantified specific property value increases due to energy efficiency. In particular, premium values of LEED and Energy Star buildings have been established to be 6% for rents and 15% for sales (DOE, 2013a,b; Fuerst and McAllister, 2009; Wiley et al., 2010). Using data for retrofitted buildings, Kok et al. (2012) measured the change in rents associated with the retrofit upgrades. They found that improving the energy efficiency of existing buildings increased the rent premiums by up to 9% and occupancy rates by 2% compared to nonretrofitted buildings. Based on these benefits, the asset value added to the retrofitted buildings is determined to be $250/m^2 well above the implementation cost of the upgrades (Kok et al., 2012).

Real estate represents a significant economic activity for GCC countries as illustrated in Fig. 9.25. The residential building sector dominates the real estate market in the GCC and represents over 70% of the property value in KSA. In particular, the value added of real estate in KSA represents $60 billion or 9.2% of its GDP based on compiled data for 2014 (JLL, 2016). In comparison, real estate contributes 6% and 5.4% in, respectively, the US and UK economies. However, the value of rental and sale within the GCC region has been generally decreasing since 2008 as noted in Fig. 9.26 for residential properties for UAE. In contrast, real estate asset value has recovered noticeable especially after 2012 within the US and UK after the 2008 depression. As noted earlier from reported studies in the US and Europe, improvements in energy efficiency and sustainability of buildings can boost real estate value.

Fig. 9.27 quantifies the impact of higher asset value of real estate in the KSA building sector's energy productivity when combined with energy efficiency retrofits. In this analysis, energy efficiency premiums are estimated by increasing the real estate value added provided in Fig. 9.25 by 1%, 2%, and 5% for, respectively, level 1, level 2, and level 3 retrofit programs applied to the existing

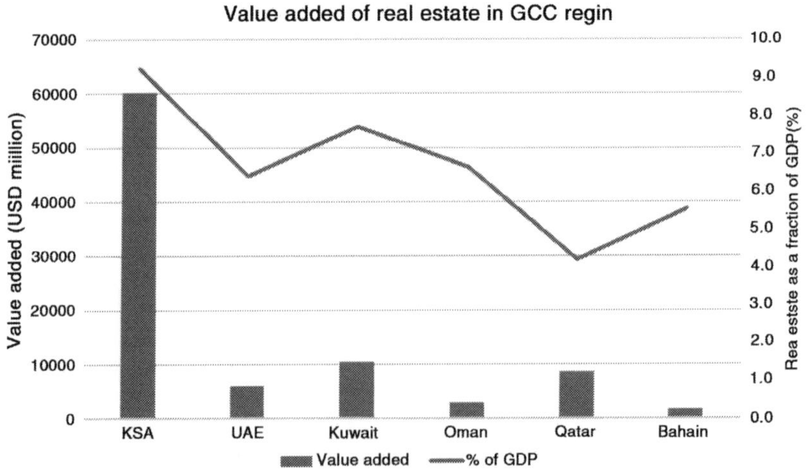

FIGURE 9.25 Value added of real estate in GDP of GCC countries for 2014.

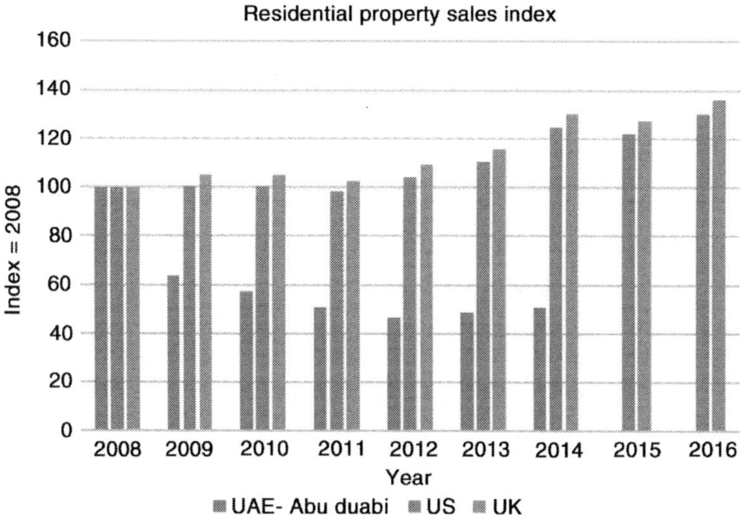

FIGURE 9.26 **Variation of residential properties in the UAE, US, and UK during 2008–2016.**

building stock. The analysis considers both applying the retrofit programs to the entire building stock and to only households. As indicated in Fig. 9.27, when the increased asset value is considered in the analysis, the building sector's energy productivity increases by 13% and 16% for the level 3 retrofit program when it is applied to households only and the entire building stock, respectively.

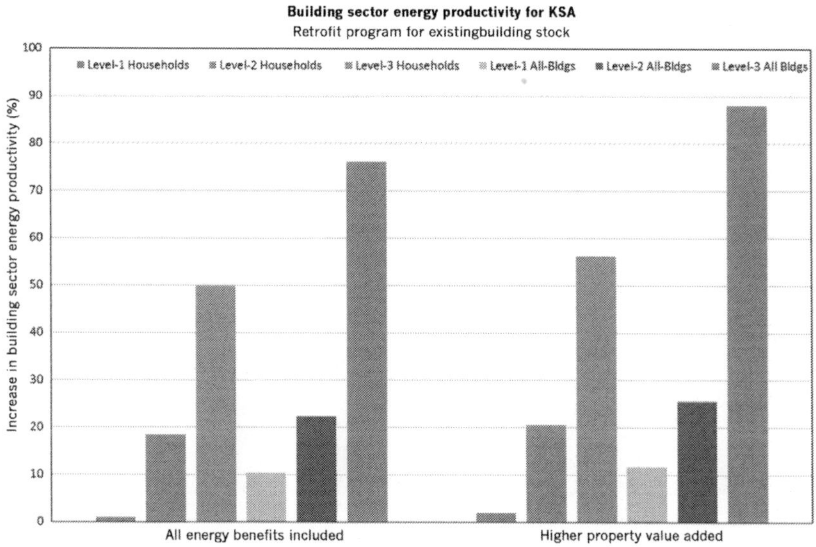

FIGURE 9.27 **Impact of property value added from energy retrofit programs on KSA building sector's energy productivity.**

9.6.5 Benefit Analysis of EE and RE Strategies

The energy productivity analysis approach can be applied to evaluate various energy strategies and alternatives. In this section, a comparative evaluation is carried out to assess the impacts of energy efficiency and renewable energy programs. The implementation costs and the benefits of installing PV systems on roofs of existing KSA housing stock have been analyzed in a reported study (Khan et al., 2017). Table 9.26 summarizes the main findings of the study that assumes that PV systems can be installed on available roof areas for the entire housing stock within KSA (Khan et al., 2017).

Fig. 9.28 presents the comparative analysis results to assess the impact of the evaluated energy programs on the energy productivity for the building

TABLE 9.26 Size and Benefits of PV Systems Installed on Roofs of the Entire Existing KSA Residential Buildings

PV roof area available for KSA housing units (million m^2)	82.032
Size of roof-PV systems (MW)	13,400
Annual electricity generated by PV (GWh/yr)	19,904
Carbon emissions avoided (million tons/yr)	15.1
Implementation costs ($ billion)	33.5
Levelized cost of energy ($/kWh)	0.144

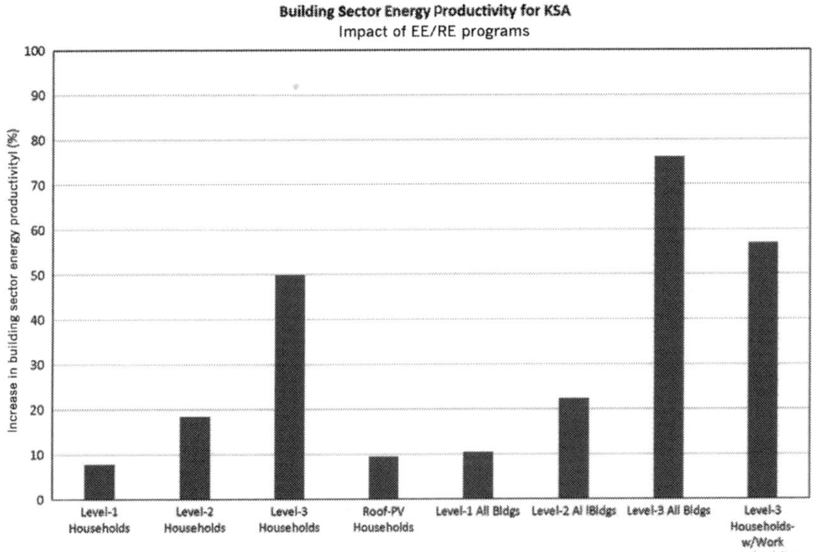

FIGURE 9.28 Impact of various energy efficiency and renewable energy programs on the KSA building sector's energy productivity.

sector. The comparative results are expressed in terms of the percentage increase of the baseline building energy productivity (i.e., relative to the business as usual scenario). It is clear that all the energy retrofit programs except level 1 when applied to only households outperforms the strategy of installing a roof PV system. As noted earlier, level 3 retrofit applied to the entire building stock has the highest effect on the energy productivity with an increase of over 75% relative to the baseline. The installation of PV systems on household's roofs results on only 9% increase of the building sector's energy productivity. The implementation costs as well as the potential job creation for all the evaluated energy programs are illustrated in Fig. 9.29. Again, level 3 retrofit program applied to the entire building sector provides the highest job creation option even though it requires the highest capital investments. Several financing mechanisms for the delivery of the energy retrofit programs can be considered for KSA and the GCC region (Krarti et al. 2017). It should be noted that a level 2 retrofit program applied to households require less capital and create more jobs than the strategy of installing a roof PV system.

It should be noted that the impact of the roof-PV installation program depends on the installation costs. Fig. 9.30 illustrates the effect of the PV system installation cost per kilowatt on the levelized cost of energy and the percent increase in the building sector's energy productivity relative to the baseline scenario. As indicated in Fig. 9.30, while the roof PV system installation costs affect significantly the levelized cost of energy, it has only little impact on the building sector's energy productivity since the considered PV capacity has a limited impact of the KSA final energy consumption specific for buildings.9.6.6 Main Findings

In the case study presented in this section, the energy productivity-based analysis was applied to account for energy and nonenergy benefits arising from

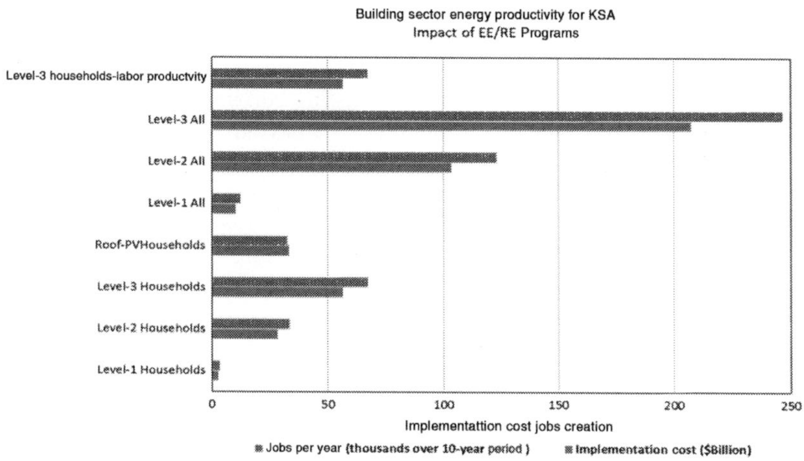

FIGURE 9.29 Implementation cost and job creation for KSA building retrofit and PV-roof programs.

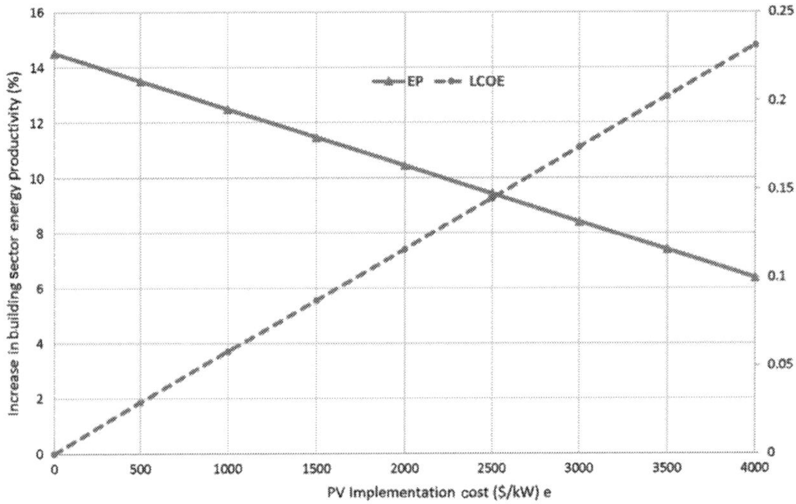

FIGURE 9.30 Impact of roof-PV installation cost on levelized energy cost and building sector's energy productivity increase.

a large-scale adoption of energy efficiency and renewable energy strategies by GCC households. First, the effect of adoption of individual retrofit strategies such as the use of high-efficiency air conditioning units and lighting fixtures on the energy efficiency of residential buildings in the GCC is evaluated. The results indicate that the adoption of high-efficiency conditioning systems can boost the energy efficiency of residential buildings by 28%. In particular, it is found that energy efficiency adoption can increase the energy affordability for households, that is, the floor area that a block of 10,000 kWh/yr can serve. In KSA, the housing floor area served by10,000 kWh/yr can increase from 52 m2 to 72 m2 when a high-efficiency air conditioning system is installed.

Then, the energy productivity-based analysis is applied to assess the impact of individual energy efficiency strategies on the direct and quantifiable benefits of large-scale energy retrofits for the existing building stock including the following: (1) reducing energy consumption, (2) avoiding future power plant construction, (3) lowering carbon emissions, and (4) creating skilled jobs. In addition, the analysis considers indirect benefits from building energy efficiency programs such as improving work productivity and increasing the asset value of real estate properties. Finally, the analysis compares the energy and economic impacts of energy retrofit programs to roof-PV installation strategies.

The analysis results clearly indicate that the energy retrofit programs proposed for the GCC households are highly cost-effective even when the governments provide the required capital investments. The implementation costs for the large-scale retrofit programs range vary widely with the desired improvement level of the household energy efficiency. For the entire GCC region, the capital needed to retrofit the entire residential building stock range from $4 billion for a small incremental improvement (i.e., level 1) to $60 billion for a deep retrofit (i.e., level 3). When all the quantifiable benefits are considered, it is found that all the retrofit programs are highly cost-effective for the GCC region since they can pay for themselves from the avoided costs of future increase in power plant capacities.

Moreover, the energy retrofit programs when fully implemented to the existing housing stock can increase the energy productivity of the building sector up to 50% in all GCC countries. As expected, the level 3 retrofit program has the highest impact in improving the energy productivity for the building sector within the GCC and creating highly skilled jobs. Over 70,000 jobs per year can be created over a 10-year implementation period of the retrofit programs to the entire GCC households. When NEBs are considered, it is found that improved work productivity associated with healthier buildings and lower absenteeism can add 5% in the energy productivity of the building sector. In addition, estimates indicated that improved GCC household's energy efficiency increases the asset value of real estate, which can result in a 13% increase in energy productivity. Higher increases are found when the retrofit programs are applied to the entire GCC building stock including residential and commercial buildings.

Finally, the energy productivity methodology has been applied to compare the benefits of the proposed energy retrofits to installation of PV systems in available roof areas for the entire housing stock. For the KSA, installing roof PV systems required almost $40 billion investments and lead to 9% increase in the energy productivity of the building sector. On the other hand, a lower capital of $28 billion needed to implement level 2 retrofits of existing housing stock results in an 18% increase in the building sector's energy productivity.

Future analysis of energy and nonenergy benefits of specific energy efficiency options and policies such as the use of district cooling in urban centers or the adoption of net-zero buildings can be carried out using the energy productivity framework.

9.7 SUMMARY

In this chapter, a general analysis approach for large-scale energy efficiency and renewable energy programs is outlined. In particular, the top–down and bottom–up methods to model building stocks are discussed. Prototypical energy models for US residential and commercial buildings are presented. The impacts of energy efficient and sustainable buildings on the well-being and work productivity of occupants as well as on the asset value of properties are also outlined. A detailed case study is used to show the applications of the bottom–up analysis method as well as of the concept of energy productivity to assess the benefits of energy retrofits of existing housing stocks in the GCC countries. The results indicate that improving the energy efficiency levels of existing buildings represent a significant strategy available for states and countries to not only their energy consumption and thus lower their carbon emissions but also to create skilled jobs and make available assets and financial capitals to invest in more productive sectors in their economies. Moreover, the energy productivity analysis has been proven to be valuable to compare the merit of various energy policies to evaluate their economic, environmental, and social benefits.

REFERENCES

Albertsen, S., Krarti, M., Bianchi, M., 2011. Estimating optimal energy savings from retrofitting US existing house stock. ASME Energy Sustainability Conference.

Al Horr, Y., Arif, M., Katafygiotou, M., Mazroei, A., Kaushik, A., Elsarrag, E., 2016. Impact of indoor environmental quality on occupant well-being and comfort: a review of the literature. Int. J. Sustain. Built Environ. 5, 1–11.

Andersson, B., Carroll, W.L., Martin, M.R., 1985. Aggregation of US population centers using climate parameters related to building energy use. J. Clim. Appl. Meteorol. 25, 596–614.

Bakó-Biró, Z.S., Kochhar, N., Clements-Croome, D.J., Awbi, H.B., Williams, M., 2008. Ventilation rates in schools and pupils' performance using computerised assessment tests. In: Proceedings of the 11th International Conference on Indoor Air Quality and Climate. Copenhagen, Sweden.

Bakó-Biró, Z., Clements-Croome, D.J., Kochhar, N., Awbi, H.B., Williams, M.J., 2012. Ventilation rates in schools and pupils' performance. Build. Environ. 48, 215–223.

Booth, A.T., Choudhary, R., Spiegelhalter, D.J., 2012. Handling uncertainty in housing stock models. Build. Environ. 48 (1), 35–47.

Caputo, P., Costa, G., Ferrari, S., 2013. A supporting method for defining energy strategies in the building sector at urban scale. Energy Policy 55, 261–270.

Clements-Croome, D., 2015. Creative and productive workplaces: a review. Intell. Build. Int. 7 (4), 164–183.

Davila, C.C., Reinhart, C., Bemis, J., 2016. Modeling Boston: a workflow for the generation of complete urban building energy demand models from existing urban geospatial datasets. Energy 117, 237–250.

Deru, M., Field, K., Studer, D., Benne, K., 2011. US Department of Energy commercial reference building models of the national building stock. , https://doi.org/NREL Report No. TP-5500-46861.

DOE, 2017. Scout, Office of Energy Efficiency and Renewable Energy (EERE). Department of Energy, Washington, DC, https://energy.gov/eere/buildings/scout.

DOE, 2013a. Business Case for Energy Efficient Building Retrofit and Renovation, Smart Market Report. McGaw-Hill Construction, New-York, NY.

DOE, 2013b. Valuing Energy Efficient Buildings. Report by Department of Energy (DOE) EEB Hub, Washington, DC.

DOE, 2005. A Look at Residential Energy Consumption in 1997. DOE/EIA-0632 (97). U.S. Department of Energy.

DOE, 2009. Building Technologies Program: Building America. DOE.

ECRA, 2015. Electricity and Cogeneration Regulatory Authority, Activities and Achievements of the Authority in 2014. Electricity and Cogeneration Regulatory Authority, Kingdom of Saudi Arabia.

Edmonds, J.E., Wise, M.A., MacCracken, C.N., 1994. Advanced Energy Technologies and Climate Change: An Analysis Using the Global Change Assessment Model (GCAM). Fondazione ENI Enrico Mattei.

Eisenhower, B., O'Neill, Z., Fonoberov, V.A., Mezic´, I., 2012. Uncertainty and sensitivity decomposition of building energy models. J. Build. Perform. Simul. 5 (3), 171–184.

EIA, 2009. The National Energy Modeling System: An Overview. Energy Information Administration, US Department of Energy, Washington, DC.

EIA, 2003. Commercial Building Energy Consumption Survey. Energy Information Administration, U.S. Department of Energy.

EIA, 2005. Residential Energy Consumption Survey. Energy Information Administration, U.S. Department of Energy, Washington, DC.

Energy Star, 2009. US Environmental Protection Agency, Information Provided in the Energy Star Website. US Environmental Protection Agency, http://www.energystar.gov.

Energy Trust of Oregon, 2009. Home Energy Solutions-Existing Homes, http://www.energytrust.org/residential/existinghomes/index.html.

Fanger, P.O., 1989. The new comfort equation for indoor air quality. ASHRAE J. 31 (10), 33–38.

Fels, M.F., 1986. PRISM: an introduction. Energy Build. 9 (1–2), 5–18.

Field, K., Deru, M., Studer, D., 2010. Using DOE commercial reference buildings for simulation studies. Conference Paper NREL/CP-550-48588. August 11–13. presented at SimBuild-2010, New York, NY.

Fishbone, L.G., Abilock, H., 1981. Markal, a linear-programming model for energy systems analysis: technical description of the BNL version. Int. J. Energy Res. 5 (4), 353–375.

Fisk, W.J., Black, D.R., Brunner, G., 2012. Changing ventilation rates in US offices: implications for health, work performance, energy, and associated economics. Build. Environ. 47, 368–372.

Fuerst, F., McAllister, P., 2009. An investigation of the effect of eco-labeling on office occupancy rates. J. Sustain. Real Estate 1 (1), 49–64.

Girardin, L., Marechal, F., Dubuis, M., Calame-Darbellay, N., Favrat, D., 2010. EnerGis: a geographical information based system for the evaluation of integrated energy conversion systems in urban areas. Energy 35 (2), 830–840.

Hopfe, C.J., Hensen, J.L.M., 2011. Uncertainty analysis in building performance simulation for design support. Energy Build. 43 (10), 2798–2805.

Howard, B., Parshall, L., Thompson, J., Hammer, S., Dickinson, J., Modi, V., 2012. Spatial distribution of urban building energy consumption by end use. Energy Build. 45, 141–151.

Huang, J., Gu, L., 2002. Prototype Residential Buildings to Represent the U.S. Housing Stock. Draft LBNL Report. Lawrence Berkeley National Laboratory, Berkeley, CA.

IEA Statistics: Energy Balance 2016. Available from http://www.iea.org/statistics.

IEA, 2014. Capturing the Multiple Benefits of Energy Efficiency, Available from http://www.iea.org/statistics.

IMF, 2016. World Economic Outlook. International Monetary Fund, Available from http://www.imf.org.

Jensen, K.L., Toftum, J., Friis-Hansen, P., 2009. A Bayesian Network approach to the evaluation of building design and its consequences for employee performance and operational costs. Build. Environ. 44, 456–462.

JLL, 2016. Riyadh Real Estate Market Overview 2014, Available from http://www.jll.com.

Kang, Y., Krarti, M., 2016. Bayesian-Emulator based parameter identification for calibrating energy models for existing buildings. Build. Simul. 9 (4), 411–428.

Kavgic, M., Mavrogianni, A., Mumovic, D., Summerfield, A., Stevanovic, Z., Djurovic-Petrovic, M., 2010. A review of bottom-up building stock models for energy consumption in the residential sector. Build. Environ. 45 (7), 1683–1697.

Kawamura, A., Tanabe, S., Nishihara, N., Haneda, M., Ueki, M., 2007. Evaluation method for effects of improvement of indoor environmental quality on productivity. In: Proceedings of the Clima 2007 Well-Being Indoors. 10-14 June 2007, Helsinki, Finland.

Khan, M.M.A., Asif, M., Stach, E., 2017. Rooftop PV potential in the residential sector of the Kingdom of Saudi Arabia. Buildings 7 (2), 46.

Kim, Y.J., 2016. Comparative study of surrogate models for uncertainty quantification of building energy model: Gaussian process emulator vs. polynomial chaos expansion. Energy Build. 133, 46–58.

Kok, N., Miller, N.G., Morris, P., 2012. The economics of green retrofits. J. Sustain. Real Estate 4 (1), 4–22, http://www.CoStar.com/uploadedFiles/JOSRE/JournalPdfs/01_4-22-J4_The_Economics_of_Green_Retrofits.pdf.

Kolter, J., Ferreira, J., 2011. A large-scale study on predicting and contextualizing building energy usage. In: Proceedings of 25th AAAI Conference. 7–11 August 2011, San Francisco, CA. pp. 1349–1356.

Korolija, I., Marjanovic-Halburd, L., Zhang, Y., Hanby, V.I., 2013. UK office buildings archetypal model as methodological approach in the development of regression models for predicting building energy consumption from heating and cooling demands. Energy Build. 60, 152–162.

Kosonen, R., Tan, F., 2004. Assessment of productivity loss in air-conditioned buildings using PMV index. Energy Build. 36, 987–993.

Krarti, M., Dubey, K., Howarth, N., 2017. Evaluation of building energy efficiency investment options for the Kingdom of Saudi Arabia. Energy 134, 595–610.

Krarti, M., Dubey, K., 2017. Energy productivity evaluation of large scale building energy efficiency programs for Oman. Sustain. Cities Soc. 29, 12–22.

Krarti, M., 2012. Weatherization of Residential Buildings. Taylor and Francis Publishing, New York, NY.

Krarti, M., 2015. Evaluation of large scale building energy efficiency retrofit program in Kuwait. Renew. Sustain. Energy Rev. 50, 1069–1076.

Kyle, P., Clarke, L., Rong, F., Smith, S.J., 2010. Climate policy and the long-term evolution of the U.S. buildings sector. Energy J. 31 (2), 301–314.

Lan, L., Wargocki, P., Wyon, D.P., Lian, Z., 2011. Effects of thermal discomfort in an office on perceived air quality, SBS symptoms, physiological responses, and human performance. Indoor Air 21, 376–390.

Lorsch, H.G., Abdou, O.A., 1994. The impact of the building indoor environment on occupant productivity. ASHRAE Trans. 100, 895–901.

Loulou, R., Remme, U., Kanudia, A., Lehtila, A., Goldstein, G., 2005. Documentation for the TIMES Model. Energy Technology Systems Analysis Programme (ETSAP).

Loveday, D.L., Hanby, V.I., Woodward, W.B., 1995. A software tool for relating the comfort and productivity of occupants in commercial buildings. In: Proceedings of the CIBSE National Conference. 1–3 October 1995, Eastbourne, UK. pp. 69–75.

Menberg, K., Heo, Y., Choudhary, R., 2016. Sensitivity analysis methods for building energy models: comparing computational costs and extractable information. Energy Build. 133, 433–445.

Moura, M.C.P., Smith, S.J., Belzer, D.B., 2015. 120 years of U.S. residential housing stock and floor space. PLOS ONE 10 (8), e0134135.

Nishihara, N., Tanabe, S., 2010. Management of indoor climate (Productivity). Kuukichouwa Eiseikougaku Binran 5, 562.

Nishio, K., Asano, H., 2006. A residential end-use demand model for analyzing the energy conservation potential of new energy efficient technologies. In: Proceedings of Energy Efficiency in Domestic Appliances and Lighting. London, England. .

Oladokun, M.G., Odesola, I.A., 2015. Household energy consumption and carbon emissions for sustainable cities—a critical review of modelling approaches. Int. J. Sustain. Built Environ. 4 (2), 231–247.

Pianosi, F., Beven, K., Freer, J., Hall, J.W., Rougier, J., Stephenson, D.B., Wagener, T., 2016. Sensitivity analysis of environmental models: a systematic review with practical workflow. Environ. Model. Softw. 79, 214–232.

Reddy, T., Maor, I., Panjapornpon, C., 2007. Calibrating detailed building energy simulation programs with measured data—Part II: Application to three case study office buildings (RP-1051). HVAC&R Res. 13 (2), 243–265.

Ritschard, R.L., Handford, J.W., Sezgen, A.O., 1992. Single-Family Heating and Cooling Requirements: Assumptions, Methods, and Summary Results. Topical Report LBL-30377, Gas Research Institute, Chicago, IL.

Roelofsen, P., 2001. The design of the workplace as a strategy for productivity enhancement. In: Proceedings of the 7th World Congress Clima 2000. 15–18 September 2001, Naples, Italy.

Rhodium Group, 2013. American Energy Productivity: The Economic, Environmental, and Security Benefits of Unlocking Energy Efficiency. Rhodium Group, New York, NY.

Russell, C., Baatz, B., Cluett, R., Amann, J., 2015. Recognizing the Value of Energy Efficiency's Multiple Benefits. American Council for an Energy-Efficient Economy, Washington, DC.

Sailor, D.J., Lu, L., 2004. A top–down methodology for developing diurnal and seasonal anthropogenic heating profiles for urban areas. Atmos. Environ. 38 (17), 2737–2748.

Satish, U., Mendell, M.J., Shekhar, K., Hotchi, T., Sullivan, D., Streufert, S., Fisk, J.W., 2012. Is CO_2 an indoor pollutant? Direct effects of low-to-moderate CO_2 concentrations on human decision-making performance. Environ. Health Perspect. 120 (12), 1671–1677.

Seppänen, O., Fisk, W.J., Faulkner, D., 2005. Control of temperature for health and productivity in offices. ASHRAE Trans. 111, 680–686.

Seppänen, O., Fisk, W.J., Lei, Q.H., 2006. Room temperature and productivity in office work. In: Proceedings of the Healthy Buildings, Volume 1. 4–8 June 2006, Lisbon, Portugal. pp. 243–247.

Seppänen, O., Fisk, W.J., 2006. Some quantitative relations between indoor environmental quality and work performance or health. Int. J. HVAC R Res. 12, 957–973.

Sherman, M., Dickerhoff, D., 1998. Air-Tightness of U.S. Dwellings. Energy Performance of Buildings Group, Energy and Environment Division. Lawrence Berkeley National Laboratory, Berkeley, CA.

Skumatz, L.A., 2014. Non-Energy Benefits/Non-Energy Impacts (NEBs/NEIs) and Their Role and Values in Cost-Effectiveness Tests: State of Maryland. Skumatz Economic Research Associates, Inc., Superior, CO.

Smith, S.T., 2009. Modelling thermal loads for a non-domestic building stock: associating a priori probability with building form and construction—using building control laws and regulations. Knowledge Creation Diffusion Utilization. University of Nottingham.

Swan, L.G., Ugursal, V.I., 2009. Modeling of end-use energy consumption in the residential sector: a review of modeling techniques. Renew. Sustain. Energy Rev. 13 (8), 1819–1835.

Tanabe, S., Nishihara, N., Haneda, M., 2007. Indoor temperature, productivity, and fatigue in office tasks. HVAC&R Res. 13, 623–633.

Tarantini, M., Pernigotto, G., Gasparella, A., 2017. A co-citation analysis on thermal comfort and productivity aspects in production and office buildings. Buildings 7, 36.

Tham, K.W., 2004. Effects of temperature and outdoor air supply rate on the performance of call center operators in the tropics. Indoor Air Suppl. 14 (7), 119–125.

Tham, K.W., Willem, H.C., 2010. Room air temperature affects occupants' physiology, perceptions and mental alertness. Build. Environ. 45, 40–44.

Vivid Economics, 2013. Energy Efficiency and Economic Growth: Report Prepared for the Climate Institute, London, UK.

Wargocki, P., Wyon, D.P., 2017. Ten questions concerning thermal and indoor air quality effects on the performance of office work and schoolwork. Build. Environ. 112, 359–366.

Wargocki, P., 2007. Improving indoor air quality improves the performance of office work and schoolwork. In: Proceedings of the 6th International Conference on Indoor Air Quality, Ventilation & Energy Conservation in Buildings. 28–31 October 2007, Sendai, Japan.

Wargocki, P., Wyon, D.P., Sundell, J., Clausen, G., Fanger, O., 2000. The effects of outdoor air supply rate in an office on perceived air quality, sick building syndrome (SBS) symptoms and productivity. Indoor Air 10, 222–236.

WGBC, 2013. World Green Building Council. The Business Case for Green Building: A Review of the Costs and Benefits for Developers, Investors and Occupants. World Green Building Council, Ontario.

Wiley, J., Benefield, J., Johnson, K., 2010. Green design and the market for commercial office space. J. Real Estate Finance Econ. 41 (2), 228–243, https://datapro.fiu.edu/campusedge/files/articles/johnsonk1590.pdf.

Winters, J., 2017. By the numbers: keeping the lid resources prices. ASME Eng. J. 139 (9), 28–29.

Wong, L.T., Mui, K.W., Hui, P.S., 2008. A multivariate-logistic model for acceptance of indoor environmental quality (IEQ) in offices. Build. Environ 43, 1–6.

Wyon, D.P., Wargocki, P., 2005. Room temperature effects on office work. In: Clements-Croome, D. (Ed.), Creating the Productive Workplace. 2nd ed. Taylor and Francis, London, UK.

Yamaguchi, Y., Shimoda, Y., Mizuno, M., 2007. Proposal of a modeling approach considering urban form for evaluation of city level energy management. Energy Build. 39 (5), 580–592.

Index

D

E

F

K

L

M

N

R

S

T

U

X

Z

FSC
www.fsc.org
MIX
Papier | Fördert
gute Waldnutzung
FSC® C083411

Kolibri 360 GmbH
Ferdinand-Jühlke-Straße 7
99095 Erfurt, Deutschland
produktsicherheit@kolibri360.de